Benchmark Papers in Organic Chemistry

Series Editor: Calvin A. VanderWerf
University of Florida

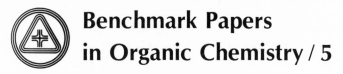

**Benchmark Papers
in Organic Chemistry / 5**

A **BENCHMARK**® Books Series

SINGLET MOLECULAR
OXYGEN

Edited by

A. PAUL SCHAAP
Wayne State University

**Dowden, Hutchinson
& Ross, Inc.**

STROUDSBURG, PENNSYLVANIA

Distributed by

HALSTED
PRESS

A Division of
John Wiley & Sons, Inc.

LIBRARY OF CONGRESS CATALOGING IN PUBLICATION DATA

Main entry under title:
Singlet molecular oxygen
    (Benchmark papers in organic chemistry / 5)
    Includes bibliographical references and indexes.
    1. Oxidation—Addresses, essays, lectures. 2. Oxygen—Addresses,
essays, lectures.   I.   Schaap, Arthur Paul, 1945–
QD281.09S52   547'.23   76–3496
ISBN 0–87933–064–3

Exclusive Distributor: **Halsted Press**
A Division of John Wiley & Sons, Inc.
ISBN: 0–470–15052–1

PERMISSIONS

The following papers have been reprinted and/or translated with the permission of the authors and copyright holders.

ACADEMIC PRESS, INC.—*Analytical Biochemistry*
  A Photoelectric Method for the Measurement of Spectra of Light Sources of Rapidly Varying Intensities

AMERICAN ASSOCIATION FOR THE ADVANCEMENT OF SCIENCE—*Science*
  Singlet Molecular Oxygen from Superoxide Anion and Sensitized Fluorescence of Organic Molecules

AMERICAN CHEMICAL SOCIETY
  Advances in Chemistry Series
    Chemistry of Singlet Oxygen: V. Reactivity and Kinetic Characterization
    Relaxation and Reactivity of Singlet Oxygen
  Biochemistry
    Dye-Sensitized Photooxidation of Tocopherols: Correlation Between Singlet Oxygen Reactivity and Vitamin E
  Chemical Reviews
    The Magnetochemical Theory
  The Journal of Physical Chemistry
    The Photoperoxidation of Unsaturated Organic Molecules: II. The Autoperoxidation of Aromatic Hydrocarbons
  Journal of the American Chemical Society
    Chemiluminescence Arising from Simultaneous Transitions in Pairs of Singlet Oxygen Molecules
    Chemistry of Singlet Oxygen: IV. Oxygenations with Hypochlorite–Hydrogen Peroxide
    Chemistry of Singlet Oxygen: VII. Quenching by β-Carotene
    Chemistry of Singlet Oxygen: IX. A Stable Dioxetane from Photooxygenation of Tetramethoxyethylene
    Chemistry of Singlet Oxygen: X. Carotenoid Quenching Parallels Biological Protection
    Chemistry of Singlet Oxygen: XIII. Solvent Effects on the Reaction with Olefins
    Chemistry of Singlet Oxygen: XVI. Long Lifetime of Singlet Oxygen in Carbon Disulfide
    Detection of the Naphthalene-Photosensitized Generation of Singlet ($^1\Delta_g$) Oxygen by Paramagnetic Resonance Spectroscopy
    The Determination of Rate Constants of Reaction and Lifetimes of Singlet Oxygen in Solution by a Flash Photolysis Technique
    Electron Paramagnetic Resonance of $^1\Delta$ Oxygen from a Phosphite–Ozone Complex
    Electron Paramagnetic Resonance of $^1\Delta$ Oxygen Produced by Gas-Phase Photosensitization with Naphthalene Derivatives
    Excited Singlet Molecular Oxygen in Photooxidation
    Olefin Oxidations with Excited Singlet Molecular Oxygen
    Photosensitized Oxygenation of Mono-Olefins
    Polymer-Based Sensitizers for Photooxidations
    Quenching of Photophysically Formed Singlet ($^1\Delta_g$) Oxygen in Solution by Amines

Quenching of Singlet Oxygen by Tertiary Aliphatic Amines: Effect of DABCO

Radiationless Decay of Singlet Molecular Oxygen in Solution: An Experimental and Theoretical Study of Electronic-to-Vibrational Energy Transfer

Remarkable Solvent Effects on the Lifetime of $^1\Delta_g$ Oxygen

The Self-Reaction of *sec*-Butylperoxy Radicals: Confirmation of the Russell Mechanism

Singlet Oxygen: A Probable Intermediate in Photosensitized Autoxidations

Singlet Oxygen ($^1\Delta_g$) Quenching in the Liquid Phase by Metal (II) Chelates

Singlet Oxygen Reactions from Photoperoxides

Singlet Oxygen Sources in Ozone Chemistry: Chemical Oxygenations Using the Adducts Between Phosphite Esters and Ozone

Solvent Effects in Dye-Sensitized Photooxidation Reactions

Stereospecific Formation of 1,2-Dioxetanes from *cis*- and *trans*-Diethoxyethylenes by Singlet Oxygen

A Study of the Peroxidation of Organic Compounds by Externally Generated Singlet Oxygen Molecules

AMERICAN INSTITUTE OF PHYSICS
  The Journal of Chemical Physics
    Metastable Oxygen Molecules Produced by Electrical Discharges
    Red Chemiluminescence of Molecular Oxygen in Aqueous Solution
    Some New Emission Bands of Molecular Oxygen
    Spectroscopic Detection of the Photosensitized Formation of $^1\Sigma$ Oxygen in the Gas Phase
  Reviews of Modern Physics
    The Interpretation of Band Spectra: Part III. Electron Quantum Numbers and States of Molecules and Their Atoms

THE CHEMICAL SOCIETY, LONDON
  Chemical Communications
    Oxidation by Photochemically Produced Singlet States of Oxygen
  Journal of the Chemical Society
    Lifetime of Singlet Oxygen in Liquid Solution
    The Oxidation of Organic Compounds by "Singlet" Oxygen

THE FARADAY SOCIETY
  Discussions of the Faraday Society
    Reactions in the Liquid Phase. Photochemistry of Anthracene: Part I. The Photo-oxidation of Anthracenes in Solution
  Transactions of the Faraday Society
    Quenching of Luminescence by Oxygen

GAUTHIER-VILLARS—*Comptes Rendus Hebdomadaires des Séances de l'Académie des Sciences*
    Un Peroxyde organique dissociable: le peroxyde de rubrène
    Photooxydation sur cycle pentagonal: photooxydiphénylisobenzofuran

INTERNATIONAL UNION OF PURE AND APPLIED CHEMISTRY—*Pure and Applied Chemistry*
    Mechanism and Stereoselectivity of Photosensitized Oxygen Transfer Reactions

MACMILLAN (JOURNALS) LTD.—*Nature*
    Interpretation of the Atmospheric Oxygen Bands: Electronic Levels of the Oxygen Molecule

NATIONAL RESEARCH COUNCIL OF CANADA—*Canadian Journal of Chemistry*
    Reactivities in Photosensitized Olefin Oxidations

THE NEW YORK ACADEMY OF SCIENCES—*Annals of the New York Academy of Sciences*
    Photosensitized Oxygenation as a Function of the Triplet Energy of Sensitizers

NORTH HOLLAND PUBLISHING COMPANY—*Chemical Physics Letters*
    Direct Measurement of the Lifetime of $^1\Delta$ Oxygen in Solution
    Formation of $O_2(^1\Sigma_g^+)$ by 1-Fluoronaphthalene Sensitization
    Production of Singlet Oxygen in the Benzene Oxygen Photochemical System
    Quantum Yield of Triplet State Formation from Naphthacene Photoperoxidation
    Reaction of Chemical Acceptors with Singlet Oxygen Produced by Direct Laser Excitation

PERGAMON PRESS
    Photochemistry and Photobiology
    Chemistry of Singlet Oxygen: XVIII. Rates of Reaction and Quenching of α-Tocopherol and Singlet Oxygen
    On the Mechanism of Quenching of Singlet Oxygen in Solution
    On the Quenching of Singlet Oxygen by α-Tocopherol
    The Photoperoxidation of Unsaturated Organic Molecules: XIII. $O_2{}^1\Delta_g$ Quenching by α-Tocopherol
    Photosensitized Oxidation Through Stearate Monomolecular Films
    Tetrahedron
    Zur Photosensibilisierten Sauerstoffübertragung: Untersuchung der Terminationsschritte durch Belichtungen bei tiefen Temperaturen
    Tetrahedron Letters
    Adamantylideneadamantane Peroxide: A Stable 1,2-Dioxetane
    The Base-Induced Decomposition of Peroxyacetylnitrate
    Chemistry of Singlet Oxygen: III. Product Selectivity
    Chemistry of Singlet Oxygen: VI. Photooxygenation of Enamines: Evidence for an Intermediate
    Mikrowellenentladung von CO_2: eine neue, ergiebige Quelle für Singulett-Sauerstoff, $O_2(^1\Delta_g)$
    Photooxygenation of Enamines: A Partial Synthesis of Progesterone

THE ROYAL SOCIETY, LONDON—*Proceedings of the Chemical Society*
    Chemiluminescence from the Reaction of Chlorine with Aqueous Hydrogen Peroxide

SOCIÉTÉ CHIMIQUE DE FRANCE—*Bulletin de la Société Chimique de France*
    Photoxydation sensibilisée d'éthyléniques: étude des deux diméthoxy-3,4 diphényl-dioxétanes

SPRINGER-VERLAG, BERLIN, HEIDELBURG, NEW YORK—*Naturwissenschaften*
    Die Aufklärung der Photoluminescenztilgung fluorescierender Systeme durch Sauerstoff: Die Bildung aktiver, diffusionsfähiger Sauerstoffmoleküle durch Sensibilisierung
    Die Synthese des Ascaridols
    Zur Theorie der photosensibilisierten Reaktion mit molekularem Sauerstoff

VERLAG CHEMIE GMBH
    Chemishe Berichte
    Photo-sensibilisierte Oxydation als Wirkung eines Aktiven, Metastabilen, Zustandes des Sauerstoff-Molekuls
    Liebig's Annalen der Chemie
    Notiz über die photochemische Bildung von Biradikalen
    Photochemische Reaktionen III. Über die Bildung von Hydroperoxyden bei photosensibilsierten Reaktionen von O_2 mit geeigneten Akzeptoren, insbesondere mit α- und β-Pinen
    Über die photochemische Oxydation des Ergosterins
    Zeitschrift für Elektrochemie
    Chemismus und Kinetik der durch fluoreszierende Farbstoffe photosensibilisierten Reaktionen mit O_2 und Primärakt der Photosynthese

SERIES EDITOR'S PREFACE

During the past fifteen years there has been an explosion of interest in singlet molecular oxygen and its reactions.

Historically, the story of singlet oxygen—from the first general proposal of the concept by G. N. Lewis, through the experimental proof of its existence, to its present-day implication in a host of oxidation reactions—constitutes one of the fascinating chapters of organic chemistry. Philosophically, there is no more elegant example in all organic chemistry of the development, testing, and application of a fundamental reaction mechanism. Practically, the potential application of the chemistry of this reactive intermediate, which is involved in reactions ranging from the most subtle physiological oxidations to the production of smog, promises to be almost unlimited.

To tell the story, in all its historical, philosophical, and practical significance, we are proud to present a contemporary pioneer and leading expert. Paul Schaap's *Singlet Molecular Oxygen* tells crisply and concisely, directly from the original literature, what every organic chemist—theorist, practitioner, and general student—should know about this exploding field.

C. A. VANDERWERF

PREFACE

In 1924 G. N. Lewis correctly interpreted the paramagnetism of molecular oxygen in terms of an electronic structure with unpaired electrons. He also recognized the possibility of a species in which the electrons were spin-paired. Molecular orbital calculations by R. S. Mulliken in the early 1930s led to the prediction of two relatively low-lying excited states for molecular oxygen, $^1\Delta_g$ and $^1\Sigma_g^+$. Although Kautsky provided unambiguous evidence in 1931 for the intermediacy of a volatile, metastable species in dye-sensitized photooxygenation, singlet molecular oxygen remained for many years only a curiosity of astrophysicists. However, the proposal in 1963 by Khan and Kasha that the chemiluminescence produced by the reaction of $NaOCl$ and H_2O_2 was mediated by singlet oxygen served to rekindle interest in the possible intermediacy of singlet oxygen in photooxygenation. We now recognize singlet oxygen as a metastable, excited form of O_2 with enhanced reactivity toward a wide variety of substrates. As we better understand the reactions of $^1O_2$ through continued research in this area, the synthetic organic chemist will find singlet oxygen to be a useful reagent for various types of oxidative transformations. The possibility that singlet oxygen may be involved in biological processes such as photocarcinogenesis and enzymatic oxidations with dioxygenases will also prompt further investigations.

The years from 1963 to the present have seen a virtual explosion of interest in singlet oxygen and its reactions. Included in this volume are some of those papers which provided new and significant insights into the nature of singlet oxygen. I have not sought to review any particular area of singlet oxygen chemistry but rather to assist the reader in understanding the historical development of the field and thereby provide a basis for reading the current literature.

I would like to thank the original authors for permitting their papers to be reprinted in this volume. I regret that because of space limitations many equally important papers could not be included. My thanks go to M. F. Mattison and Wilfried Heller for their assistance in translating several papers and to Christopher S. Foote, Michael Kasha, Robert W. Murray, Elmer A. Ogryzlo, and Thérèsa Wilson for reading the manuscript and making several valuable suggestions. I also want to thank Carol A. Craig, Kristine S. Curtis, and Bashir Kaskar for their assistance in preparing this volume. It is also a pleasure to express appreciation to Carl R. Johnson for many helpful discussions.

A. PAUL SCHAAP

CONTENTS

Contents

Contents

Contents

Contents

SOLVENT EFFECTS ON SINGLET OXYGEN REACTIONS:
LIFETIME OF $^1\Delta_g\,O_2$

Contents

CONTENTS BY AUTHOR

Contents by Author

SINGLET MOLECULAR OXYGEN

INTRODUCTION

The oxidation of organic and biological substrates under the influence of light, oxygen, and a sensitizer has been under investigation since the report by Fritzsche in 1867 that photooxygenation of naphthacene yields a peroxide. Two general types of photosensitized oxygenation are observed: (1) the excited sensitizer serves to initiate a free-radical propagated autoxidation, and (2) the reactive intermediate is an electronically excited state of molecular oxygen (singlet oxygen) produced by the transfer of energy from the excited sensitizer to oxygen. An example of photochemically initiated autoxidation is the benzophenone-sensitized oxidation of isopropyl alcohol in the presence of oxygen, initially investigated by Bäckström.[1] Early photooxygenation reactions, which were subsequently shown to involve singlet oxygen, include the photooxygenation of rubrene investigated by Moureu, Dufraisse, and Dean (Paper 5) and the dye-sensitized photooxygenation of ergosterol investigated by Windaus and Brunken (Paper 6). However, it was the classic synthesis of ascaridole from a-terpinene by Schenck and Ziegler in 1944 (Paper 10) that prompted extensive preparative and mechanistic investigations of photooxygenation.

The nature of the reactive intermediate in photooxygenation has been the subject of considerable controversy for many years. It was suggested by Straub in 1904 and later by Noack in 1920 that the dye sensitizer combined with oxygen to form an intermediate peroxide which could transfer oxygen to a substrate with regeneration of the dye. In 1935, Schönberg (Paper 12) proposed a general mechanism for sensitized photooxygenation that involved the formation of an unstable complex between the excited sensitizer and oxygen. He suggested that this

1

complex could donate oxygen to an organic substrate molecule to yield the oxygenation products. G. O. Schenck (Paper 13) was a strong proponent of the sensitizer–oxygen complex (moloxide) mechanism.

In 1931, Kautsky and de Bruijn (Paper 7) proposed that dye-sensitized photooxygenation involved the transfer of electronic excitation energy from the excited sensitizer to oxygen to produce a "reactive, metastable state of the oxygen molecule." At the time of Kautsky's proposal, only the $^1\Sigma_g^+$ excited state of oxygen had been observed spectroscopically, and this was assumed to be the reactive oxygen species. Following the report by Ellis and Kneser[2] in 1933 of the $^1\Delta_g$ state of oxygen, both states of oxygen were considered as possible reactants in photooxygenation (Paper 9). Kautsky supported his proposed mechanism with a series of elegant experiments that should have put to rest the sensitizer–oxygen complex mechanism. It was observed that photooxygenation was possible even when the sensitizer and the acceptor were physically separated on different grains of silica gel, which demonstrated that only a diffusible molecule such as $^1O_2$ could be the reactive species. In spite of these results, the Kautsky mechanism was almost totally disregarded and was not revived until the independent generation of singlet oxygen with NaOCl and H_2O_2 and with the electrodeless discharge (Papers 18–28).

Molecular oxygen, a ground-state triplet with paramagnetic and diradical-like properties, has two low-lying singlet excited states, $^1\Delta_g$ and $^1\Sigma_g^+$. As the transition of $^1\Delta_g$ to $^3\Sigma_g^-$ is spin-forbidden, $^1\Delta_g$ is a relatively long-lived species. The $^1\Sigma_g^+$ state is relatively short-lived with a spin-allowed transition to $^1\Delta_g$. The lifetime of the $^1\Sigma_g^+$ is sufficiently short that all singlet oxygen chemistry in solution involves the $^1\Delta_g$ state.

In addition to photosensitization, several alternative methods for the generation of singlet oxygen have been developed: the reaction of so-

Electronic states of molecular oxygen[3-5]

State	Energy above ground state (kcal)	Lifetime in solution(s)	Radiative lifetime at zero pressure
$^1\Sigma_g^+$	37.5	10^{-9} to 10^{-12}	7.1 s
$^1\Delta_g$	22.5	10^{-3} to 10^{-6}	45 min
$^3\Sigma_g^-$	0		

dium hypochlorite with hydrogen peroxide, the thermal decomposition of phosphite ozonides, the decomposition of epidioxides, the reaction of potassium superoxide in water, the self-reaction of *sec*-peroxy radicals, and microwave discharge through gaseous oxygen.

The reactions of singlet oxygen with a wide variety of organic substrates are discussed in the reviews listed in the Selected Readings. Singlet oxygen exhibits three modes of reaction with alkenes: 1,4-cycloaddition with conjugated dienes to yield cyclic peroxides, an "ene" -type reaction to form allylic hydroperoxides, and 1,2-cycloaddition with olefins to give 1,2-dioxetanes, which cleave thermally to carbonyl-containing products. Other reactions of singlet oxygen include oxidation of sulfides to sulfoxides and sulfones and addition to heterocycles such as pyrroles, furans, oxazoles, imidazoles, and thiophenes. Singlet oxygen also reacts with such biologically important substrates as fatty acids, purines, pyrimidines, DNA, RNA, amino acids (tyrosine, tryptophan, methionine, cystine, histidine) and various proteins. [6a-f] The possible role of singlet oxygen in biological oxidations has been considered by several investigators.[7]

NOTES AND REFERENCES

1. H. L. J. Bäckström, *Z. Physik. Chem.*, **B25**, 99 (1934).
2. J. W. Ellis and H. O. Kneser, *Z. Physik*, **86**, 583 (1933).
3. G. Herzberg, *Molecular Spectra and Molecular Structure, I. Spectra of Diatomic Molecules*, 2nd ed., Van Nostrand Reinhold Company, New York, 1950, p. 560.
4. S. J. Arnold, M. Kubo, and E. A. Ogryzlo, "Relaxation and Reactivity of Singlet Oxygen, *"Advan. Chem. Ser.*, **77**, 133 (1968); Paper 43.
5. The electronic structure of oxygen is considered in papers by Ogryzlo and Kasha: E. A. Ogryzlo, *Photophysiology*, **5**, 35 (1970); and M. Kasha and A. U. Khan, *Ann. N.Y. Acad. Sci.*, **171**, 5 (1970).
6. See, for example, (a) J. D. Spikes and B. W. Glad, *Photochem. Photobiol.*, **3**, 471 (1964); (b) J. D. Spikes and M. L. MacKnight, *Ann. N.Y. Acad. Sci.*, **171**, 149 (1970); (c) R. Nilsson, P. B. Merkel, and D. R. Kearns, *Photochem. Photobiol.*, **16**, 117 (1972); (d) F. H. Doleiden, S. R. Fahrenholtz, A. A. Lamola, and A. M. Trozzolo, *ibid.*, **20**, 519 (1974); (e) D. A. Lightner and D. C. Crandall, *Tetrahedron Lett.*, No. 953 (1973); (f) A. F. McDonagh, *Biochem. Biophys. Res. Commun.*, **44**, 1306 (1971).
7. See the following review: C. S. Foote, "Photosensitized Oxidation and Singlet Oxygen: Consequences in Biological Systems," in W. A. Pryor (ed.), *Free Radicals in Biology*, Academic Press, New York, 1975.

Editor's Comments
on Papers 1 Through 3

ELECTRONIC STRUCTURE OF MOLECULAR OXYGEN

Prior to 1924, molecular oxygen was thought to have an electronic structure similar to its carbon analog, ethylene. However, the paramagnetic properties of oxygen and its unique chemical reactivity toward carbon free radicals led G. N. Lewis (Paper 1) to suggest that "the assumed double bond in oxygen is broken in a symmetrical manner so as to leave to each atom an odd electron," a suggestion that in qualitative terms still adequately describes the ground state of molecular oxygen.

The predictions by R. S. Mulliken (Papers 2 and 3) of the existence of metastable, electronically excited singlet states for molecular oxygen stand as one of the early triumphs of quantum mechanics. Mulliken formulated the electronic configuration for the ground state of molecular oxygen as $(K)\ (K)\ (\sigma_g 2s)^2\ (\sigma_u 2s)^2\ (\sigma_g 2p)^2\ (\pi_u 2p)^4\ (\pi_g 2p)^2$. This formulation suggested that there should be three states for oxygen: ^3S, ^1D, and ^1S (in order of increasing energy). The spectroscopic notation for these states was later changed to $^3\Sigma_g^-$, $^1\Delta_g$, and $^1\Sigma_g^+$, respectively (Paper 3). As a ground-state triplet, molecular oxygen would be expected to exhibit paramagnetic properties. Mulliken correctly identified an atmospheric absorption band at 1.62 eV with the transition from $^3\Sigma_g^-$ to $^1\Sigma_g^+$. He also suggested that there should exist a metastable $^1\Delta_g$ state at approximately 0.81 eV (19 kcal/mol). He noted that, if this prediction was correct, there should be a weak absorption for molecular oxygen in the infrared region corresponding to the $^3\Sigma_g^- \longrightarrow {}^1\Delta_g$ transition.

The first unambiguous evidence for the $^1\Delta_g$ state of molecular oxygen was subsequently provided by Ellis and Kneser in 1933.[1] The infrared absorption spectrum of liquid oxygen exhibited a band at 1.261 μm (22.7 kcal/mol), which was correctly attributed to the $^3\Sigma_g^- \rightarrow {}^1\Delta_g$ transition.

NOTES AND REFERENCES

1. J. W. Ellis and H. O. Kneser, *Z. Physik,* **86,** 583 (1933).

1

Reprinted from *Chem. Rev.*, 1, 243-245 (1924)

THE MAGNETOCHEMICAL THEORY

G. N. Lewis

[*Editor's Note:* In the original, material precedes this excerpt.]

THE CASE OF OXYGEN

If we consider the three substances, ethylene, formaldehyde and oxygen, with the conventional formulae $H_2C = CH_2$, $H_2C = O$, and $O = O$, we see that they form a continuous series, and we might expect to be able to predict the magnetic properties of oxygen from those of the first two substances. However, a careful determination of the magnetic properties of ethylene and formaldehyde will probably show that they have small negative susceptibilities, while the susceptibility of oxygen is large and positive. This difference must correspond to a radical difference in molecular structure. In my paper of 1916 the possibilities of tautomerism in molecules containing a double bond were discussed. In one tautomer the double bond consists of two pairs of electrons shared between two atoms. In another tautomer the bond is broken in a polar manner, so that one of the electron pairs becomes the sole property of one of the atoms. This is the tautomeric form which Lowry has recently considered as responsible for many of the reactions of unsaturated compounds. But it is the third type of tautomerism which I suggested that interests us here. In this tautomer one of the bonds is broken in such a way as to be equally divided between the two atoms, so that each atom possesses an odd electron.

It appears that this type of rupture of the double bond occurs to a large extent in the oxygen molecule, so that the predominant form of molecules in oxygen is quite different from that in ethylene, as can be represented in the two formulae, $H : \overset{..}{C} : : \overset{..}{C} : H$ and $: \overset{..}{\underset{.}{O}} : \overset{..}{\underset{.}{O}} :$. We may thus regard the paramagnetic properties of oxygen as due to the existence in its molecule of two odd atoms.

Now if this is the correct explanation of the magnetic properties of oxygen it implies the existence of great differences between

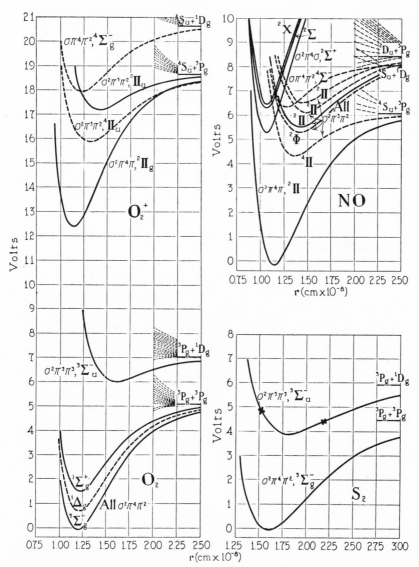

Fig. 48. $U(r)$ curves of known and (dashed lines) some predicted states of O_2, O_2^+, NO, and S_2. There is a possibility that the predicted $^4\Sigma^-$ and $^4\Pi$ states of O_2^+ are the upper and lower states of the visible O_2^+ bands, whose structure has not yet been analyzed. Other predicted curves for O_2^+ analogous to those shown for NO could also have been drawn. The numerous predicted curves derived from $O(^3P)+O(^3P)$, $O(^3P)+O(^1D)$, $O^+(^4S)+O(^3P)$, $O^+(^4S)+O(^1D)$, $N(^4S)+O(^3P)$, $N(^4S)+O(^1D)$ and $N(^2D)+O(^3P)$ are all indicated qualitatively in the figure. Predicted curves of S_2, which would be analogous to those of O_2, are not shown. Two points of intersection of one,—or perhaps two,—of these predicted curves with the known $^3\Sigma^-_u$ curve are, however, shown by crosses, Their existence is indicated by predissociation phenomena.

11

and unexcited O_2 is then $\cdots (v\pi)^2$, $^3\Sigma^-_g$. Since ξ must be fairly large even for the outer orbits, we presumably have the binding order indicated by

$$\cdots (x\sigma)^2(w\pi)^4(v\pi)^2.$$

Probably ξ is so large for the orbits $z\sigma$ and $y\sigma$ in O_2 that they can definitely be correlated with $2s$ orbits of the separated atoms and that they can best be described as $\sigma_g 2s$ and $\sigma_u 2s$. For the normal state of O_2 we may then write:

$$(K)(K)(\sigma_g 2s)^2(\sigma_u 2s)^2(x\sigma)^2(w\pi)^4(v\pi)^2, \; ^3\Sigma^-_g.$$

This description implies that the first two shells in O_2 are essentially unchanged atomic K shells, the next two are somewhat modified atomic $2s$ shells, while the remaining electron orbits are strongly modified as compared with their condition in the separated atoms.

Granting that the normal state of O_2 is $\cdots (v\pi)^2$, $^3\Sigma^-_g$, there should be two other stable states $^1\Delta_g$ and $^1\Sigma^+_g$ with the same electron configuration and slightly higher energy (cf. section $A6$). These two states should be metastable, since transitions from Δ_g or Σ^+_g to Σ^-_g would violate the usual (dipole) selection rules (Part IIc, p. 149) and since it is extremely improbable that other electron states exist, between the $^3\Sigma^-_g$ and the $^1\Delta_g$ and $^1\Sigma^+_g$, which would remove the metastability. Actually, however the atmospheric absorption bands probably correspond to the forbidden transition $^1\Sigma^+_g \leftarrow ^3\Sigma^-_g$. The $^1\Sigma^+_g$ is 1.62 volts above the $^3\Sigma^-_g$; according to theoretical considerations,[118] the $^1\Delta_g$ is probably about halfway between $^1\Sigma^+_g$ and $^3\Sigma^-_g$ (cf. Fig. 48).

Childs and Mecke[141] have shown that the absorption intensities in the atmospheric bands of O_2 are of the order of 10^{-8} times as large as in ordinary bands. This is so weak as to represent really an extraordinarily high degree of obedience to the dipole selection rules. As a matter of fact, the observed intensity would be about right for a quadrupole transition. The observed band-lines, which, if the upper level is $^1\Sigma^+_g$, go only from $+$ rotational levels to $+$ rotational levels and so violate the dipole selection rules, are in agreement with the selection rules for a quadrupole transition $^1\Sigma^+_g \leftarrow ^3\Sigma^-_g$. There is, however, one difficulty, namely that this would be an *intersystem* quadrupole transition and its intersystem character should almost certainly, according to the theory,[114] reduce the intensity by a further factor of probably about 10^{-4}, making 10^{-12} altogether, as compared with the observed 10^{-8}. The most probable way out of this difficulty is to suppose that we are after all not really dealing with a quadrupole absorption, but that the observed intensity owes its existence almost entirely to disturbing effects, such as collisions between gas molecules, which tend to break down the usual selection rules and to induce an otherwise forbidden dipole transition. [In the discussion of these bands in Part IIc, section $D1$, p. 96, it was concluded that they are probably $^1\Sigma^-_u \leftarrow ^3\Sigma^-_g$ or possibly $^1\Sigma^+_g \leftarrow ^3\Sigma^-_g$. The former should, however, have a transition probability probably of the order of 10^{-4} times that of an ordinary transition, as compared with the observed 10^{-8}. (This last figure was not known when Part IIc was written.) Also, it is very difficult in terms of electron configurations to explain the existence of a $^1\Sigma^-_u$ state so near the $^3\Sigma^-_g$.]

The most important excited states of O_2^+ can probably be understood by considering the result of removing an electron from each of the various outer orbits of O_2, just as we were able to explain the three observed states of CO^+ by considering the removal of an electron from each of the three outermost orbits in CO. If we remove a $v\pi$ electron from $\cdots (w\pi)^4(v\pi)^2$, we get the normal state of O_2^+, $\cdots (w\pi)^4 v\pi$, $^2\Pi_g$. If we remove a $w\pi$ electron from $\cdots (w\pi)^4(v\pi)^2$, $^3\Sigma^-_g$, we get $\cdots (w\pi)^3(v\pi)^2$, $^4\Pi_u$ or $^2\Pi_u$, of which the latter may reasonably be identified with the upper state of the ultraviolet O_2^+ bands. Additional states $\cdots (w\pi)^3(v\pi)^2$, $^2\Pi_u$ and $^2\Phi_u$ of O_2^+ should be obtained by removing a $w\pi$ electron from the metastable state $\cdots (w\pi)^4(v\pi)^2$, $^1\Delta_g$ of O_2. From $\cdots (v\pi)^2$, $^1\Sigma^+_g$ of O_2, still a third $^2\Pi_u$ of O_2^+ with the same electron configuration must be obtained. From $\cdots (x\sigma)^2(w\pi)^4(v\pi)^2$, $^3\Sigma^-_g$ by removal of one $x\sigma$ electron a $^4\Sigma^-_g$ and a $^2\Sigma^-_g$ state must be obtained. Possibly the visible O_2^+ bands represent a transition from this $^4\Sigma^-_g$ to the $^4\Pi_u$ mentioned above (cf. Fig. 48), but it has not yet been possible to analyze their structure and so determine their nature.

It may seem strange that of the three predicted $^2\Pi_u$ states belonging to $\cdots (w\pi)^3(v\pi)^2$, only one is known, since all three should combine with the normal state of the molecule. One plausible explanation is that while the known $^2\Pi_u$ is obtained by direct ionization of a $^3\Sigma^-_u$ molecule, the unknown ones could be obtained directly only by ionization of the metastable states $^1\Delta_g$ and $^1\Sigma^+_g$ of O_2. Evidence that the $^2\Pi_u$ of O_2^+ is really related to $^3\Sigma^-_g$ of O_2 in the way suggested is given by the small positive A value (cf. Table XIV) of the observed $^2\Pi$. By a consideration of the interaction of the two groups of electrons $(w\pi)^3$ and $(v\pi)^2$, by means of the vector model, it can be shown that (1) when π^2 is constituted as in $^3\Sigma^-_g$, an A value much less than a of $w\pi$ or $v\pi$, but probably positive, is to be expected, while (2) when π^2 is as in $^1\Delta_g$, $A \sim +a$ of $w\pi$ is expected, and (3) when π^2 is as in $^1\Sigma^+_g$, $A \sim -a$ of $w\pi$ is expected. Probably a of $w\pi$ is not far different from a of $v\pi$, whose value is 195 according to data on the normal state of O_2^+, which is $\cdots v\pi$, $^2\Pi$, with $A = a$ (cf. Table XIV). The observed $A = +8$ of $^2\Pi_u$ of O_2^+ is evidently in agreement with the interpretation given here.

[*Editor's Note:* Material has been omitted at this point.]

Editor's Comments
on Papers 4 Through 6

FIRST REPORTS OF PHOTOOXYGENATION

One of the earliest descriptions of a photooxygenation was given by Fritzsche (Paper 4) in 1867; he observed that exposure of a solution of naphthacene (*1*) to air and light produced a crystalline material which regenerated the hydrocarbon when heated. It has subsequently been shown that this reaction involves the addition of singlet oxygen to (*1*) to yield the epidioxide (*2*).[1]

(1)   (2)

In 1926, Moureu, Dufraisse, and Dean (Paper 5) reported similar experiments with a red aromatic hydrocarbon of unknown structure, rubrene (*3*). They observed that irradiation (sunlight) of a solution of rubrene in benzene in the presence of air produced a colorless solution from which a crystalline peroxide could be isolated in quantitative yield. They described several experiments which demonstrated that thermal decomposition of the rubrene peroxide at 150° liberated molecular oxygen in 87 percent yield, with concomitant formation of rubrene.

The structure of the rubrene peroxide was established several years later as the epidioxide (4).[1] However, the authors initially suggested that the peroxide was a molecular complex "in which the molecules have preserved their individuality." They considered this system as a possible organic chemical analog to the respiratory function of oxyhemoglobin in the reversible binding of oxygen.

Moureu, Dufraisse, and Butler subsequently reported that artificial light may be used in the photooxygenation of rubrene.[2] They also observed that the thermal decomposition of the rubrene peroxide is attended with chemiluminescence (see Papers 22 and 51). A translation of three pertinent paragraphs from their 1926 paper follows:

> Another method of demonstrating the necessity for the action of light which can be used as a routine test is the following: one prepares a suspension of finely pulverized rubrene in a very slowly evaporating solvent (of the type of tetrahydronaphthalene); a white plate is painted with this ink and, after having partially masked the plate with black paper in which designs have been cut out, the plate is exposed to strong irradiation. When the masking is removed, the design is found reproduced in white and the masked part remains red. If the solvent is then removed by evaporation, the rubrene that has not been oxidized will remain unchanged even under the light; i. e., the image will be fixed. This is in fact a genuine process of photographic reproduction.
>
> The formation of rubrene peroxide requires light; however, the easy dissociation of this compound indicates that in the act of combination a great loss of potential energy does not take place. These considerations lead us to conclude that rubrene peroxide retains a considerable portion of the luminous energy used in its formation, and that during the dissociation this energy would be liberated with the oxygen in the form of light. We have put this assumption to the test.
>
> In the dark, a tube containing a small quantity of rubrene peroxide was immersed in a paraffin oil bath (contained in a glass vessel) heated to 180°C. Several observers who had stayed in the dark for a sufficient time simultaneously followed the experiment. A short time after introduction of the tube into the bath, a brilliant greenish yellow light appeared at the bottom of the well and on the walls (because of the particles adhering); this light gradually diminished and was extinct after several minutes. After the test, the tube contents were examined in the light and exhibited the characteristic red color of the regenerated ru-

15

brene. Therefore, as we had assumed, rubrene peroxide dissociates with the emission of light.

In Paper 6, Windaus and Brunken described the first example of a *dye*-photosensitized reaction of oxygen to yield an isolable, crystalline peroxide. As part of their investigations of the photochemical conversion of the plant hormone ergosterol *(5)* to vitamin D, the Windaus group observed that irradiation of *(5)* in the presence of a dye and oxygen gave a stable peroxide, which was later shown to be the epidioxide *(6)*.[3] We now recognize this reaction as an example of the 1,4-cycloaddition of singlet oxygen to a conjugated diene. This reaction was later extended to simple 1,3-dienes by G. O. Schenck with his synthesis of ascaridole (Paper 10).

$$\xrightarrow[O_2]{\text{dye, } h\nu}$$

(5)   (6)

NOTES AND REFERENCES

1. K. Gollnick and G. O. Schenck, in J. Hamer (ed.), *1, 4-Cycloaddition Reactions*, Academic Press, Inc., New York, 1967.
2. C. Moureu, C. Dufraisse, and C. -L. Butler, *Compt. Rend.*, **183**, 101 (1926).
3. W. Bergmann and M. J. McLean, *Chem. Rev.*, **28**, 367 (1941).

4

Reprinted from *Compt. Rend.,* **64,** 1035–1037 (1867)

NOTE SUR LES CARBURES D'HYDROGÈNE SOLIDES, TIRÉS DU GOUDRON DE HOUILLE

M. Fritzsche

« En continuant mes recherches sur les carbures d'hydrogène solides tirés du goudron de houille, je suis arrivé à des résultats que je crois dignes d'être portés à la connaissance de l'Académie, et je profite de mon séjour à Paris pour lui présenter en même temps quelques échantillons des corps que j'ai obtenus récemment.

» La plus grande difficulté de mon travail consistait dans la préparation de substances sur la pureté desquelles on ne pourrait plus être en doute. Ayant acquis la conviction que le produit brut était un mélange de plusieurs corps très-semblables les uns aux autres, je cherchai des réactions spéciales pour chacun d'eux, et j'ai été assez heureux pour trouver un réactif qui donne avec les corps en question des réactions aussi caractéristiques que l'hydrogène sulfuré avec les métaux. Grâce à ce réactif, qui résulte de l'action de l'acide azotique sur un de ces corps, nommément le corps $C^{14}H^{10}$ (1), j'ai reconnu l'existence de cinq corps bien distincts, dont le point de fusion est situé entre 180 ou 190 et 235 degrés centigrades environ, et dont je vais brièvement énoncer les réactions.

» Le corps $C^{14}H^{10}$, le seul que j'ai obtenu jusqu'ici en état de parfaite pureté, donne avec mon réactif une combinaison d'un beau rouge violet qui cristallise en tables rhomboïdales; un second corps, dont le point de fusion est de 235 degrés centigrades environ, donne avec le même réactif des tables rectangulaires d'un bleu violacé très-foncé. La combinaison du même réactif avec un troisième corps représente des tables rectangulaires d'une couleur vert foncé, et celle d'un quatrième corps des prismes aciculaires d'une belle couleur orangée.

---

(1) $C = 12,\ H = 1.$

» Le cinquième corps enfin a une très-grande ressemblance avec le corps $C^{14}H^{10}$, et retient avec tant de ténacité des traces de ce dernier, que je ne suis pas encore en état de signaler exactement la couleur de sa combinaison avec mon nouveau réactif; mais en tout cas elle est beaucoup plus foncée que celle du corps $C^{14}H^{10}$. Et comme en même temps sa solubilité est beaucoup moindre, il est hors de doute qu'il y a là une substance particulière. Je ne connais pas encore la composition des quatre derniers corps, et voilà pourquoi je n'ose encore me prononcer sur la question de savoir auquel de ces cinq corps on doit donner le nom d'*anthracène*. M. Anderson a désigné en 1862, sous ce nom, le corps $C^{14}H^{10}$, et lui attribue un point de fusion de 213 degrés centigrades; mais comme M. Dumas a donné pour son anthracène un point de fusion de 180 degrés centigrades seulement, et que le point de fusion de mon cinquième corps se rapproche beaucoup plus de ce dernier chiffre que de celui de M. Anderson, il me paraît probable que le nom d'*anthracène* devra être accordé à ce dernier corps.

» Le précieux réactif dont je viens de parler donne aussi des combinaisons caractéristiques avec d'autres corps solides, provenant non-seulement du goudron de houille, mais aussi d'autres sources. Ce sont d'abord le chrysène de M. Laurent, et un corps incolore qui accompagne ce dernier, mais qui ne paraît pas être le pyrène; puis des corps qui se trouvent dans le goudron de houille et dont le point de fusion est proche de 100 degrés, et enfin l'idrialène de M. Dumas et le rétène. La naphtaline au contraire a résisté à toutes mes tentatives de la combiner avec ledit réactif.

» Le corps $C^{14}H^{10}$ que j'ai l'honneur de présenter à l'Académie en état de parfaite pureté est très-remarquable par plusieurs qualités. Il montre une fluorescence très-belle qui le fait paraître dans la lumière réfléchie coloré en violet très-brillant, surtout quand on fait adhérer ses paillettes aux parois d'un ballon dans lequel on a versé quelques gouttes de benzine. Cette fluorescence n'a pas été remarquée par M. Anderson, probablement parce qu'il n'avait pas éliminé les dernières traces de coloration jaune qui l'altèrent avec ténacité. Je suis parvenu à cela seulement par l'exposition des solutions du corps en question à la lumière directe du soleil, qui en peu de temps détruit le chrysogène, dont les plus légères traces suffisent à donner une couleur jaune très-prononcé à tous les corps que j'ai énumérés, et même à la naphtaline.

» Une seconde qualité non moins remarquable du corps $C^{14}H^{10}$ est la manière dont agit sur lui la lumière directe du soleil. En exposant aux rayons solaires des dissolutions de ce corps saturées à la température ambiante, il se dépose bientôt des cristaux d'une substance qui ne donne plus du tout de combinaison avec mon réactif, et qui non-seulement est presque insoluble dans tous les dissolvants, mais aussi presque inattaquable par les acides sulfurique et azotique concentrés, qui agissent énergiquement sur le corps $C^{14}H^{10}$. En soumettant cependant ce nouveau corps à une température assez élevée pour le fondre, il se transforme entièrement dans le corps auquel il doit sa formation, et il paraît évident que la lumière produit dans l'agrégation des molécules un changement que la chaleur détruit.

» L'action de l'acide azotique sur le corps $C^{14}H^{10}$ est très-remarquable par la diversité des produits qui en résultent. Outre le nouveau réactif dont j'ai déjà parlé, j'ai obtenu des traces d'un autre réactif, qui donne avec le corps $C^{14}H^{10}$ non pas des tables rhomboïdales rouge violacé, mais des prismes verdâtres, et je ne désespère pas d'obtenir encore d'autres réactifs qui m'aideront à séparer entre eux entièrement les carbures d'hydrogène solides.

» Ayant trouvé que l'action de l'acide azotique sur le corps $C^{14}H^{10}$ est très-énergique même à froid, j'ai cherché un corps qui, tout en diluant l'acide azotique, ne lui cédât cependant pas de l'eau, et j'ai trouvé que l'acide acétique cristallisable remplit parfaitement ces conditions. En versant sur le corps $C^{14}H^{10}$ assez de cet acide pour former un mélange pas trop épais, et en ajoutant à ce mélange goutte à goutte de l'acide azotique, on obtient bientôt une dissolution complète sans le moindre dégagement de vapeurs nitreuses, si on empêche l'élévation de température en plongeant le vase dans l'eau. De cette solution, de laquelle l'eau précipite un corps jaune résineux, on peut obtenir plusieurs substances jouissant de propriétés remarquables, suivant qu'on la laisse en repos, ou qu'on la porte à une température plus ou moins élevée, avec ou sans addition ultérieure du corps $C^{14}H^{10}$. J'ai l'honneur de présenter à l'Académie deux de ces corps, dans lesquels la combinaison nitreuse qui a opéré leur formation est encore très-mobile, puisqu'il suffit de faire bouillir ces substances avec de l'acide acétique pour provoquer un développement de vapeurs nitreuses. Le nouveau réactif au contraire est très-stable et supporte une chaleur de 200 degrés centigrades sans se décomposer. J'ai lieu de croire que c'est un corps trinitré, mais sa composition n'a pas encore pu être déterminée d'une manière exacte. »

5

Reprinted from *Compt. Rend.*, **182**, 1584–1587 (June 1926)

UN PEROXYDE ORGANIQUE DISSOCIABLE : LE PEROXYDE DE RUBRÈNE

Charles Moureu, Charles Dufraisse et Paul Marshall Dean

1. Dans une Note précédente (³), nous avons indiqué que le rubrène pouvait fixer l'oxygène libre, sous l'influence de la lumière, en se décolo-

---

(²) Séance du 21 juin 1926.
(³) *Comptes rendus*, **182**, 1926, p. 1440.

20

rant. Nous avons constaté que cette oxydation donnait naissance à un produit blanc, bien cristallisé, contenant du solvant de constitution, et qui présente la remarquable propriété de se dissocier sous l'action de la chaleur en régénérant ses constituants : solvant, rubrène, oxygène libre.

2. *Préparation et propriétés.* — Une solution de rubrène dans le benzène, par exemple, est exposée à la lumière, en présence d'une quantité suffisante d'air, soit qu'on opère dans un récipient de grande capacité, soit qu'on renouvelle le gaz au-dessus de la solution soumise à l'agitation. La teinte s'affaiblit peu à peu, en même temps qu'on voit s'atténuer la fluorescence, et, au bout d'un temps variable suivant les conditions de l'expérience, la liqueur est devenue complètement incolore et non fluorescente. La vitesse dépend surtout de l'intensité d'éclairement et de la concentration en oxygène. A la lumière solaire directe, la réaction est rapide. On peut observer une décoloration complète en quelques minutes, ce qui prouve que, même aux grandes dilutions, l'oxydation est intense. Après décoloration, le solvant ayant été chassé par distillation dans le vide à basse température, on obtient régulièrement un corps blanc, cristallisé en belles aiguilles.

Le rendement est sensiblement quantitatif. Tout au plus observe-t-on la formation d'une très petite proportion d'une substance jaune, facile à éliminer par lavage.

Le peroxyde de rubrène est un corps relativement peu soluble dans les solvants usuels, et, en particulier, dans le benzène. A cause de sa sensibilité à l'action de la chaleur, on ne peut déterminer son point de fusion qu'au bloc; il est situé aux environs de 190°.

3. *Dissociation.* — *a.* Dans une petite ampoule de verre on introduit, soigneusement desséché dans le vide, un peu de peroxyde pulvérisé, et l'on soude l'ampoule à une canalisation en relation avec une trompe à mercure, qui permettra l'extraction des gaz. Sur le trajet de la canalisation on dispose un tube en U où l'on pourra condenser les vapeurs par refroidissement.

Tout l'appareil étant à la température ambiante, on commence par y faire le vide (au millième de millimètre de mercure), pour éliminer l'air et les traces de solvant non combiné. On refroidit ensuite le tube en U dans le mélange neige carbonique-acétone, et l'on chauffe l'ampoule dans un bain d'huile de paraffine (en vase de verre, qui permettra d'observer le contenu pendant le chauffage). Peu après que l'on a dépassé la température de 100°, on voit apparaître des bulles dans la colonne de mercure : un gaz commence à se dégager. Le dégagement augmente à mesure que la température monte; il est important vers 150°.

Tandis que le gaz se dégage, la masse pulvérulente, qui était blanche au début, prend peu à peu une teinte rose de plus en plus foncée, et il se condense dans le tube en U des cristaux incolores très réfringents.

On élève la température lentement, en plusieurs heures, jusqu'au-dessus de 200°; à la fin, il n'y a plus de dégagement gazeux appréciable.

b. Dans une de nos expériences, par exemple, effectuée avec $0^g,50$ de peroxyde, nous avons recueilli $17^{cm^3},7$ de gaz, ainsi composé : $16^{cm^3},2$ d'oxygène, 1^{cm^3} de gaz carbonique, $0^{cm^3},5$ de gaz résiduel. Le tube en U avait condensé $0^g,026$ de vapeurs, et l'ampoule chauffée retenait $0^g,450$ de matière. Si l'on tient compte, outre l'oxygène libre, de l'oxygène combiné (sous forme de CO^2), on retrouve, au total, $0^g,498$ de matière sur $0^g,50$ dont on était parti, ce qui est un bilan très satisfaisant.

D'autre part, si nous comparons le poids $0^g,0216$ d'oxygène dégagé (libre ou sous forme de CO^2) au poids théorique $0^g,0249$, calculé en admettant 1^{mol} de benzène comme solvant de constitution (voir ci-dessous), nous trouvons, pour la décomposition, la proportion de 87 pour 100. Il est donc resté fixé sur le résidu 13 pour 100 de l'oxygène du peroxyde, soit par suite d'une dissociation incomplète, soit par évolution du peroxyde vers une forme oxydée stable. Étant donné qu'il a été impossible d'éviter la fusion d'une partie de la masse, celle-ci a nécessairement dissous du peroxyde, ce qui a pu limiter la dissociation.

A supposer, au surplus, qu'une réaction secondaire se soit effectuée pour une proportion de 13 pour 100, il n'en reste pas moins que la dissociation intégrale de notre peroxyde, avec formation d'oxygène *libre*, a porté sur la proportion considérable de 82 pour 100 de la quantité traitée.

Un tel rendement, même pour des réactions régulières et simples, est considéré comme excellent en Chimie organique. Or il s'agit ici d'une réaction vraiment extraordinaire, où l'oxygène se sépare à l'état libre, sous l'action de la chaleur, du carbone et de l'hydrogène qui l'avaient avidement fixé.

c. Nous nous sommes naturellement assurés que le résidu solide resté, après l'opération, dans notre ampoule, était en majeure partie constitué par du rubrène, dont il avait d'ailleurs la couleur (un peu plus foncée toutefois).

Déjà le point de fusion du produit brut, à 313° (au bloc), au lieu de 331°, était une sérieuse indication. Mais, de plus, ses solutions, comme celles du rubrène pur, étaient rouges et fluorescentes, et subissaient la même décoloration sous l'action de l'oxygène et de la lumière, ce qui achevait l'identification.

d. En ce qui concerne le poids $0^g,026$ de vapeurs condensées dans le tube en U, nous ne croyons pas devoir tirer une conclusion ferme sur sa signification pour la teneur en solvant de constitution, pour les raisons suivantes : 1° nous avions employé comme solvant, dans l'espèce, pour faire cristalliser le produit, un mélange de benzène et de ligroïne, dont nous ignorons le poids moléculaire moyen; 2° il n'est pas impossible qu'une partie du solvant ait été retenu par la masse, et, pour l'en chasser, il faudrait peut-être chauffer fortement et longtemps. Indiquons, à toutes fins utiles, que, pour le benzène, le poids trouvé s'accorderait mieux avec $0^{mol},5$ de solvant de constitution (théorie, pour 1^{mol}, $0^g,0607$, et, pour $0^{mol},5$, $0^g,0323$).

D'autres expériences ont été faites et ont donné des résultats analogues, et d'autres sont en cours.

4. Quelle peut être la structure du peroxyde de rubrène? La formule devra obligatoirement rendre compte de la propriété caractéristique : la séparation à l'état libre de l'oxygène et du rubrène. On conçoit que nous ne soyons pas encore en mesure de proposer une solution définitive à ce difficile problème. Il est hautement probable que dans le complexe rubrène-solvant-oxygène, les trois molécules qui doivent se séparer dans la réaction

de dissociation ont gardé leur individualité, comme les pierres d'un édifice et le ciment qui les unit. Ici le ciment serait le solvant. Nous aurions donc affaire à une combinaison du type dit *moléculaire*. Et il semble bien difficile, en effet, d'admettre qu'il ait pu y avoir entre le carbone et l'oxygène saturation mutuelle des valences chimiques ordinaires, ce qui exclurait, à nos yeux, la possibilité d'une dissociation intégrale du produit d'une telle union.

Nous continuons nos recherches sur ce point.

5. La peroxydation du rubrène, ainsi que la dissociation du peroxyde, soulèvent d'importants problèmes d'énergétique chimique et photochimique. Nous les envisageons avec d'autant plus d'intérêt qu'ils viennent rejoindre les recherches que par ailleurs nous poursuivons activement sur la catalyse d'autoxydation (actions antioxygènes et prooxygènes), où les peroxydes jouent, selon nous, un rôle de première importance. Bornonsnous, pour le moment, à souligner le fait que la dissociation du peroxyde de rubrène, avec libération de l'oxygène combiné, réfute péremptoirement une objection sérieuse qui pouvait être faite à notre théorie de la catalyse antioxygène, dans laquelle nous supposons qu'un peroxyde $A[O^2]$, où A peut être, en particulier, un corps organique, est susceptible de perdre, sous certaines influences, tout son oxygène, en régénérant le mélange $A + O^2$. Il pouvait sembler inadmissible que le carbone et l'hydrogène, après s'être combinés avec l'oxygène, pussent le laisser ainsi se détacher d'eux à l'état libre. Le cas du peroxyde de rubrène montre que cette objection est sans fondement.

6. La dissociation du peroxyde de rubrène n'est pas sans présenter certaines analogies avec celles de l'oxyhémoglobine, avec, toutefois, cette différence capitale, que l'hémoglobine renferme un élément minéral, le fer. Or l'aptitude à donner des peroxydes dissociables était considérée jusqu'ici comme l'apanage de corps entièrement minéraux. Pour l'hémoglobine, qui semblait faire exception, on rapportait généralement à l'atome de fer cette propriété, qui restait ainsi propriété minérale. Notre travail, en établissant qu'elle peut appartenir à des corps purement organiques, composés uniquement de carbone et d'hydrogène, autorise à penser que la fonction respiratoire de l'hémoglobine peut parfaitement être une propriété de la partie organique de la molécule, sans être nécessairement liée à la peroxydabilité de l'atome de fer.

23

6

Reprinted from *Liebigs Ann. Chem.*, **460**, 225–230 (Jan. 1928)

Über die photochemische Oxydation des Ergosterins;

von *A. Windaus* und *J. Brunken.*

Mit 1 Figur im Text.

[Aus dem Allgem. Chem. Universitätslaboratorium Göttingen.]

(Eingelaufen am 7. Januar 1928.)

In früheren Untersuchungen[1]) ist festgestellt worden, daß Ergosterin bei der Bestrahlung mit *ultraviolettem* Licht in einen antirachitisch sehr wirksamen Stoff übergeht. *Sichtbares* Licht ist dagegen ohne Einwirkung auf Ergosterin. Wir haben nun geprüft, ob es gelingt, durch Zusatz von *Sensibilisatoren* das Ergosterin für sichtbares Licht zu aktivieren. Dies ist tatsächlich der Fall, und zwar sind die Umwandlungsprodukte verschieden, je nachdem ob die Luft bei der Bestrahlung fern gehalten wird oder nicht. Bei Anwesenheit von Sauerstoff findet eine *Photooxydation* statt; hierbei erweisen sich *fluorescierende* Farbstoffe wie *Eosin, Erythrosin, Methylenblau, Chlorophyll* und *Hämatoporphyrin* als wirksam, während *Acridon* sehr wenig wirksam und *Fuchsin* ohne Wirkung ist. Ein Zusatz von *Kaliumcyanid* verhindert die Photooxydation nicht, im *Dunkeln* unterbleibt dagegen jede Reaktion.

Bei der Photooxydation lagert sich 1 Mol. Sauerstoff an Ergosterin an, und es entsteht ein ziemlich beständiges *Peroxyd* $C_{27}H_{42}O_3$ vom Schmelzp. 178°. Bei Verwendung von *Ergosteryl-acetat* erhält man nach der gleichen Reaktion das Peroxyd des Ergosteryl-acetats, das bei der Verseifung das Ergosterin-peroxyd liefert.

Das *Ergosterin-peroxyd* krystallisiert sehr schön und gibt ein krystallisiertes *Acetylderivat*, das mit dem Photooxydationsprodukt des Ergosteryl-acetats identisch ist.

---

[1]) A. Windaus u. A. Hess, Nachr. Ges. d. Wissenschaften Göttingen. Mathem.-Phys. Klasse **1926**, S. 175. — O. Rosenheim u. T. A. Webster, Lancet, i. 306 (1927, 1); Biochem. Journ. **21**, 397 (1927). — György, Klin. Wochenschr., VI. Jahrg. 1927, Nr. 13.

Aus Kaliumjodid scheidet es schon in der Kälte Jod aus,
mit Digitonin liefert das Ergosterin-peroxyd eine in heißem
Alkohol lösliche *Additionsverbindung*. Es ist nicht unzer-
setzt destillierbar. Während es von alkoholischer Salz-
säure schnell verändert wird, erweist es sich als wider-
standsfähig gegenüber alkoholischer Kalilauge. Bei der
Reduktion mit Natrium und Alkohol spaltet es überraschenderweise den Peroxydsauerstoff ab; das hierbei zunächst ge-
bildete Ergosteri nimmt langsam 1 Mol. Wasserstoff auf
und geht in *Dihydro-ergosterin* über. Letzteres kann dem-
gemäß auch bei der Behandlung des *Ergosterins* mit Natrium und Alkohol erhalten werden. Bei der *katalytischen
Hydrierung* des Dihydro-ergosteryl-acetats und des Ergosteryl-acetats mit Platin und Eisessig bei 70° entsteht
sofort das γ-Ergostylacetat von F. Reindel.[1])

Über zahlreiche andere Umwandlungsprodukte des
Ergosterin-peroxyds werden wir später berichten.

Die glatte photochemische Bildung des Ergosterin-
peroxyds aus Ergosterin und Sauerstoff besitzt darum
einiges Interesse, weil es nur selten gelungen ist, die bei
der Oxydation ungesättigter Kohlenwasserstoffe mit Luft
primär entstehenden Moloxyde in krystallisiertem Zustande
zu isolieren.[2]) Auch bei den von Gaffron[3]) untersuchten
Photooxydationen sind die Peroxyde nur durch Reaktionen
nachgewiesen, nicht aber in Substanz gefaßt worden.

Unter den Sterinen wird nur das Ergosterin leicht
photochemisch verändert, während *Cholesterin*, *Sitosterin*,
Stigmasterin und *Dihydro-ergosterin* unter *denselben Bedingungen* unverändert bleiben. Bemerkenswert erscheint das
Ergosterin-peroxyd auch dadurch, daß es mit Arsenchlorid
und Antimonchlorid ähnliche *Farbenreaktionen* zeigt wie sie
für Vitamin A als charakteristisch gelten.[4])

Indessen besitzt das Ergosterin-peroxyd weder bestrahlt

---

[1]) A. **452**, 45 (1927).
[2]) Kohler, Am. **36**, 177 (1906). — C. Moureu u. C. Dufraisse,
Ph. Ch. **130**, 472 (1927).
[3]) B. **60**, 2229 (1927).
[4]) Rosenheim u. Webster, Lancet 1926, 806.

noch unbestrahlt eine antixerophtalmische oder eine antirachitische Wirkung. Sehr bemerkenswert erscheint uns
die Tatsache, daß das aus dem *nicht aktivierbaren Peroxyd
zurückgewonnene Ergosterin wiederum aktivierbar ist.* Das
deutet darauf hin, daß das Ergosterin selbst und nicht eine
Verunreinigung desselben das antirachitische Provitamin
darstellt. Das *Dihydro-ergosterin* ist unbestrahlt und bestrahlt physiologisch unwirksam.

Darstellung des Ergosterin-peroxyds.

Zur Photooxydation haben wir die folgende Apparatur
verwendet: Eine Osram-Nitralampe von 200 Watt befand
sich in einem mit Wasser gefüllten und mit Zufluß und
Abfluß versehenen Becherglase; dieses war umgeben von
einem zweiten, etwas weiteren
Becherglase, das die zu be-
strahlende Lösung enthielt (Fig.).
Um eine möglichst gute Ausnutzung der Lichtenergie zu er-
zielen, wurde der Apparat in
einen mit Spiegelglasscheiben
ausgekleideten Kasten gesetzt.

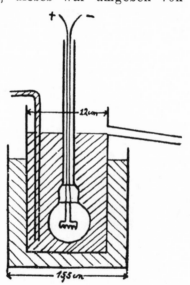

Eine auf 60° erwärmte Lö-
sung von 4 g Ergosterin in
1200 ccm 95-proc. Alkohol wurde
mit 6 mg Eosin versetzt und in
den Apparat gefüllt; während
der Belichtung wurde ein langsamer Strom von Sauerstoff
durch die Reaktionsflüssigkeit
geleitet. Nach etwa 3 Stunden war die Oxydation beendet,
was daran erkannt wurde, daß eine Probe der Lösung nicht
mehr mit einer alkoholischen Digitoninlösung reagierte. Nun-
mehr wurde die Lösung im Vakuum auf ein kleines Volum
eingedampft, die sich abscheidenden Krystalle wurden abfiltriert und wiederholt aus Alkohol oder Aceton umkrystallisiert, bis der Schmelzpunkt bei 178° konstant blieb. Ausbeute 70%.

An sonnigen Tagen sind wir auch so verfahren, daß wir große flache Glasschalen, in die wir eine 2—3 cm hohe Schicht einer alkoholischen Ergosterinlösung und etwas Eosin gefüllt hatten, lose bedeckt auf das Dach des Instituts gestellt haben. Unter diesen Bedingungen wird das Ergosterin ebenfalls in wenigen Stunden oxydiert.

Die *Sauerstoffabsorption* bei der Photooxydation des Ergosterins wurde volumetrisch mittels kleiner Schüttelgefäße bestimmt; die Belichtung erfolgte hier mit einer 1000 Wattlampe; als Sensibilisator wurde Eosin verwendet. Der Verlauf der Reaktionskurve zeigte, daß die Sauerstoffabsorption nach Aufnahme von 1 Mol. Sauerstoff für 1 Mol. Ergosterin nahezu zum Stillstand kam.

Eigenschaften. Das Ergosterin-peroxyd ist sehr schwer löslich in Petroläther, ziemlich leicht löslich in Äther, Aceton und Alkohol. Aus wenig Aceton oder Äther krystallisiert es in sehr großen Prismen. Gegen Säuren ist es äußerst unbeständig. Beim Erhitzen einer Eisessiglösung des Peroxyds erfolgt nach kurzer Zeit Braunfärbung. Eine Chloroformlösung färbt sich auf Zusatz von Arsentrichlorid erst grünblau, dann tiefblau und schließlich violett. Mit Antimonchlorid erhält man eine braun-violette, mit Zinntetrachlorid eine weinrote Lösung. Schüttelt man eine Chloroformlösung des Peroxyds mit konz. Schwefelsäure, so färbt sich die Säure tiefrot.

4,948, 4,852 mg Subst.: 14,240, 13,960 mg CO_2, 4,70, 4,56 mg H_2O

$C_{27}H_{42}O_3$　　Ber. C 78,20　　　　H 10,22
　　　　　　　　Gef. „ 78,50, 78,49　　„ 10,63, 10,57.

Molekulargewichtsbestimmung. 0,1986 g Subst. erniedrigten den Schmelzpunkt von 11,987 g Naphtalin um 0,28⁰. — 13,4 mg Subst. erniedrigten den Schmelzpunkt von 142,2 mg Campher um 9,0⁰.

Molgew. $C_{27}H_{42}O_3$　　Ber. 414　　　　Gef. 411, 419.

Drehung: 0,2805 g Subst. mit Chloroform zu 10 ccm gelöst gaben $\alpha_D = -1,00^0$; $l = 1$ dm, $t = 19^0$. $[\alpha]_D^{19} = -35,7^0$. 0,3243 g Subst. mit Chloroform zu 10 ccm gelöst, gaben $\alpha_D = -1,15^0$; $l = 1$ dm, $t = 18^0$ $[\alpha]_D^{18} = -35,5^0$.

Wie die *Bestimmung der Hydroxylgruppen* nach Zerewitinoff ergibt, enthält das Ergosterin-peroxyd noch *eine* Hydroxylgruppe.

0,1119 g Subst. entwickelten 7,2 ccm Methan (754 mm, 20°).

$C_{27}H_{42}O_3$: 1 OH. Ber. 4,10 Gef. 4,54.

Bestimmung des aktiven Sauerstoffs. 0,1363 g Subst. wurden in 20 ccm Eisessig gelöst und mit überschüssigem Kaliumjodid versetzt; unter gleichen Bedingungen wurde eine Blindprobe angesetzt. Die nach 24 stündigem Stehen vorgenommene Titration ergab gegenüber der Blindprobe einen Mehrverbrauch von 5,4 ccm $^n/_{10}$-Natriumthiosulfat-lösung. Der für den aktiven Sauerstoff gefundene Wert ist zu niedrig, und zwar aus dem Grunde, weil die Lösung noch unzersetztes Peroxyd enthielt.

Thermisches Verhalten des Peroxyds: Bei der Destillation im Hochvakuum gibt das Peroxyd ein krystallisiertes Destillat, das noch nicht genau untersucht worden ist. Bis zu Temperaturen von 180—185° ist das Peroxyd dagegen ziemlich beständig.

0,5 g Subst. wurden 30 Minuten auf 180° erhitzt; hierbei erfolgte eine leichte Gelbfärbung, aber beim Abkühlen erstarrte die Schmelze krystallin und das Drehungsvermögen hatte sich kaum geändert: $[\alpha]_D^{21} = -34,3°$. (0,0437 g Subst. wurden mit Chloroform auf 2 ccm gelöst, $\alpha_D = -0,75°$, $l = 1$ dm, $t = 21°$). Nach zweimaligem Umkrystallisieren der Schmelze aus Aceton wurde reines Ergosterin-peroxyd vom Schmelzp. 178° in einer Ausbeute von 75 Proc. zurückgewonnen.

Ergosterinperoxyd-digitonid.

Eine Lösung von Ergosterin-peroxyd in Alkohol wurde in der Hitze mit einer 1-proc. alkoholischen Digitonin-lösung versetzt. Die über Nacht entstandenen weißen, amorphen Flocken wurden abfiltriert und aus Alkohol umgelöst. Das Digitonid ist in Pyridin und heißem Alkohol leicht löslich, unlöslich in Wasser.

$[\alpha]_D^{15} = -38°$ (0,1221 g Subst. mit Pyridin zu 10 ccm gelöst. $\alpha_D = -0,47°$, $l = 1$ dm, $t = 15°$).

Acetylderivat des Ergosterin-peroxyds.

Eine geringe Menge Peroxyd wurde 10 Minuten mit Essigsäureanhydrid gekocht. Beim Erkalten schieden sich aus der stark braun gefärbten Lösung Krystalle ab, die nach häufigem Umkrystallisieren aus Alkohol und Aceton bei 202° schmolzen. Aus Aceton wurde das Acetylderivat in schönen, langgestreckten Blättchen erhalten.

5,223 mg Subst.: 14,610 mg CO_2, 4,70 mg H_2O.

$C_{29}H_{44}O_4$ Ber. C 76,22   H 9,72
            Gef. „ 76,31   „ 10,07.

Photooxydation des Ergosteryl-acetats.

5 g Ergosteryl-acetat wurden in 1500 ccm Alkohol gelöst und mit 5 mg Eosin versetzt. Die mit Luft gesättigte Lösung wurde in einer flachen Schale dem Sonnenlicht ausgesetzt, und zwar die gleiche Zeitdauer, welche unter denselben Bedingungen zur Photooxydation von Ergosterin erforderlich ist. Falls sich im Laufe der Bestrahlung Blättchen von unverändertem Ergosteryl-acetat ausschieden, wurden diese durch gelindes Erwärmen wieder in Lösung gebracht. Der größte Teil des Alkohols wurde im Vakuum abdestilliert und das in Nadeln auskrystallisierte *Ergosteryl-acetat-peroxyd* durch Umkrystallisieren aus Alkohol und Essigester gereinigt, bis der Schmelzpunkt konstant bei 202⁰ lag. Der Stoff erwies sich durch den Mischschmelzpunkt als identisch mit dem Acetat, das durch Acetylieren des Ergosterin-peroxyds gewonnen war. Die Ausbeute betrug 40 Proc. Das Acetat ist leicht löslich in Chloroform und Äther, ziemlich leicht in Aceton, schwer in Alkohol. Mit Arsentrichlorid und Schwefelsäure gibt es die gleichen Farbreaktionen wie das Ergosterin-peroxyd.

0,3986 g Subst. mit Chloroform zu 10 ccm gelöst, $\alpha_D = -0,69^0$, $l = 1$ dm, $t = 15^0$.

0,3577 g Subst. mit Chloroform zu 10 ccm gelöst, $\alpha_D = -0,64^0$, $l = 1$ dm, $t = 15^0$.

$[\alpha]_D^{15} = -17,3^0, -17,9^0$.

Verseifung des Ergosterin-peroxyd-acetats.

1 g Ergosterin-peroxyd-acetat wurde mit 35 ccm 5-proc. alkoholischer Kalilauge 2 Stunden gekocht. Die Lösung wurde mit Wasser angespritzt; die beim Erkalten sich abscheidenden Krystalle wurden aus Aceton umkrystallisiert.

Der Schmelzpunkt des Verseifungsproduktes lag bei 178⁰, und das spezifische Drehungsvermögen betrug $[\alpha]_D^{20} = -36,4^0$.

(0,3677 g Subst. mit Chloroform zu 10 ccm gelöst, $\alpha_D = -1,34^0$, $l = 1$ dm, $t = 20^0$).

[*Editor's Note:* Material has been omitted at this point.]

Editor's Comments
on Papers 7 Through 9

7   KAUTSKY and DE BRUIJN
    Die Aufklärung der Photoluminescenztilgung fluorescierender Systeme durch Sauerstoff: Die Bildung aktiver, diffusionsfähiger Sauerstoffmoleküle durch Sensibilisierung

8   KAUTSKY et al.
    Photosensitized Oxidation Mediated by a Reactive, Metastable State of the Oxygen Molecule

9   KAUTSKY
    Quenching of Luminescence by Oxygen

THE KAUTSKY MECHANISM FOR DYE-SENSITIZED PHOTOOXYGENATION

In 1931, Kautsky and de Bruijn (Paper 7) first proposed that singlet molecular oxygen was the reactive intermediate in dye-sensitized photooxygenations. Quenching of the fluorescence and phosphorescence of dyes (see Papers 34 and 35) by oxygen was interpreted as resulting from energy transfer from an excited state of the dye to O_2 to form a metastable, reactive oxygen molecule. Because of the historical importance of this classic work of Kautsky, a translation of the paper describing the experiments is also included (Paper 8).

In an incisive series of experiments, Kautsky and coworkers demonstrated that photooxygenation could occur even when the dye sensitizer and the substrate were physically separated on silica gel particles. Therefore, the reactive intermediate in photooxygenation had to be a gaseous species capable of diffusing from the particles with adsorbed sensitizer to the particles with adsorbed substrate. Kautsky originally identified the reactive intermediate as the $^1\Sigma_g^+$ state of O_2. This proposal was challenged by Gaffron on the basis of energy considerations; that is, dyes with less than the 37.5 kcal/mol required for the formation of the $^1\Sigma_g^+$ state were able to sensitize the photooxygenation of allyl thiourea. However, following the spectroscopic observation of $^1\Delta_g$ O_2 at 22.5 kcal/mol by Ellis and Kneser, Kautsky revised his mechanism to include this species as the reactive intermediate (Paper 9).

Although the ingenious experiments of Kautsky unambiguously demonstrated the intermediacy of a diffusible species, the singlet oxygen mechanism was generally disregarded in favor of a mechanism first proposed by Schönberg (Paper 12), which involved an unstable sensitizer-oxygen complex. This mechanism was also advocated by Schenck in 1948 (Paper 13). The Kautsky mechanism was later revived by experiments which showed that chemically produced singlet oxygen would oxygenate various organic substrates to give products identical to those obtained from dye-sensitized photooxygenation (Papers 23–28).

Reprinted from *Naturwiss.*, **19**(52), 1043 (1931)

Die Aufklärung der Photoluminescenztilgung fluorescierender Systeme durch Sauerstoff: Die Bildung aktiver, diffusionsfähiger Sauerstoffmoleküle durch Sensibilisierung.

Fluorescenz und Phosphorescenz vieler Farbstoffmoleküle, die an Grenzflächen adsorbiert sind, werden durch molekularen Sauerstoff geschwächt bzw. völlig getilgt[1]. Wir haben jetzt diesen Sauerstoffeinfluß, den wir als die Grunderscheinung der photodynamischen Oxydationswirkungen durch belichtete fluorescierende Farbstoffe erkannt haben, aufgeklärt, und zwar: als Bildung einer nur kurze Zeit beständigen, sehr aktiven, diffusionsfähigen Form der Sauerstoffmoleküle, die durch die Übertragung der Anregungsenergie fluorescierender Farbstoffe auf den Sauerstoff zustande kommt[2]. Der zur Aktivierung nötige Energiebetrag ist verhältnismäßig gering, da selbst die tiefrote Fluorescenz des Chlorophylls durch Sauerstoff weitgehend ausgelöscht wird.

Wir stellen getrennt zwei Kieselsäureadsorbate her, das eine von einem fluorescierenden Farbstoff (z. B. Trypaflavin), das andere von einem Acceptor (z. B. p-Leukanilin). Beide werden durch gemeinsames Verreiben innig vermischt. Auf diese Weise sind Sensibilisator und Acceptor räumlich vollkommen voneinander getrennt. Belichtet man die Mischung mit sichtbarem Licht, so färbt sich die belichtete Stelle durch Oxydation der Leukoverbindung rot. Das farblose Adsorbat der Leukoverbindung allein wird bei den gegebenen Versuchsbedingungen in keiner Weise verändert. Wir verstehen dieses merkwürdige Ergebnis nur so, daß die einzelnen, adsorptiv festgelegten, energetisch isolierten, langlebigen, angeregten Farbstoffmoleküle die absorbierte Lichtenergie in der Grenzfläche auf die Sauerstoffmoleküle übertragen. Der so aktivierte, einige Zeit beständige Sauerstoff diffundiert aus der Grenzfläche des Trypaflavinadsorbates durch den Gasraum an die Grenzfläche des Acceptoradsorbates und oxydiert dort die adsorbierten Moleküle des p-Leukanilins zum roten p-Rosalinin. Der Vorgang ist äußerst druckabhängig. Beobachtet man die Veränderung des Adsorbatgemisches bei verschiedenen Gasdrucken, immer nach je 10 Minuten Belichtungszeit, so findet man ein ausgeprägtes Maximum der Rotfärbung bei etwa 0,001 mm Sauerstoffdruck. Druckerhöhung auf etwa 0,005 mm läßt in 10 Minuten nur noch eine sehr geringe Rotfärbung entstehen; ebenso wirkt eine Verminderung des Druckes unter 0,001 mm. Die Druckabhängigkeit ist nicht merkbar, wenn die Diffusionswege vom fluorescierenden Farbstoff zum Acceptor in der Grenzfläche sehr gering werden, weil dann die Lebensdauer der aktivierten Sauerstoffmoleküle keine so ausschlaggebende Rolle mehr spielt. Kieselsäureadsorbate, die gleichzeitig fluorescierenden Farbstoff und Acceptor in derselben Grenzfläche adsorbiert enthalten, färben sich, auch bei normalem Luftdruck belichtet, sehr rasch intensiv rot.

Wir finden hier also einen ganz neuartigen, besonders klaren und durchsichtigen Fall von Energieumwandlung an Grenzflächen, bei der die einzelnen Teilprozesse zeitlich und räumlich vollkommen getrennt werden können. Wir werden auch weiterhin unsere Aufmerksamkeit diesen experimentellen Befunden zuwenden.

Wir vermuten, daß der von uns durch seine Eigenschaften charakterisierte aktivierte Sauerstoff im Zusammenhang mit dem von W. H. J. CHILDS und R. MECKE[1] rein spektroskopisch ermittelten metastabilen Zustand ($^1\Sigma$ Term) des Sauerstoffmoleküls steht.

Heidelberg, Chemisches Institut der Universität, den 22. November 1931.   H. KAUTSKY.   H. DE BRUIJN.

[1] H. KAUTSKY u. A. HIRSCH, Ber. dtsch. chem. Ges. **64**, 2677 (1931).
[2] An anderer Stelle wird demnächst ein ausführlicher Bericht erscheinen.

[1] W. H. J. CHILDS u. R. MECKE, Z. Physik **68**, 344 (1931).

8

PHOTOSENSITIZED OXIDATION MEDIATED BY A REACTIVE, METASTABLE STATE OF THE OXYGEN MOLECULE

H. Kautsky, H. de Bruijn, R. Neuwirth, and W. Baumeister

Chemical Institute of the University of Heidelberg

*This article was translated expressly for this Benchmark
volume by M. F. Mattisson, Detroit, Michigan, from
"Energie-Umwandlung an Grenzflächen: VII.
Photo-sensibilisierte Oxydation als Wirkung eines
aktiven, metastabilen Zustandes des Sauerstoff-Moleküls,"
Chem. Ber., **66,** 1588–1600 (1933)*

In this article, we report *experimental evidence for a metastable, reactive state of the oxygen molecule,* which is formed by the interaction of oxygen with excited molecules of fluorescent materials. A short preliminary communication on this subject was published by Kautsky and de Bruijn in Naturwissenschaften.[1] The experiments have been continued in partial collaboration with Dr. Baumeister using improved techniques. Several questions of Mr. Neuwirth were thoroughly examined.

Prior to our observation of the reactive, metastable form of oxygen, its strong oxidizing effects were known as the "photo-dynamic effect of fluorescent dyestuffs."[2] We suggest that it would be useful to change this still widely used expression to "photosensitized oxidation." This expresses more clearly what is involved: oxidizable materials (acceptors) in the living organism or in the test tube, which in the dark or when exposed to light are not changed by atmospheric oxygen, absorb molecular oxygen when irradiated in the presence of fluorescent materials (sensitizers).

The use of simplified experimental techniques has enabled us to establish that the fundamental mechanism of photosensitized oxidation involves a mutual interaction between excited fluorescent molecules and oxygen.[3] This interaction is immediately manifested visibly by the partial extinction of the fluorescence of various fluorescent systems upon the addition of oxygen. The experiments reported in this article provide a better understanding of *photosensitized oxidation, which involves the transfer of excitation energy from the fluorescent system exposed to light to oxygen molecules which are converted into a diffusible, metastable, active form that is more reactive in oxidations.*

Our experiments were based on the assumption that, during the interaction between excited molecules of fluorescent materials and oxygen, activated oxygen molecules are formed that can diffuse a certain distance before they are converted back to normal oxygen with a loss of energy. Therefore, we spatially isolated the molecules of the sensitizer from those of the acceptor; that is, we separated by a given distance the place at which the oxygen is activated and the place at which the oxygen exerts those special oxidizing effects that distinguish it from normal oxygen. If, in fact, an active form of gaseous oxygen with a long lifetime is produced by the sensitizer, we would expect that, despite the spatial separation of the sensitizer from the acceptor, oxidation of the acceptor would be observable. This would exclude an oxidation mechanism that requires binding with the sensitizer, as has been hitherto assumed.

The sensitizer as well as the acceptor have to be molecularly distributed in order to be effective; they must be spatially separated by depriving the molecules of their free mobility. This immobilization can be achieved by the polar adsorption of the molecules at the expanded inner boundary surfaces of a gel.[4] We prepared separately two adsorbates; one contained the sensitizer and the other, the acceptor. When portions of the sensitizer adsorbate are mixed with those of an acceptor adsorbate, the various gel particles are in contact with each other, but the molecules of the sensitizer cannot interact with the molecules of the acceptor as almost all adsorbed molecules are uniformly distributed and fixed to the interior of the gel particles. Therefore, the sensitizer molecules are close to the acceptor molecules but completely separated from them.

The critical experiment for the detection of the diffusible, activated oxygen involved irradiating a thoroughly mixed sample of both adsorbate granules in the presence of oxygen, with the result that the acceptor is oxidized despite the total separation from the sensitizer. It is to be expected that the *lifetime of the energy-rich, metastable oxygen molecules depends upon the frequency with which they collide with other molecules. The oxidation of the acceptor when exposed to light, therefore, depends on the oxygen pressure and on the distance of the acceptor from the sensitizer.* These opinions guided us in the design of our experiments.

DESCRIPTION OF THE EXPERIMENTS

The type and arrangement of the individual components of the system and the specific experimental conditions are discussed below.

Preparation and Nature of the Individual Adsorbates

Adsorbent: As adsorbent, we used exclusively Silica Gel B with a particle size of 1–3 mm, great quantities of which are manufactured in a very uniform and pure state by Agfa (Aktiengesellschaft fur Annilinfarbenfabrikation) to whom we are indebted for this material. The gel is evacuated until it is as clear as glass. The gel has large pores, which is advantageous for the gas exchange. It contains small quantities of Fe(III) compounds, which have an oxidizing effect upon the acceptor during the adsorption of the latter. Therefore, we specially purified the gel by treating a large quantity of it with sulfuric acid (1 vol. H_2SO_4 : 1 vol. H_2O) until a sample of the gel did not give any blue coloration or only a weak blue coloration when tested with diphenylamine sulfuric acid or with potassium ferrocyanide. This takes about three weeks. The gel is then continuously washed with water to which a small amount of ammonia has been added and finally, with pure water until no trace of ammonia can be detected with Nessler's reagent.

Sensitizer adsorbate: As sensitizer, we used the strongly fluorescent dyestuff trypaflavine, which is adsorbed by silica gel as a basic dyestuff.

Trypaflavine standard solution: 5 mmol (1.2946 g pure trypaflavine) in 1 liter water.

Preparation of the adsorbate: Powdered silica gel (10 g) in a solution containing 5 cc of trypaflavine standard solution is introduced into water, stirred and allowed to

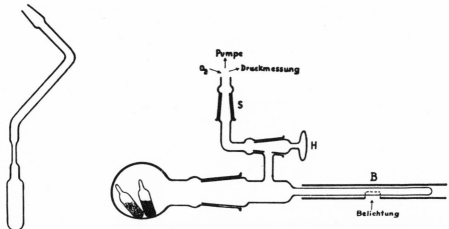

Figure 1   Device for pumping out the adsorbate.

Figure 2   Apparatus for detection of the activated metastable form of the oxygen molecule. *Pumpe,* pump; *Druckmessung,* pressure measurement; *Belichtung,* irradiation.

stand overnight, and washed with water. The adsorbate is then dried in a drying oven at about 50°C. To carry out the experiments in the apparatus of Figure 2, the adsorbate must be in sealed, very thin vials that are highly evacuated and free from gases and vapors. This is achieved in the simple, small device depicted in Figure 1. It consists of an angular glass tube, with a ground joint at one end for connection to the high-vacuum pump and carries a test tube at the other end with a length of about 4 cm. Close to the point of attachment, the tube, after having been filled with the predried adsorbate, is constricted so that subsequently, it can easily be sealed under the high vacuum. The constriction also offers protection against the considerable dusting of the adsorbate during the degassing. The evacuation lasts at least 4 hours at a bath temperature of 110°C.

The properties of the trypaflavine adsorbates (yellow color, intense yellow-green fluorescence and phosphorescence, and extinction of the luminescence by oxygen, which is of special importance for the experiments carried out here) have been described in detail in papers II and IV of this series.[5] However, the coloration of the trypaflavine adsorbates upon prolonged intensive irradiation has to be emphasized. The coloration becomes more pronounced with increasing oxygen pressure and manifests itself in that the irradiated spot becomes rust brown while the oxygen absorption becomes gradually smaller and that this spot completely loses its capacity of luminescence. This behavior will be discussed later.

Acceptor adsorbate: The acceptor has to be a compound that is polarly adsorbed by silica gel and which, under the given conditions, is not oxidized by normal oxygen but by activated oxygen. These requirements are met by leuco compounds of triphenylmethane dyestuffs. They possess the special advantage that, when oxidized, they are converted into intensely colored dyestuffs; therefore, they can serve as visible indicators for very small amounts of activated oxygen. The first experiments were carried out with *p*-leucaniline, and the later experiments with leucomalachite green, which is more suitable because the blue color of malachite green formed during the oxidation of this compound is easier to distinguish from the above-mentioned rust-red coloration of the trypaflavine adsorbate arising upon extended irradiation. We describe here only experiments in which leucomalachite green was used as the acceptor.

Preparation of leucomalachite green: Malachite green is dissolved in hydrochloric acid and reduced with (not too much) zinc dust until the solution is almost colorless. The leuco base is precipitated with sodium hydroxide solution, filtered off by suction, dissolved in alcohol, and crystallized repeatedly from alcohol. The leucomalachite green thus obtained is pure white.

Leucomalachite green standard solution: Two mmol (0.660 g) leucomalachite green is dissolved in a small amount of dilute hydrochloric acid and diluted with water to 100 cc (to be stored in the dark).

Preparation of the adsorbate: Powdered silica gel (5 g) is introduced into a solution containing 5 cc of leucomalachite green standard solution and 95 cc of water. The mixture is stirred for 15 min (longer adsorption times are ineffective since moist adsorbates are gradually oxidized in the air), washed four times with water, and, finally, rapidly rinsed once with a small quantity of pure acetone to accomplish faster drying of the adsorbate. The adsorbates should not be stored in the moist state in the air since they take on a bluish color; therefore, they have to be evacuated rapidly (in the apparatus depicted in Figure 1). One evacuates first for 2 hours without heating, and then for an additional 3-4 hours at 50-60°C. The leucomalachite green adsorbate sealed in the vial is almost completely colorless and stable for an indefinite time.

It is important to note that the acceptor adsorbate should not show any color when irradiated with oxygen in the absence of a sensitizer. The dry, evacuated leucomalachite green adsorbate ideally meets this requirement in that the adsorbate does not change in any way when intensively irradiated for 10 hours in the apparatus described in Figure 2 when varying oxygen pressures are applied.

Experimental apparatus (Figure 2): Into the thick-walled round flask (with a capacity of about 300 cc), which is attached by a wide ground joint, one introduces, depending upon the type of the experiment, one or two evacuated vials containing the adsorbates. After attaching the greased ground joint, the whole apparatus is evacuated. Stopcock H is then closed and the apparatus removed at ground joint S. The vials are shattered by shaking. If two vials are used, the contents of which have to be well mixed, one continues shaking until uniform mixing has been achieved. By inclining the apparatus, one allows the adsorbates to flow into the narrow, thin-walled irradiation tube B (diameter of 6 mm) and attaches again the apparatus to the ground joint S. This connects the apparatus to the high-vacuum pump, a mercury manometer, and a McLeod gauge, which makes it possible to measure the adjustable oxygen pressures. The experiments are carried out with the adsorbates or adsorbate mixtures contained in the small irradiation tube B. This tube is masked by a loosely fitting, movable tube of black paper with a square window. The light passes through this window to the contents of the small tube B. In this manner, one obtains sharply defined, photochemically changed spots that are easily distinguishable from the unchanged surroundings. The light source was a New Liliput Arc Lamp (5 A, d. c.) from the company *Leitz*, Wetzlar. The light is directly focused on the window by means of a lens at the arc lamp. An interposed copper sulfate filter protects the irradiated system against heat.

Effect of Oxygen Pressure on the Photosensitized Oxidation
in the Adsorbate Mixture

The powdered, thoroughly mixed adsorbates of trypaflavine and leucomalachite green were irradiated in the small irradiation tube B at different oxygen pressures. Oxygen or air is allowed to enter the apparatus and is determined with the McLeod apparatus or, in the case of higher pressures, with the mercury manometer. The oxygen pressures thus measured are the pressures at the boundary surface of the adsorbates, but only the pressures that are in equilibrium with the quantity of oxygen adsorbed at the adsorbate boundary surfaces. For each newly adjusted oxygen pressure, the irradiation window of the small tube B is laterally displaced by more than a window width so that the results of the individual irradiations over the same length of time for the different oxygen pressures can be evaluated. Results of several irradiations are summarized in the following table:

O_2 pressure in 0.001 mm	Irradiation time (min)	Coloration
0.4	10	None
1.4	10	None
12	10	Little
18	10	Blue
24	10	Little
40	10	None

When pressures higher than those listed in the table (up to the partial pressure of atmospheric oxygen) were employed, the blue coloration of the irradiated spot was no longer observed. The oxygen pressure decreases measurably with increasing time of irradiation; that is, oxygen is consumed. The duration of the phosphorescence of the trypaflavine adsorbate makes it possible to determine how much oxygen is present within a range of very small pressures from below 0.001 to about 0.005 mm.[3]

These experiments gave the following results. Despite spatial separation of sensitizer and acceptor, oxidation of the acceptor is observable as the formation of blue color when the system is exposed to light. Therefore, the presence of oxygen is required. Under the experimental conditions, photosensitized oxidation takes place only in a very limited, low range of pressure; this result is understandable in terms of the intermediacy of a metastable, excited state of molecular oxygen. It is not suprising that with low pressures coloration does not appear within the irradiation periods as the quantity of oxygen used is too small for effecting visible changes.

Effect of the Distance of the Acceptor Molecules from
the Sensitizer Molecules on the Photosensitized Oxidation

The upper limit of the oxygen pressure range in which the photosensitized oxidation can still be observed is determined essentially by the distance of the sensitizer molecules from the acceptor molecules: the shorter the distance, the higher the pressure at which the blue coloration will still occur under the same experimental conditions. An increase of this distance has the opposite effect. This limit will finally drop to such small oxygen pressures that, even with experimental times that are considerably longer, only such a small quantity of acceptor is oxidized that no visible color is produced. These considerations have been examined experimentally.

Reduction of the distance between acceptor and sensitizer: We reduced the length of the path over which the active oxygen diffuses from the sensitizer to the acceptor by adsorbing the sensitizer and the acceptor on the same gel particles. A double adsorbate is formed on which both molecule species are held in close proximity.

Preparation of the double adsorbate: Trypaflavine adsorbate (5 g) (the same as in the previous experimental series) is added to a solution containing 95 cc of leucomalachite green standard solution and 95 cc of water. The trypaflavine + leucomalachite green silica gel is then treated as the above-described simple leucomalachite green adsorbate, except that light must be excluded during the preparation.

The irradiation of the double adsorbate in the small tube B of the apparatus depicted in Figure 2 exhibited a pressure dependence which differed considerably from that of a mixture of the two individual adsorbates. The irradiated double adsorbate became blue at oxygen pressures adjusted from 0.003 mm to 1 atm. Therefore, there is no effect on photosensitized oxidation by oxygen pressure with this arrangement of the molecules. The path that the activated oxygen has to traverse from the sensitizer to the acceptor is so shortened in the double adsorbate that the number of deactivating collisions on this path is greatly decreased as compared to those in the system of the separated adsorbates.

Increase of the distance between sensitizer and acceptor: We separated a layer of trypaflavine adsorbate from a layer of leucomalachite green adsorbate by a distance of a few millimeters. This can be experimentally accomplished in various ways (which we shall not describe here as the description is too lengthy). Even when we used very long irradiation times, the result was always negative. A blue coloration of the acceptor was visible neither at low nor high pressures. The distance caused such a decrease in the concentration of activated oxygen that the photosensitized oxidation of the acceptor failed to occur.

EXAMINATION OF VARIOUS QUESTIONS

The experiments described provide support for the proposal that an excited, metastable state of oxygen is formed during photosensitized oxidation. However, several questions must be answered experimentally to prove unequivocally this proposal. In those experiments in which the place of formation of the activated oxygen and the place at which it exerts its effect are separated spatially (i. e., separately prepared adsorbates were mixed), conclusive proof is lacking that neither the molecules of trypaflavine migrate onto the leucomalachite green adsorbate nor the molecules of leucomalachite green adsorbate diffuse onto the trypaflavine adsorbate within the period of time in which the experiments were conducted. Only if we can unambiguously demonstrate that the acceptor molecules were completely separated from the sensitizer molecules during the irradiation does the mechanism of the sensitized oxidation require a gaseous intermediate.

To determine whether the adsorbed molecules of the sensitizer or the acceptor diffuse, one must be able to differentiate the particles of the sensitizer adsorbate

Figure 3

from the particles of the acceptor adsorbate both before and after the experiments. The finely powdered mixtures that have been employed so far are not suitable for such a separation. To distinguish between the two types of adsorbates, we have used a uniform, characteristic particle size for each, obtained by fractionally screening crushed silica pieces. The preparation of the adsorbates was, otherwise, carried out as has been described for the finely powdered gels. To obtain an exit surface for the activated oxygen that is as large as possible, we chose an especially small particle size (0.15–0.3 mm) for this adsorbate and, by contrast, we chose a large particle size (1.0–1.5 mm) for the acceptor adsorbate. As a result, the large acceptor particles in the adsorbate mixture are surrounded by the small sensitizer particles, as shown in Figure 3.

In such a system, it would be easy to verify the migration of the adsorbed molecules from one gel particle to the other. If the intense fluorescence of trypaflavine does not remain restricted to the sensitizer particles and if the acceptor particles also fluoresce after the experiment, the trypaflavine would then have migrated into the acceptor adsorbate. If, after irradiation under conditions leading to a blue coloration, not only the acceptor adsorbate particles but also the sensitizer adsorbate particles are colored blue, then acceptor molecules have migrated onto the sensitizer adsorbate.

To carry out this examination, one cuts off the small irradiation tube B (Figure 2), very carefully removes portions of irradiated and nonirradiated spots by means of a spatula, and spreads the granules on glass. Although, in general, visual inspection of the large and small particles is sufficient, an exact microscopic examination provides a more definitive evaluation.

Strength of the adsorptive bond of trypaflavine on silica gel: Gel particles that have adsorbed trypaflavine fluoresce intensively and are green under an analytical quartz lamp. This fluorescence makes it possible to detect very small quantities of trypaflavine.

Experiment: Finely divided, preevacuated trypaflavine adsorbate together with silica gel particles of 1.0–1.5 mm (the same as those used for the preparation of the above-mentioned malachite green adsorbate) were evacuated at 90° C for 3 hours and then were left standing mixed for 14 days. A fluorescence test of the dyestuff-free coarser particles gave completely negative results after this time period. As with the pure silica gel, particles of the malachite green adsorbate that had been in contact with trypaflavine adsorbate for a long time did not fluoresce, regardless of whether they had been previously irradiated or not. Therefore, trypaflavine molecules adsorbed on silica gel do not show any detectable mobility. They are completely fixed at the inner boundary surfaces.

Strength of the bond of the adsorbed malachite green: If photosensitized oxidation occurs via direct contact of the acceptor with the sensitizer, the leuco com-

pound would have to migrate onto the trypaflavine adsorbate to give a blue coloration to the irradiated particles of the trypaflavine adsorbate, because we know now
that trypaflavine molecules do not migrate from the gel particles in which they are
adsorbed. Irradiation experiments with a mixture consisting of fine-grained sensitizer adsorbate and coarse-grained acceptor adsorbate have answered this question.

Experiment: Both adsorbates having different particle sizes are mixed immediately before the irradiation. The irradiation is carried out for 1½ hours at 0.005 mm
of oxygen pressure. After this time, the spot exposed to light is strongly but not
uniformly colored blue. One can see with the naked eye, but more clearly with a
microscope, that only particles of the leucomalachite green adsorbate are colored
blue. The small gel particles containing trypaflavine preserve their appearance unchanged. They do not show even small signs of oxidation of trypaflavine, which
would clearly appear as a rust-brown coloration.

These experiments provide unambiguous evidence that photosensitized oxidation
takes place with complete spatial separation of the sensitizer from the acceptor.

We considered it important to conduct experiments in which the sensitizer and
the acceptor adsorbates were mixed and in contact for prolonged periods before the
irradiation. Even small mobility of the adsorbed molecules of the malachite green
might cause migration into the trypaflavine adsorbate when the contact times were
long.

Experiment: Acceptor and sensitizer adsorbates of different particle sizes were
mixed in the apparatus and were left undisturbed for 10 days. After this period, we
irradiated a given spot. The time required for the appearance of the blue color depends on the oxygen pressure: after 15 min at an oxygen pressure of 0.004 mm and
25 min at an oxygen pressure of 0.015 mm. It was noted that in this test, the effect
of the oxygen pressure is considerably smaller than in tests in which the adsorbates
were irradiated immediately after the mixing. This suggests that the sensitizer and
acceptor molecules were pushed into closer proximity by the long contact of the
two adsorbates. After irradiation for 1½ hours with the above oxygen pressures
(which caused a strong blue coloration), we investigated the contents of the small
irradiation tube. Besides large, blue-colored acceptor particles, a large number of
small trypaflavine particles which show spots that are intensely blue colored was
found. The type of coloration indicates that the leuco compound has diffused into
the trypaflavine adsorbate only at the points of contact between the different gel
particles during long periods after mixing (Figure 4). The same phenomenon is observable, although to a much lesser extent, when the contact time is 1½ days.

These results show that the molecules of the malachite green which are adsorbed
on silica gel possess a significant, although small, mobility. *Although short contact
times must be used, these experiments also demonstrate that for photosensitized*

Figure 4

oxidation a collision of molecules of the sensitizer with those of the acceptor dur-ing the irradiation is not required.

We now know that a gaseous species with a strong oxidizing effect is formed with a sensitizer upon irradiation which diffuses to the acceptor molecules and oxidizes them. However, the question remains as to whether this gaseous intermediate is an activated, metastable form of oxygen. Although the presence of molecular oxygen is a prerequisite for photosensitized oxidation, other possibilities have to be taken into account. In addition to the nonreactive silica gel, the sensitizer adsorbate con-tains trypaflavine molecules and possibly also water molecules, not removed from the surface by the evacuation. Volatile, unstable photochemical products of trypa-flavine and water that may be formed with oxygen might lead to oxidation of leu-comalachite green on the adsorbate to form malachite green. These two possibili-ties will be discussed separately.

Significance of the decomposition of the sensitizer: If the photosensitized oxida-tion of the leucomalachite green depends on photooxidation of the trypaflavine, we expect the blue coloration of the acceptor to be much stronger as more brown-red color of the oxidized trypaflavine appears. This is contradicted by facts. The most intense blue coloration of the acceptor adsorbate was always observed when the trypaflavine adsorbate was not colored or barely colored (i. e., at low oxygen pressures). At higher oxygen pressures at which the trypaflavine became intensely red-brown, the photosensitized oxidation of the acceptor completely failed to ap-pear. In the decisive irradiation experiments with adsorbate mixtures at low oxygen pressure and short irradiation times in which the adsorbate was colored a deep blue, the color of the trypaflavine adsorbate was not changed at all.

The oxidation of leucomalachite green by light is not only sensitized by trypafla-vine but also by various fluorescent materials that have the same effect. Studies by H. Kautsky and R. Neuwirth on the relation of photosensitized oxidation to the extinction of fluorescence show that there is no correlation between the photosen-sitized oxidation of the acceptor and the photosensitized oxidation of the sensiti-zing dyestuffs. The same is very clearly evident from the measurements of H. Gaf-fron[6] regarding the absorption of oxygen in irradiated solutions of fluorescent dye-stuffs containing an acceptor. A result which is important to our studies is that a slow oxidation of the sensitizer is only observed in acceptor-free solutions. How-ever, when the acceptor concentration is sufficient, the acceptor is oxidized exclu-sively and the fluorescent dyestuff remains unchanged (i. e., acts only as a sensiti-zer). *Therefore, photosensitized oxidation is not mediated by decomposition pro-ducts of the sensitizing dyestuff.*

Significance of water: Experiments by Gaffron[7] also answer the question as to whether water is absolutely necessary for photosensitized oxidation. He found that it is not only possible to exclude water, but also that water often considerably re-duces the yield of the absorbed light energy for the photosensitized oxidation. The results of Gaffron's quantitative measurements (summarized below) clearly support our proposal for the mechanism of photosensitized oxidation and are inconsistent with an oxidation mechanism involving radicals derived from water.

Allyl thiourea dissolved in a solution of hematoporphyrin or chlorophyllide is ox-idized when exposed to light. In an approximately 10 percent solution of this ac-ceptor, one molecule of oxygen is consumed for the oxidation of the acceptor for each quantum of absorbed light.

41

Water molecules cannot be responsible for the photosensitized oxidation described above or in another reaction studied by Gaffron. Using pure anhydrous isoamylamine in which a fluorescent dyestuff was dissolved, Gaffron found molecular oxygen upon irradiation. In the dark, he was able to liberate practically all the bound oxygen molecules. How can one better explain this peculiar reaction than by the formation and the subsequent binding of an especially reactive form of molecular oxygen? Also, water was not needed in these experiments for the formation of the oxygen bond; on the contrary, addition of water caused secondary changes that reduced the catalytically cleavable quantity of oxygen to half its amount.

RESULTS OF THE EXPERIMENTS

Photosensitized oxidation is mediated by photosensitized activation of oxygen and consists of two reactions, a reaction in the light, followed by a reaction in the dark. The light reaction involves the activation of the oxygen molecule by transfer to oxygen of the light energy absorbed by the sensitizer. The following reaction, for which irradiation might not be necessary, results in the oxidation of the acceptor by oxygen that has been activated in the light reaction.

What is the nature of this activated state of oxygen? It is formed by the transfer to the oxygen molecules of the light energy absorbed by the fluorescent dyestuff molecules. The amount of energy stored in the oxygen molecule can only be relatively small. Dyestuffs, such as hematoporphyrin and chlorophyll that fluoresce in the very long wavelength red range are particularly capable of activating oxygen as their own fluorescence is quenched. The above-mentioned photosensitized oxidation of allyl thiourea with chlorophyll as sensitizer was measured by Gaffron, who used a substantially monochromatic light having a wavelength of 6600 Å. Therefore, the energy transferred to oxygen presumably will not exceed 40,000 cal. Under these conditions, a cleavage of the oxygen molecule into atoms cannot be envisioned, so the formation of ozone is also excluded. In view of the "photosensitized oxidation dependence on distance," our first thought is an excitation state of the oxygen molecule. This excitation state has to be metastable. This is clearly shown by the evidence for the diffusion of the activated oxygen. This activated oxygen has the additional property of exerting a much stronger oxidizing effect than normal oxygen. All experiments regarding photosensitizing oxidation are examples for this effect.

In consulting the literature for low-energy excited states of oxygen, we find one such state that meets our requirements. It was detected by Heurlinger[8] based on purely spectroscopic measurements. Oxygen possesses an absorption band in the long-wavelength red at 7623 Å (= 37,257 cal). The absorption of light of this wavelength by the oxygen molecule is so small that it was concluded that a metastable state is involved. W. Childs and R. Mecke,[9] using quantitative intensity measurements in the absorption band, succeeded in determining that the lifetime of this $^1\Sigma$ state of the oxygen molecule was several seconds.

One is tempted to identify the oxygen activated by photosensitization with the $^1\Sigma$ state. We do not have any direct proof for this. The concentration of oxygen obtained in the $^1\Sigma$ state by direct light absorption is so small that its chemical properties as well as its special oxidizing effects cannot be investigated and therefore cannot be compared with the effects of the active state of the oxygen molecule that

42

we have investigated. However, the probability that the two activation states are identical is quite high in view of the agreement of important physical properties (such as the excitation energy below 40,000 cal and the metastable state).

We do not know as yet whether this "activated oxygen" is involved in the well-known processes of autoxidation; we consider it reasonable that this activated state is also attained during the transfer of chemical energy to the oxygen molecule.

The main importance of this active oxygen would seem to be in the biological field. When one considers that in each green assimilating plant the fluorescence of chlorophyll is extinguished by oxygen to a large extent, and that this extinction of the fluorescence is directly correlated with energy conversion in the assimilation process, [10] one can judge how important activated oxygen may be for the biological processes on the surface of the earth.

We want to express our thanks to the *van't Hoff Foundation* of the Royal Academy of Sciences in Amsterdam for support for this research.

LITERATURE REFERENCES

1. H. Kautsky and H. de Bruijn, *Naturwiss.* **19**, 1043 (1931).
2. H. Gaffron, Literature Review, *Ber. Deut. Chem. Ges.,* **60**, 2229(1927).
3. H. Kautsky and A. Hirsch, *ibid.,* **64**, 2677 (1931).
4. H. Kautsky and W. Baumeister, *ibid.,* **64**, 2446 (1931).
5. H. Kautsky, A. Hirsch, and W. Baumeister, *ibid.,* **64**, 2053 (1931); H. Kautsky, and A. Hirsch, *ibid.,* **64**, 2677 (1931).
6. H. Gaffron, *Biochem. Z.,* **179**, 171 (1926).
7. H. Gaffron, *loc. cit.,* ref. 2.
8. Heurlinger, Dissertation, Lund, 1919.
9. W. H. J. Childs and R. Mecke, *Z. Physik,* **68**, 344 (1931).
10. H. Kautsky, A. Hirsch, and Davidshofer, *Ber. Deut. Chem. Ges.,* **65**, 1763 (1932).

9

Reprinted from *Trans. Faraday Soc.*, 35, 216–219 (1939)

QUENCHING OF LUMINESCENCE BY OXYGEN.

By Hans Kautsky.

Received in German on 18th *July*, 1938.

A connection exists between fluorescence and the action of sensitisers. Fluorescent substances are able temporarily to store up energy quanta in excited states. They can emit the excitation energy as light, but in appropriate systems, they can also transfer it, to other molecules. In the latter case, the fluorescence is quenched. My experiments show that oxygen has a quenching effect upon the fluorescence of many dye-stuffs. The question as to what happens in the interaction of excited dyestuffs and oxygen is of importance in connection with many natural processes. I need only call attention to some biologically important fluorescent colouring matters, chlorophylls, porphyrins, carotinoids, lactoflavin, and so on.

Organic substances of various kinds, illuminated in presence of fluorescent dyes, take up molecular oxygen. There is an extensive literature relating to this sensitised photo-oxidation.

I have tried to clear up two general questions:[1] 1. What process underlies the quenching of fluorescence by oxygen? 2. Is this process connected with photo-oxidation sensitised by fluorescent dyes?

Experiments in solution seemed to me ill-adapted to the purpose of explaining qualitatively the sequence of the separate steps involved in reactions, since in solution all molecules, like and unlike, are freely mobile, and thus are simultaneously in continuous interaction. I aimed at fixing the molecules of appropriate sensitisers in space, so that the influence of molecular oxygen upon their fluorescence could be examined without any disturbance by secondary circumstances. This was achieved in the following way: suitable dyestuffs, such as Trypa-flavin, chlorophyll, porphyrins, etc., were adsorbed from their dilute solutions on surfaces of solid silica gel or aluminium oxide gel. The adsorbates, especially those of silica gel, are quite transparent. The dye molecules, as was shown experimentally, are firmly fixed, but are freely accessible to all molecules diffusing up to the surface. The most carefully evacuated adsorbates not only fluoresce, but in many cases show a strong phosphorescence which even at ordinary temperatures lasts for many seconds. At very low temperatures, new phosphorescence bands of longer wave-length appear, while the excited states corresponding to the normal short wave emission, become frozen in. They can subsequently be made to give out light by warming. Silica gel must be a specially good medium for insulating the energy, since the excited states of the adsorbed dye molecules may have a very long life.

[1] H. Kautsky, *Biochem. Z.*, 1937, **291**, 271 ; Papers by H. Kautsky and collaborators in the *Ber.*, from 1931 to 1935.

The fluorescence and phosphorescence of the adsorbates are quenched by oxygen : the process is reversible. The degree of fluorescence quenching depends upon the oxygen pressure. The phosphorescence, however, is completely quenched independently of the oxygen pressure, down to very small values of the latter—a few thousandths mm. O_2. (In this way, the minutest quantities of oxygen—less than 0·001 mm. O_2 can be detected by their action in quenching phosphorescence.) The greater sensitivity to oxygen of the phosphorescence is understandable, since with increasing life-period of the excited states the probability of collisions leading to quenching of luminescence at very low oxygen pressures becomes greater.

The extinction of luminescence is the expression of an energy transfer between excited dyestuff molecules and oxygen, in which activated oxygen O_2^e is produced. Corresponding to the small fluorescence energy of many sensitisers (red or even infra-red fluorescence) the activation energy of the O_2 can only be small. Essentially, it is only the meta-stable oxygen states of long life, $1\Sigma O_2$ (37·3 Cal.) and $1\Delta O_2$ (22·5, 26·8, 30·9 Cal.) which need be considered. The activated oxygen has been detected by its oxidising actions on organic molecules which are spatially separated from the molecules of the fluorescent dye. The O_2^e must have diffused from the place where it was formed (sensitizer) to the place where it was used up (oxygen acceptor).

As acceptor was employed the colourless leuco malachite green which is transformed by oxidation into the greenish blue dye (indicator of photo-oxidation). A silica gel acceptor adsorbate has been prepared in which the grain size considerably exceeded that of the adsorbate of sensitiser. Under the conditions of the experiment, the acceptor adsorbate by itself is not changed in light. But if it is mixed with the adsorbate of the sensitiser, so that the larger particles of the acceptor adsorbate are surrounded by the smaller particles of the sensitiser adsorbate, and illuminated in presence of O_2, then the particles of the acceptor adsorbate become coloured blue by photo-oxidation. This is the connection between quenching of luminescence by oxygen and photo-oxidation. The sensitiser remains chemically unchanged during the photo-oxidation. For the success of the experiment, the oxygen pressure should be small enough just to quench the phosphorescence.

The experimental results lead to the conclusion, that we can only suppose an active form of oxygen of long life, and not any other oxidising agent which dif-

O_2 Pressure in 10^{-4} mm.	Time of Illumination. Minutes.	Coloration.
0·4	10	None.
1·4	10	,,
12	10	Faint.
18	10	Strong blue.
24	10	Faint.
40	10	None.

fuses from sensiter to acceptor. Water plays no part in the processes occurring, and its influence on the photo-oxidation is an adverse one.

Shortening the diffusion path must make the photo-oxidation less dependent upon the oxygen pressure. Particles of silica gel, which contain both sensitiser and acceptor molecules adsorbed on the same surface are rapidly coloured blue in light, even at much higher oxygen pressures—up to an atmosphere. This mixed adsorbate forms a transition to the case of solutions.

The degree of fluorescence quenching in solutions depends upon the oxygen concentration. The duration of fluorescence in solutions has been measured (by Gaviola) in presence of air, and, presumably on this account, has generally been found to be very short (about 10^{-8} sec.). If one examines the luminescence of oxygen-free solutions of dyes with the phosphoroscope, many of them (*e.g.*, chlorophyll, hæmatoporphyrin, eosin, erythrosin, benzoflavin, etc.), depending upon the solvent, show a persistence of the fluorescent light, often up to about 10^{-2} sec. This phosphorescence is extinguished by the smallest quantities of oxygen. The solutions thus behave in a way which shows a marked analogy to the case of the adsorbates.

Phosphorescence only appears in a region of concentration where the dyestuff molecules are slightly associated (presumably to double molecules). The total luminescence yield of such sensitiser solutions is usually only small, since with increasing lifetime, there is also an increasing probability, that the excitation energy will be dispersed as random thermal motion before emission, although this dispersal in itself is a rarely occurring process. (Self quenching.) In this state of slight association, fluorescent dyes are peculiarly well adapted to be sensitisers on account of the long life of the absorbed energy in the solution.

The quantum yield of sensitised photo-oxidations must, according to these results, be dependent upon the degree of association of the dye (dyestuff concentration, nature of the solvent) and upon the acceptor concentration (in connection with the life of O_2^e), but it must be independent of the oxygen pressure when there is the right sensitiser concentration.

This is in agreement with H. Gaffron's quantitative measurements [2] of the velocity of photo-oxidation in solution in relation to the concentration of the reacting substances, and the adsorbed light quanta. He found under favourable conditions a quantum yield of 1, independently of the oxygen pressure: Gaffron neglected the quenching of luminescence by oxygen and formed the idea that the excitation energy is transferred from the sensitiser, not to oxygen, but directly to the oxygen acceptor, as a result of which the latter is transformed into an active form, of long life, which can add on normal oxygen directly. The untenability of this view follows from experiments on the influence of the oxygen acceptors which Gaffron used (allyl thiourea and *iso*-amylamine) on the fluorescence and phosphorescence of the sensitiser dyes. These acceptors do not decrease the luminescence of the dissolved dyes in the slightest degree so long as they do not change them chemically. In an evacuated solution of chlorophyll, for example, in *iso*amylamine, in which chlorophyll molecules are in continual collision with the acceptor, the chlorophyll shows a particularly fine phosphorescence. These oxygen acceptors thus act purely as solvents, and are rather energy-insulators than energy-acceptors. In solutions containing acceptors, however, the fluorescence and phosphorescence are quenched in just the same way as in acceptor-free solutions. Thus the quenching of luminescence and photo oxidation in solution must be interpreted, as with the adsorbates, by a primary formation of O_2^e.

I should like also to mention shortly my observations on the numerous regular changes with time of the intensity of chlorophyll fluorescence in green plants, and the dependence of these changes upon various internal and external factors. Free and bound oxygen is most probably the cause

[2] H. Gaffron, *Biochem. Z.*, 1936, **287**, 130 (other references here).

of the decrease of fluorescence with time which occurs in the green particles of the chloroplasts (grana). By transfer to these compounds, the excitation energy of the chlorophyll may be stabilised in the form of chemical energy, and become concentrated and available at appropriate places in the assimilatory apparatus. Whether carotinoids take part in these processes cannot yet be said from my experiments, though at all events, their fluorescence is also strongly quenched by oxygen and they are capable of adding on oxygen.

University of Leipzig.

Editor's Comments
on Papers 10 and 11

1,4-CYCLOADDITION OF OXYGEN TO
CONJUGATED DIENES

Ascaridole (2), the active constituent in chenopodium oil, is a naturally occurring transannular peroxide. Paper 10 is the report of Schenck and Ziegler describing the first dye-sensitized photooxygenation of a simple 1,3-diene. Irradiation of α-terpinene in the presence of chlorophyll and oxygen produced a peroxide that was identical with the natural material. Schenck and Ziegler found a quantum yield of 1 for the formation of ascaridole. The authors also indicate that the peroxide (norascaridole) from 1,3-cyclohexadiene has been prepared by this procedure.

In 1946, Dufraisse and Ecary (Paper 11) reported the isolation of the transannular peroxide from the photochemical addition of oxygen to 1,3-diphenylisobenzofuran (3) (DPBF). The peroxide that might be considered as the ozonide of 1,2-diphenyl-3,4-benzocyclobutadiene is relatively stable at –78°C but explodes at +18°C. DPBF has subsequently been shown to be an extremely reactive singlet oxygen acceptor and has been used in many investigations (see Papers 32 and 72).

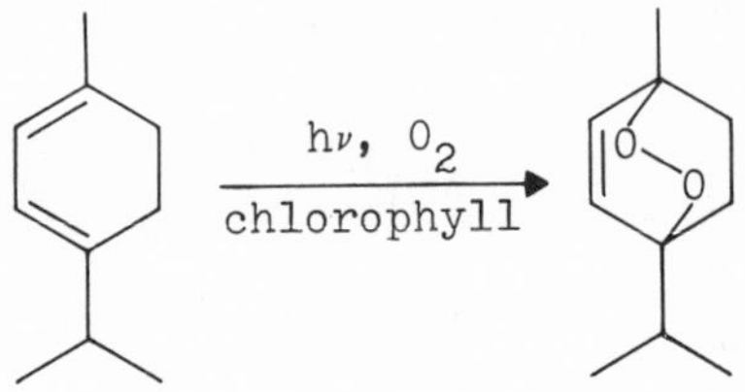

$$h\nu, O_2 \quad \text{chlorophyll}$$

(1)   (2)

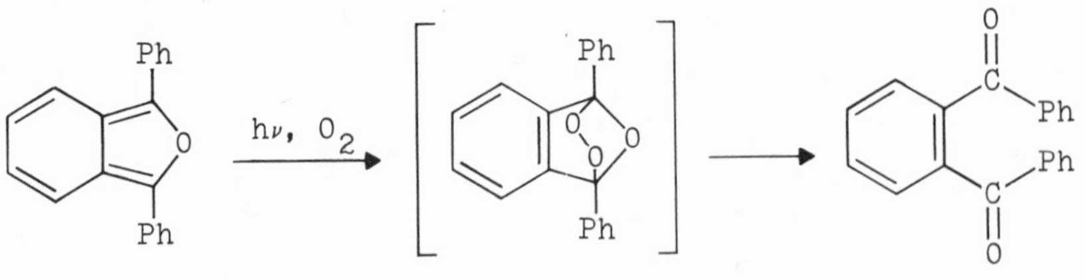

(3)   (4)   (5)

48

10

Copyright © 1944 by Springer-Verlag, Berlin, Heidelberg, New York

Reprinted from *Naturwiss.*, 32(14/26), 157 (Apr.–June 1944)

Die Synthese des Ascaridols.

Das Ascaridol, das wirksame Prinzip des amerikanischen Wurmsamenöls (Ol. chenopodii anth. var. Gray) besitzt nicht allein wegen seiner bekannten wurmwidrigen Eigenschaften großes Interesse. Es ist das einzige in der Natur vorgefundene stabile Peroxyd, dem nach den grundlegenden Untersuchungen von WALLACH[1]) die Formel des 1,4-Peroxido-p-menthens-(2) (I) zukommt. Bisherige Versuche zur Synthese dieser Verbindung waren erfolglos. Obwohl gelegentlich in der Literatur die Vermutung geäußert wurde, daß die Bildung des Ascaridols in der Pflanze als eine 1,4-Addition von einem Molekül Sauerstoff an ein Molekül α-Terpinen (II) erfolgen dürfte, kann beim Schütteln von α-Terpinen mit Sauerstoff kein Ascaridol erhalten werden. Zwar wird hierbei nach BODENDORFF[2]) der Sauerstoff an den Enden des konjugierten Systems aufgenommen, doch sind die hierbei gebildeten Peroxyde polymer und nach III zu

formulieren; die Reaktion verläuft also nach Art einer Mischpolymerisation, und die Polymerisate lassen sich nicht zu Ascaridol depolymerisieren.

Verbindungen mit Anthrazenstruktur[3]) sind bei Belichtung zur Diensynthese mit Sauerstoff bereit; das gleiche gilt für Diene der Sterinreihe[4]), wie z. B. Ergosterin, die unter Sauerstoff in Gegenwart von fluoreszierenden Farbstoffen wie Eosin belichtet werden. Von der letztgenannten Reaktion war jedoch nicht bekannt, ob sie etwa nur auf Diene der Sterinreihe beschränkt sei. Wir vermuten nun, daß als Voraussetzung für die Entstehung des Ascaridols in der Pflanze folgende Bedingungen anzusehen sind:

1. Die Anlagerung des Sauerstoffs an das α-Terpinen erfolgt im Licht unter der photosensibilisierenden Wirkung des fluoreszierenden Chlorophylls.

2. Das Terpinen liegt während der Reaktion in einer solchen Verdünnung vor, daß die Bildung des monomeren Peroxyds, also des Ascaridols, gegenüber der polymerisierenden Autoxydation stark in Vordergrund tritt.

Unsere Vermutung gewinnt einen hohen Grad von Wahrscheinlichkeit dadurch, daß es uns gelang, den Übergang des α-Terpinens in das Ascaridol unter obigen Bedingungen in vitro zu verwirklichen. Das reine synthetische Ascaridol (Totalsynthese, da α-Terpinen synthetisch zugänglich) schmolz bei + 2,5° und erwies sich mit dem natürlichen Produkt in jeder Weise identisch.

Beim eingehenden Studium der Reaktion ergab sich, daß außer der Wahl geeigneter Lösungsmittel und Sensibilisatorfarbstoffe (Chlorophyll liefert z. B. bessere Resultate als Eosin) auch gewisse Zusätze (wie Toluol) für die Bildung des Ascaridols förderlich sind. Ein Ascaridolmolekül benötigt zu seiner Bildung ein Lichtquant.

Wir haben unsere Erfahrungen bereits mit Erfolg auf eine ganze Reihe cyclischer Diene übertragen und behalten uns weitere Untersuchungen mit der angegebenen Methodik auf breitester Basis ausdrücklich vor. Von den bisherigen Ergebnissen interessiert hier im Zusammenhang, daß wir auch mehrere bisher nicht bekannte Isomere des Ascaridols, z. B. aus α-Phellandren das Peroxyd IV sowie aus Cyclohexadien-(1,3) den Grundkörper V erhalten haben. Diese Verbindungen besitzen mit Ascaridol große Ähnlichkeit.

Ausführliche Mitteilung unserer Ergebnisse wird später erfolgen.

Halle a. d. S., Chemisches Institut der Universität, den 24. Oktober 1943.

GÜNTHER O. SCHENCK. K. ZIEGLER.

[1]) WALLACH, Liebigs Ann. **392**, 67 (1912).
[2]) BODENDORF, Arch. Pharm. **1933**, 1.
[3]) Z. B. DUFRAISE und GERARD, Compt. rend. **201**, 428 (1935).
[4]) Erstmals WINDAUS und BRUNKEN, Liebigs Ann. **460**, 225 (1928).

11

Reprinted from *Compt. Rend.*, **223**, 735–737 (Nov. 1946)

PHOTOOXYDATION SUR CYCLE PENTAGONAL: PHOTOXYDIPHÉNYLISOBENZOFURAN

Charles Dufraisse et Serge Ecary

a. La théorie de l'union labile de l'oxygène au carbone donnée antérieurement ([1]) a pour caractéristique une addition 1.4 sur deux liaisons éthyléniques conjuguées. Dans les exemples connus (acènes : anthracène, naphtacène, etc.), le système conjugué fait partie d'un mésonoyau, mais on imagine qu'il pourrait appartenir aussi bien à tout autre cycle, hexagonal, pentagonal, etc., ou même à une chaîne linéaire. En fait, la structure diénique hexagonale paraît être un réceptacle bien adapté à la molécule d'oxygène. comme il ressort de l'existence de peroxydes *transannulaires* hexagonaux, relativement stables, comme l'ascaridole et les similaires, ou les peroxydes. stéroïdes. Par contre l'incertitude règne sur le comportement des diènes pentacycliques.

Par ailleurs des différences considérables se manifestent entre l'oxygénation des acènes et celle des autres diènes, ainsi qu'entre les propriétés des peroxydes formés. Elles tiennent assurément à l'influence de la mésomérie qui s'exerce dans les acènes. Aussi avons-nous entrepris d'étudier l'oxygénation de dispositifs diéniques autres que ceux des mésonoyaux, mais en maintenant dans les nouvelles structures au moins l'essentiel de ce qui paraît favorable dans celle des acènes, c'est-à-dire un accolement benzénique au diène (III) (partie renforcée). C'est un dispositif orthoquinonique, on le trouve, par exemple, dans le squelette isobenzocyclopentadiénique (I), où X représente un élément quelconque, carboné ou non. Un aryle aux points d'attache de la molécule d'oxygène étant supposé propice à la réactivité ([2]), on a envisagé d'abord la formule (II).

b. Le nombre de corps décrits, appartenant à ce type est assez restreint.

([1]) Ch. Dufraisse, *Bull. Soc. Chim.*, 4ᵉ série, **53**, 1933. p. 837.

([2]) Ch. Dufraisse et A. Etienne, *Comptes rendus*, **201**, 1935, p. 280.

Ils comprennent : le diphénylisobenzofuran (V) ([3]), le dérivé sulfuré correspondant, ou diphénylisobenzothiofène ([4]), la phènecyclone (IV) ([5]). Les uns et les autres, comme les acènes, sont photooxydables au sens strict du terme, c'est-à-dire stables à l'air dans l'obscurité, mais les photooxydes n'ont pas été signalés jusqu'ici.

Pour le diphénylisobenzofuran (V), on avait déjà recherché le photooxyde dans des essais préliminaires ([6]), et aucun indice de sa présence n'avait été relevé. Il y avait là un point de doctrine assez important pour justifier de nouveaux essais.

Nous venons de constater que le photooxyde attendu se forme bien et sans doute avec de bons rendements, mais qu'il est extrêmement instable : à l'état cristallisé il explose brutalement à des températures qui n'atteignent pas 20°, et, en solution, il se transforme rapidement dès la température ordinaire en orthodibenzoylbenzène (VII); ceci explique que Guyot n'ait trouvé que ce terme final dans l'oxydation à la lumière du diphénylisobenzofuran.

Par analogie, nous attribuons au photooxyde la formule (VI). Pour l'obtenir il faut opérer prestement, ce qui fixe les conditions les meilleures. On expose par grand soleil, pendant 70 secondes et en agitant vivement, du diphénylisobenzofuran dans du sulfure de carbone, à raison de 2^g par litre. On évapore aussitôt sous vide, en laissant la température de la liqueur s'abaisser en dessous de 0°. Le photooxyde se dépose en beaux cristaux par amorçage spontané. La température d'explosion est aux environs de 18°; mais elle est plus élevée (jusqu'à 50°) quand l'opération à été défectueuse, parce que le peroxyde subsistant est dilué dans les produits de sa transformation. Les cristaux paraissent être stables pendant plusieurs heures à la température de la neige carbonique. La dilution retarde aussi l'altération, quoique moins que le froid; elle est cependant assez efficace pour qu'il n'y ait pas avantage à irradier les solutions diluées à —40° plutôt qu'à la température ordinaire, pourvu que ce soit pendant un temps très court.

La décomposition du photooxyde dégage peu d'un gaz formé surtout d'anhydride carbonique, avec une trace d'oxygène. Ce résultat apparente le nouveau photooxyde avec ceux de certains acènes, spécialement ceux qui sont dépourvus d'aryles sur les mésosommets, anthracène ou naphtacène. Il comporte cependant un certain imprévu, puisque l'oxygène est en excès d'un atome dans le photooxyde (VI) relativement au terme final de l'autoxydation, le dibenzoylbenzène (VII); la demi-molécule qui ne se dégage pas à l'état libre attaque le

---

([3]) A. Guyot et J. Catel, *Bull. Soc. Chim.*, 3ᵉ série, 35, 1906, p. 1124. Divers dérivés de l'isobenzofuran ont été décrits par Guyot et surtout par Adams [U. S. P. 2. 325. 757 (1944)]; leur photooxydation n'a pas été étudiée.

([4]) Ch. Dufraisse et D. Daniel, *Soc. Chim.*, 5ᵉ série, 4, 1937, p. 2063.

([5]) W. Dilthey, *J. Prak. Ch.*, 151, 1938, p. 97.

([6]) Ch. Dufraisse et L. Enderlin, *Comptes rendus*, 190, 1930, p. 1229.

support carboné, d'où l'origine de l'anhydride carbonique. En fait la décomposition explosive donne notablement mo.ıs de la dicétone (VII) que la transformation lente dans un solvant, qui est intégrale : peut-être dans ce dernier cas l'atome d'oxygène part-il à l'état libre en s'unissant à un autre atome voisin de même origine. Ce dégagement, s'il a lieu, est trop faible pour être décelable en solution.

Editor's Comments
on Papers 12 and 13

12   SCHÖNBERG
     Notiz über die photochemische Bildung von Biradikalen

13   SCHENCK
     Zur Theorie der photosensibilisierten Reaktion mit molekularem Sauerstoff

MOLOXIDE MECHANISM FOR PHOTOOXYGENATION

In 1935, Schönberg (Paper 12) proposed that an unstable sensitizer–oxygen complex was the reactive intermediate in both direct photooxygenation of aromatic hydrocarbons such as rubrene[*] and in dye-sensitized photooxygenations. It was suggested that this intermediate could transfer oxygen to an acceptor with concomitant regeneration of the sensitizer. However, Schönberg[†] noted that this oxygen transfer might be effected either by transfer of oxygen in a collision of the complex with an acceptor or by liberation of oxygen which was trapped by an acceptor. Schönberg indicated that the "oxygen that is liberated by the transition from III (the complex) to II may be especially active."

Paper 13 is a discussion by G. O. Schenck of the mechanism of photooxygenation sensitized by dyes such as eosin, fluorescein, and chlorophyll. The reactions of oxygen with dienes to form endoperoxides and with olefins to give allylic hydroperoxides were considered. The proposed mechanism involves the excitation of the sensitizer to a metastable state with biradical character (S^{rad}), which reacts with O_2 to form a labile sensitizer–oxygen complex (moloxide). This reactive intermediate transfers oxygen to an acceptor in a collision complex to give the oxygenation product. Support for this mechanism came from several photooxygenation experiments that seemed to be highly dependent on the sensitizer used.

[*]The structures I and II given for rubrene in Paper 12 are incorrect. See Editor's Comments on Papers 4 Through 6.

[†] See footnote 3 on Page 300 of Paper 12.

12

Reprinted from *Liebigs Ann. Chem.*, **518**(2-3), 299-302 (1935)

Notiz über die photochemische Bildung von Biradikalen;

von *Alexander Schönberg.*

[Aus dem Institut für medizinische Chemie der Universität Edinburgh.]

(Eingelaufen am 11. Mai 1935.)

1. In dem soeben erschienenen Annalenheft schreiben E. Müller und Müller-Rodloff[1]): „Das Rubren zeigt die merkwürdige Eigenschaft, bei Belichtung, seiner roten benzolischen Lösungen den Luftsauerstoff unter Bildung eines dissoziierbaren Peroxyds zu absorbieren. Ch. Dufraisse[2]) sowie A. Schönberg[3]) erklären diese Eigenschaft durch Übergang des chinoiden Systems in ein Biradikal, das nun den Sauerstoff unter Peroxydbildung aufnimmt."

Die Nennung von Dufraisse (mit einer Quellenangabe aus dem Jahre 1933) vor mir (mit einer Quellenangabe aus dem Jahre 1934) *muß* bei dem Leser den Eindruck erwecken, daß obige Anschauung von Herrn D. *vor* mir vertreten worden wäre. Demgegenüber sei betont, daß vor Erscheinung meiner Publikation[4]) „Über die Diradikalformel des Rubrens …" von Herrn D. der Übergang von I in II (unter irgendwelchen Versuchsbedingungen) niemals in Betracht gezogen wurde[5]) — da er auch in seinen späteren oben erwähnten Mitteilungen[6]) zu der von mir angenommenen Umlagerung I → II nicht in positivem Sinne Stellung genommen hat, so scheint es mir, daß Herr Dufraisse in den oben angeführten Sätzen zu Unrecht genannt worden ist[7]).

---

[1]) A. **517**, 134 (1935).

[2]) Bl. [4] **53**, 837 (1933); **67**, 1021, 2018 (1934).

[3]) B. **67**, 633, 1404 (1934).   [4]) B. **67**, 633 (1934).

[5]) Ch. Dufraisse hat dagegen als erster die Ansicht vertreten, daß beim thermischen Zerfall von Rubrenperoxyd IV die Verbindung II entsteht, welche dann in I übergeht.

[6]) B. **67**, 1021, 2018 (1934).

[7]) Vgl. auch A. Schönberg, B. **68**, 162 (1935).

Ich lege auf die alleinige Priorität hinsichtlich der oben erwähnten Diradikaltheorie solchen Wert, weil sie sich mit Erfolg auch
auf andere Verbindungen übertragen läßt, z. B. auf Chromanorufen VI
und Chlorophyll. Zur Wahrung der Priorität sei folgendes schon jetzt
mitgeteilt.

2. Über den Chemismus der Einwirkung von O_2 auf Rubren und Chlorophyll.

Triphenylmethyl reagiert bekanntlich[1]) mit O_2 zuerst unter Bildung eines Peroxyds mit Radikalcharakter (V), welches seinen Sauerstoff auf andere Substanzen übertragen kann. Vorläufige Versuche
scheinen zu beweisen, daß das *erste* Reaktionsprodukt zwischen Rubren
und O_2, die Verbindung III, ein Radikalperoxyd ist. III stabilisiert
sich entweder durch Übergang in IV oder durch Sauerstoffabspaltung
unter Rückbildung von Rubren (I bzw. II). — Bei Gegenwart eines
Sauerstoffacceptors kann eine Stabilisierung von III sich auch wie folgt
vollziehen: der Sauerstoffacceptor wird oxydiert, III geht selbst in
Rubren über.

H. Gaffron hat gefunden[2]), daß eine bestrahlte Lösung von
Rubren bei Gegenwart von O_2 schnell ausbleicht [Bildung des farb-
losen Rubrenperoxyds (IV)], setzt man einer solchen Lösung den Sauer-
stoff-acceptor Thiosinamin zu, so bleicht das Rubren nur langsam oder
gar nicht aus. Dabei wird von der Lösung aber ebensoviel oder mehr
Sauerstoff verbraucht als ohne den schützenden Zusatz. — Im Lichte
der oben erwähnten Radikaltheorie lassen sich diese Beobachtungen
zwanglos deuten[3]).

Die oxydierende Wirkung bestrahlter Chlorophyllösungen[4]) aut
einen Acceptor, z. B. Isoamyl-amin, bei Gegenwart von O_2 läßt sich auf
Grund vorläufiger Versuche analog erklären: die bestrahlten Chlorophyllösungen enthalten Moleküle mit Biradikalcharakter, welche mit O_2

---

[1]) K. Ziegler u. Ewald, A. 504, 162 (1933).

[2]) H. Gaffron, Bio. Z. 264, 260 (1933).

[3]) Wie der Chemismus der Übertragung des Sauerstoffs von III
auf den Acceptor ist, soll hier nicht näher erörtert werden; 2 Mög-
lichkeiten kommen besonders in Frage. Entweder stoßen Moleküle
von III mit denen des Acceptors zusammen und die Sauerstoffübertragung erfolgt im Reaktionsknäuel, oder der Sauerstoff, welcher beim
Übergang von III in II frei wird, ist besonders „aktiv". (Oxydierende
Wirkung des Sauerstoffs in statu nascendi.) — Analoges gilt hinsichtlich der oxydierenden Wirkung des Chlorophylls (im Licht bei Gegenwart von O_2).

[4]) H. Gaffron, B. 60, 755, 2229 (1927).

unter Bildung von Radikalperoxyden reagieren (Analoga zu III und V),
welche ihrerseits ihren Sauerstoff auf Sauerstoffacceptoren, z. B. Isoamyl-amin, übertragen können unter Rückbildung von Chlorophyll [1]).

3. Untersuchungen am Chromanorufen (VI).

Nach vorläufigen Versuchen [2]) kann VI als Biradikal VII reagieren [3]),
wodurch die schon beschriebene Fähigkeit des Chromanorufens im gelösten Zustand bei Gegenwart von Licht mit O_2 zu reagieren [4]) seine
Erklärung findet; es bildet sich auch hier zuerst ein Radikalperoxyd
(Analogon von III bzw. V).

[1]) Vgl. auch R. Willstätter, Naturwiss. 21, 252 (1933); F. Hauro-
witz, Klin. Wochenschr. 13, 321 (1934); H. Kautsky u. A. Hirsch,
Bio. Z. 277, 250 (1935); diese 3 Mitteilungen enthalten zahlreiche
Quellenangaben.

[2]) Diese Versuche wurden gemeinsam mit R. Michaelis durchgeführt.

[3]) Das gleiche gilt auch vom Tetraphenyl-p-chinodimethan.

In seinen Lösungen stellt sich ein photochemisches Gleichgewicht ein.

[4]) H. Liebermann u. J. Barrollier, A. 509, 45 (1934).

H. Liebermann und Barrollier[1]) geben an, daß z. B. bei der Einwirkung von Salpetersäure auf das Chromanorufen VI eine Verbindung entsteht, welche sich von VI durch den Mehrgehalt von 2 Sauerstoffatomen unterscheidet; sie wird als Analogon des Rubrenperoxyds IV aufgefaßt[2]). — Es hat sich ergeben[3]), daß diese Analogie nicht besteht; die in Rede stehende Verbindung ist ein Glycol [5,12-Dioxychromano-rufan (VIII)].

VIII

[1]) H. Liebermann u. Barrollier, A. 509, 45 (1934).

[2]) Vgl. auch Jean Barrollier, Diss. Berlin 1932, S. 13 u. 14.

[3]) Diese Versuche wurden gemeinsam mit R. Michaelis durchgeführt.

13

Reprinted from *Naturwiss.*, 35(1), 28–29 (1948)

Zur Theorie der photosensibilisierten Reaktion mit molekularem Sauerstoff.[1])

Die eigenartige Wirkungsweise der Sensibilisatoren bei den durch fluoreszierende Farbstoffe photosensibilisierten Reaktionen zwischen O_2 und Akzeptoren A in Lösung kann meines Erachtens nicht als einfache Übertragung von Anregungsenergie auf A oder O_2 aufgefaßt werden. Zur Klärung dieser Frage unternahm ich vergleichende Untersuchungen über im Lichte erfolgende Reaktionen zwischen A und O_2 in Gegenwart und in Abwesenheit von Photosensibilisatoren (z. B. Eosin, Fluoreszein, Chlorophyll usw.). Dabei ergab sich die prinzipielle Verschiedenheit der verglichenen Reaktionen, für deren Eintreten nicht die gleichen Voraussetzungen der Konstitution gelten, und die nur in besonderen Fällen zu den gleichen Reaktionsprodukten führen.

Zum Beispiel ergab sich folgendes Bild:

Autoxydation von 2,5-Dimethylfuran.

	unsensibilisiert	sensibilisiert mit Eosin
O_2-Aufnahme	0,5 O_2/Mol	
Reaktionsprodukte	Fur. $\dot{O}$; 50 % α-, β-Diacetyläthylen, wenig Peroxyd (FurO_2).	nur ozonidartiges Peroxyd durch Diensynthese mit O_2, Diacetyläthylen auch in Spuren nicht nachweisbar.
Kinetik	typische Kettenreaktion. Quantenausbeute über 35.	Quantenausbeute < 1.

Soweit die an über 50 untersuchten organischen Verbindungen gewonnenen Erfahrungen bereits gewisse Regelmäßigkeiten erkennen lassen, kann über die photosensibilisierte Oxydation folgendes gesagt werden:

1. Konjugene addieren im allgemeinen O_2 unter Bildung von Peroxyden vom Typ C-O-O-C. Diese Addition erfolgt sehr häufig nach Art einer Diensynthese, z. B. bei Furanen unter Bildung von Peroxyden vom Ozonidtyp.

2. Mono-Olefine bzw. Olefine mit isolierten Doppelbindungen reagieren zu Hydroperoxyden C-O-O-H, die durch Addition des O_2 an die Doppelbindung und Wanderung eines H-Atoms der Allylstellung an den Sauerstoff unter Verschiebung der Doppelbindung entstehen. So sind die Hydroperoxyde der sensibilisierten und nicht sensibilisierten Autoxydation des Cyclohexens miteinander identisch, die α-Methylcyclohexens dagegen nicht miteinander identisch, während so leicht autoxydable Verbindungen wie Tetralin, Indan, auch Dekalin usw. in Abwesenheit des Sensibilisators (Fehlen der olefinischen Doppelbindung) nicht mit O_2 reagieren.

3. Die photosensibilisierte Reaktion mit O_2 ist nicht auf nur eine bestimmte Atomkombination spezifisch eingestellt. Analoge Umsetzungen, z. B. mit Maleinsäureanhydrid an Stelle von O_2, werden durch Belichtung in Gegenwart von Sensibilisatoren nicht gefördert.

4. Es besteht eine sehr hohe Spezifität der Sensibilisatorwirkungen für Reaktionen mit O_2. Soweit Reaktionen angeregter Sensibilisatoren in Abwesenheit von O_2 zu beobachten sind, werden diese durch O_2 inhibiert. Fast alle photosensibilisierten Reaktionen in Lösung sind solche mit O_2; zu nennen sind außer den Reaktionen mit Konjugenen, Furanen, Pyrrolen, Olefinen, Aldehyden, Aminen, Phenolen usw. die sogenannten photodynamischen Erscheinungen, die durch fluoreszierende Farbstoffe hervorgerufenen Lichtkrankheiten und die Biosynthese des Askaridols in Chenopodium anthelminticum.

Alle Beobachtungen vereinigen sich zum Beweise der Forderung, daß der angeregte Sensibilisator auf O_2 und nicht (in Ggw. v. O_2) auf A einwirkt. Den gleichen Schluß hat H. Kautsky[2]) aus der Tilgung der Fluoreszenz durch Sauerstoff gezogen und als Aktivierung von O_2 zu einem metastabilen O_2^{ε} interpretiert.

Für die entgegengesetzte, auf H. Gaffron zurückgehende, von K. Weber[3]) vertretene Auffassung, daß bei den fluoreszeinsensibilisierten Oxydationen nicht O_2, sondern immer A durch die vom Sensibilisator übertragene Lichtenergie aktiviert werde, spricht kein Experiment.

Die Wirkung des Photosensibilisators und zugleich die Fluoreszenzlöschung durch O_2 beruht auf einer chemischen Reaktion des angeregten Sensibilisators mit O_2[4]), d. h. der normalerweise gegen O_2 indifferente Sensibilisator geht durch Licht in eine gegenüber O_2 reaktionsfähige Substanz über.

Bereitschaft zur Dunkelreaktion mit O_2 ist für organische Radikale charakteristisch. Diese können nach K. Ziegler und L. Ewald[5]) auf andere Verbindungen O_2 übertragen. Entsprechend meiner Erwartung geht der Sensibilisator durch das absorbierte Lichtquant in ein gegenüber O_2 reaktionsfähiges Radikal, nach seiner Entstehung ein Diradikal, über. Diese Vorstellung wird vorzüglich gestützt:

1. Durch die Beobachtung, daß beim Belichten von Anthrazen in Abwesenheit von O_2 ein Dimeres[6]), in Gegenwart von O_2 ein Photoperoxyd[7]) entsteht. Diese Reaktionen sind nur als Reaktionen photochemisch gebildeter Radikale zu deuten.

2. Durch die Beobachtung von G. N. Lewis und M. Calvin[8]), daß Fluoreszein durch Belichtung (nach G. N. Lewis und M. Kasha[9]) durch Übergang in ein Diradikal!) paramagnetisch wird, und daß dieser Effekt durch O_2 gestört wird.

3. Durch quantentheoretische Überlegungen von Th. Förster[10]), wonach fluoreszierende Farbstoffe durch Licht in einen metastabilen Diradikalzustand übergehen.

So ergibt sich folgendes Bild der photosensibilisierten Reaktion mit molekularem Sauerstoff[11]):

1. Der Sensibilisator S_{norm} *geht durch Aufnahme eines Lichtquants* hν *in das phototrop-isomere Diradikal* S_{rad} *über.*

2. S_{rad} *addiert* O_2 *zu einem labilen* $S_{rad} \cdots O_2$[12]).

3. $S_{rad} \cdots O_2$ *gibt mit A den Stoßkomplex* $[S_{rad} \cdots O_2 \cdots A]$.

4. Das im Stoßkomplex enthaltene S_{rad} *geht unter Energieabgabe an* AO_2 *in* S_{norm} *über und wird abgespalten.*

5. Das so entstandene energiereiche AO_2 *stabilisiert sich unter Abgabe von Wärme zu* AO_2, *soweit nicht* AO_2 *endotherm ist.*

Da S_{norm} gegen O_2 indifferent ist, wird ein weiteres O_2 durch den Sensibilisator erst nach neuerlicher Aufnahme eines Lichtquants übertragen. Die photosensibilisierte Reaktion zwischen A und O_2 kann also gegenüber einer zwischen diesen Reaktionsteilnehmern gedachten Dunkelreaktion um den Betrag der absorbierten, photochemisch nutzbaren Energie, also etwa um den Betrag der Fluoreszenz- bzw. wahrscheinlicher der Phosphoreszenzenergie des Sensibilisators endotherm sein.

Wie weit das hier zugrunde liegende Prinzip einer gekoppelten Reaktion beim endothermen Aufbau organischer Verbindungen in der belebten Natur Bedeutung besitzt, bedarf der Prüfung. Die von H. Kautsky[13]) geforderte Mitbeteiligung des Sauerstoffs im Geschehen der Assimilation der Kohlensäure wird durch meine Versuche gestützt.

Ausführliche Mitteilung meiner Ergebnisse erfolgt an anderer Stelle. Von den skizzierten Ergebnissen ausgehend, werden auch andere Probleme, insbesondere der Photochemie und der Reaktionen mit O_2, bearbeitet.

Heidelberg.   Günther O. Schenck.

Eingegangen am 4. März 1948.

[1]) 5. Mitteilung über Autoxydation in der Furanreihe; 4. Mitteilung: Chem. Ber. im Druck.

[2]) Kautsky, H., Ber. Dtsch. Chem. Ges. **64**, 2677 (1931). Kautsky H. u. de Bruijn, H., Naturwiss. **19**, 1043 (1931).

[3]) Weber, K., Ber. Dtsch. Chem. Ges. **69**, 1026 (1936). Weber, K., Inhibitorwirkungen, Ferd. Enke Verlag, Seite 78, 108, Stuttgart (1938). Gaffron, Ber. Dtsch. Chem. Ges. **60**, 2229 (1927). Gaffron, H., Biochem. Zeitschr. **264**, 251 (1933).

[4]) Als Vermutung ausgesprochen in meiner Habilitationsschrift Halle a. S. (1943).

[5]) Ziegler K. u. Ewald, L., Annalen **504**, 162 (1933).

[6]) Fritzsche, J., Journ. prakt. Chemie **101**, 337 (1867); **106**, 274 (1869).

[7]) Dufraisse, Ch. u. Gerard, M., Compt. rend. **201**. 428 (1935). Dufraisse, Ch. u. Etienne, A., Compt. rend. **201**, 280 (1935).

[8]) Lewis, G. N. u. Calvin, M., J. Amer. Chem. Soc. **67**, 1232 (1945).

[9]) Lewis, G. N. u. Kasha, M., J. Amer. Chem. Soc. **66**, 2100 (1944); **67**, 994 (1945). Eistert, Hr. B. machte mich freundlicherweise auf die zitierten Arbeiten von Lewis und Mitarbeitern aufmerksam in einer Diskussionsbemerkung zu meinem Vortrag am 17. Februar 1948 vor der Chemischen Gesellschaft Heidelberg.

[10]) Förster, Th., Naturwiss. **33**, 220 (1946).

[11]) Schenck, G. O., vorläufige Mitteilung für FIAT-Review September (1946).

[12]) Spaltung in S_{norm} + O_2^{ε} könnte die Bildung des von Kautsky H.[2]), nachgewiesenen gasförmigen metastabilen Sauerstoffs erklären.

[13]) Erstmals Kautsky, H., Hirsch, A. u. Davidshöfer, F., Ber. Dtsch. Chem. Ges. **65**, 1762 (1932).

Editor's Comments
on Papers 14 and 15

14 SCHENCK
Chemismus und Kinetik der durch fluoreszierende Farbstoffe photosensibilisierten Reaktionen mit O_2 und Primärakt der Photosynthese

15 BOWEN
Reactions in the Liquid Phase. Photochemistry of Anthracene: Part I. The Photo-oxidation of Anthracenes in Solution

KINETICS OF PHOTOOXYGENATION

In Paper 14, G. O. Schenck described the first detailed investigation of the kinetics of dye-sensitized photooxygenation. The results were interpreted in terms of the intermediacy of a sensitizer–oxygen complex. Schenck indicated that reactivity of various oxygen acceptors could be expressed as a ratio of two rate constants, k_V/k_{III}, where k_V is the rate constant for the decomposition of the sensitizer–oxygen complex to ground-state sensitizer and oxygen, and k_{III} is the rate constant for the reaction of the complex with the acceptor (see Paper 70).

The kinetics of the direct photooxygenation of several substituted anthracenes were described by E. J. Bowen in Paper 15. He found that the kinetics were consistent with either the Kautsky singlet oxygen or a sensitizer–oxygen complex (AO_2^*) mechanism.

14

Reprinted from *Z. Elektrochem.*, 55(6), 505–511 (1951)

Chemismus und Kinetik der durch fluoreszierende Farbstoffe photosensibilisierten Reaktionen mit O₂ und Primärakt der Photosynthese

Von *GÜNTHER O. SCHENCK*, Göttingen

Aus dem Organisch-Chemischen Institut der Universität Göttingen

(Vortrag, gehalten auf der 50. Hauptversammlung der Deutschen Bunsengesellschaft
in Göttingen am 4. Mai 1951)

Der Sensibilisator bildet im Primärakt energiereicheres sauerstoffaffines Isomeres, das O₂ addieren und so auf geeignete Akzeptoren übertragen kann (Photobiosynthese des Askaridols). Die Kinetik weist kurzlebiges Zwischenprodukt „Sensibilisator . . . O₂" nach und bestimmt die Quantenausbeute. Vorstellungen über den Primärakt führen zu neuen Photoreaktionen. Zur Methodik: Automatische Registrierung des Gasverbrauches. Warburg-Apparatur mit Leuchtstoffröhren.

Belichtet man Lösungen geeigneter Akzeptoren A in Gegenwart fluoreszierender Farbstoffe, wie Chlorophyll, Methylenblau oder Eosin, unter Sauerstoff, so vereinigen sich A und O₂ zu Produkten AO₂ — das sind häufig organische Peroxyde oder Hydroperoxyde, die isoliert werden können —, während die Sensibilisatoren unverändert bleiben.

Für solche photosensibilisierte Reaktionen mit molekularem Sauerstoff kennt man bisher anscheinend nur ein natürlich vorkommendes Beispiel: die Photobiosynthese des Askaridols[1] in einigen Pflanzen, vor allem in Chenopodium anthelminticum. α-Terpinen (I) und O₂ vereinigen sich hier im Sonnenlicht unter der sensibilisierenden Wirkung des Chlorophylls zum Askaridol (II).

In vitro gelingt diese Photosynthese etwa auch in der Weise[2], daß man in eine verdünnte alkoholische Lösung von α-Terpinen einige Brennessel- oder Spinatblätter bringt und die Flüssigkeit, die sich bald grün färbt, in einem Glaskolben unter Luft oder Sauerstoff in die Sonne stellt.

Die Sensibilisatorleistung des Chlorophylls bei der Askaridolsynthese und die des Chlorophylls bei der Assimilation der Kohlensäure stehen nach meinen weiteren Untersuchungen[3] dadurch in engstem Zusammenhang, daß die photochemischen Primärakte der beiden Reaktionen identisch sein müssen, wenn die beiden Chlorophylle miteinander identisch sind. Es zeigte sich nämlich, daß hier die Wirkung des Photosensibilisators nicht auf einer reinen Übertragung von Anregungsenergie auf einen der Reaktionsteilnehmer beruhen kann[4]. Vielmehr besteht der photochemische Primärakt in beiden Fällen in der Bildung eines kurzlebigen energiereicheren Isomeren des Sensibilisators mit den Eigenschaften eines Diradikals, das ich auf Grund seiner Bildung als phototrop-isomeres Diradikal[5] bezeichne. Diese mit dem Sensibilisator im photochemischen Anregungszustand iden-

---

[1] G. O. Schenck, Habilitationsschrift, Halle a. S. 1943, S. 45; G. O. Schenck und K. Ziegler, Naturw. *32* (1944), 157.

[2] G. O. Schenck und K. G. Kinkel, unveröffentlicht.

[3] G. O. Schenck, Angew. Chemie *61* (1949), 332.

[4] G. O. Schenck, Naturw. *35* (1948), 28.

[5] Siehe hierzu auch Eugen Müller, Fortschritte der chem. Forschung Bd. *1* (1949/50), S. 387.

tische Verbindung greift dann in einer Dunkelreaktion in den Assimilationsprozeß bzw. in den Vorgang der O_2-Übertragung ein.

Bei der Bildung von AO_2 sind dann 3 aufeinanderfolgende Schritte zu unterscheiden:

1. Der Sensibilisator im Grundzustand ($Sens_{norm}$) geht durch Aufnahme eines Lichtquants $h\nu$ in das phototropisomere Diradikal $Sens^{rad}$ über.

2. $Sens^{rad}$ addiert O_2 zu einer labilen und eigentümlich reaktionsfähigen Verbindung $Sens^{rad} \ldots O_2$.

3. $Sens^{rad} \ldots O_2$ setzt sich mit A um unter Bildung von AO_2 und $Sens_{norm}$. Dabei wird Wärme frei, soweit nicht etwa die Bildung von AO_2 endotherm ist.

Mit Reaktion II konkurrieren aber alle übrigen Prozesse (IV), die $Sens^{rad}$ in $Sens_{norm}$ (z. B. durch Strahlung, Löschvorgänge usw.) umwandeln. Ferner führen nicht alle gebildeten $Sens^{rad} \ldots O_2$ nach Reaktion III zu AO_2, da hier noch der exotherme Zerfall (V) von $Sens^{rad} \ldots O_2$ zu $Sens_{norm} + O_2$ in Konkurrenz steht.

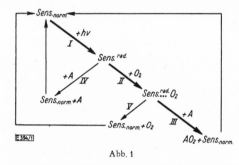

Abb. 1

Wegen der prinzipiellen Bedeutung des Befundes, daß $Sens^{rad}$ *nur* durch Wirkung auf O_2 zur Bildung von AO_2 beiträgt[4]), haben wir uns eingehend mit der Wechselwirkung zwischen $Sens^{rad}$ und O_2 bzw. $Sens^{rad}$ und A beschäftigt. Dabei leitete uns folgender Gedanke:

Setzt sich $Sens^{rad}$ unter gleichzeitiger Selbstinaktivierung mit A (Fall 1) oder mit O_2 (Fall 2) oder in jedem der beiden Fälle (Fall 1 + 2) unter Bildung von AO_2 um, so muß sich die Quantenausbeute der AO_2-Bildung mit zunehmender Konzentration von A und O_2 asymptotisch einem maximal bei 1 liegenden Wert nähern. Setzt sich dagegen $Sens^{rad}$ unter gleichzeitiger Selbstinaktivierung mit A *und* mit O_2 um, wobei jedoch nur *entweder* (Fall 1 + 4) die Umsetzung mit A *oder* (Fall 2 + 3) die Umsetzung mit O_2 zur Bildung von AO_2 führt, so muß die Quantenausbeute mit zunehmender Konzentration des *nur inaktivierenden* Stoßpartners durch ein unter 1 liegendes Maximum gehen und dann wieder abfallen:

1) $Sens^{rad} + A \rightarrow Sens_{norm}$,   Bildung von AO_2

2) $Sens^{rad} + O_2 \rightarrow Sens_{norm}$,   Bildung von AO_2

3) $Sens^{rad} + A \rightarrow Sens_{norm}$,   + Wärme

4) $Sens^{rad} + O_2 \rightarrow Sens_{norm}$,   + Wärme

Wir fanden nun regelmäßig, daß die Geschwindigkeit der Sauerstoffaufnahme bzw. die Quantenausbeute der AO_2-Bildung mit zunehmender Konzentration des Akzeptors (z. B. Diene, Thioharnstoff) ansteigt und nach Erreichung eines Maximums wieder abfällt[2]). Dies beweist erstmalig für jeden einzelnen untersuchten Fall, daß $Sens^{rad}$ sowohl durch A als auch durch O_2 inaktiviert wird, daß aber nur Stöße von $Sens^{rad}$ mit O_2 zur Bildung von AO_2 (z. B. Askaridol) führen. Es gilt also Fall $2 + 3$.

Nun ist zu unterscheiden zwischen den Affinitäten von $Sens^{rad}$ gegenüber A (ausgedrückt durch k_{IV}) und gegenüber O_2 (ausgedrückt durch k_{II}), und man darf nicht erwarten, daß diese etwa einander parallel laufen. Auch wird bei verschiedenen phototrop-isomeren Diradikalen das Verhältnis der Akzeptor- und Sauerstoffaffinitäten variieren, so daß man auch Fälle mit nur Akzeptor- oder mit nur Sauerstoffaffinität erwarten kann. Letztere wird aber nun als wichtige Voraussetzung für die hier untersuchten Sensibilisatorwirkungen erkannt.

Für die Bildung des zu AO_2 führenden Zwischenproduktes $Sens^{rad} \ldots O_2$ gilt nun, wenn wir mit J den wirksam absorbierten Quantenstrom bezeichnen und wenn alle übrigen Umwandlungsreaktionen von $Sens^{rad}$ zu $Sens_{norm}$ gegenüber der Inaktivierung durch A und durch O_2 vernachlässigt werden können, die Beziehung:

$$\frac{d[Sens^{rad} \ldots O_2]}{dt} = J \cdot \frac{k_{II}[O_2]}{k_{II}[O_2] + k_{IV}[A]} = \frac{J}{1 + \frac{k_{IV}[A]}{k_{II}[O_2]}} = J \cdot \varphi .$$

Hier gibt der Quotient φ das Verhältnis der allein zu AO_2 führenden inaktivierenden Stöße mit O_2 zur Gesamtzahl der $Sens^{rad}$ inaktivierenden Stöße an. Die Bildung von $Sens^{rad} \ldots O_2$ erfolgt demnach mit der Quantenausbeute φ. Diese wird sich beim Übergang von Sauerstoff auf Luft so lange nur geringfügig ändern, als in

$$\varphi = \frac{1}{1 + \frac{k_{IV}[A]}{k_{II}[O_2]}} \quad \text{der Summand} \quad \frac{k_{IV}[A]}{k_{II}[O_2]} \quad \text{sehr klein ist}$$

gegenüber 1. Nach diesen Überlegungen muß das Verhältnis $\dfrac{\varphi \text{ unter Luft gemessen}}{\varphi \text{ unter } O_2 \text{ gemessen}}$ mit zunehmender Konzentration des Akzeptors abnehmen. Mit Methylenblau in Pyridin fanden wir[6]) bei 26° für

	2%	3%	4% Thioharnstoff
$\dfrac{\varphi_{Luft}}{\varphi_{O_2}} =$	0,83	0,77	0,68 .

Eine Überraschung erlebten wir, als wir das Zyklooktatetraen (III), dessen Sauerstoffempfindlichkeit[7]) bekannt war, als Akzeptor verwenden wollten. Wir erwarteten, daß es in formaler Analogie zur Reaktion mit Maleinsäureanhydrid ein Endoperoxyd bilden würde. Indessen wurde überhaupt kein Sauerstoff aufgenommen, und das Zyklooktatetraen benahm sich unter den Bedingungen der photosensibilisierten Autoxydation wie eine aromatische Verbindung!

[6]) G. O. Schenck und E. Theiling, unveröffentlicht.
[7]) W. Reppe, O. Schlichting, K. Klager und T. Toepel, Annalen *560* (1948), 1.

Wenn im Zyklooktatetraen ein nennenswerter Anteil der gelegentlich diskutierten isomeren Form IV vorhanden wäre, so hätten wir sicherlich eine Reaktion beobachtet. Wir fanden nämlich an einer Reihe von Verbindungen der strukturell nahe verwandten Norkaradienreihe (z. B. V, VI), daß hier sogar recht glatt Endoperoxyde (Va, VIa) gebildet werden[8]).

auf die mit Millimeterpapier bespannte Schreibtrommel eines Kymographions (12). Unsere Vorrichtungen erwiesen sich als sehr zeitsparend und genauer als die subjektive Einstellung und Ablesung. Abb. 4 gibt einige solcher direkt beschriebener Kurven wieder, die eine eigentümliche Hemmung der Sauerstoffaufnahme bei der durch Methylenblau photosensibilisierten Autoxydation des Terpinens durch von Kurve 1–4 steigende kleine Mengen von Zyklooktatetraen zeigen.

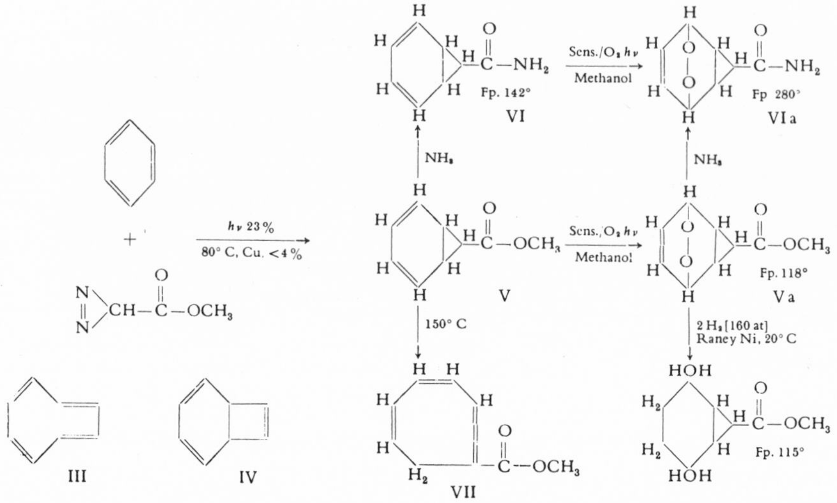

Durch eine photochemische Arbeitsweise konnten wir die Ausbeuten an den benötigten Norkaradienderivaten gegenüber Buchner[9]) und anderen vervielfachen und erstmals reine Norkaradienkarbonsäureester (V) darstellen. Diese nehmen bei der photosensibilisierten Autoxydation sehr rasch Sauerstoff auf im Gegensatz zu den isomeren Zykloheptatrienkarbonsäureestern (VII) (Abb. 2). Die Photooxydationsprodukte der Zykloheptatrienderivate sind mit denen der Norkaradienreihe nicht identisch, aber noch nicht näher untersucht. Die in einigen Kurven erkennbare verfrühte Verlangsamung der Sauerstoffaufnahme läßt Rückschlüsse auf den Gehalt an Norkaradienderivaten der verwendeten Präparate zu.

Bei unseren Versuchen bedienten wir uns einer vielseitig verwendbaren Methode zur automatischen Registrierung des Gasverbrauchs[2]), für die wir in der uns zugänglichen Literatur kein Vorbild fanden. Die durch das Schema erläuterte Arbeitsweise (Abb. 3) liefert eine direkt geschriebene Kurve des zeitlichen Verlaufes der Sauerstoffaufnahme, indem der durch ein empfindliches Kontaktbarometer (6) gesteuerte Antrieb (8), z. B. über einen von der Spindel (9) geführten Wagen (10), Niveaugleichheit der Sperrflüssigkeit im Gasometer (2) und im beweglichen Schenkel mit dem Niveaugefäß (3) einstellt. Gleichzeitig schreibt eine mit dem Wagen verbundene Barographenfeder dessen Höhe

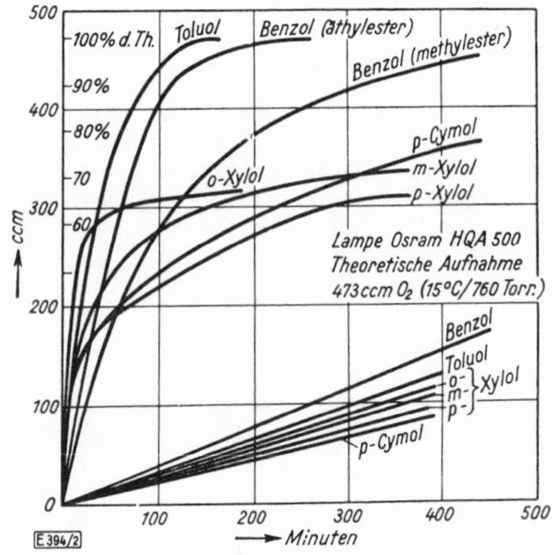

Abb. 2

Sauerstoffaufnahme bei der Photooxydation der homologen Norkaradienkarbonsäuremethylester und Zykloheptatrienkarbonsäuremethylester in Methanol. (Bezifferung der Ester nach den Ausgangs-Kohlenwasserstoffen, Einwaage 0,02 Mol in 75 cm³ Sensibilisatorlösung entspricht 0,267 Mol/l. Sensibilisator Methylenblau [500 mg im l]

---

[8]) G. O. Schenck und H. Ziegler, unveröffentlicht.
[9]) E. Buchner und Th. Curtius, Ber. 18 (1885), 2377; A. Loose, J. pr. Ch. (N. F.) 79 (1909), 509.

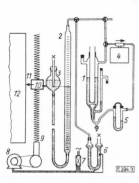

Abb. 3

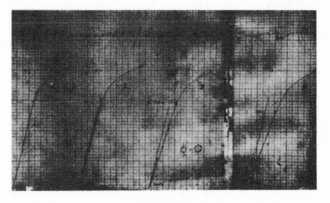

Abb. 4

Wir haben uns nämlich mit der Feststellung, daß Zyklooktatetraen bei der photosensibilisierten Autoxydation keinen Sauerstoff aufnimmt, nicht begnügt, sondern haben jeweils nach einiger Zeit eine kleine Menge eines anderen bewährten Akzeptors zugegeben. Dabei fand dann stets die erwartete Sauerstoffaufnahme statt, sie war aber in jedem Falle etwas langsamer als erwartet. Diese beobachtete Verlangsamung führte sofort zur Umkehrung des Versuches mit dem obigen Ergebnis.

Wir vermuteten, daß das Zyklooktatetraen zwar wie andere Akzeptoren zur Inaktivierung von $Sens^{rad}$ befähigt sei, indessen aber gegenüber dem Zwischenprodukt $Sens^{rad}\ldots O_2$ indifferent sei. Um dies zu beweisen, arbeiteten wir mit einer für diesen Zweck besonders konstruierten Warburg-Apparatur, in der durch Verwendung von Leuchtstoffröhren jedem Gefäß innerhalb 1% der gleiche Quantenstrom zugeführt wird[10]. Außer den Versuchen lassen wir stets eine aktinometrisch bewährte Reaktion mitlaufen und haben so die Kontrolle über die Konstanz der Versuchsbedingungen während der Messung.

Die Tabelle[11] zeigt für Methylenblau in Isopropanol bei 26° den Einfluß von Zyklooktatetraen auf φ unter O_2 und unter Luft.

Depression von φ durch Zyklooktatetraen unter O_2 und unter Luft.

	[C₈H₈]	0,1	0,3	0,5	Mol/Liter
100 φ_{O_2}	gef.	78,0	56,7	44,6	% von $\varphi_{optimal} = 1$
	ber.	77,5	55,9	43,7	
100 φ_{Luft}	gef.	48,8	24,6	15,5	
	ber.	46,3	24,0	16,2	

Sämtliche gefundenen φ sind Mittelwerte aus 4 Meßreihen mit α-Terpinen, α-Pyronen, Zyklopentadien und 2-Methylfuren als Akzeptoren und bezogen auf extrapoliertes $\varphi_{optimal} = 1$. Messungen bei 26° C in Isopropanol mit 0,5 g Methylenblau je l mit [A] = 0,2.

[10] Der Fa. B. Braun, Melsungen, haben wir für entgegenkommenden Umbau eines Modells zu danken.
[11] G. O. Schenck, K. G. Kinkel und E. Theiling, unveröffentlicht.

$$\varphi_{O_2}\,\text{ber.} = \frac{1}{1 + 0,2\,[A] + 2,5\,[C_8H_8]}$$

$$\varphi_{Luft}\,\text{ber.} = \frac{1}{1 + 4\,(0,2\,[A] + 2,5\,[C_8H_8])}$$

Es zeigte sich, daß die inaktivierende Wirkung von 1 Mol Zyklooktatetraen gleich der von etwa 12,5 Mol der untersuchten Akzeptoren ist.

Die gute Übereinstimmung zwischen Experiment und Theorie zeigt, daß die Depression der Quantenausbeute durch Zyklooktatetraen nicht auf einer inneren Lichtfilterwirkung beruht. Wir haben vielmehr hier isoliert die inaktivierende Wirkung der Akzeptoren ohne deren Affinität gegen $Sens^{rad}\ldots O_2$. Im Gegensatz zur Wirkungsweise der Inhibitoren, die bei Radikalkettenreaktionen durch Abfangen von Gliedern der Kette wirksam werden, dabei aber zugleich in jedem einzelnen Falle selbst chemisch verändert werden, kehrt das Zyklooktatetraen in den von uns untersuchten Fällen jedesmal aus dem Hemmvorgang unverändert zurück; man könnte das Zyklooktatetraen auch als einen besonders wirksamen Katalysator zur Umwandlung photochemischer Anregungsenergie in Wärme ansehen.

Aus der Gültigkeit der Gleichung (1) ergibt sich, daß außerhalb der Fehlergrenze (höchstens 5%) in den von uns bisher untersuchten Konzentrationen keine sonstigen Umwandlungen $Sens^{rad} \rightarrow Sens_{norm}$ stattfinden. Hieraus folgt, daß $Sens^{rad}$ eine relativ hohe Lebensdauer besitzt und daß eine Umwandlung $Sens^{rad} \rightarrow Sens_{norm} + Wärme$ ohne Mitwirkung einer hierzu spezifisch geeigneten Substanz (hier z. B. O_2, Zyklooktatetraen, Akzeptor) nicht möglich ist. Dies bedeutet umgekehrt, daß eine direkte Umwandlung $Sens^{rad} \rightarrow Sens_{norm}$ nur unter Strahlung erfolgen kann. Wir haben also in $Sens^{rad}$ einen relativ langlebigen, strahlungsfähigen Anregungszustand des Sensibilisators, den phosphoreszenzfähigen Anregungszustand, vor uns. In diesem Ergebnis sehe ich einen von den physikalischen Überlegungen[12], die die Gleich-

[12] Über die Notwendigkeit weiterer Beweise s. Th. Foerster, Fluoreszenz organischer Verbindungen, 1951 bei Vandenhoeck & Ruprecht, Göttingen, S. 281.

setzung von Triplett- und phosphoreszenzfähigem Anregungszustand rechtfertigen, unabhängigen Beweis hierfür.

Trotz dieser Gleichsetzung müssen wir uns vor einer ausreichenden Klärung der hier behandelten Zusammenhänge vor der Vorstellung hüten, als seien hier nun Phosphoreszenz und Triplett- bzw. Diradikalzustand notwendig miteinander gekoppelte Begriffe. Einmal ist denkbar, daß in manchen Fällen der fluoreszenzfähige Anregungszustand erst durch Einwirkung anderer hierfür spezifisch geeigneter Molekeln in den Diradikalzustand übergeht. Andererseits scheint möglich, daß manche phototrop-isomere Diradikale infolge besonderer Reaktionsfähigkeit unter normalen Bedingungen keinerlei Photolumineszenz zeigen: Man wird daher gerade der chemischen Seite bei der Erforschung des Triplettzustandes besondere Aufmerksamkeit zuwenden müssen.

Für die Richtigkeit und den heuristischen Wert dieser Vorstellungen spricht folgendes an einer nicht phosphoreszierenden Verbindung gefundene Beispiel:

Beim Phenanthrenchinon müßten die beiden Radikalqualitäten des photochemischen Anregungszustandes an den O-Atomen sich in 1,4-Stellung befinden und chemisch etwa Halogenatomen entsprechen. Da Chloratome, nicht Chlormolekeln, an SO₂ zu Sulfurylchlorid addiert werden können, sollte man Phenanthrenchinon analog photochemisch an SO₂ zu zyklischen o-Arylen-Sulfaten addieren können. Wir fanden diese Reaktion[13]), die zu einer neuen Verbindungsklasse führt, tatsächlich bei Phenanthrenchinon und konnten sie inzwischen auch in der Benzol-, Naphthalin- und Chrysenreihe durchführen. Auch eine im einzelnen noch ungeklärte Photoaddition von NO haben wir beobachtet.

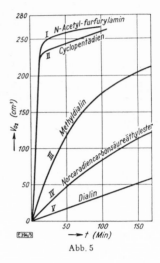

Was geschieht mit dem kurzlebigen Zwischenprodukt ,Sens^rad... O₂ bei verschiedenen Konzentrationen der Akzeptoren? Wie leicht festzustellen, muß es auch in Abwesenheit von A sehr schnell zerfallen. Andernfalls müßte hier nämlich beim Belichten ein gewisser, im Dunkeln reversibler Verbrauch an O₂ zu beobachten sein. Ebensowenig ist eine Verminderung der Extinktion der Sauerstoff enthaltenden Sensibilisatorlösung beim Belichten unter unseren Versuchsbedingungen zu erkennen. Aber die unter völlig übereinstimmenden Bedingungen erhaltenen Kurven der Sauerstoffaufnahme[14]) verschiedener Akzeptoren sind sehr aufschlußreich (Abb.5): Sie zeigen erhebliche und, wie aus der Übereinstimmung der Kurven für N-Acetyl-furfurylamin und Zyklopentadien zu entnehmen, auch sehr charakteristi-

sche Unterschiede. Bei N-Azetyl-furfurylamin und Zyklopentadien finden wir praktisch einen linearen Verlauf, während beim Methyldialin in gleicher molarer Konzentration (0,03 m) eine Reaktion 1. Ordnung vorzuliegen scheint.

Abb. 5

Wie wir immer wieder fanden, nähert sich der Charakter der Kurven mit zunehmender Konzentration von A der 0. Ordnung, mit abnehmender Konzentration von A der 1. Ordnung. Dies ist aber leicht zu verstehen: Das betrachtete System erhält einen durch den vorhin erläuterten Ausdruck $J \cdot \dfrac{1}{1 + \dfrac{k_{IV}[A]}{k_{II}[O_2]}} = \alpha$ gegebenen Zustrom an Sens^rad... O₂. Dieser Zustrom wird aber nun laufend durch die beiden konkurrierenden Reaktionen III und V verbraucht, und da Zulieferung und Verbrauch hier einander gleich sind, so gilt:

$$\alpha = k_{III}\,[\text{Sens}^{\text{rad}}\ldots O_2]\,[A] + k_V\,[\text{Sens}^{\text{rad}}\ldots O_2].$$

Für die Ausnutzung von α zur Bildung von AO₂ ergibt sich damit der Ausnutzungsquotient q

$$q = \frac{k_{III}\,[\text{Sens}^{\text{rad}}\ldots O_2]\,[A]}{k_{III}\,[\text{Sens}^{\text{rad}}\ldots O_2]\,[A] + k_V\,[\text{Sens}^{\text{rad}}\ldots O_2]} = \frac{[A]}{[A] + \dfrac{k_V}{k_{III}}},$$

der sich mit zunehmender Konzentration von A an 1 nähert, so daß wir dann eine Reaktion 0. Ordnung finden. Wird dagegen die Konzentration von A sehr klein gegen $\dfrac{k_V}{k_{III}}$, so wird α praktisch entsprechend der Konzentration von A ausgenutzt, und zwar im Verhältnis $\dfrac{[A]}{\dfrac{k_V}{k_{III}}}$ bzw. $\dfrac{[A]\,k_{III}}{k_V}$. Hier finden wir einen Verlauf nach der 1. Ordnung.

---

[13]) G. O. Schenck und G. A. Schmidt, unveröffentlicht.
[14]) G. O. Schenck und Hrch. Mertens, unveröffentlicht.

Für die Bildung von AO$_2$ gilt also in Übereinstimmung mit unseren Experimenten die Geschwindigkeitsgleichung

$$\frac{d\,[AO_2]}{dt} = \frac{\alpha \cdot [A]}{[A] + \dfrac{k_V}{k_{III}}}\,.$$

Wir ermittelten α und $\frac{k_V}{k_{III}}$ aus dem weiteren Verlauf der Sauerstoffaufnahme und fanden in niedrigen Konzentrationen von A (wo α noch annähernd konstant), die zu erwartenden Verhältnisse[14]): Während α von der Art des Akzeptors praktisch unabhängig ist, liegt in $\frac{k_V}{k_{III}}$ eine für jeden Akzeptor charakteristische Konstante vor, deren Abhängigkeit von der chemischen Konstitution der Akzeptoren theoretisches Interesse besitzt.

einer Reihe erfolgloser Stöße reagieren, wobei die für die Bildung von AO$_2$ erfolglosen Stöße jedoch das Sens^{rad}... O$_2$ unverändert lassen. Ferner ist beim α-Pinen eine sterische Selektivität der Sauerstoffaddition zu beobachten, die zum bisher unbekannten Pinocarveylhydroperoxyd[15]) führt. Analog erhielten wir durch photosensibilisierte Autoxydation von β-Pinen (VIII) das Myrtenylhydroperoxyd[15]) (IX), das durch Reduktion in den Naturstoff Myrtenol (X) übergeht[15]). Möglicherweise können also noch weitere natürlich vorkommende photosensibilisierte Reaktionen mit O$_2$ gefunden werden. Als ein theoretisch wie organisch präparativ besonders interessantes Beispiel sei schließlich noch ein photosensibilisierte Autoxydation des 1,2-Dimethylzyklohexans (XI) genannt, das praktisch quantitativ (90% des aufgenommenen Sauerstoffes) in das tertiäre Hydroperoxyd XII übergeht[16]).

Mit den zuletzt skizzierten Ergebnissen und der eingangs behandelten Möglichkeit φ und damit auch $\frac{k_{IV}}{k_{II}}$ experimentell zu bestimmen, finden wir für die von uns untersuchte photosensibilisierte Reaktion mit O$_2$ folgende Geschwindigkeitsgleichung gültig:

$$\frac{d\,[AO_2]}{dt} = \frac{J}{1 + \dfrac{k_{IV}[A]}{k_{II}[O_2]}} \cdot \frac{[A]}{[A] + \dfrac{k_V}{k_{III}}}\,.$$

Wir sind daher auch in der Lage, J unabhängig von Energiemessungen mit reaktionskinetischen Methoden zu bestimmen.

Für Sens^{rad}... O$_2$ ergibt sich aus unseren Untersuchungen eine gewisse Beständigkeit, und es wird eine Aufgabe weiterer Untersuchungen sein, die Abhängigkeit von k_V von der chemischen Konstitution verschiedener Sensibilisatoren zu eigen. Wir fanden, daß manche Akzeptoren, z. B. α-Pinen mit Sens^{rad}... O$_2$ erst nach

Zusammenfassend kann gesagt werden: Unsere Ergebnisse sind bereits im Grundsätzlichen mit den bisher als gültig angesehenen Auffassungen über photosensibilisierte Reaktionen der von uns untersuchten Art nicht vereinbar. Dies gilt vor allem für die aus anderen Gründen (z. B. sensibilisierte Fluoreszenz) naheliegende Vorstellung, der Photosensibilisator übertrage Anregungsenergie[17]) auf einen der Reaktionsteilnehmer, der so die charakteristische Reaktionsfähigkeit erlange. Im besonderen erfolgt die Bildung von AO$_2$ nicht über eine „Akzeptoraktivierung" im Sinne von v. Tappeiner, Gaffron, K. Weber[18]) u. a. Wie ich früher zeigte[4]), ist kon-

---

[15]) G. O. Schenck und H. Eggert, unveröffentlicht.
[16]) G. O. Schenck und Wi. Hake, unveröffentlicht.
[17]) A. Eucken, Lehrbuch II, 1, S. 415, 3. Aufl. 1950; Th. Foerster, Z. Naturf. 2b (1947), 174, insbes. 181.
[18]) v. Tappeiner, Ergebnisse der Physiologie Bd. 8, S. 698 (1909); H. Gaffron, Ber. 60 (1927), 2229; Biochem. Z. 264 (1933), 251; K. Weber, Ber. 69 (1936), 1026; Inhibitorwirkungen, Ferd. Enke Verlag, S. 78, 108. Stuttgart (1938).

sequenterweise auch die Vorstellung unzulässig, daß es für Reaktionen zwischen A und O_2 im Grunde gleichgültig sei, ob diese Reaktion direkt durch Absorption eines Lichtquants durch einen der Reaktionsteilnehmer oder indirekt durch Vermittlung eines Photosensibilisators bewirkt werde[19]).

[19]) O. Warburg und V. Schocken, Arch. of Biochemistry *21* (1949), 368.

Wir hoffen, daß die von uns studierte Chemie des photochemischen Anregungszustandes zu einem tieferen Verständnis mancher photochemischer Prozesse in Natur und Technik beitragen wird; ich denke hierbei z. B. an das Ausbleichen und die faserschädigende Wirkung von Farbstoffen im Licht, an die photodynamischen Effekte und Lichtkrankheiten und vor allem an die Assimilation der Kohlensäure.

15

Reprinted from *Discussions Faraday Soc.*, No. 14, 143–146 (1953)

REACTIONS IN THE LIQUID PHASE

PHOTOCHEMISTRY OF ANTHRACENE

PART I.—THE PHOTO-OXIDATION OF ANTHRACENES IN SOLUTION

By E. J. Bowen

Physical Chemistry Laboratory, Oxford

Received 25th March, 1952

The kinetics of photo-oxidation by dissolved oxygen of solutions of various substituted anthracenes in three solvents have been measured by oxygen up-take. The quantum efficiencies depend on the anthracene but not on the oxygen concentrations. Very small additions of hexane, benzene, or alcohol to carbon disulphide solutions greatly increase the dependence on anthracene concentration. The reaction is interpreted in terms of a reactive intermediate for which anthracene and molecules with C—H bonds compete.

Studies of the photo-oxidation by dissolved oxygen of the aromatic hydro-carbons, rubrene and anthracene, in solution have shown that the products are cyclic peroxides of considerable stability formed by non-chain mechanisms which, however, are unexpectedly complex.[1, 2, 3, 4] The reaction of anthracene has not hitherto been studied in detail, but the following facts have been reported.

(i) In solvents such as benzene or hexane, anthracene shows fluorescence in dilute solutions and forms dianthracene in concentrated ones; photo-oxidation is very slow and is a function of the anthracene but not of the oxygen concentration. The quenching effect of dissolved oxygen on the fluorescence appears to be unrelated to the oxidation processes.

(ii) In solvents such as carbon disulphide or chloroform, anthracene fluor-escence is more or less quenched and photo-oxidation is rapid; when no oxygen is present, however, dianthracene formation does not occur readily.[1] It is natural to associate the quenching and reaction with the production of triplet levels ^3A of anthracene by spin-reversal due to nuclei of the heavy atoms in the solvent molecules.[5]

EXPERIMENTAL

Anthracene was purified by double distillation with ethylene glycol;[6] the substituted anthracenes were synthesized and purified by chromatography until impurities could not be detected by examination of the absorption spectra. The solvents were A.R. specimens, except chloroform (B.P.) which contained a small quantity of ethyl alcohol.

Photo-oxidation was carried out in a modified Barcroft differential gas-absorption apparatus [7] with oxygen above the solutions, the whole being shaken with an amplitude and frequency adjusted to give rapid equilibration between gas and solution. The oxygen taken up in the illuminated limb (which had a fixed aperture below the broken surface of the liquid) was followed by changes of level of the manometer connecting it to the unilluminated limb. The light used was the 3650 Å line of the mercury arc, and quantum efficiencies were estimated by the use of the uranyl oxalate actinometer. When air was used instead of oxygen the rates were found to be only slightly slower.

In the concentration range covered the light absorption by the solutions was complete; corrections were applied to the results for carbon disulphide as this liquid itself absorbs somewhat at 3650 Å; linear reaction curves with time were obtained, and from these quantum efficiency against concentration graphs were plotted. Fig. 1 is a plot of the reciprocals of quantum efficiency Q and concentration [A] for B.P. chloroform solutions showing that a linear relation exists between them. The results for all the solvents are expressible by the equation,

$$\text{quantum efficiency} = \frac{k[A]}{[A] + K},$$

(see later paper for modifications due to solvent reaction for anthracene in carbon tetra-chloride solution).

Table 1 gives values of k and K for different solvents and substituted anthracenes. In earlier work by Mr. D. T. G. Morgan [8] it was found that small additions of hexane, benzene or alcohol to carbon disulphide had a remarkable effect in diminishing the rate of oxidation.   Fig. 2 and table 2 show that this arises from changes in the constant K

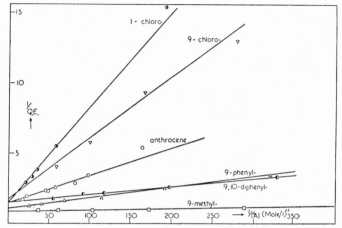

FIG. 1.—Plot of reciprocals of quantum efficiency of photo-oxidation and of con-centrations of several anthracenes for solutions in B.P. chloroform.

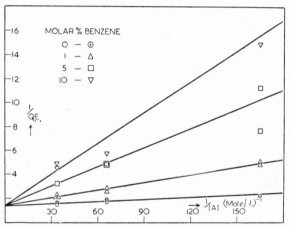

FIG. 2.—Plot of reciprocals of quantum efficiency and of anthracene concentration for solutions in carbon disulphide containing varying additions of benzene.

TABLE 1

	values of k in			values of $K \times 10^4$ in		
	CCl$_4$	CS$_2$	CHCl$_3$	CCl$_4$	CS$_2$	CHCl$_3$
anthracene	0·75	0·70	0·69	40·0	36·0	136
1-chloro	0·45	0·64	0·66	41·0	49·0	452
9-chloro	0·63	0·57	0·56	32·0	30·0	242
9-10 dichloro	0·89	2·0	1·6	101·0	170·0	603
9-methyl	1·3	—	1·2	33·0	—	12
9-phenyl	0·85	0·91	0·94	7·4	8·4	72
9 : 10-diphenyl	0·75	1·8	0·69	6·0	32·0	40

rather than in k. Addition of benzene to carbon disulphide makes the photo-oxidation much more dependent on the anthracene concentration; the mixed solvent resembles the B.P. chloroform in this respect. Alcohol and hexane act very similarly.

TABLE 2

molar % benzene	0	1	5	10
$K \times 10^4$	36	150	380	620

DISCUSSION

The form of the dependence of the photo-oxidation on anthracene concentration and its practical independence on oxygen concentration can be interpreted by at least three mechanisms. Since photo-oxidation is characteristic of those solvents which quench the fluorescence by bringing about singlet-triplet conversions, it is natural to begin these by assuming the participation of triplet anthracene levels ^3A, which have an energy 45 kcal/mole above the ground state.

reaction constant

$$(1) \quad ^3A + O_2 \rightarrow {}^1\Sigma O_2 + A \quad \text{(complete)}$$
$$^1\Sigma O_2 + A \rightarrow AO_2 \quad k_1$$
$$^1\Sigma O_2 \rightarrow {}^3\Sigma O_2 \quad k_2$$

$$(2) \quad ^3A + O_2 \rightarrow AO_2{}^* \quad \text{(complete)}$$
$$AO_2{}^* + A \rightarrow AO_2 + A \quad k_1$$
$$AO_2{}^* \rightarrow A + O_2 \quad k_2$$

$$(3) \quad ^3A + A \rightarrow A_2{}^* \quad k_1$$
$$^3A \rightarrow A \quad k_2$$
$$A_2{}^* + O_2 \rightarrow AO_2 + A \quad \text{(complete)}$$

The ratios k_2/k_1 represent the K of tables 1 and 2. These three mechanisms are associated with views put forward by Kautsky (1),[9] Koblitz and Schumacher (2),[2] and Dufraisse (3).[1] None of them, however, explains the large effect of small quantities of hexane or benzene on the value of K; it is difficult to see by any physical mechanism how the k_1 reactions can be greatly retarded or the k_2 reactions accelerated. It seems necessary to postulate a kind of chemical effect involving a hydrogen atom from a C—H link which is the common feature of the molecules giving large K values. The hypothetical intermediate $AO_2{}^*$ might be regarded as a combination between triplet levels of anthracene and of oxygen. Possibly by the temporary appropriation of a hydrogen atom to give a radical with an odd number of electrons, spin-reversal may occur leading to dissociation of the intermediate. The fact that 9-methyl anthracene does not show a steep concentration dependence in B.P. chloroform may be significant; its methyl radical may serve to retain the oxygen within the molecule. It does not seem possible on present facts to speculate further. The scheme below sets out the various photo-reactions of anthracene in solution up to the point of the oxidation stages.

$$A + h\nu \rightarrow {}^1A \qquad \text{(74 kcal/mole)} \qquad (1)$$
$$^1A \rightarrow A + h\nu^1 \qquad \text{(fluorescence)} \qquad (2)$$
$$^1A \rightarrow A \qquad \text{(internal conversion;} \qquad (3)$$
$$\text{constant} = k_a)$$
$$^1A + A \rightarrow (^3A + {}^3A - 16 \text{ kcal/mole}) \rightarrow A_2 \qquad (4)$$
$$^1A + {}^3\Sigma O_2 \rightarrow {}^3A + {}^1\Delta O_2 + 5 \cdot 5 \text{ kcal/mole} \qquad (5)$$
$$^1A + S \rightarrow {}^3A + S \qquad \text{(constant} = k_b) \qquad (6)$$
$$^3A + S \rightarrow A + S \qquad (7)$$

Reactions (2), (3), (4), and (5) apply particularly to solutions in benzene or hexane, (3) explaining why the fluorescence yield is only about 0·24, (4) being dianthracene formation, and (5) interpreting fluorescence quenching by oxygen as a mutual electron-spin change on collision. With the exception of (3), these reactions may be neglected for the solvents used in this paper. Reactions (6) and (7) represent singlet-triplet changes caused by collision with solvent molecules such as carbon disulphide, (7) being negligible in view of the small oxygen dependence of the rate. The oxidation constant k is then given by $k_b/(k_a + k_b)$.

The experimental measurements in this paper were carried out by Messrs. B. W. Forgham and K. J. Heritage.

[1] Dufraisse, *Compt. rend.*, 1935, **201**, 428 ; *Bull. Soc. chim.*, 1939, **6**, 422.
[2] Koblitz and Schumacher, *Z. physik. Chem. B*, 1937, **35**, 11.
[3] Bowen and Steadman, *J. Chem. Soc.*, 1934, 1098.
[4] Bowen and Williams, *Trans. Faraday Soc.*, 1939, **35**, 765.
[5] Kasha, *Chem. Rev.*, 1947, **41**, 401.
[6] Vogel, *Practical Organic Chemistry* (Longmans, 1948), p. 826.
[7] Barcroft, *J. Physiol.*, 1908, **37**, 16.
[8] D. T. G. Morgan, *Thesis* (Oxford, 1951).
[9] Kautsky, *Trans. Faraday Soc.*, 1939, **35**, 216.

Editor's Comments
on Papers 16 and 17

PHOTOSENSITIZED FORMATION OF ALLYLIC
HYDROPEROXIDES: THE ENE REACTION

In 1953, Schenck, Eggert, and Denk (Paper 16) reported that the dye-sensitized photooxygenation of alkyl-substituted alkenes gave allylic hydroperoxides in which the double bond is shifted. This type of photo-oxygenation reaction has been identified as the "ene" reaction of singlet oxygen (Papers 29 and 30). Nickon and Bagli (Paper 17) found that the oxygen is added to the same face of the molecule from which the hydrogen atom is removed. While the mechanism of this reaction is still the subject of considerable investigation, most evidence to date is consistent with a concerted reaction via a cyclic transition state.

16

Reprinted from *Liebigs Ann. Chem.*, **584**(2-3), 177–181 (1953)

Photochemische Reaktionen III

Über die Bildung von Hydroperoxyden bei photosensibilisierten Reaktionen von O_2 mit geeigneten Akzeptoren, insbesondere mit α- und β-Pinen

Von *Günther O. Schenck*[1]), *Hans Eggert*[2]) und *Walter Denk*[3])

Die vorliegende Untersuchung entstand, weil der eine von uns wissen wollte, wie die in den beiden vorangegangenen Mitteilungen[4]) beschriebenen, durch Eosin, Methylenblau oder Chlorophyll photosensibilisierten Reaktionen von O_2 mit einfachen Dien-Kohlenwasserstoffen und Furanen zustande kommen. Hierzu war es notwendig, durch Variation der Akzeptoren A die Voraussetzung für das Eintreten der photosensibilisierten Reaktionen mit O_2 kennen- und verstehen zu lernen.

Als erstes Beispiel untersuchten wir Cyclohexen, dessen Autoxydation zum Cyclohexenyl-hydroperoxyd (I)[5]) führt. Die O_2-Aufnahme erfolgt nach H. Hock und O. Schrader[5]a) bei Belichtung in Gegenwart von Eosin doppelt so schnell wie ohne den Zusatz. Wir belichteten nun Cyclohexen in verdünnter alkoholischer Lösung in Gegenwart von Eosin und von O_2 unter den Bedingungen der ersten Askaridol-Synthese mit einer Glühlampe, beobachteten eine langsame O_2-Aufnahme und erhielten ebenfalls das bekannte Cyclohexenyl-hydroperoxyd (I).

Niemand wird sich darüber wundern, daß die durch Eosin photosensibilisierte Autoxydation des Cyclohexens zum gleichen Produkt führte wie die nichtsensibilisierte Autoxydation. Vielmehr erfüllt dieses Ergebnis die mit dem Begriff einer photosensibilisierten Reaktion verbundene Erwartung. Dieser entsprechend bildet sich bei der Autoxydation des Benzaldehyds die Benzoesäure, gleichgültig, ob in Gegenwart oder in Abwesenheit eines Photosensibilisators wie Eosin belichtet wird. Ebenso entsteht das latente

[1]) Ein Teil der Ergebnisse stammt noch aus dem Chemischen Institut Halle a. S., 1942—1945, aus Untersuchungen über Spezifität und Mechanismus der photosensibilisierten Autoxydation, über die (eingegangen Sept. 1946) in FIAT-Review of German Science 1939—1946, Preparative organic Chemistry, Part II, Senior Author K. Ziegler, S. 197 (1948) referiert wurde.

[2]) Diss. Göttingen 1952.

[3]) Diss. TH Darmstadt 1945.

[4]) Liebigs Ann. Chem. **584** 125, 156 (1953); für einen Überblick siehe G. O. Schenck, Naturwissenschaften **40**, 205, 229 (1953).

[5]) a) H. Hock u. O. Schrader, Naturwissenschaften **24**, 159 (1936); b) R. Criegee, Liebigs Ann. Chem. **522**, 75 (1936); c) H. Hock, Öl und Kohle **13**, 697 (1937); d) H. Hock u. K. Gänicke, Ber. dtsch. chem. Ges. **71**, 1430 (1938); e) R. Criegee, H. Pilz u. H. Flygare, Ber. dtsch. chem. Ges. **72**, 1799 (1939).

Bild, gleichgültig ob das Silberhalogenid selbst oder ein zugesetzter Photosensibilisator (z. B. auch Eosin) das Lichtquant wirksam absorbiert. Dennoch können wir zeigen, daß das obige Ergebnis ein durch die zufällige Wahl des untersuchten Objektes bestimmter Zufall war, und daß die Erwartung identischer Produkte bei der photosensibilisierten und der nichtsensibilisierten Autoxydation nicht berechtigt ist.

„Die photochemischen Sensibilisationsvorgänge sind allgemein dadurch gekennzeichnet, daß in dem reagierenden System zwei Stoffgruppen vorliegen, von denen die eine (Sens) die strahlende Energie aufnimmt, ohne eine dauernde chemische Veränderung zu erfahren, während die übrigen Substanzen den eigentlichen chemischen Umsatz verrichten, ohne selbst zu absorbieren" [6f] Hierbei erfolgt entweder eine Übertragung der vom Sensibilisator absorbierten Anregungsenergie auf einen der Reaktionsteilnehmer, (z. B. A oder O_2), wie bei der sensibilisierten Fluoreszenz, und die hierdurch reaktionsbereit gemachten „angeregten" Substanzen (z. B. A* oder O_2*) reagieren dann chemisch weiter. Oder aber führt die Absorption des wirksamen Lichtquants zu einer vorüber-gehenden chemischen Veränderung des Sensibilisators, der dann im Verlauf einer Zwischenreaktionskatalyse die bleibenden Veränderungen bewirkt und selbst wieder regeneriert wird.

In der von vielen Autoren [6a-n] angenommenen Formulierung einer reinen Energieübertragung durch den Sensibilisator steckt die Vor-stellung, daß es prinzipiell gleichgültig sei, ob eine Molekel M direkt durch Absorption eines Lichtquants oder indirekt durch Energie-übertragung vom angeregten Sensibilisator in den Anregungs-zustand M* übergeführt wird. Hiernach wären für beide denkbaren Reaktionen identische Umsetzungsprodukte zu erwarten.

Bei der Durchmusterung der in Tab. 1 aufgeführten Verbin-dungen, die wir in gleicher Weise in Gegenwart von Eosin in Alkohol unter O_2 belichteten, beobachteten wir nun charakteristische Unter-schiede, die folgende 4 Gruppen erkennen lassen:

1. Verbindungen, die in beiden Fällen **nicht** mit O_2 reagieren.
2.   „   die in beiden Fällen mit O_2 reagieren.
3.   „   die nur unsensibilisiert mit O_2 reagieren.
4.   „   die nur photosensibilisiert mit O_2 reagieren.

[6a] H. v. Tappeiner, Ergebn. Physiol., biol. Chem. exp. Pharmakol. **8**, 698 (1909); b) H. Gaffron, Ber. dtsch. chem. Ges. **60**, 2229 (1927); c) A. Berthoud, Photochimie, Paris, Librairie Octave Doin, 1928, S. 228; d) A. Windaus u. H. Brunken, Liebigs Ann. Chem. **460**, 225 (1928); e) H. Kautsky, Ber. dtsch. chem. Ges. **64**, 2677 (1931); f) J. Eggert, Lehrbuch d. Physik. Chem., S. Hirzel, Leipzig 1931; g) H. Gaffron, Biochem. Z. **264**, 251 (1933); h) K. Weber, Ber. dtsch. chem. Ges. **69**, 1026 (1936); i) ders., Inhibitorwirkungen, S. 78 u. 108, Enke, Stuttgart 1938; j) A. Eucken, Lehrbuch der Chemischen Physik, 3. Aufl. 1948, Bd. II, 1, S. 415; k) G. S. Egerton, J. Soc. Dyers Colourists **65**, 764 (1949); l) O. Warburg u. V. Schocken, Arch. Biochem. **21**, 363 (1949); m) W. A. Waters, The Chemistry of free Radicals, University Press, Oxford 1950, S. 109; n) Holleman-Wiberg, Lehrbuch der Anorganischen Chemie, Berlin 1951, S. 83 und 445.

Tab. 1

1. Unsensibilisiert und photosensibilisiert keine Reaktion mit O_2:
   Benzol, Naphthalin, Malonsäurediäthylester, Azodicarbonsäure-diäthyl-ester, Maleinsäure-diäthylester, Acetylen-dicarbonsäure-diäthylester, Crotonsäure-methylester, Methanol, Äthanol, Isopropanol, Aceton, Pyridin.

2. Unsensibilisiert und photosensibilisiert reagieren mit O_2:
   Cyclohexen, Cyclopenten, 1-Methylcyclopenten-(1), 1-Methylcyclohexen-(1), 1,2-Dimethylcyclohexen-(1), α-Pinen, β-Pinen, Limonen, Terpinolen, Carvomenthen.

3. Es reagieren mit O_2 unsensibilisiert, aber nicht durch Eosin sensibilisiert:
   Toluol, Xylol, Cumol, p-Cymol, Tetralin, Dekalin, Tetrahydrofuran.

4. Es reagieren mit O_2 nur sensibilisiert:
   3,4-Dihydrofuran

Aus der Auffindung der Gruppen 3 und 4 ergibt sich, daß für das Eintreten der photosensibilisierten und der unsensibilisierten Reaktion mit O_2 nicht die gleichen Voraussetzungen der Konstitution gelten.

Bei der orientierenden Aufarbeitung der aus den Verbindungen der Gruppe 2 durch photosensibilisierte Autoxydation entstandenen Umsetzungsprodukte erhielten wir in sämtlichen Fällen Hydroperoxyde, die durch Aufnahme einer Molekel O_2 pro Molekel Akzeptor (A) entstanden waren und die allgemeine Zusammensetzung AO_2 hatten. Überraschenderweise waren aber die auf den beiden verschiedenen Wegen erhaltenen Reaktionsprodukte AO_2, die sich vom gleichen Akzeptor ableiteten, nicht miteinander identisch; eine Ausnahme davon bildeten Cyclopenten und Cyclohexen, die auf beiden Wegen zu identischen Hydroperoxyden führen.

Hier bestätigt sich die schon in den beiden vorangegangenen Arbeiten getroffene Feststellung, daß die Produkte der photosensibilisierten und der unsensibilisierten Autoxydation nicht oder nur in Sonderfällen miteinander identisch sind.

In Gruppe 3 sind diejenigen autoxydablen Verbindungen zur photosensibilisierten Reaktion mit O_2 nicht bereit, bei denen die Bildung des Hydroperoxyds, z. B. II im Falle des Cumols, nur

durch einen Substitutionsprozeß erfolgen kann, der nach dem RH-
Schema formulierbar ist. Umgekehrt fällt auf, daß sämtliche zur
photosensibilisierten Reaktion mit O_2 bereiteten Verbindungen
(Gruppe 2 und 4) z. B. α-Pinen und β-Pinen olefinische Doppelbindungen enthalten. Nach dieser Gegenüberstellung ist für die
photosensibilisierte Reaktion zu vermuten, daß hier der Bildung der
Hydroperoxyde kein Dehydrierungsvorgang am Akzeptor voraus-
gehen kann, wie dies bei den Verbindungen der Gruppe 3 notwendig
ist, sondern daß die Doppelbindung in besonderer Weise
an dem Zustandekommen der Hydroperoxyde beteiligt
ist. Dies ist in der Weise denkbar, daß bei der photosensibilisierten
Autoxydation der Sauerstoff an ein C-Atom der Doppelbindung
addiert wird, worauf ein H-Atom aus der Allylstellung an den
Sauerstoff wandert und die Doppelbindung verlegt wird:

$$\begin{array}{ccc}
\text{H} \quad R_2 \; R_3 \quad \text{H} & & \text{H} \quad R_2 \; R_3 \quad \text{H} \\
R_1\text{C}\!-\!\text{C}\!=\!\text{C}\!-\!\text{C} \; R_4 & \dashrightarrow & R_1\text{C}\!-\!\text{C}\!-\!\text{C}\!=\!\text{C}R_4 \\
\text{H} \qquad\quad \text{H} & & \text{H} \\
 & & \text{OOH}
\end{array}$$

K. Alder[7]) hat solch einen Reaktionstyp auch bei Reaktionen des Maleinsäureanhydrids an Stelle von O_2 gefunden und als
„indirekte substituierende Addition in der Allylstellung" bezeichnet.
Wegen der formalen Analogie schließen wir uns dem Vorschlag an.

Nimmt man nun für die unsensibilisierte Autoxydation der Verbindungen der Gruppe 2 ausschließlich eine substituierende Autoxydation, für die photosensibilisierte Reaktion die oben erläuterte
addierende Hydroperoxydbildung an, so kommen wir nur im Falle
des Cyclopentens und Cyclohexens zu identischen Verbindungen der
Zusammensetzung AO_2. Die geschilderten gegensätzlichen Vorstellungen vermögen also das Auftreten identischer wie nicht identischer Umsetzungsprodukte auf den beiden verschiedenen Wegen
der Reaktionen mit O_2 richtig vorauszubestimmen.

Die Richtigkeit dieser Aussage kann nur durch sorgfältige präparative und analytische Durcharbeitung der photosensibilisierten Reaktion eines für die Entscheidung der Frage geeigneten
Studienobjektes, wie z. B. der beiden Pinene, festgestellt werden.
Dabei muß nämlich auch ausgeschlossen werden können, daß die
zunächst nur durch addierende Autoxydation formulierbaren Verbindungen nicht doch noch auf einem anderen Wege gebildet werden
können. Wenn bei der normalen Autoxydation der Anlagerung von
Sauerstoff eine Dehydrierung des Akzeptors in der Allylstellung zu
einem Radikal vorangeht, so sollte aus α-Pinen (III) durch Dehydrierung in der Myrtenylstellung das Radikal IVa und durch Dehydrierung in der Verbenylstellung das Radikal Va entstehen können.
Zwischen IVa und Va besteht keine Mesomerie; wohl aber ist mit

---

[7]) Liebigs Ann. Chem. **565**, 73 (1949).

einer Allylmesomerie mit den Grenzformeln IVa⟷IVb und Va⟷Vb zu rechnen. In diesem Falle wäre das Radikal IVa⟷IVb identisch mit dem aus β-Pinen (VI) durch Dehydrierung in der Pinocarveylstellung gebildeten Radikal. Erfolgt nun im Falle des α-Pinens eine Addition von O_2 an den Grenzformeln IVb und Vb, so entstehen die Hydroperoxyde XIII und XIV, deren Bildung zunächst nur mit der Vorstellung einer addierenden Hydroperoxydbildung zu erklären war.

Derartige Reaktionen haben H. Farmer[8]) und Mitarbeiter an mehreren Beispielen erstmals beschrieben. Bei der Autoxydation des Ölsäuremethylesters konnten sämtliche vier zu erwartenden Autoxydationsprodukte nachgewiesen werden[9]).

[8]) J. chem. Soc. [London] **1943**, 541 u. D. A. Sutton, ebenda **1946**, 10.
[9]) J. Ross, A. J. Gebhart u. J. F. Gerecht, J. Amer. chem. Soc. **71**, 282 (1949).

Reprinted from *J. Amer. Chem. Soc.*, **81**, 6330 (1959)

PHOTOSENSITIZED OXYGENATION OF MONO-OLEFINS

Sir:

Oxidations of olefins with molecular oxygen conducted photochemically in the presence of a sensitizing dye are proving most useful in synthetic work.[1] With mono-olefins Schenck, *et al.*, have established that the initial products are hydroperoxides and that the double bond always undergoes an allylic shift during the process.[1] We have studied the geometric requirements of photosensitized oxidations and have found that the reaction (a) is stereospecific, (b) is markedly subject to steric hindrance, (c) may have specific conformational (*i.e.*, stereoelectronic) requirements.

Photochemical oxygenation of various Δ^6-cholestenes (Ia,b,c,) in pyridine in the presence of hematoporphyrin gave the corresponding Δ^5-cholestene-7α-hydroperoxides (II), but no isolable amounts of the 7β-epimers.[2] For characterization the hydroperoxides were reduced without purification to the known allylic alcohols, which were identified as such and by conversion to known benzoates. As a typical result: Ia gave Δ^5-cholestene-$3\beta,7\alpha$-diol (*ca.* 60% isolated), some $\Delta^{3,5}$-cholestadien-7-one (*ca.* 5–10%), and some starting material (*ca.* 5–10%). In one case (IIb) the hydroperoxide was isolated separately and purified. Similar oxygenation of cholesterol-7α-d gave us 3β-hydroxy-5α-hydroperoxy-Δ^6-cholestene[1] (IIIa) that retained only 8.5% of the original deuterium, whereas cholesterol-7β-d gave IIIa that retained 95% of the original deuterium.[3] We conclude that in hydroperoxide formation the new C–O bond bears a *cis* relationship to the C–H bond that suffers cleavage.

The effect of steric blocking is exemplified with $3\beta,5\alpha$-dihydroxy-Δ^6-cholestene (IIIb), which we find is largely unchanged even on prolonged photosensitized oxygenation.

The operation of a conformational factor is suggested by studies with Δ^6-coprosten-3-one (VI) and Δ^6-coprostene (VII), where the β-hydrogen at C-5 is *quasi*-equatorial on the (non-flexible) B ring. Peracid epoxidation of $\Delta^{4,6}$-cholestadien-3-

one gave Δ^4-cholesten-3-one-6,7-epoxide (IV), hydrogenated at $-27°$ (Pd/C) to coprostan-3-one-6,7-epoxide (V). Treatment with HBr, then acetylation and the action of zinc gave VI, which provided VII on Wolff–Kishner reduction. For characterization VI and VII were hydrogenated to coprostan-3-one and to coprostane, respectively. Both VI and VII proved inert to photosensitized oxygenation even on prolonged treatment.

Our findings suggest a cyclic mechanism (concerted or not) for the olefin–oxygen combination, after the system has been suitably energized. The reaction is of special interest as a possible pathway for biological oxidations, particularly in plants, and may even represent a pathway for non-photochemical processes where the reactants can be activated enzymatically.

Constants[4] for the new compounds mentioned are: IIb m.p. 142–142.5°; $\alpha-137°$; λ(chf) 3540, 3300 cm.$^{-1}$. IV m.p. 138.5–139°; $\alpha-59°$ λ 1684, 1621, 870 cm.$^{-1}$; λ (EtOH) 241 mμ (ϵ 12,010). V, m.p. 122–123°; $\alpha-46°$; λ 1724, 892 cm.$^{-1}$. VI, m.p. 109–110°; $\alpha-52°$; λ 1727; 1656 cm.$^{-1}$. VII, m.p. 44–45°; $\alpha-7°$; λ 1647 cm.$^{-1}$.

(1) See G. O. Schenck, *Angew. Chem.*, **69**, 579 (1957), for a review and leading references.

(2) Mother liquors also were thoroughly investigated.

(3) The deuterated cholesterols were kindly provided by Dr. E. J. Corey.

(4) Optical rotations in chloroform; infrared spectra in CS$_2$. All compounds gave satisfactory C and H analyses.

(5) This work was supported by the National Science Foundation and by the Alfred P. Sloan Foundation.

(6) Alfred P. Sloan Foundation Fellow.

DEPARTMENT OF CHEMISTRY[5] ALEX NICKON[6]
THE JOHNS HOPKINS UNIVERSITY
BALTIMORE 18, MARYLAND JEHANBUX F. BAGLI

RECEIVED OCTOBER 7, 1959

Editor's Comments
on Papers 18 Through 22

SPECTROSCOPIC DETECTION OF SINGLET OXYGEN
PRODUCED BY THE REACTION OF SODIUM
HYPOCHLORITE AND HYDROGEN PEROXIDE

In 1927, Mallet observed that a weak red chemiluminescence attends
the formation of molecular oxygen from the reaction of sodium hypochlorite and hydrogen peroxide.[1] In a subsequent "rediscovery" of this
chemiluminescent reaction, Seliger (Paper 18) reported that the luminescence consists primarily of a sharp band centered at 634 nm.

$$NaOCl + H_2O_2 \longrightarrow O_2 + NaCl + H_2O + h\nu$$

In 1963, Khan and Kasha (Paper 19) published a particularly insight-
ful paper in which they attributed this chemiluminscence to emission
from electronically excited molecular oxygen.[2] Although the two observed emission bands at 633.4 and 703.2 nm were incorrectly assigned to
solvent shifted (0,0) and (0,1) bands of the $^1\Sigma_g^+ \longrightarrow {}^3\Sigma_g^-$ transition of molecular oxygen, Khan and Kasha must be credited with recognizing the

78

importance of a chemical reaction that produced excited molecular oxygen. Their work served to revive interest in Kautsky's original proposal that singlet oxygen was the reactive intermediate in dye-sensitized photooxygenations:

> The existence of a metastable excited molecular oxygen in solution is of outstanding interest in several problems. Originally Kautsky about 1935 had speculated on the existence of such a species in connection with dye-photosensitized oxidations, but the suggestion was lost for want of direct evidence. We believe that in many biological peroxide systems the chemiluminescence may be found as an indication of the breakdown mechanism; and in addition, similar observations may be made in radiobiological and radiation chemical reactions. Applications of these ideas deserve further investigation.

Ogryzlo and coworkers (Papers 20 and 21) demonstrated that simultaneous emission from two $^1\Delta_g$ oxygen molecules would account for the bands reported by Khan and Kasha. Browne and Ogryzlo (Paper 21) also reported emission bands at 1070 and 1270 nm, which were assigned to the (1,0) and (0,0) transitions of $^1\Delta_g$ to $^3\Sigma_g^-$.

NOTES AND REFERENCES

1. L. Mallet, *Compt. Rend.*, **185**, 352 (1927).
2. See also Paper 22.

18

Reprinted from *Anal. Biochem.,* **1**(1), 60–65 (1960)

A Photoelectric Method for the Measurement of Spectra of Light Sources of Rapidly Varying Intensities

H. H. SELIGER

*From the McCollum-Pratt Institute and Department of Biology,
Johns Hopkins University, Baltimore, Maryland*

Received February 16, 1960

INTRODUCTION

The emission spectra of bioluminescent and chemiluminescent reactions exhibit rapidly decreasing light intensities due to the kinetics of the reactions. Heretofore the photographic technique was required to obtain a reasonably accurate spectrum of a varying intensity light source since the entire spectrum is displayed at one time and the photographic plate is essentially a light integrator. However, the assumption of reciprocity is not too well justified at high and at low light intensities and further the range of intensity that can be measured with reasonable accuracy on a single exposure seldom reaches 10:1 (1).

In this paper a simple technique is described which combines the much higher sensitivity and linearity of the electron multiplier phototube with the integrating "principle" of the photographic plate to measure the relative emission spectra of light sources independent of their time variation in intensity. This relative emission spectrum we shall designate by $f(\lambda)$.

METHOD

The experimental arrangement is shown in Fig. 1. Light from the source, S, after passing through the beam-splitting glass plate, P, is focused by the spherical mirror onto the entrance slit of a spectrometer. The phototube, PMT-1, at the exit slit receives quanta in the wavelength interval $\lambda + d\lambda$. A small fraction of the emitted light is reflected by P onto the face of the second phototube PMT-2. The light source is masked so that the image just fills the entrance slit of the spectrometer and so that its linear dimensions are smaller than those of the photocathode of PMT-2. This is a rather important point since both phototubes should "see" the same effective area of the source. (With chemiluminescent solutions the intensity usually varies over the effective surface of the solution cell due to convection when the reactants are combined.) The electrons collected at the respective anodes of the phototubes are delivered to separate capacitor current-integrating circuits so that the total charge collected by each phototube during the light flash is recorded. At a wavelength setting λ_i the charge in coulombs collected by PMT-1 is given by

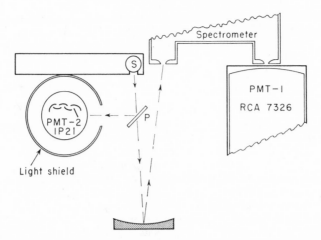

FIG. 1. Schematic drawing of the experimental method of beam-splitting in the charge normalization technique.

$$(q_1)_{\lambda_i} = \int_0^t dt I(\lambda_i, t) G_1 T(\lambda_i) \epsilon_1(\lambda_i) A_1 \times 1.60 \times 10^{-19} \tag{1}$$

where $I(\lambda_i, t)$ is the number of quanta of wavelength λ_i and at time t emitted per second by the source; G_1 is the fraction of the light focused by the mirror; $T(\lambda_i)$ is the transmission of the spectrometer which can be assumed to be flat over the wavelength interval involved; $\epsilon_1(\lambda_i)$ is the spectral efficiency of the photocathode of PMT-1 for a quantum of wavelength λ_i; and A_1 is the gain of the phototube. The total charge collected by PMT-2 is given by

$$(q_2)_i = \int_0^t \int_{\lambda_{min}}^{\lambda_{max}} dt\, d\lambda\, I(\lambda, t) R \epsilon_2(\lambda) A_2 \times 1.60 \times 10^{-19} \tag{2}$$

where R is the fraction of the light reflected by the beam-splitting plate P. Equation (2) can be simplified by separation of variables and by performing the time integration over the entire flash. Then, integrating Eq. (1) and dividing by Eq. (2), we obtain

$$\frac{(q_1)_{\lambda_i}}{(q_2)_i} = \frac{N(\lambda_i) G_1 T \epsilon_1(\lambda_i) A_1}{\int_{\lambda_{min}}^{\lambda_{max}} d\lambda\, N(\lambda) \epsilon_2(\lambda) R A_2} \tag{3}$$

where $N(\lambda_i)$ is the total number of photons emitted in the wavelength interval between λ_i and $\lambda_i + d\lambda_i$. If we now lump together all of the constants of the experiment, recognizing that $\int_{\lambda_{min}}^{\lambda_{max}} d\lambda\, N(\lambda) \epsilon_2(\lambda)$ is just the average efficiency of the phototube PMT-2 multiplied by N_0, the total number of quanta emitted by the source during the measured time interval, we arrive at the final equation

$$\frac{(q_1)_{\lambda_i}}{(q_2)_i} = K \epsilon_1(\lambda_i) \frac{N(\lambda_i)}{N_0} = K \epsilon_1(\lambda_i) f(\lambda_i). \tag{4}$$

Here the subscripts i are retained to indicate that this ratio is to be computed at each wavelength setting λ_i, and $f(\lambda_i)$ is the desired relative emission spectrum.

APPARATUS

The capacitor current integration circuits were the same as those used in previous measurements of the quantum yield of firefly bioluminescence (2). By placing the input capacitor in the negative feedback loop of a d.c. amplifier, the voltage change across it is never more than about 0.1 v so that the charging of the capacitor does not affect the phototube anode voltage to more than 0.1%. The circuit permits variation of either the voltage applied to the phototube or the size of the input capacitor used. There is a sensitive relay at the output of the VTVM portion of the d.c. amplifier which shorts out the input capacitor when the charging voltage reaches 10 v and an electrical message register then tallies the number of times this discharge has occurred. This further extends the useful range of the circuit and has the advantage that for most measurements only one capacitor need be used over all integrated intensities, thus increasing the precision. The instruments have separate settings with resistors in place of the input capacitors so that current measurements can also be made.

The spectrometer was an $f/3$ grating instrument used in first order. The dispersion was 20 A/mm at the exit slit. PMT-1 was an experimental model of the RCA 7326 phototube with an S-20 response. PMT-2 was a conventional RCA 1P-21 phototube.

SPECTRAL MEASUREMENTS

1. Chemiluminescence of Luminol (3-Aminophthalhydrazide)

One of the important criteria in the determination of the emission spectra of chemiluminescent solutions is that the concentration of the reactants be low so that the spectrum will not be distorted by self-absorption. In those instances where the quantum yield of chemiluminescence is high as in the case of the firefly (2), the problem is not so severe since relatively low concentrations of reactants will yield appreciable light. However, in the case of luminol in aqueous solution the quantum yield is low, of the order of several times 10^{-3}, and one must be careful to work with as dilute a solution as practicable. In the present case concentrations of approximately 10^{-5} M of luminol in 0.1 N sodium hydroxide were oxidized by the injection of small amounts of hydrogen peroxide and sodium hypochlorite. The light flash was just visible in a darkened room and lasted no more than 1 or 2 sec. It was necessary to repeat this procedure several times for each wavelength setting in order to obtain a reasonable precision. The control phototube PMT-2 always obtains sufficient light since the light is not dispersed before striking its photocathode. The chemiluminescence spectrum of luminol is shown in Fig. 2, together with the fluorescence spectrum of luminol at pH 7.5. The coincidence of the chemiluminescence and the fluorescence spectra would indicate that the unreacted molecule, rather than the oxidized product, is the light emitter since the final products are not fluorescent. This point has been discussed by Pringsheim (3) and by Spruit and Spruit-Van Der Burg (4). The peak positions agree almost exactly with the earlier data of Spruit and Spruit-Van Der Burg.

2. *Chemiluminescence of Hydrogen Peroxide in the Presence of Chlorine*

In the course of preliminary measurements of the quantum yield of the luminol chemiluminescence, it was found that the mixing of hydrogen peroxide and sodium hypochlorite themselves gave rise to a very small amount of light. Thus far, the following facts have been established qualitatively: (*a*) $H_2O_2 + NaOCl$ give rise to a concentration-dependent red light-emission; (*b*) the efficiency of the reaction is increased in acid solution; (*c*) the light emission appears to be temperature dependent, becoming less intense as the temperature is increased; (*d*) the reaction $H_2O_2 + Cl_2$ gives rise to the same red chemiluminescence.

The rather sharp band spectrum of this red chemiluminescence is shown in Fig. 3*a*. The light consisted of a rapid but weak flash each time NaOCl

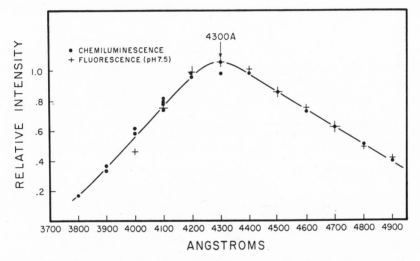

FIG. 2. Chemiluminescence and fluorescence spectrum of luminol.

was added to the H_2O_2 solution. The duration of the light flash was approximately 0.1 sec. Again it was necessary to repeat the reaction several times at each wavelength setting in order to obtain sufficient total light. It was also necessary to widen the spectrometer slits somewhat in order to increase the sensitivity. The effect of the slit width on the spectral resolution is shown in Fig. 3*b* which is a scan of the 5461A line from a low-pressure mercury discharge tube. The close agreement of the different points at each wavelength setting indicates the precision of the technique for even this extreme case of intensity variation.

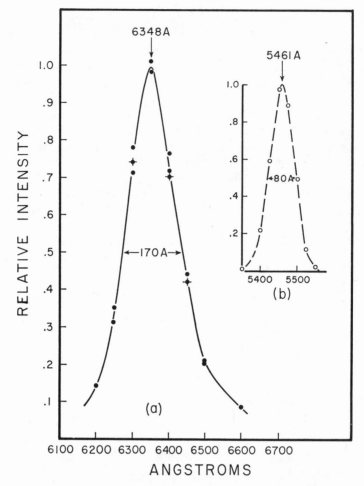

FIG. 3. (a) Chemiluminescence of the hydrogen peroxide–chlorine reaction; and (b) the effect of slit width on the resolution of the mercury 5461A line.

DISCUSSION

The charge normalization technique described above should have wide application in the measurement of spectra of intermittent light sources in chemiluminescence and in flame spectroscopy where a great deal of time and effort has been expended in devising experimental procedures to give constant light intensities. The phototube in general has higher sensitivity, greater spectral range, and gives much higher precision than the photographic plate.

It must be noted that while the application of this combination of current integration and beam-splitting normalization for spectral measurement appears to be new, techniques based on one or the other of these principles have been used for some time. A rate-of-charge balance technique for precise measurement of currents in ionization chambers was used by Taylor (5) in 1931. More recently, the author has used this technique with ionization chambers (6) and with phototubes (2). Current integration has also become rather standard in astronomical photometry (7, 8). A variation of the normalization method was used several years ago in the gamma-gamma coincidence-counting standardization of cobalt-60, (9) as a control over possible time variations in the counting efficiencies of the different counting channels. Several recording spectrophotometers use a variation of the normalization technique by splitting a light beam at high speed alternately through the sample cell and the reference cell and electrically measuring the ratio of the two transmission intensities. The charge normalization technique is thus a logical extension of this principle.

The chemiluminescence spectrum of the hydrogen peroxide–chlorine system has not been reported previously. More detailed reports on quantum yields and spectra of chemiluminescence and fluorescence of luminol and the peroxide system will be reported in appropriate journals.

ACKNOWLEDGMENTS

This work has been supported in part by the U. S. Atomic Energy Commission and the National Institutes of Health. The author would like to thank Dr. E. F. MacNichol, Jr. for the original design of the d.c. current integrator and Mr. W. Fastie for the design and construction of the grating spectrometer used.

SUMMARY

A method is presented for the measurement of spectra of light sources of rapidly varying intensities using phototubes in a charge-normalization technique. The method is applied to the luminol chemiluminescence and to a newly found chemiluminescence of hydrogen peroxide and chlorine.

REFERENCES

1. Gaydon, A. G., in "The Spectroscopy of Flames," p. 32. John Wiley and Sons, New York, 1957.
2. Seliger, H. H., and McElroy, W. D., Arch. Biochem. Biophys. **88**, 136 (1960).
3. Pringsheim, P. in "Luminescence of Liquids and Solids," p. 33. Interscience Publ., New York, 1946.
4. Spruit, C. J. P., and Spruit-Van Der Burg, A., in "The Luminescence of Biological Systems" (F. H. Johnson, ed.), p. 107. Amer. Assoc. Adv. Sci., Washington, D. C. 1955.
5. Taylor, L. S., Radiology **17**, 294 (1931).
6. Seliger, H. H., and Schwebel, A., Nucleonics **12**, 54 (1954).
7. Gardiner, A. J., and Johnson, H. L., Rev. Sci. Instr. **26**, 1145 (1955).
8. Weitbrecht, R. H., Rev. Sci. Instr. **28**, 883 (1957).
9. Mann, W. B., and Seliger, H. H., J. Research Natl. Bur. Standards **50**, 197 (1953).

Reprinted from *J. Chem. Phys.*, 39(8), 2105–2106 (1963)

Red Chemiluminescence of Molecular
Oxygen in Aqueous Solution*

Ahsan Ullah Khan and Michael Kasha

*Institute of Molecular Biophysics and Department of Chemistry,
Florida State University, Tallahassee, Florida*

(Received 5 August 1963)

THE extremely interesting observation[1] that a red-orange chemiluminescence is produced during the reaction of hydrogen peroxide and sodium hypochlorite in aqueous solution was reported by Seliger. Using a photomultiplier and a light integrating device, he was able to record one relatively narrow emission band, giving its peak wavelength as 6348 Å and half-width as 170 Å. Since Seliger's apparatus possessed low resolution and was insensitive to wavelengths longer than 6700 Å, we decided to extend the spectroscopic study of this chemiluminescence.

A cell was designed which permitted a mixing of rapidly flowing 1-cm×1-mm jets, of 0°C 30% H_2O_2 and commerical sodium hypochlorite (or reagent-grade) solutions, upon impingement on a 3-cm quartz window. The red chemiluminescence was photographed by means of a Steinheil Universal GH three-prism glass spectrograph with $f3.9$ optics in an infrared setting, giving a dispersion of 108 Å per mm in the region of 7000 Å. Eastman Kodak I–N water-hypersensitized plates were used. Exposure times from 1 to 15 min were used at various slit settings. Neon discharge-tube lines were used for calibration. Two principal bands were observed in the spectral range 4000 to 9000 Å, as depicted in Fig. 1. Extreme overexposure of these two bands failed to yield additional bands of any comparable intensity. However two extremely weak bands, estimated at $\frac{1}{10\,000}$ of those shown, were recorded on the long-wavelength side of the two main bands; these are under further investigation.

Our best calibration yields the values for the two peaks: 6334 Å (15 788 cm^{-1}) and 7032 Å (14 221 cm^{-1}). The two bands have half-widths of 100 and 125 Å, respectively, or approximately identical energy half-widths 250 cm^{-1}.

In view of the nature of the reaction and the bands observed we tentatively assign the red chemiluminescence bands as the 0,0 and 0,1 bands of the $^1\Sigma_g^+ \rightarrow {}^3\Sigma_g^-$ transition of molecular oxygen. The band separation we measure is 1567 cm^{-1}, which is within experimental

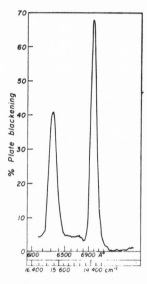

Fig. 1. Chemiluminescence spectrum of reaction of hydrogen peroxide with sodium hypochlorite in aqueous solution at 20°C.

error of the ground state vibrational frequency (1580 cm^{-1}) of molecular oxygen.[2] The study of the mechanism of the reaction by Cahill and Taube[3] using isotopic labeling of the peroxide indicates that the O–O bond remains intact, reaffirming the possibility of excited molecular oxygen rather than atomic oxygen as a primary product.

Our assignment of a 0,0 band at 15 788 cm^{-1} in water requires a solvent shift of this band of 2593 cm^{-1} to higher frequencies compared[2] with the gas phase 0,0 band (13 195.2 cm^{-1}) for this transition. This is not an unusual solvent shift, but it is not possible to account for it by means of a simple molecular orbital model, since the $^1\Delta_g$, $^1\Sigma_g^+$, and $^3\Sigma_g^-$ states arise from the same state configuration. We rule out the $^1\Delta_g$ state on grounds of energy discrepancy.

The observed chemiluminescence lines in aqueous

solution bear a striking resemblance to those found by Kaplan[4] in gas-discharge luminescence experiments on oxygen mixtures, except for the wavelength shift noted. He also reported a 0,0 and 0,1 band pair. Higher quantum number transitions could be observed with other plates, although some of these could be Franck–Condon intensity limited; we believe the absence of a 0,2 band could be due to accidentally low vibronic eigenfunction overlap. High-resolution carbon monoxide–oxygen flame spectra show[5] the same 0,0 band at 7593.7 Å.

A kinetic study of this chemiluminescence was made by Stauff and Schmidkunz[6]; they expressed the opinion that van der Waals O_2–O_2* complex was responsible for the emission. Physical evidence for the existence of O_4 is doubtful even in liquid oxygen,[7] and we believe it to be an unlikely species in the water system.

The existence of a metastable excited molecular oxygen in solution is of outstanding interest in several problems. Originally Kautsky[8] about 1935 had speculated on the existence of such a species in connection with dye-photosensitized oxidations, but the sugges-

tion was lost for want of direct evidence. We believe that in many biological peroxide systems the chemiluminescence may be found as an indication of the breakdown mechanism; and in addition, similar observations may be made in radiobiological and radiation chemical reactions. Applications of these ideas deserve further investigation.

* Work supported in part by a contract between the Division of Biology and Medicine, U. S. Atomic Energy Commission, and the Florida State University; and in part by a contract between the Physics Branch, Office of Naval Research and the Florida State University.

[1] H. Seliger, Anal. Biochem. **1**, 60 (1960).
[2] G. Herzberg, *Molecular Structure and Molecular Spectra I. Diatomic Molecules* (D. Van Nostrand Inc., New York, 1957), p. 560.
[3] A. E. Cahill and H. Taube, J. Am. Chem. Soc. **74**, 2312 (1952).
[4] J. Kaplan, Nature **159**, 673 (1947); Phys. Rev. **71**, 274 (1947).
[5] R. C. Herman, H. S. Hopfield, G. A. Hornbeck, and S. Silverman, J. Chem. Phys. **17**, 220 (1949).
[6] J. Stauff and H. Schmidkunz, Z. physik. Chem. N. F. **35**, 295 (1962).
[7] C. M. Knobler, *Dissertation: On the Existence of a Molecular Oxygen Dimer* (Drukkerij Pasmans, 'S-Gravenhage, 1961).
[8] H. Kautsky, in *Luminescence, A General Discussion of the Faraday Society* (Gurney and Jackson, London, 1938), p. 216.

Erratum: Red Chemiluminescence of Molecular Oxygen in Aqueous Solution

[J. Chem. Phys. **39**, 2105 (1963)]

AHSAN ULLAH KHAN AND MICHAEL KASHA

The Department of Chemistry and The Institute of Molecular Biophysics, The Florida State University Tallahassee, Florida 32306

IN our communication we attributed the observation of a red chemiluminescence, in the reaction of hydrogen peroxide with hypochlorite or chlorine in alkaline aqueous solution, to Seliger.[1] Seliger did pub-

lish the fullest account of this observation, together with the first spectrum. Much earlier, Mallet[2] made a passing observation of this chemiluminescence, noting its color qualitatively. Also, a brief communication by Gattow and Schneider[3] gave some variations on the observaion and quoted the rough wavelength range for the chemiluminescence. The authors thank Dr. H. Seliger and Dr. G. Schenck, respectively, for calling these papers to our attention.

[1] H. H. Seliger, Anal. Biochem. **1**, 60 (1960).
[2] L. Mallet, Compt. Rend. **185**, 352 (1927).
[3] G. Gattow and A. Schneider, Naturwiss. **41**, 116 (1954).

20

Reprinted from *J. Chem. Phys.*, **40**(6), 1769–1770 (1964)

Some New Emission Bands of Molecular Oxygen

S. J. ARNOLD, E. A. OGRYZLO, AND H. WITZKE

Chemistry Department, University of British Columbia,
Vancouver, Canada

(Received 1 November 1963)

CALORIMETRIC,[1] spectroscopic,[2] and mass spectrometric[3] studies of electrically discharged oxygen have shown that the gas stream contains about 10% $O_2(^1\Delta_g)$ and about 0.1% $O_2(^1\Sigma_g^+)$. In a spectroscopic search for other excited molecules present in the gas stream we have recorded two bands shown in Fig. 1

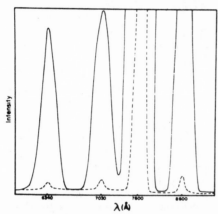

FIG. 1. Emission spectrum of electrically discharged oxygen with all atoms removed. Total O_2 pressure=2 mm Hg. $f/4.6$ monochromator with an RCA-7102 photomultiplier. Solid line—slit 750 μ; broken line—slit 100 μ.

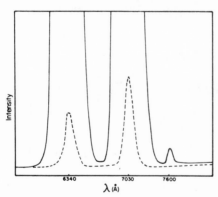

FIG. 2. Emission spectrum from the reaction of Cl_2 and H_2O_2 in alkaline solution. Solid line—slit 250 μ; broken line—slit 100 μ (and less amplification).

with peaks at 6340 and 7030 Å. When oxygen atoms are removed by the addition of Hg to the discharge, the 6340 and 7030 Å peaks remain unaltered while the 7600 and 8600 Å peaks [(0, 0) and (0, 1) transitions in the $^1\Sigma_g^+-^3\Sigma_g^-$ system] are only slightly lowered.

The two new bands appear diffuse even with a spectral slitwidth of 0.6 Å (Jarrell–Ash $f/6.3$ grating spectrograph). This fact together with their positions makes it obvious that they do not belong to the $^1\Sigma_g^+-^3\Sigma_g^-$ system.

Two impurities could conceivably give rise to emissions in this region. Water absorption peaks have been observed at 6340 and 7000 Å, and NO_2 emission bands are observed throughout this region.[4] Hence specially dried, nitrogen-free oxygen was tried, and then both water and NO_2 were added before and after the discharge. The two new bands were essentially unaltered by these changes.

An entirely different observation makes it even less likely that this could be an "impurity" emission. The reaction between Cl_2 and H_2O_2 gives rise to a red emission which was first reported by Seliger.[5] The spectrum of the emission is shown in Fig. 2. There is no doubt that the emitter is the same as that in the discharge experiment. The extremely small 7600 Å peak is consistent with the rapid quenching of $O_2(^1\Sigma_g^+)$ by water in the discharge products.

Ozone cannot be the emitter since the addition of NO has no effect. Any excited molecules such as $O_2(^3\Sigma_u^+)$ with a lifetime shorter than 0.1 sec would not be observed in our flow system. The $O_2(^1\Sigma_u^-$ and $^3\Sigma_u^+)$ states would not give rise to only two bands spaced 1550 cm^{-1} apart.

In our opinion there remains only one other possible source for these bands. The clue to the identification comes from the parallel behavior of the new bands and the $O_2(^1\Delta_g)$ concentration. Neither is affected by fewer than 10^8 collisions with any gas we have added. We therefore proposed the following reaction in which O_4^* and O_4 may be stabilized by van der Waals forces:

$$2O_2(^1\Delta_g)\overset{(M)}{\rightleftharpoons}O_4{}^*\to O_4\overset{(M)}{\rightleftharpoons}2O_2(^3\Sigma_g^-)\\+\\h\nu$$

The $O_2(^1\Delta_g)$ dimer would then give rise to a band at 6343 Å (when only ground vibrational levels are involved) and one at 7038 Å [when an O_2 ($^3\Sigma_g^-$) molecule is left with one vibrational quantum]. Evidence for transitions such as this comes from studies of the absorption spectrum of compressed oxygen.[6] In an attempt to make the identification more certain we are undertaking a kinetic study of the relation between $O_2(^1\Delta_g)$ concentration and the new emission bands.

Note added in proof: Since this paper was submitted A. U. Khan and M. Kasha [J. Chem. Phys. **39**, 2105 (1963)] have tentatively assigned the 6340 and 7030 Å bands to solvent-shifted $^1\Sigma_g^+-^3\Sigma_g^-$ transitions in oxygen. Since we have obtained these bands in the gas phase it is obvious the assignment must be in error.

* The research of this paper was supported by Defence Research Board of Canada, Grant Number 9530–31.

[1] L. Elias, E. A. Ogryzlo, and H. I. Schiff, Can. J. Chem. **37**, 1680 (1959).
[2] M. A. A. Clyne, B. A. Thrush, and R. P. Wayne, Nature **199**, 1057 (1963).
[3] S. N. Foner and R. L. Hudson, J. Chem. Phys. **25**, 601 (1956).
[4] A. Fontijn and H. I. Schiff, *Chemical Reactions in the Lower and Upper Atmosphere* (Interscience Publishers, Inc., New York, 1961), p. 239.
[5] H. H. Seliger, Anal. Biochem. **1**, 60 (1960).
[6] V. I. Dianov-Klokov, Opt. Spectry. **6**, 290 (1959)[Opt. i Spektroskopiya **6**, 457 (1959)]. J. W. Ellis and H. O. Kneser, Z. Physik **86**, 583 (1933).

Reprinted from *Proc. Chem. Soc.*, 117 (Apr. 1964)

Chemiluminescence from the Reaction of Chlorine with Aqueous Hydrogen Peroxide

By R. J. BROWNE and E. A. OGRYZLO*

IN 1960 Seliger[1] reported that the reaction between hydrogen peroxide and sodium hypochlorite (in acidic solution) or chlorine (in basic solution) gives rise to a red chemiluminescence. He recorded one emission band at 6348 Å, but made no attempt to identify the source of the emission. More recently Khan and Kasha[2] reported a second band at 7032 Å for this reaction. In view of the 1567 cm.$^{-1}$ spacing between the bands they tentatively assigned the bands as solvent shifted (0,0) and (0,1) bands of the $^1\Sigma_g^+ \longleftrightarrow {}^3\Sigma_g^-$ system in oxygen. Arnold, Witzke, and Ogryzlo[3] showed that this assignment could not be correct since the same two bands are observed in the products of electrically discharged pure gaseous oxygen.

We have recorded the emission spectrum from this reaction with a cooled RCA-7102 photomultiplier and an f 4·6 Hilger-Watts spectrometer. Chlorine was bubbled into an alkaline solution of 30% hydrogen peroxide in water. The spectrum of the emission (which comes from the bubbles) is shown in Fig. 1. The spectral sensitivity of the photomultiplier is indicated with the dotted line.

The overall reaction can be written:

$$Cl_2 + HO_2^- + OH^- \rightarrow 2Cl^- + H_2O + O_2. \quad (1)$$

All the bands observed can be assigned to known transitions of $O_2(^1\Sigma_g^+)$ and $O_2(^1\Delta_g)$. Since the photomultiplier sensitivity beyond 12,000 Å is about 5×10^{-3} of the maximum value, the emission at 12,700 Å (peak 8) is probably the most intense detected. This is undoubtedly the (0,0) transition in the $O_2(^1\Delta_g \longleftrightarrow {}^3\Sigma_g^-)$ system. Peak (7) at 10,700 Å is then the (1,0) transition in the same system, indicating the presence of vibrationally excited $O_2(^1\Delta_g)$. In view of the known efficiency of water in deactivating vibrationally excited oxygen,[4] the large (1,0), intensity is surprising. Peak (2) is due to a weakly bound complex of two $O_2(^1\Delta_g)$ molecules.[5] Peak (3) arises from the same complex when one of the oxygen molecules ends up in the first excited vibrational level. Band (1) at 5800 Å confirms the presence of vibrationally excited $O_2(^1\Delta_g)$ since it arises from a complex between $O_2(^1\Delta_g)_{v=0}$ and $O_2(^1\Delta_g)_{v=1}$. Peak (4) at 7619 Å and peak (6) at 8645 Å are the (0,0) and (0,1) transitions in the $O_2(^1\Sigma_g^+ \longleftrightarrow {}^3\Sigma_g^-)$ system. Since water

rapidly deactivates $O_2(^1\Sigma_g^+)$,[5] these peaks are most intense when the reaction mixture is cooled to reduce the vapour pressure of water. Peak (5) at 7700 Å which we assign as the (1,1) transition in the same system provides further evidence for vibrational excitation in oxygen.

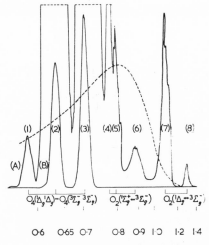

$O_4(^1\Delta_g^1\Delta_g)-O_4(^3\Sigma_g^{-3}\Sigma_g)$ $O_2(^1\Sigma_g^{+3}\Sigma_g^-)$ $O_2(^1\Delta_g^{-3}\Sigma_g^-)$

0·6   0·65 0·7   0·8   0·9 1·0   1·2 1·4

Wavelength (μ)

FIG. 1. *Emission spectrum from the reaction of chlorine and hydrogen peroxide, obtained with a 500 μ slit width. The broken line gives the relative spectral sensitivity of the photomultiplier. The amplification for curve (A) ≈ 1000 times the amplification for curve (B).*

The production of oxygen in excited "singlet" states can probably be traced to the spin conservation requirement in reaction 1 (or one of its more elementary steps). Since molecular oxygen seems essential for bioluminescent processes, and peroxide intermediates appear to be formed,[6] such processes may, in some ways, resemble the present reaction.

(*Received, February 7th, 1964.*)

* Department of Chemistry, University of British Columbia, Vancouver 8, Canada.
[1] Seliger, *Analyt. Biochem.*, 1960, **1**, 60.
[2] Khan and Kasha, *J. Chem. Phys.*, 1963, **39**, 2105.
[3] Arnold, Witzke, and Ogryzlo, *J. Chem. Phys.*, in press.
[4] Knotzel and Knotzel, *Ann. Physik*, 1948, **2**, 393.
[5] Bader and Ogryzlo, *Discuss. Faraday Soc.*, 1964, April.
[6] White, "Light and Life," The Johns Hopkins Press, Baltimore, 1961, p. 183.

22

Reprinted from *J. Amer. Chem. Soc.*, **92**(11), 3293–3300 (1970)

Chemiluminescence Arising from Simultaneous Transitions in Pairs of Singlet Oxygen Molecules[1]

Ahsan Ullah Khan and Michael Kasha

Contribution from the Institute of Molecular Biophysics and the Department of Chemistry, The Florida State University, Tallahassee, Florida 32306. Received September 23, 1969

Abstract: A complete spectroscopic comparison is made of the numerous chemiluminescence bands observed in the 20° aqueous solution reaction of hydrogen peroxide with hypochlorite ion, with gaseous molecular oxygen absorption bands at high pressure. Both single molecule and simultaneous transitions involving molecular pairs are observed, all involving the highly metastable singlet molecular oxygen states $^1\Delta_g$ and/or $^1\Sigma_g^+$. The visible chemiluminescence bands observed are simultaneous transitions; the red pair at 6334 and 7032 Å corresponding to the $(^1\Delta_g)(^1\Delta_g) \rightarrow (^3\Sigma_g^-)(^3\Sigma_g^-)$ transition and a green band at 4780 Å corresponding to the $(^1\Delta_g)(^1\Sigma_g^+) \rightarrow (^3\Sigma_g^-)(^3\Sigma_g^-)$ transition. These correlate spectroscopically with the high-pressure (150 atm) gaseous oxygen absorption bands. No emission was observed corresponding to the $(^1\Sigma_g^+)(^1\Sigma_g^+) \leftarrow (^3\Sigma_g^-)(^3\Sigma_g^-)$ absorption, but evidence for the production of this pair state is obtained from sensitized luminescence studies. An infrared chemiluminescence band centered on 7620 Å corresponding to the single molecule transition $^1\Sigma_g^+ \rightarrow {}^3\Sigma_g^-$ for molecular oxygen is shown to reveal rotational structure. Sensitized liminescence of organic fluorescent molecules present in the peroxide–hypochlorite reacting system is interpreted by means of a physical energy transfer from simultaneous transition singlet molecular oxygen pair states to the organic acceptor.

Oxygen molecules are among the most ubiquitous in nature and play a unique role in many photophysical, photochemical, and photobiological processes.[2] Yet in spite of the early observations on absorption spectroscopy of oxygen, interpreted in terms of its comparatively well understood excited electronic states, it is only in the last few years that the importance of these excited states has been recognized in the many systems in which oxygen participates as a physical or chemical agent. Especially have the lowest metastable singlet states of oxygen ($^1\Delta_g$ and $^1\Sigma_g^+$) appeared prominent in giving rise to visible region chemiluminescence phenomena.[3–8] As an immediate consequence of this demonstration, important new interpretations involving excited singlet oxygen have appeared, offering applications to photosensitized oxidations, reversible quenching of excited states of polyatomic molecules, and unique chemical reactions of the singlet oxygen species produced chemically.

Not only has the facile chemical production and the unusual metastability of the two singlet oxygen species resulted in this renascence of interest, but also attention has been focused as a consequence on oxygen molecule-pair interactions in their simultaneous transitions.

Simultaneous transitions as a one-photon two-molecule process were first invoked as an idea by Ellis and Kneser[9] in the interpretation of liquid oxygen absorption spectra in the visible and near-ultraviolet regions, based on a suggestion by Lewis[10] that an "(O₂)₂ species" could account for the paramagnetic behavior of liquid oxygen, although this latter problem is still not perfectly resolved. The observation of visible chemiluminescences in peroxide decompositions yielding molecular oxygen has similarly required invoking simultaneous transitions in pairs of excited singlet oxygen molecules as a corollary to the absorption process in condensed oxygen.

It is altogether surprising how slowly the recognition of the unusual chemiluminescences of molecular oxygen produced in the simple aqueous reaction

$$H_2O_2 + OCl^- \longrightarrow O_2^* + H_2O + Cl^- \qquad (1)$$

(where O_2^* represents $^1\Delta_g$ and $^1\Sigma_g^+$ oxygen) was established. In 1927 Mallet reported[11] a red chemiluminescence accompanying this reaction, but made no further observations on it except in a chemiluminescence sensitizing technique. In 1938, Groh duplicated[12] the simple observation (using alkaline Br₂ instead of hypochlorite), and in 1942 Groh and Kirrmann made an approximate measurement[13] of the prominent visible band associated with this chemiluminescence as occurring at 6320 Å. More importantly, these authors made the observation from thermochemical criteria that: "It is necessary to acknowledge that the emission results from an accumulation of energy of two reacting molecules, *e.g.*, by the collisions of two activated molecules of oxygen, which supposes a relatively long lifetime." Groh and Kirrmann also noted the correspondence of this red chemiluminescence emission with some astrophysical observations on atmospheric oxygen. These first observations did not come to a fruition worthy of the subject. Rediscoveries of these results were given in the qualitative observations by Gattow and

(1) Work supported in part by a contract between the Division of Biology and Medicine, U. S. Atomic Energy Commission, and the Florida State University; and in part by a contract between the Physics Branch, Office of Naval Research, and the Florida State University.

(2) (a) *Cf.* C. Reid, "Excited States in Chemistry and Biology," Butterworth and Co., Ltd., London, 1957, especially pp 98–106; (b) H. H. Seliger and W. D. McElroy, "Light: Physical and Biological Action," Academic Press, New York, N. Y., 1965.

(3) A. U. Khan and M. Kasha, *J. Chem. Phys.*, **39**, 2105 (1963); **40**, 605 (1964).

(4) A. U. Khan and M. Kasha, *Nature*, **204**, 241 (1964).

(5) A. U. Khan and M. Kasha, *J. Amer. Chem. Soc.*, **88**, 1574 (1966).

(6) S. J. Arnold, E. A. Ogryzlo, and H. Witzke, *J. Chem. Phys.*, **40**, 1769 (1964).

(7) R. J. Browne and E. A. Ogryzlo, *Proc. Chem. Soc. London*, 117 (1964).

(8) J. S. Arnold, R. J. Browne, and E. A. Ogryzlo, *Photochem. Photobiol.*, **4**, 963 (1965).

(9) J. W. Ellis and H. O. Kneser, *Z. Phys.*, **86**, 583 (1933).

(10) G. N. Lewis, *J. Amer. Chem. Soc.*, **46**, 2027 (1924).

(11) L. Mallet, *C. R. Acad. Sci., Paris*, **185**, 352 (1927).

(12) P. Groh, *Bull. Soc. Chim. Fr.*, **5**, 12 (1938).

(13) P. Groh and K. A. Kirrmann, *C. R. Acad. Sci., Paris*, **215**, 275 (1942).

Schneider[14] in 1954, and the quantitative observations of Seliger[15] in 1960, who reported a single emission band at 6348 Å with a half-band width of 170 Å, but gave no interpretation. Seliger's quantitative report focused new attention on the oxygen chemiluminescence phenomena.

Stauff and Schmidkunz[16] interpreted the "6348-Å band" as the direct emission corresponding to the liquid oxygen 6300-Å absorption band interpreted by Ellis and Kneser[9] as the double molecule 0,0 band for the $2(^1\Delta_g) \leftarrow 2(^3\Sigma_g^-)$ absorption. This possibly coincidental interpretation has been established subsequently by 'more rigorous spectroscopic investigation. However, other assignments of comparatively broad luminescences in reacting chemiluminescent systems made by Stauff and Schmidkunz[16] and by Stauff and Lohmann[17] on the basis of Ellis and Kneser absorption band positions must be regarded as mere coincidences on the basis of later work by the present authors and others.

Khan and Kasha discovered that the red chemiluminescence in the aqueous peroxide–hypochlorite reaction yields two principal bands[3] instead of the one previously noted, at 6334 Å (15788 cm^{-1}) and 7032 Å (14221 cm^{-1}), as well as two other faint bands in the near-infrared. The 1567-cm^{-1} spacing between these comparatively narrow (approximate half-band width of 100–125 Å; or 250 cm^{-1}) bands, compared with the ground state lowest vibrational spacing (1556 cm^{-1}) of molecular oxygen, confirmed that a molecular oxygen species was directly involved in the emission. The uncorrected spectrographic results showed surprisingly that the 0,0 and 0,1 bands were of comparable intensity. Seliger's determination[18] of the relative intensities of these two bands gave an integrated intensity ratio of 1/0.6 for the 0,0 vs. the 0,1 strong chemiluminescence bands, but other measurements (see below) yield ratios closer to 1/1. Seliger also observed an extremely weak band at 5780 Å (17,300 cm^{-1}). Khan and Kasha[3] considered two interpretations of the two main chemiluminescent bands, tentatively favoring the idea that the bands represented the oxygen $^1\Sigma_g^+ \leftarrow {}^3\Sigma_g^-$ transition shifted by hydration, considering the Stauff and Schmidkunz comparison[16] with liquid oxygen double-molecule transitions as a possible coincidence.

However, in 1964 Arnold, Ogryzlo and Witzke,[6] pursuing research on molecular species in electrically discharged gaseous oxygen, noted that the red bands observed by Khan and Kasha were present although very weakly even in the gas-discharge spectrum. Moreover, the band envelopes and relative intensities (uncorrected) for the two bands (reported as 6340 and 7030 Å) were very similar to the bands determined by them for the peroxide–hypochlorite aqueous system. Equally significant was their observation of an extremely weak band at 7600 corresponding to the single-molecule $^1\Sigma_g^+ \to {}^3\Sigma_g^-$ emission. Browne and Ogryzlo reported[7] more general spectroscopic observations at low resolution for the peroxide–hypochlorite reaction.

The strong bands at 6340 and 7030 Å were definitely assigned as the $2(^1\Delta_g) \to 2(^3\Sigma_g^-)$ transition, with Seliger's weak 5800-Å band as a companion 1,0 band. Weak bands at 10,700 and 12,700 Å were identified as 1,0 and 0,0 bands of the $^1\Delta_g \to {}^3\Sigma_g^-$ transition. Weak bands in the region 7600–8700 Å were assigned to the $^1\Sigma_g^+ \to {}^3\Sigma_g^-$ single molecule transition of O_2.

The unusual and unexpected appearance of a double-molecule simultaneous transition of molecular oxygen (at one atmosphere) in aqueous chemiluminescence spectra requires a complete spectroscopic investigation. In the following section a complete correlation is made of chemiluminescence spectra for the peroxide–hypochlorite aqueous system with high-pressure gaseous oxygen absorption spectra in the corresponding regions.

Correlation of Oxygen Chemiluminescence Spectra for Aqueous Solutions with Absorption Spectra of Gaseous Oxygen at High Pressure

In polyatomic molecular luminescence studies the correlation of the lowest absorption band for each multiplicity change with the corresponding luminescences (normally two) is the established method for testing authenticity of the luminescences in one-photon, one-molecule excitations. The molecular oxygen spectroscopy case is more complex phenomenologically for two reasons. First, the absence of radiationless transitions[19] would permit any excited state of a diatomic molecule to exhibit an intrinsic luminescence spectrum, and, second, the absence of intramolecular vibrational relaxation mechanisms in diatomics[20] makes metastable states of diatomics exceedingly susceptible to intermolecular perturbations.

For molecular oxygen, several luminescences are observed in chemiluminescence experiments in aqueous peroxide oxidation. Not only is emission observable from single-molecule oxygen metastable excited states (i.e., $^1\Sigma_g^+ \to {}^3\Sigma_g^-$) but also from O_2–O_2 molecular pair excited states (e.g., $[^1\Delta_g][^1\Delta_g] \to [^3\Sigma_g^-][^3\Sigma_g^-]$) as simultaneous transitions, arising from one-photon, two-molecule nonresonance interactions. (We use the wave function convention $[^1\Delta_g][^1\Delta_g]$ to describe simultaneous transition states, and the energy convention $[^1\Delta_g + {}^1\Delta_g]$ in reference to transition frequencies.)

For luminescence observations on O_2–O_2 pairs, low pressures such as 1 atm suffice (the "solution" chemiluminescences arise from metastable singlet oxygen molecules trapped in gas bubbles). All of the chemiluminescences observed involve pairs of metastable $^1\Delta_g$ and/or $^1\Sigma_g^+$ molecules in binary collisions. Owing to the extreme metastability of these two lowest singlet states of molecular oxygen, pair interaction leading to simultaneous emission of one photon can occur even at very low partial pressures of the metastable species.

For absorption observations on O_2–O_2 pairs, effectively a rare ternary event is required for a simultaneous transition to occur: a photon must interact with an O_2–O_2 collision pair undergoing electron exchange interaction. Thus, high gas pressure (here 140–150 atm) in 6.5-cm optical cells must be used to find simultaneous transitions in the molecular gas. It is understandable then why the spectrum of liquid (or solid)

(14) G. Gattow and A. Schneider, Naturwissenschaften, 41, 116 (1954).
(15) H. H. Seliger, Anal. Biochem., 1, 60 (1960).
(16) J. Stauff and H. Schmidkunz, Z. Phys. Chem. (Frankfurt am Main), 35, 295 (1962).
(17) J. Stauff and F. Lohmann, ibid., 40, 123 (1964).
(18) H. H. Seliger, J. Chem. Phys., 40, 3133 (1964).

(19) B. R. Henry and M. Kasha, Ann. Rev. Phys. Chem., 19, 161 (1968).
(20) P. J. Gardner and M. Kasha, J. Chem. Phys., 50, 1543 (1969).

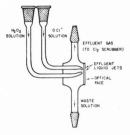

Figure 1. Continuous flow liquid–liquid mixing cell for spectral study of peroxide–hypochlorite chemiluminescence.

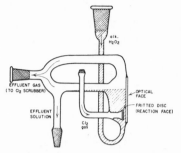

Figure 2. Hirtian apparatus: continuous mixing gas–liquid cell for spectral study of peroxide–hypochlorite chemiluminescence.

oxygen would exhibit simultaneous optical absorption transitions which could correlate with chemiluminescence simultaneous transitions for low-pressure systems.

In this section we shall present the full correlation of all the presently observed oxygen solution chemiluminescence spectra with the high-pressure gaseous absorption spectra.

Experimental Section

Reports in the literature indicated that a red chemiluminescence could be observed by mixing alkaline aqueous hydrogen peroxide with sodium hypochlorite or hypobromite. Initially the mixing apparatus for two liquid stream shown in Figure 1 was developed. The reagent grade hydrogen peroxide (10–30%) was kept chilled to 0°, and introduced into a separatory funnel above the apparatus just before use. Commercial hypochlorite bleach solution as well as reagent grade dilute sodium hypochlorite was similarly disposed. Constant slow flow of the two solutions produced a bright red-orange glow on the optical face, disposed within the cone of vision of the spectrograph.

A Steinheil Universal GH spectrograph was used, with f 3.9 three-prism glass short-focus optics in a special infrared setting with a dispersion of 108 Å/mm at 7000 Å; the exposure times were 1–15 min at various slit settings using water-hypersensitized Kodak I-N infrared spectrographic plates. Neon discharge tube lines were used for calibration.

For high resolution studies and for the observation of very weak bands, an apparatus involving gaseous chlorine proved more convenient and efficient for very long observation periods. This specially designed Hirtian apparatus[21] is illustrated in Figure 2. The surface of the fritted filter glowed brilliantly as the flow of gaseous Cl_2 into alkaline 10–30% chilled aqueous H_2O_2 was optimized. The Steinheil GH three-prism glass spectrograph with f 10 long-focus optics was used, yielding a dispersion of 41 Å/mm at 7000 Å.

For extremely weak band emissions ultra-high-speed Polaroid film (Type 410, ASA 10000) was used with the Steinheil GH three-prism spectrograph with f 3.9 glass optics with visible setting.

Absorption spectra of gaseous oxygen at high pressure were determined with a Cary Model 15 spectrophotometer, using enhanced optical density recording with 0.1 absorbance unit full scale. An Aminco high-pressure quartz optics absorption cell was used with an optical path of 6.5 cm. Purified tank oxygen at 140–150 atm at 20° was used.

Discussion of Spectral Observations

Red Chemiluminescence Bands. The $2[^1\Delta_g] \rightarrow 2[^3\Sigma_g^-]$ Transition. The spectrum for the prominent red chemiluminescence bands (Figure 3c) with prominent peaks at 6334 Å (15,788 cm^{-1}) and 7032 Å (14,221 cm^{-1}) was reported previously[3] for the hydrogen peroxide–aqueous hypochlorite reaction. The chemiluminescence is brilliant enough to be visible in strong daylight, and required exposure times of 1 sec or less

under the best conditions. As indicated, the separation between these two main peaks is 1567 cm^{-1}, which should represent the ground state lowest vibrational spacing of molecular oxygen, which is known[22] to be 1556 cm^{-1}. In the experiments it is evident (see below) that the chemiluminescence is emitted from metastable excited oxygen molecules within the gas bubble. Comparison was then thought desirable with gas-phase visible region absorption of pure oxygen. Figure 4c shows the absorption spectrum of gaseous O_2 at 150.0 atm and 20° in a 6.5-cm optical path in the 5000–6300-Å region, corresponding to the red chemiluminescence region from 6100 to 7500 Å. The absorption envelope and wavelength correspond well to those reported by Salow and Steiner,[23] who accepted Ellis and Kneser's assignment[9] of this absorption as the simultaneous transition $2[^1\Delta_g] \leftarrow 2[^3\Sigma_g^-]$ for O_2 pairs. We shall discuss this interpretation in the next section. The comparison of the red chemiluminescence of Figure 3c and the corresponding absorption spectrum of Figure 4c shows a classical correlation of 0,0 bands and vibrational envelopes for luminescence and absorption. The 0,2 band for the $2[^1\Delta_g] \rightarrow 2[^3\Sigma_g^-]$ band appears very weakly at 7860 Å in Figure 3d.

The Franck–Condon envelopes for the $2[^1\Delta_g] \leftrightarrow 2[^3\Sigma_g^-]$ absorption and chemiluminescence are in good approximate correspondence; although the absorption contours are accurately given by the dual-beam spectrophotometer, the emission results require plate correction, so are approximate. Seliger's determination[18] yielded a value of 0.6 for the integrated ratio for the 0,1/0,0 solution chemiluminescence bands. Whitlow and Findlay[24] measured 0.93 ± 0.1 for the 6340/7030 Å or 0,0/0,1 band ratio for the electrical discharge gaseous oxygen luminescence. The lack of agreement of these intensity ratios is puzzling; since our absorption data (Figure 4c) were obtained from a dual beam spectrophotometer (1 atm air path blank), any source of error connected with luminescence measurements would be absent. We obtain as integrated areas for 0, 0/1, 0/2,0 bands (Figure 4c) the values 1.00/1.21/0.09; however, the small optical densities involve an error of ca. 10% in these values. It is clear, however, that in the $2[^1\Delta_g] \leftrightarrow 2[^3\Sigma_g^-]$ transition, the 1,0 or 0,1 band is of

(21) A. Hirt, *Vieux Carré*, New Orleans, La.

(22) G. Herzberg, "Molecular Spectra and Molecular Structure I. Spectra of Diatomic Molecules," 2nd ed, D. Van Nostrand Co., Inc., New York, N. Y., 1950.

(23) H. Salow and W. Steiner, *Nature*, **134**, 463 (1934); *Z. Phys.*, **99**, 137 (1936).

(24) S. H. Whitlow and F. D. Findlay, *Can. J. Chem.*, **45**, 2087 (1967).

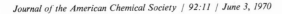

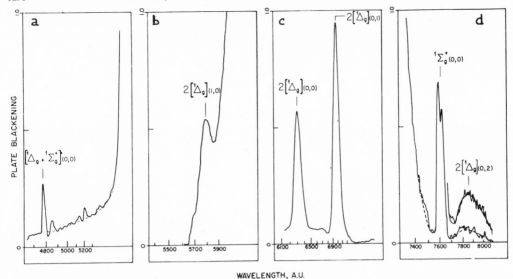

Figure 3. Chemiluminescence bands for the aqueous reaction at 20° of hydrogen peroxide with hypochlorite. Bands labeled according to upper electronic state by energy convention, with vibronic components.

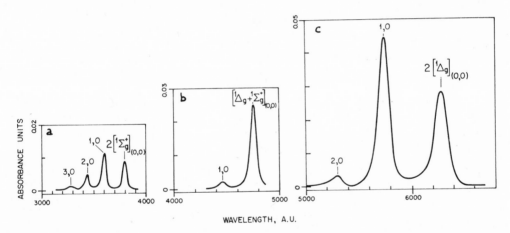

Figure 4. Enhanced optical density absorption spectra of gaseous molecular oxygen at 150 atm in a 6.5-cm cell at 20°. Bands labeled according to upper electronic state by energy convention, with vibronic components.

comparable intensity to the 0,0 band; a comment on this Franck–Condon anomaly of the simultaneous transition compared with the single-molecule Franck–Condon intensities will be made in the next section.

The 6334- and 7032-Å chemiluminescence bands for the $2[^1\Delta_g] \rightarrow 2[^3\Sigma_g^-]$ transition, in spite of their being effectively gas-phase emissions, proved to be not further resolvable. Using medium focus glass optics at $f/10$ in the Steinheil spectrograph, a setting which revealed[4] rotational structure in the $^1\Sigma_g^+ \rightarrow {}^3\Sigma_g^-$ solution chemiluminescence bands at 7600 Å, failed to reveal any further detail in the 6334- and 7032-Å bands. This lack of further structure has been confirmed by other workers.[24,25]

The vibrationally excited band 1,0 for the $2[^1\Delta_g] \rightarrow 2[^3\Sigma_g^-]$ reported by Seliger[18] at 5780 Å and assigned by Browne and Ogryzlo[7] is shown in Figure 3b. This appears as an extremely weak shoulder on the high-frequency side of the intense 0,0 band at 6334 Å.

An additional weak but persistent band was found on all plates at about 6700 Å, but is not assigned. It cannot correspond to the 6900-Å 1,0 hot band for $^1\Sigma_g^+ \rightarrow {}^3\Sigma_g^-$ reported by Whitlow and Findlay,[24] since their resolved structure clearly shows it to be on the high-frequency side of the 7032-Å band; sighting along

(25) L. W. Bader and E. A. Ogryzlo, *Discuss. Faraday Soc.*, No. 37, 46 (1964).

this side of the 7032-Å band in our figure is a very slight convexity corresponding to this 1,0 band.

Green Chemiluminescence Bands. The $[^1\Sigma_g^+ + {}^1\Delta_g] \rightarrow 2[^3\Sigma_g^-]$ Transition. The spectrum for a green chemiluminescence was determined by searching under spectroscopic conditions of extreme sensitivity in the region of the 4770-Å band of gaseous molecular oxygen identified by Salow and Steiner[23] as the 0,0 band of the Ellis and Kneser[9] $[^1\Sigma_g^+ + {}^1\Delta_g] \leftarrow 2[^3\Sigma_g^-]$ oxygen pair transition. A very sharp green chemiluminescence band was found[5] (Figure 3a) for the Cl_2-alkaline hydrogen peroxide reaction at 4780 Å (20,920 cm^{-1}); only a single band was recorded, which must be the 0,0 band for the oxygen pair $[^1\Sigma_g^+ + {}^1\Delta_g] \rightarrow 2[^3\Sigma_g^-]$ emission. The corresponding absorption spectrum (Figure 4b) for gaseous molecular oxygen at 150.0 atm in a 6.5-cm optical path, at 20° indicates the close correspondence of the 0,0 bands in absorption and emission, and also that the 0,1 band is expected to be relatively weak, explaining our failure to observe it in the very feeble emission.

The Ultraviolet Bands. The $2[^1\Sigma_g^+] \leftarrow 2[^3\Sigma_g^-]$ Absorption. The absorption spectrum of gaseous oxygen at 150 atm, in a 6.5-cm optical path, at 20° in the region 3000–4000 Å was determined and is shown in Figure 4a. It is in agreement with the early work of Salow and Steiner,[23] who identified the four bands observed with the 0,0, ..., 3,0 vibronic components of the $2[^1\Sigma_g^+ \leftarrow 2[^3\Sigma_g^-]$ oxygen pair transition of Ellis and Kneser.[9] An exhaustive search for chemiluminescence with the alkaline peroxide–Cl_2 reaction in the expected 4000–5000-Å region failed to reveal any bands. Dissolved and gaseous Cl_2 (liberated as a by-product in the exothermic reaction) absorption could possibly screen any feeble emission which would be present. Moreover, studies of the quenching of metastable $^1\Sigma_g^+$ oxygen by water[7] indicate that a very small fraction of the $^1\Sigma_g^+$ molecules would survive nucleation into bubbles from the water phase, so that emission for a $2[^1\Sigma_g^+]$ pair state would be even less likely than from the $[^1\Sigma_g^+ + {}^1\Delta_g]$ pair state, which we have found to be extremely weak. Nevertheless, luminescence sensitization experiments clearly indicate that even the $2[^1\Sigma_g^+]$ double molecule state has a finite probability of being excited in the peroxide–hypochlorite reactions, as discussed in the next section.[25a]

The Single Molecule $^1\Sigma_g^+ \rightarrow {}^3\Sigma_g^-$ Infrared Chemiluminescence. The electrical discharge single-molecule infrared $^1\Sigma_g^+ \rightarrow {}^3\Sigma_g^-$ luminescence in gaseous oxygen was first studied by Kaplan[26] at low resolution. We observed this emission as a very weak band in our first peroxide–hypochlorite chemiluminescence studies. The spectrum published by Browne and Ogryzlo[7] revealed weak bands of comparable intensity at 7619 and 7700 Å assigned as 0,0 and 1,1 bands of the $^1\Sigma_g^+ \rightarrow {}^3\Sigma_g^-$ transition, and a much weaker band at 8645 Å as a 0,1 band. Using the same medium-resolution equipment described above for the red chemiluminescence bands, the infrared $^1\Sigma_g^+ \rightarrow {}^3\Sigma_g^-$ chemiluminescence 0,0 band at 7620 Å (band center) was partially resolved (Figure 3d) into two *rotational* branches[4] even though

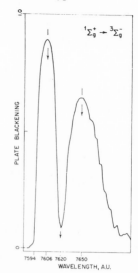

Figure 5. Gas-phase luminescence of molecular oxygen at room temperature in electrical discharge indicating rotational resolution.

the chemiluminescence was generated from an aqueous reaction of alkaline peroxide with chlorine.

The high resolution rotationally resolved spectrum obtained by Branscomb[27] for the $^1\Sigma_g^+ \rightarrow {}^1\Sigma_g^-$ emission produced in electrodeless discharge in gaseous oxygen reveals a missing Q branch near 7621 Å as expected for the $^1\Sigma$–$^3\Sigma$ band of a diatomic. A Q-branch gap appeared clearly near 7620 Å in the solution chemiluminescence spectrum at medium resolution (Figure 3d). A spectrum of the gas-discharge $^1\Sigma_g^+ \rightarrow {}^3\Sigma_g^-$ luminescence of O_2 at similar medium resolution is presented in Figure 5. This observation confirmed the qualitative interpretation that the observed chemiluminescence in the aqueous hydrogen peroxide–hypochlorite reaction comes from metastable oxygen molecules which survive as excited molecules upon nucleation into gas bubbles.

A comparison of the absorption envelope for the first vibrational band of the 7620-Å $^1\Sigma_g^+ \leftarrow {}^3\Sigma_g^-$ transition (Figure 7) and the corresponding emission envelope for the 0,0 band at 7620 Å of the same transition (Figure 3d) reveals identical contours, as is expected for rotational envelopes.

Mechanism of Singlet Oxygen Formation, Simultaneous Transition in Molecular Oxygen Pairs, and Singlet Oxygen Sensitization of Chemiluminescence

Metastable Molecular Oxygen Singlet States, $^1\Delta_g$ and $^1\Sigma_g^+$. In molecular oxygen, starting with a single configuration molecular orbital wave function in zeroth order, configurational splitting as a result of electron repulsion leads to three distinct molecular states.

$$[(K)(K)(\sigma_g)^2(\sigma_u)^2(\sigma_g)^2(\pi_u)^4](\pi_g)^1(\pi_g)^1 \equiv {}^1\Sigma_g^+ + {}^1\Delta_g + {}^3\Sigma_g^- \quad (2)$$

The highest multiplicity state of a given configuration lying lowest leads to a triplet ground state for the even-electron molecule O_2. The ground state of O_2 is thus

(25a) NOTE ADDED IN PROOF. E. W. Gray and E. A. Ogryzlo, *Chem. Phys. Lett.*, **3**, 658 (1969), have now found gas discharge bands at 3800 and 4000 A corresponding to the O_2-pair uv transition.

(26) J. Kaplan, *Nature*, **159**, 673 (1947); *Phys. Rev.*, **71**, 274 (1947).

(27) L. Branscomb, *ibid.*, **86**, 258 (1952).

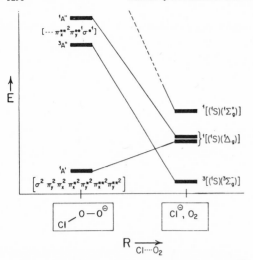

Figure 6. Mechanism of singlet oxygen formation in peroxide oxidation. Spin-state correlation diagram for chloroperoxy ion and its products.

assist in deciding that singlet oxygen should be uniquely evolved by the peroxide–hypohalite reaction (eq 1). The spectroscopic evidence is definite that both $^1\Delta_g$ and $^1\Sigma_g^+$ molecular oxygen are produced, and chemical reactivity studies[32] with singlet oxygen show that the yields of product may exceed 70% and approach unity. The kinetic mechanism of the alkaline–peroxide hypohalite reaction was studied by Cahill and Taube,[33] who used isotopic labeling to prove that the O–O bond of the peroxide did not undergo fission, and that the rate-determining intermediate in the reaction was the chloroperoxy ion OOCl⁻. It is clear that ionic fission of this species to yield chloride ion must *perforce leave an oxygen molecule with all electrons paired*, i.e., $^1\Delta_g$ oxygen, as deduced from the spin-state correlation diagram (Figure 6). It appears that the reaction

$$^1\Delta_g + {}^1\Delta_g \longrightarrow {}^1\Sigma_g^+ + {}^3\Sigma_g^- \tag{3}$$

may account for the presence of $^1\Sigma_g^+$ in the peroxide system in small amounts, the ratio $[^1\Sigma_g^+]/[^1\Delta_g]$ being[34] less than 10^{-6}.

Simultaneous Transition in Molecular Oxygen Pairs. A spectroscopic summary of the optical absorption spectrum from 2600 to 13,400 Å of gaseous molecular oxygen at 20° and 150 atm (6.5-cm cell) is given in

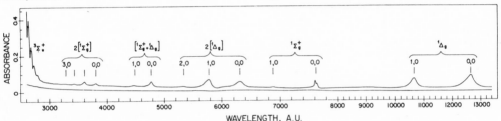

Figure 7. Complete absorption spectrum from 2600 to 13,400 Å of gaseous molecular oxygen at 150 atm in a 6.5-cm cell at 20°. Bands labeled according to upper electronic state by energy convention, with vibronic components.

$^3\Sigma_g^-$, with $^1\Delta_g$ observed[22] at 7882.4 cm⁻¹ (0.9772 eV, 22.544 kcal) and $^1\Sigma_g^+$ at 13,120.9 cm⁻¹ (1.6266 eV, 37.527 kcal) above the ground state, respectively.

The optical transitions $^1\Sigma_g^+ \leftrightarrow {}^3\Sigma_g^-$ and $^1\Delta_g \leftrightarrow {}^3\Sigma_g^-$ are strictly electric dipole forbidden. The transition $^1\Sigma_g^+ \leftrightarrow {}^3\Sigma_g^-$ represents a spin-forbidden, symmetry-forbidden magnetic dipole transition, with an intrinsic mean radiative lifetime of 7 sec.[28] The transition $^1\Delta_g \leftrightarrow {}^3\Sigma_g^-$ is a spin-forbidden, symmetry-forbidden, orbitally forbidden magnetic dipole transition with an intrinsic mean radiative lifetime of 45 min.[29] Both of these transitions are pressure sensitive, and acquire electric dipole character through intermolecular perturbation.

The electronic structure and spectra of molecular oxygen have been studied in exhaustive detail.[22,30,31] The *electronic mechanism of chemical production* of singlet molecular oxygen from hydrogen peroxide poses a special problem. It might appear since the three lowest states, $^3\Sigma_g^-$, $^1\Delta_g$, and $^1\Sigma_g^+$, all arise from the same molecular orbital configuration, that a quantum mechanical analysis based on MO theory could not directly

Figure 7. In *energy* labeling are designated the vibronic components of the three *simultaneous transitions for molecular oxygen pairs* in the near-uv and visible regions. In the infrared are the vibronic components of the *single molecule* O₂ transitions.

In Figure 8 is given an energy level diagram (wave function labeling) for molecular oxygen using experimental data for the vibronic levels observed by us in absorption at high pressure (~150 atm) in gaseous O₂, and in solution chemiluminescence studies. It is striking to note the richness of electronic and vibronic states of molecular oxygen in the near-uv, visible, and near-ir regions.

The possibility of *simultaneous transition* from pairs of metastable singlet molecular oxygen molecules offers a mechanism of "quantum doubling" in the sense that the electron spin coupling collision of two excited molecules can lead to the emission of one photon of additive excitation energy. We may describe the simultaneous transition as a *transition for a pair of atomic or molecular species to an (algebraically) additive composite excited state*.

(28) W. H. J. Childs and R. Mecke, *Z. Phys.*, **68**, 344 (1931).
(29) R. M. Badger, A. C. Wright, and R. F. Whitlock, *J. Chem. Phys.*, **43**, 4345 (1965).
(30) W. Moffitt, *Proc. Roy. Soc., Ser. A*, **210**, 224 (1951).
(31) J. Slater, "Quantum Theory of Molecules and Solids. I. Electronic Structure of Molecules," McGraw-Hill Book Co., New York, N. Y., 1963, *cf.* Appendix 11, p 294 ff.

(32) (a) C. S. Foote and S. Wexler, *J. Amer. Chem. Soc.*, **86**, 3879, 3880 (1964); (b) C. S. Foote, *Accounts Chem. Res.*, **1**, 104 (1968).
(33) A. E. Cahill and H. Taube, *J. Amer. Chem. Soc.*, **74**, 2312 (1952).
(34) S. J. Arnold, M. Kubo, and E. A. Ogryzlo, Advances in Chemistry Series, No. 77, American Chemical Society, Washington, D. C., 1968, p 133.

Experiments on simultaneous transitions in oxygen molecules clearly show that these one-photon two-molecule processes are collisionally induced. The transitions have low oscillator strength and spectrally appear to correspond almost exactly to the sum of the energy of the states of the colliding partners. Robinson,[35] Rettschnick and Hoytink,[36] and Krishna[37] have treated this problem theoretically, and have successfully explained a number of the qualitative aspects for transitions in oxygen pairs. They have shown that forbidden transitions in the single molecules can acquire enhanced transition probability in the complex by borrowing intensity from an allowed (uv) transition under the influence of intermolecular interactions.

We may describe the *intermolecular* enhancement[35-37] of the highly forbidden transitions of molecular oxygen in simple terms involving combined multiplicities for molecular oxygen pair states. The spin-forbidden transition $^1\Delta_g \leftrightarrow {}^3\Sigma_g^-$ becomes spin-allowed for the simultaneous transition

$$^1[(^1\Delta_g)(^1\Delta_g)] \longleftrightarrow {}^{1,3,5}[(^3\Sigma_g^-)(^3\Sigma_g^-)] \qquad (4)$$

The superscript prefix denotes the combined multiplicities for the molecular pairs; in the molecular pair, a singlet–singlet component exists. The Schumann–Runge pair state$^{1,3,5}[(^3\Sigma_u^-)(^3\Sigma_u^-)]$ could provide the perturbation intensity source. Similarly, the spin-forbidden single molecule transition $^1\Delta_g \leftrightarrow {}^3\Sigma_g^-$ may become spin-allowed in the collision pair involving a second normal oxygen molecule

$$^3[(^1\Delta_g)(^3\Sigma_g^-)] \longleftrightarrow {}^{1,3,5}[(^3\Sigma_g^-)(^3\Sigma_g^-)] \qquad (5)$$

in which a triplet–triplet component now appears. Thus, as the spectra of Figure 7 emphasize, both the simultaneous transitions and the forbidden single-molecule transitions appear with comparable intensity, and all become pressure- or collision-dependent electric dipole transitions. The intrinsic lifetimes of the metastable singlet states of molecular oxygen may become shortened many orders of magnitude by O_2–O_2 collisions, as well as by the influence of other intermolecular collisions. It is characteristic of *vibrationally deficient*[20] molecules to be unusually collision-sensitive, owing to the absence of intramolecular relaxation mechanisms.

The Franck–Condon envelopes observed for the simultaneous transitions in molecular oxygen pairs in some cases differ widely from those of the single molecule transitions. For example, the relative intensities of the (0,0), (1,0), and (2,0) vibronic components in the $(^1\Delta_g)(^1\Delta_g) \leftarrow (^3\Sigma_g^-)(^3\Sigma_g^-)$ transition have ratios (integrated absorption) of 1/1.21/0.1, respectively, whereas both experimental and theoretical values[38] for the single-molecule Franck–Condon factors of the corresponding vibronic components in the $^1\Delta_g \leftarrow {}^3\Sigma_g^-$ transition have ratios of 1/0.013/0.000, respectively. Similar behavior is shown also in the vibronic intensities of the $(^1\Sigma_g^+)(^1\Sigma_g^+) \leftarrow (^3\Sigma_g^-)(^3\Sigma_g^-)$ transition. Such a change of vibronic intensities could come from O_2 distortion due to strong interaction between the two molecules, but for the weakly interacting O_2–O_2 pairs this is not applicable.

(35) G. W. Robinson, *J. Chem. Phys.*, **46**, 572 (1967).
(36) R. P. H. Rettschnick and G. J. Hoytink, *Chem. Phys. Lett.*, **1**, 145 (1967).
(37) V. G. Krishna, *J. Chem. Phys.*, **50**, 792 (1969).
(38) R. W. Nicholls, P. A. Fraser, W. R. Jarmin, and R. P. McEachran, *Astrophys. J.*, **131**, 399 (1960).

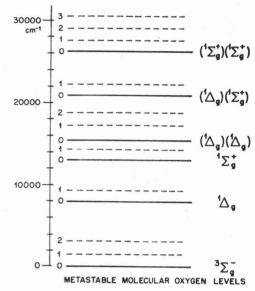

Figure 8. Complete experimental energy level diagram of molecular oxygen up to the near-uv, showing vibronic components observed for single molecule states and simultaneous transition molecular pair states.

Moreover, in the mixed simultaneous transition $(^1\Sigma_g^+)$-$(^1\Delta_g) \leftarrow (^3\Sigma_g^-)(^3\Sigma_g^-)$ the vibronic components (0,0) and (1,0) have ratios of about 1/0.06, respectively, which resemble those of the single molecule transitions. The Franck–Condon envelopes for the simultaneous transitions indicate that the Born–Oppenheimer separation of the nuclear and electronic wave functions may not be valid. A detailed investigation of the vibronic coupling is needed. Dianov-Klokov has suggested loss of center of symmetry with vibrational excitation in one member of the molecular pair as a qualitative explanation.[39]

Singlet Oxygen Sensitization of Chemiluminescence. Chemiluminescence is a widely occurring phenomenon which can arise by a variety of mechanisms. The case described by Mallet[11] is of particular interest since it involves observation of organic molecule luminescence in the peroxide–hypochlorite aqueous reaction system. Further studies of this type were described by Kurtz, who studied especially violanthrone luminescence sensitized by the peroxide reaction.[40] It is clear that in all of the cases studied the excitation energy of the metastable singlet oxygen separate molecules would be insufficient to sensitize the luminescence of the organic acceptor molecule. However, the simultaneous transition states of singlet molecular oxygen pairs would possess enough excitation energy to sensitize the luminescence of a suitable organic molecule.

Khan and Kasha[5] have proposed a physical theory of sensitized chemiluminescence for such cases, in which it is assumed that the singlet molecular oxygen produced in the peroxide reaction could transfer excitation energy from simultaneous transition pair states to the organic acceptor. In this mechanism it is assumed that a pene-

(39) V. I. Dianov-Klokov, *Opt. Spectrosc.*, **6**, 290 (1959).
(40) R. B. Kurtz, *Ann. N. Y. Acad. Sci.*, **16**, 399 (1954).

Figure 9. Mechanism of fluorescence sensitization by the O_2^*–O_2^* simultaneous transition pair state.

trating electron exchange collision of the metastable oxygen molecular pairs is required, but that the acceptor molecule may be at a considerable distance as an excitation acceptor (Figure 9), avoiding the usual unfavorable statistics on ternary processes. It is interesting to note that the efficiency of the sensitized "chemiluminescence" in the peroxide systems is much greater than the efficiency of direct oxygen (pair) chemiluminescence. Evidently, the energy transfer step successfully competes with quenching mechanisms for the metastable oxygen.

Ogryzlo and Pearson[41] have suggested an alternative mechanism involving effectively a two-step successive excitation of a very low-lying triplet of the organic acceptor, utilizing the energy of the $^1\Delta_g$ state. But the low energy of this molecular oxygen state (7882.4 cm^{-1}, 0.9772 eV) makes this mechanism unacceptable, since there are almost no organic molecules with such low-lying triplet states. It is possible that the Ogryzlo and Pearson "disproportionation pooling" mechanism would fit the rare cases of molecules with lowest singlet–triplet transitions below 7800 cm^{-1}. Recently Wilson[42] reported a study of rubrene luminescence excited by energy transfer from singlet oxygen. Her studies do not permit an absolute distinction between (a) the two-step excitation of the lowest singlet excited state of the acceptor (rubrene) with the lowest triplet excited state as an intermediate, and (b) the near-triple collision between the acceptor and a "dimol" $[^1O_2]_2$. The latter description conforms to the proposal made by Khan and Kasha[5] as elaborated above.

The study of excitation energy transfer for organic molecules to and from singlet molecular oxygen, or the various simultaneous transition pair states, deserves

much additional study. The detailed theoretical mechanisms[43] proposed for such processes may become refined only when the physical criteria become more clearly delineated. A complete spectroscopic study of the resonance conditions for efficient energy transfer in the singlet oxygen molecular and pair states is needed, with a complete correlation with the full electronic energy manifold of the interacting organic molecule.

Summary

Singlet molecular oxygen, once neglected as of mere astrophysical interest, is now being found to be a possibly ubiquitous intermediate in physical, chemical, and biological processes involving molecular oxygen. The many methods now available for its generation and detection allow extensive studies to be made in systems in which it was previously assumed normal ground state oxygen ($^3\Sigma_g^-$) was involved. The extreme metastability of the singlet species ($^1\Delta_g$ and $^1\Sigma_g^+$) and their unique reactivity at normal temperatures make them species of exceptional chemical interest. It is well established now that these excited states of oxygen participate in ordinary chemical steps involving molecular oxygen.

In excitation phenomena, involving photo- and radiation-induced processes, it could also be expected that the metastable singlet molecular oxygen states would be especially prominent. It is now established that in the reversible quenching of excited states of molecules by normal oxygen, and in chemiluminescences accompanying the evolution of molecular oxygen, physical energy transfer steps involve singlet molecular oxygen either as single molecules or as molecular-pair states. These processes are of great importance in photochemistry and photobiology, and their applicability to radiation events should be investigated.

Acknowledgment. We are indebted to Drs. A. Llewellyn and V. G. Krishna for the electrical discharge gas-phase luminescence spectrum of molecular oxygen used to make the microdensitometer trace in Figure 5. We thank Dr. G. Holzwarth for repeating and refining some of the high-pressure oxygen gas-phase absorption curves used in this paper.

(41) E. A. Ogryzlo and A. E. Pearson, *J. Phys. Chem.*, **72**, 2913 (1968).
(42) T. Wilson, *J. Amer. Chem. Soc.*, **91**, 2387 (1969).

(43) K. Kawaoka, A. U. Khan, and D. R. Kearns, *J. Chem. Phys.*, **46**, 1842 (1967).

Editor's Comments
on Papers 23 Through 26

OXYGENATION OF ORGANIC SUBSTRATES BY SINGLET OXYGEN FROM SODIUM HYPOCHLORITE AND HYDROGEN PEROXIDE

The spectroscopic investigations of the reaction of NaOCl and H_2O_2 had demonstrated that singlet molecular oxygen could be produced chemically (Papers 18–22). In 1964, Foote and Wexler (Paper 23) reported that singlet oxygen produced by this reaction would oxygenate various organic substrates to give products identical to those obtained from dye-sensitized photooxygenation of these substrates.* On the basis of these results, and those of Corey and Taylor with singlet oxygen generated by a radiofrequency discharge (Paper 28), Foote and Wexler (Paper 24) concluded that the reactive intermediate in sensitized photooxygenations was probably also singlet oxygen, as originally proposed by Kautsky. Although the evidence was certainly compelling, it did not unambiguously demonstrate the intermediacy of $^1O_2$ in photooxygenations.

McKeown and Waters (Paper 25) also found that singlet oxygen generated by sodium hypochlorite and hydrogen peroxide or by alkaline hydrogen peroxide and molecular bromine would oxygenate several substrates that were known to undergo photooxygenation. They also described a two-phase system in which the singlet oxygen formed in the lower aqueous phase rises through the upper organic layer containing the acceptor. However, it should be noted that the utility of the method using Br_2 is often lessened by side reactions.

*See also Paper 26.

23

Reprinted from *J. Amer. Chem. Soc.,* **86**, 3879–3880 (1964)

Olefin Oxidations with Excited Singlet Molecular Oxygen[1]

Sir:

This communication describes a novel and useful synthetic method for the oxidation of olefins and dienoid compounds to give products identical with those of the well-studied dye-photosensitized autoxidations.[2] The active species appears to be molecular oxygen in an excited singlet state, formed *in situ* by the reaction of sodium hypochlorite and hydrogen peroxide.[3-7]

The significance of this reaction for the mechanism of the photosensitized autoxidations is discussed in the accompanying communication.[8]

The oxidations proceed exceptionally smoothly, and in preparatively useful yields. Examples of the conversions so far carried out are given below. As no attempt at optimization of conditions has been made, the yields given are minimal.

In methanol, 2,5-dimethylfuran (I) is converted to 2,5-dimethyl-2-hydroperoxy-5-methoxydihydrofuran (II) in 84% yield (based on sodium hypochlorite used; in this case, dimethylfuran was in excess). The same product is produced in the photosensitized autoxidation of I in 74% yield, presumably by reaction of methanol with the hypothetical ozonide-like peroxide III.[9]

(1) Supported by N.S.F. Grant G-25086.

(2) G. O. Schenck, *Angew. Chem.,* **69**, 579 (1957), and references therein cited; G. O. Schenck and E. Koch, *Z. Elektrochem.,* **64**, 170 (1960); W. Bergmann and M. J. McLean, *Chem. Rev.,* **28**, 367 (1941).

(3) The chemiluminescence[4] of this reaction was originally assigned to emission from the $^1\Sigma_g{}^+$ state of molecular oxygen[5]; more recent studies, however, have shown that the visible luminescence is more probably derived mainly from a dimer of the $^1\Delta_g$ state[6,7]; infrared emission from the monomeric $^1\Delta_g$ state is also observed.[7]

(4) H. H. Seliger, *Anal. Biochem.,* **1**, 60 (1960); L. Mallet, *Compt. rend.,* **185**, 352 (1927); G. Gattow and A. Schneider, *Naturwiss.,* **41**, 116 (1954).

(5) A. U. Khan and M. Kasha, *J. Chem. Phys.,* **39**, 2105 (1963).

(6) S. J. Arnold, E. A. Ogryzlo, and H. Witzke, *ibid.,* **40**, 1769 (1964).

(7) R. J. Browne and E. A. Ogryzlo, *Proc. Chem. Soc.,* 117 (1964).

(8) C. S. Foote and S. Wexler, *J. Am. Chem. Soc.,* **86**, 3880 (1964).

(9) Only polymeric peroxides had been previously isolated from the photosensitized autoxidation.[10] The structure of II (m.p. 75–76°) follows from its chemical and physical properties: *Anal.* Found: C, 52.42; H, 7.45; OCH₃, 18.83; mol. wt., 177.3 (osmometer); infrared (CCl₄, 0.005 *M*) 3512 cm.⁻¹ (OOH); n.m.r. (CDCl₃) τ 0.78, 3.98, 6.64, 8.42, and 8.47 (all sharp singlets, relative areas 1:2:3:3:3, assigned to OOH, olefinic H, OCH₃, C–CH₃, and C–CH₃, respectively). The *p*-nitrobenzoate (m.p. 91–92°) has an infrared band (CHCl₃) at 1775 cm.⁻¹ (perester). II gives the bis-2,4-dinitrophenylhydrazone of diacetylethylene, m.p. 286° dec., m.m.p. with a sample (m.p. 284° dec.) prepared by the method of Levisalles[11] (reported m.p. 291–292°) was 284° dec. Schenck has also observed product II.[12] The product formed from photosensitized autoxidation of methofuran in methanol has an analogous structure (not the one originally assigned[13]).[12]

(10) G. O. Schenck, *Ann.,* **584**, 165 (1953).

(11) J. Levisalles, *Bull. soc. chim. France,* 997 (1957).

(12) G. O. Schenck, private communication.

(13) G. O. Schenck and C. S. Foote, *Angew. Chem.,* **70**, 505 (1958).

$$CH_3 - \boxed{O} - CH_3 \xrightarrow[CH_3OH]{NaOCl/H_2O_2} CH_3 - \boxed{O} - CH_3$$

I

$$CH_3 - \boxed{O} - CH_3$$
$$HOO \quad OCH_3$$

II

$$CH_3 - \boxed{O} - CH_3$$
$$O - O$$

III

The reaction of 2,3-dimethylbutene-2 (IV) gives 2,3-dimethyl-3-hydroperoxybutene-1 (V), 63%, b.p. 54° (9 mm.), $n^{24.5}$D 1.4408; the infrared and n.m.r. spectra are identical with those of the product of the photosensitized autoxidation (formed in 54% yield; lit.[14]: 82%, b.p. 55° (12 mm.), n^{20}D 1.4428).

$$\begin{array}{ccc} CH_3 & & CH_3 \\ \diagdown & & \diagup \\ & C = C & \\ \diagup & & \diagdown \\ CH_3 & & CH_3 \end{array} \longrightarrow \begin{array}{ccc} & & OH \\ CH_2 & O & CH_3 \\ \diagdown & \diagup & \\ & C - C & \\ \diagup & & \diagdown \\ CH_3 & & CH_3 \end{array}$$

IV   V

Tetraphenylcyclopentadienone (VI) gives cis-dibenzoylstilbene (VII), 50%, m.p. 215.9–216.3°, infrared spectrum identical with that reported[15]. The photosensitized autoxidation gave VII in 65% yield.[16] The peroxide VIII is presumably intermediate.

VI   VII   VIII

Cyclohexadiene-1,3 (IX) gives 5,6-dioxabicyclo-[2.2.2]octene-2 ("norascaridol," X, 20%), with infrared and n.m.r. spectra identical with those of a sample prepared by the photosensitized autoxidation (35% yield; lit.[17] 21%).

(14) G. O. Schenck and K. Schulte-Elte, *Ann.*, **618**, 185 (1958).

(15) D. R. Berger and R. K. Summerbell, *J. Org. Chem.*, **24**, 1881 (1959).

(16) C. F. Wilcox, Jr., and M. P. Stevens, *J. Am. Chem. Soc.*, **84**, 1258 (1962); G. O. Schenck, *Z. Elektrochem.*, **56**, 855 (1952).

(17) G. O. Schenck and W. Willmund, reported by R. Criegee in Houben–Weyl, "Methoden der Organischen Chemie," Vol. VIII, 4th Ed., E. Müller, Ed., Georg Thieme Verlag, Stuttgart, 1952, p. 16. The yields referred to above are of distilled but unrecrystallized material.

(18) Corey has recently found that externally generated singlet O₂ is capable of oxidizing aromatic hydrocarbons: E. J. Corey and W. C. Taylor, *J. Am. Chem. Soc.*, **86**, 3881 (1964).

IX   X

Preliminary indications are that rubrene and other aromatic hydrocarbons also form peroxides analogous to those formed photochemically.[18]

Sample Procedure.—To a solution of 5.05 g. (60 mmoles) of 2,3-dimethylbutene-2 (IV) in 300 ml. of methanol was added 19.2 ml. of 30% H₂O₂ [180 mmoles). The solution was stirred at 10°, and 145.5 ml. of 1.03 M NaOCl solution (150 mequiv.) was added in the course of 90 min. The solution was diluted with water and extracted with ether. Ether was distilled from the dried extracts through a column to prevent loss of the volatile product. The residue was distilled, giving 4.39 g. (37.8 mmoles, 63%) of V, with the physical constants described above.

The new reaction parallels the photosensitized autoxidations in every case so far studied; it appears to be a general and practical synthetic method for oxidization of olefins which give high quantum yields in the photosensitized autoxidation (for example, s-cis dienes, furans, and tri- and tetrasubstituted olefins). The products (when they are peroxides) are, of course, readily reduced to the corresponding alcohols. Where the parallel photosensitized autoxidation goes in low quantum yield, it is likely that the peroxide–hypochlorite reaction will be less useful, because of the large quantities of reagents which would be required; this suggestion is supported by preliminary studies. The quantities of oxidant are best adjusted so that nearly 1 mole of oxygen is actually taken up; the oxygen which evolves from the solution (small amounts with good acceptors such as dimethylfuran) is collected in a gas buret so that this loss can be compensated by further addition of reagents. The use of excess H₂O₂ appears to be of value in preventing further oxidation of the products.

The reaction is being extended to other acceptors and to other systems potentially capable of yielding singlet oxygen.

Contribution No. 1689   Christopher S. Foote
Department of Chemistry   S. Wexler
University of California
Los Angeles 24, California

Received June 1, 1964

100

24

Reprinted from *J. Amer. Chem. Soc.*, **86**, 3880 (1964)

Singlet Oxygen. A Probable Intermediate in Photosensitized Autoxidations[1]

Sir:

Several mechanisms have been suggested for the dye-photosensitized autoxidations of dienes and olefins, studied in detail by Schenck,[2] and the analogous photooxidations of aromatic hydrocarbons, studied by many workers, particularly by Bowen.[3] Of these mechanisms, all but two have been ruled out by careful kinetic studies and by other criteria such as product or energy considerations. The two mechanisms which are consistent with the kinetics differ only in steps 3 and 4 (see below). In mechanism a (originally proposed by

$$\text{Sens} \xrightarrow{h\nu} {}^1\text{Sens} \qquad (1)$$

$$^1\text{Sens} \longrightarrow {}^3\text{Sens} \qquad (2)$$

$$^3\text{Sens} + {}^3O_2 \longrightarrow \cdot\text{Sens-O-O}\cdot \qquad (3a)$$

$$\cdot\text{Sens-O-O}\cdot + A \longrightarrow AO_2 + \text{Sens} \qquad (4a)$$

$$^3\text{Sens} + {}^3O_2 \longrightarrow \text{Sens} + {}^1O_2 \qquad (3b)$$

$$^1O_2 + A \longrightarrow AO_2 \qquad (4b)$$

Schönberg[4]), the postulated reactive intermediate (an adduct ($\cdot$SensOO$\cdot$) of sensitizer (Sens) and oxygen) reacts with acceptor (A) to form the product peroxide (AO$_2$). Mechanism b was originally proposed by Kautsky[5]; the reactive intermediate is an excited singlet state of molecular oxygen.

The evidence which supports mechanism a is as follows. (1) Gaffron found that with one sensitizer, at least, light of 8200 Å. is effective. Since this wave length is of insufficient energy to excite the $^1\Sigma_g{}^+$ state of O$_2$, Gaffron concluded that singlet oxygen could not be intermediate.[6] Kautsky, however, pointed out that there was ample energy in Gaffron's system to excite

oxygen to the lower-lying $^1\Delta_g$ state.[5c] (2) Livingston's kinetic studies indicate that the active intermediate has somewhat different kinetic behavior when formed with anthracene or with diphenylanthracene as sensitizer.[7] The rates were measured indirectly, and with a fairly high probable error, making this conclusion somewhat questionable. In other cases, Schenck has found that the rate of disappearance of the active intermediate does not depend on sensitizer.[2d]

Until now, the singlet oxygen mechanism has been generally disregarded despite the following observations. (1) Kautsky observed that reaction occurs even when sensitizer and acceptor are adsorbed on separate granules of silica gel, which would indicate that a reactive intermediate is formed which is capable of diffusion under vacuum.[5a,b] Similar phenomena were noted by Rosenberg and Shombert.[8] *No satisfactory explanation of these observations in terms of mechanism a has yet been made.* (2) The "moloxide" would have to have an unexpected property; for example, in the reaction with anthracene, which is both sensitizer and acceptor, the kinetics require that the intermediate (which, if it is a moloxide, is presumably of structure I) cannot collapse to product II itself, but must transfer oxygen to a second anthracene molecule.[3,9]

Since singlet oxygen has now been shown to be effective in carrying out reactions identical with the photosensitized autoxidations,[10] the weight of the evidence favors the intermediacy of singlet oxygen. Further experiments are in progress.

(1) Supported by N.S.F. Grant G-25086; we are pleased to acknowledge helpful discussions with Professor K. D. Bayes.

(2) (a) G. O. Schenck, *Angew. Chem.*, **69**, 579 (1957); (b) *Naturwiss.*, **35**, 28 (1948); (c) *ibid.*, **40**, 205, 229 (1953); (d) G. O. Schenck and E. Koch, *Z. Elektrochem.*, **64**, 170 (1960).

(3) E. J. Bowen in "Advances in Photochemistry," Vol. 1, W. A. Noyes, G. S. Hammond, and J. N. Pitts, Jr., Ed., Interscience Publishers, Inc., New York, N. Y., 1963, p. 23, and references therein cited.

(4) A. Schönberg, *Ann.*, **518**, 299 (1935).

(5) (a) H. Kautsky and H. de Bruijn, *Naturwiss.*, **19**, 1043 (1931); (b) H. Kautsky, H. de Bruijn, R. Neuwirth, and W. Baumeister, *Ber. deut. chem. Ges.*, **66**, 1588 (1933); (c) H. Kautsky, *Biochem. Z.*, **291**, 271 (1937).

(6) H. Gaffron, *ibid.*, **287**, 130 (1936), and references cited therein.

(7) R. Livingston and V. Subba Rao, *J. Phys. Chem.*, **63**, 794 (1959).

(8) J. L. Rosenberg and D. J. Shombert, *J. Am. Chem. Soc.*, **82**, 3527 (1960).

(9) G. O. Schenck, *Naturwiss.*, **41**, 452 (1954).

(10) (a) C. S. Foote and S. Wexler, *J. Am. Chem. Soc.*, **86**, 3879 (1964); (b) E. J. Corey and W. C. Taylor, *ibid.*, **86**, 3881 (1964).

Contribution No. 1690   Christopher S. Foote
Department of Chemistry   S. Wexler
University of California
Los Angeles 24, California

Received June 1, 1964

25

Reprinted from *J. Chem. Soc. (B)*, 1040–1046 (1966)

.

The Oxidation of Organic Compounds by " Singlet " Oxygen *

By E. McKeown and William A. Waters

The normal or "triplet" ground state ($^3\Sigma g^-$) of molecular oxygen is that of a biradical, ·O–O·, but "singlet" oxygen ($^1\Delta g$) O=O, which is structurally similar to ethylene, is the initial product of heterolytic decompositions of hydrogen peroxide or of per-acids. "Singlet" oxygen reverts to normal "triplet" oxygen by a bimolecular reaction that produces chemiluminescence but is sufficiently long-lived to be able to react as a dienophil; it combines with anthracenoid hydrocarbons at their *meso* positions, or adds to suitable 1,3-dienes.

Singlet oxygen has been produced (i) by the reaction of alkaline hydrogen peroxide with either sodium hypochlorite or bromine and (ii) by decompositions of alkaline solutions of per-acids, either alone or with added hydrogen peroxide. By these methods the *endo*-peroxides (the "photoperoxides") of a number of 9,10-di-substituted anthracenes have been prepared in the dark, whilst 9-substituted anthracenes have given typical hydrolysis products of the primary *endo*-peroxides. The most effective route uses the H_2O_2–Br_2 reaction, but gives side-products owing to concurrent bromine addition unless a two-phase reaction mixture is used. With the H_2O_2–NaOCl system the dienophilic addition of oxygen to lumisteryl acetate and to 2,4-cholestadiene has also been established.

Similar evidence is given to show that the chemiluminescent Trautz reaction also yields singlet oxygen.

SPECTROSCOPISTS have shown that molecular oxygen can exist for finite periods of time in a number of different electronic states. Of these the ground state ($^3\Sigma g^-$) is paramagnetic, having two unpaired electrons with parallel spins in degenerate V_π orbitals, whilst the more energy-rich $^1\Delta g$ state, with which the visible luminescence of excited oxygen molecules is probably associated,[1] contains no unpaired electron. These allotropes of oxygen differ chemically as well as physically. Thus ground-state, or "triplet," oxygen can rationally be written as ·O–O· because it is a stable biradical; its chemical reactions are of free-radical type since they normally proceed by chain mechanisms whilst again the vast majority of the reactions of molecular oxygen

* Read at the Chemical Society Meeting in Nottingham, September 1965.

[1] R. J. Browne and E. A. Ogryzlo, *Proc. Chem. Soc.*, 1960, A, **254**, 317.

are initiated by free radicals or radical-producing catalysts.

$$R-H + Cat. \longrightarrow R\cdot \qquad (1)$$

$$\left.\begin{array}{l} R\cdot + \cdot O-O\cdot \longrightarrow R-O-O\cdot \qquad (2) \\ R-O-O\cdot + H-R \longrightarrow R-O-O-H + R\cdot \qquad (3) \end{array}\right\} \text{ chain}$$

In contrast the $^1\Delta g$, or " singlet " form of oxygen can be written as O=O to indicate its structural resemblance to ethylene, and the experimental evidence which we present below shows that singlet oxygen is indeed a dienophil, capable of immediate addition to conjugated 1,3-dienes and to anthracenoid hydrocarbons, no irradiation or catalyst being needed.

$$(4)$$

Whereas triplet oxygen is produced directly from one-electron reaction of hydrogen peroxide, such as (5), (6),

$$\{Fe(CN)_6\}^{3-} + {}^-O-OH \longrightarrow \{Fe(CN)_6\}^{4-} + \cdot O-OH \qquad (5)$$

$$\{Fe(CN)_6\}^{3-} + \cdot O-OH \longrightarrow \{Fe(CN)_6\}^{4-} + \cdot O-O\cdot + H^+ \qquad (6)$$

singlet oxygen must be formed initially from its heterolytic (i.e., electron pair) decompositions such as (7),[2]

$$(7)$$

and we have already [3] drawn attention to the fact that only the heterolytic reactions are chemiluminescent. This light emission is due to the rapid conversion of singlet oxygen into triplet oxygen and most probably involves [4] a bimolecular reaction (8), for this, unlike a

$$2O_2(^1\Delta g) \longrightarrow 2O_2(^3\Sigma g-) + h\nu \qquad (8)$$

unimolecular spin inversion, would not be spectroscopically " forbidden."

Now when peroxides decompose heterolytically to give singlet oxygen in solution these active molecules are produced separately and consequently it is possible for reactions between singlet oxygen and appropriate substrates (i.e., reactions of type 4) to be kinetically competitive with the bimolecular reaction (8). We have confirmed this deduction by effecting a number of reactions of type (4). Our results substantiate and extend the scope of the work of Foote and Wexler [5] who have been independently investigating oxidations of olefins and conjugated dienes by hydrogen peroxide-hypochlorite mixtures and whom we thank for interchange of information.

The deduction that reaction (8) can be kinetically competitive with reactions of type (4) is also valid if singlet oxygen has been formed at low concentration

in the gas phase. This ($^1\Delta g$) active form of oxygen can be produced by passing an electrodeless discharge through oxygen gas and Corey and Taylor [6] have recently shown that by activating oxygen in this way it is possible to obtain anthracene endo-peroxide from anthracene in the dark. This provides further substantiation for our basic theory which, as Foote, Wexler, and Ando pointed out,[7] shows that photosensitised autoxidations, such as the formation of the endo-peroxide from anthracene and oxygen in sunlight, are in reality reactions of singlet oxygen and not addition reactions of triplet forms of organic molecules and normal triplet oxygen. Indeed our own reactions, which can be carried out in the dark, all yield the same products as those which can be obtained from the same substrates by photosensitised autoxidation.[8] One of our procedures (the use of bromine and alkaline hydrogen peroxide) provides an effective and much more rapid alternative procedure to the photosensitised autoxidation of hydrocarbons of the anthracene type.

The procedures involving the use of organic peracids are more of theoretical than practical significance for they serve to substantiate the reaction mechanism which have been advanced to explain the instability of alkaline solutions of organic peroxides. These reactions proceed too slowly for the concurrent emission of chemiluminescence to be conclusively detectable.

Use of Sodium Hypochlorite–Hydrogen Peroxide Mixtures.—This, our first procedure, has also been studied by Foote et al. The organic substrate, dissolved or slurried in methanol, is mixed with an excess of aqueous hydrogen peroxide, stirred vigorously, and cooled. Concentrated aqueous sodium hypochlorite is then added gradually from the capillary tip of a burette which dips far below the surface of the stirred mixture so that the oxygen as it is formed by reaction (7) is rapidly disseminated through the solution. When an equivalent of hypochlorite has been added the mixture is diluted with water and the organic products are separated and purified. This procedure is not very efficient, mainly because the dilution of the methanol with the aqueous reagents necessitates the use of low concentrations of organic substrates. Thus anthracene was too insoluble in aqueous methanol to yield any endo-peroxide, but from the more soluble 9-methyl-10-phenylanthracene a 13% yield of endo-peroxide based on the hydrocarbon taken was obtained. From 2,5-dimethylfuran, which had been examined by Foote and Wexler,[5] we obtained about a 10% yield of a mixture of 5-hydroxy-2,5-dimethyl-2-hydroperoxyfuran together with a little 2-hydroperoxy-5-methoxy-2,5-dimethylfuran (Foote and Wexler's main product; they had used a higher methanol–water ratio than ourselves).

Ergosterol failed to show any sign of reaction but its acetate reacted to a slight extent. From lumisteryl

[2] A. H. Khan and M. Kasha, *J. Chem. Phys.*, 1963, **39**, 2105.
[3] E. McKeown and W. A. Waters, *Nature*, 1964, **203**, 1063.
[4] A. M. Viner and K. D. Bayes, *J. Phys. Chem.*, 1966, **70**, 302.
[5] C. S. Foote and S. Wexler, *J. Amer. Chem. Soc.*, 1964, **86**, 3879, 3880.

[6] E. J. Corey and W. C. Taylor, *J. Amer. Chem. Soc.*, 1964, **86**, 3881.
[7] C. S. Foote, S. Wexler, and W. Ando, *Tetrahedron Letters*, 1965, **46**, 4111.
[8] P. F. Southern and W. A. Waters, *J. Chem. Soc.*, 1960, 4340.

acetate however the β epidioxide was isolated in 13% yield together with a small amount of 3-β-hydroxy-lumista-5,8(9),22-trien-7-one, a product which had previously been obtained by Bladon [9] from the photo-oxidation of lumisteryl acetate using eosin as a sensitiser.

With crude cholestadiene the 2,4-diene proved to be much more reactive than the 3,5-isomer and yielded 5% of epidoxide together with decomposition products which contained both hydroxyl and carbonyl groups (cf. Conca and Bergmann [10]). It must be remembered that the direct attack on steroid dienes by hydrogen peroxide cannot entirely be discounted.

Use of Bromine and Alkaline Hydrogen Peroxide.— Alternatives to sodium hypochlorite were examined to obviate the use of a high percentage of water in the re-action mixtures and it was found that the heterolytic decomposition of alkaline hydrogen peroxide by bromine (eqn. 9) was visibly chemiluminescent; the reacting

$$\overset{\curvearrowleft}{\text{O--O--H}} + \text{Br} \overset{\curvearrowright}{-}\text{Br} \longrightarrow \text{O=O} + \text{HBr} + \text{Br}^-$$

(9)

substances give even more distinctly the same orange-red glow as do hydrogen peroxide–hypochlorite mixtures. Solutions of anthracenoid hydrocarbons in alcoholic potassium hydroxide were stirred vigorously in the dark whilst, from two burettes with capillary tips, hydrogen peroxide and bromine were added simultaneously below the level of the rotating liquid, the amount of peroxide added being kept always in excess of the amount of bromine. The products and yields so obtained from 9,10-disubstituted anthracene hydrocarbons are in the Table.

Yields from H_2O_2–Br_2 and anthracenes

	(I)	(II) (%)	
	(I)	$R^3 = R^4 =$	$R^3 = OEt$;
Hydrocarbon	(%)	—OEt	$R^4 = $—OH
9,10-Dimethyl-anthracene	20—27	34	14
9,10-Diphenyl-anthracene *	46	Nil	Nil
9-Methyl-10-phenyl-anthracene	34	11	12 (mixture)
		$R^3 = R^4 = R^3 = $—OMe —OMe	
,, †	44	8	$R^4 = $—OOH(—OH) }13

* In chlorobenzene.   † In methanol.

It can be seen that in this way substantial yields of *endo*-peroxides (I) can regularly be formed. The other products of these reactions in alcohol were the alkoxy- or hydroxy-adducts (II). These are undoubtedly formed by *meso* addition of bromine to the hydro-carbons followed by its nucleophilic replacement by alkoxy- or hydroxy-groups. Tests showed that products

of type (II) are not formed by the action of alcoholic alkali on the *endo*-peroxides (I). However the isolation of product (II) necessitates the consideration of altern-ative, though less likely, mechanisms of the formation of *endo*-peroxides. A *trans*-dibromo-adduct (III) could possibly react with −O·OH anions to give a *cis*-hydro-peroxide (IV) which might cyclise to (I) but this ring-closure would require retention of configuration at the C–Br bond in (III). Alternatively some *cis*-adduct (III) might have been formed in which event both the reactions (III) ⟶ (IV) and (IV) ⟶ (I) could be inversions of S_N2 type.

Experiments with 9-methyl-10-phenylanthracene have shown however that this alternative mechanism can be discounted, for if bromine, either with or without alkali, is first added to the alcoholic solution of the hydrocarbon and the hydrogen peroxide is added sub-sequently, then no *endo*-peroxide (I) is formed though the formation of the adduct (III) was found to be rapid. Thus peroxide formation, under conditions in which the singlet form of oxygen is present, must be a one-step process (4) that is sterically similar to a Diels–Alder reaction.

An even better procedure than that outlined above is to use a two-phase system so that the singlet oxygen is formed in the lower (aqueous) phase and rises through an upper organic, but oxygen resistant, liquid containing the oxidisable substrate where oxidation but not bromination occurs. This is exemplified by the highly successful oxidation of sparingly soluble 9,10-diphenyl-anthracene to its endo-peroxide in 46% yield. A solu-tion of the hydrocarbon in chlorobenzene was floated on top of strong aqueous hydrogen peroxide made alka-line with potassium hydroxide and bromine was intro-duced gradually into the gently stirred lower phase at such a rate that the oxygen which was evolved did not carry up with it any free bromine. No adduct of type (II) was then formed.

When the hydrogen peroxide–bromine reaction (9) was extended to 9-substituted anthracenes *endo*-peroxides (V) did not eventuate, 9-hydroxyanthrone derivatives (VI) being isolated instead. We suggest

[9] P. Bladon, *J. Chem. Soc.*, 1955, 2176.
[10] R. J. Conca and W. Bergmann, *J. Org. Chem.*, 1953, **18**, 1104.

that the peroxides (V) may have been intermediates, and that in our strongly alkaline media these were rapidly attacked by base to give the final products (VI).

This reaction, and also the formation of hydroxy- or alkoxy-adducts (II), show the limitation of the hydrogen peroxide–bromine method of oxidising conjugated dienes by singlet oxygen. In view of these limitations the method was not tested for the oxidation of alicyclic dienes.

Reactions of Alkaline Solutions of Organic Peracids.— Organic peracids are less stable in alkaline than in acidic solutions. Perbenzoic acid in aqueous solutions evolves oxygen most rapidly when pH = pK_a, *i.e.*, when the acid is 50% dissociated.[11] We have confirmed this also for peracetic acid. Mechanism (10) accounts satisfactorily for this decomposition; it should yield singlet oxygen.

$$(10)$$

At higher pH's the rate of evolution of oxygen decreases again and peracids can be decomposed by strong alkali to give hydrogen peroxide (11), but in dilute

$$(11)$$

alkali, solutions of peracids containing free hydrogen peroxide give off oxygen, the pH of minimum stability being *ca.* 10, conditions in which the peracid is present almost entirely as its anion whilst hydrogen peroxide is still largely undissociated. A mechanism (12) similar to (10), can be written here, but Akiba and Simamura,[12]

$$(12)$$

who by studies with oxygen-18 have shown that the oxygen comes mainly from the hydrogen peroxide, prefer mechanism (13) though clearly the direction of the electron movements could be reversed without detriment to their concept.

[11] J. F. Goodman, P. Robson, and E. R. Wilson, *Trans. Faraday Soc.*, 1962, **58**, 1846.

$$(13)$$

Di-isoperoxyphthalic acid was selected to show that reaction (10) gives singlet oxygen. It was added to a solution of 9-methyl-10-phenylanthracene in ethanol, maintained at pH 8—9 by the addition of potassium hydroxide and the reaction mixture was kept in the dark for 48 hours. On working up, 9,10-dihydro-9-methyl-10-phenylanthracene *endo* peroxide (I; $R^1 = Me$, $R^2 = Ph$) was isolated in 21% yield calculated on the hydrocarbon or 10% yield calculated on the available oxygen. However peracetic acid at pH 8 failed to oxidise this hydrocarbon, but at pH 10 with addition of some hydrogen peroxide a small amount of (I; $R^1 = Me$, $R^2 = Ph$) was formed.

Definite success in the application of reaction (13) to the formation of *endo*-peroxides was later achieved by using *p*-peroxytoluic acid and hydrogen peroxide at pH 10 for the oxidation of 9-methyl-10-phenylanthracene in ethanol, an 18% yield of the desired oxidation product being obtained.

Other Reactions.—We reported[3] that the reaction between alkaline hydrogen peroxide and nitriles (14) is chemiluminescent, but, as yet, we have failed to use this

$$(14)$$

reaction for the addition of singlet oxygen to an anthracenoid hydrocarbon or to a diene. Possibly reaction (14) like (13) requires specific pH control.

Another reaction which has been shown spectroscopically to evolve singlet oxygen[13] is the Trautz reaction, *i.e.*, the oxidation of formaldehyde by alkaline hydrogen peroxide in the presence of pyrogallol. This is an exceedingly vigorous reaction and its orange-red chemiluminescence can easily be seen in a dimly lit room, but when it was carried out in the presence of alcoholic methylphenylanthracene no *endo*-peroxide was formed. However a number of other phenols can be used in place of pyrogallol and with resorcinol the reaction, though less vigorous, did give a 5% yield of *endo*-peroxide from 9-methyl-10-phenylanthracene. We therefore support Bowen's[13] view that the Trautz reaction does yield singlet oxygen.

We pointed out that the addition of singlet oxygen to 1,3-dienes had to be viewed as a process competitive with the chemiluminescent conversion of active singlet oxygen into stable triplet oxygen. The relative rates of

[12] K. Akiba and O. Simamura, *Chem. and Ind.*, 1964, 705.
[13] E. J. Bowen, *Pure and Appl. Chem.*, 1964, **9**, 477.

these concurrent reactions will undoubtedly be solvent-dependent as well as component-dependent and so prolonged researches would be needed to obtain the maximum yield of peroxides in the reactions which we have examined. Again our reacting mixtures contain several components (peroxide, halogen, alkali, etc.) owing to which side reactions can occur. So at this stage we consider that the significance of our study of singlet oxygen lies almost entirely in its theoretical import, though we feel that it well may be possible to improve considerably the yields in reactions such as those which we have examined. However our findings may have significance in relation to other aspects of peroxide chemistry. Thus reactions of singlet oxygen should not be excluded from consideration with reference to oxidation effected by peroxides in alkaline media amongst which some enzymic reactions could be included (cf. wider implications of the Trautz reaction). Wider implications of the now proven mechanism of photosensitised oxidations can be envisaged throughout the whole of biological chemistry.

EXPERIMENTAL

Use of Sodium Hypochlorite and Hydrogen Peroxide.—
Oxidation of 9-methyl-10-phenylanthracene. A slurry of the hydrocarbon (2 g., ca. 7·5 mmoles) in methanol (150 ml.) was stirred vigorously with sodium hypochlorite solution (14 ml., 10—14% available Cl, ca. 50 mmoles) in a blackened flask, and aqueous hydrogen peroxide (30 ml., equivalent to 20 mmoles H_2O_2) was added below the surface from a burette with a capillary tip. When the addition had been completed the yellow solid product was collected (1·76 g.) and separated by chromatography through alumina with light petroleum–benzene into unchanged 9-methyl-10-phenylanthracene (1·23 g., m. p. 114—115°) and 9,10-dihydro-9-methyl-10-phenylanthracene 9,10-*endo*-peroxide [0·29 g., m. p. 181—183° (decomp.)] (Found: C, 84·0; H, 5·4. Calc. for $C_{21}H_{16}O_2$: C, 84·0; H, 5·3%). The infrared spectrum had no band corresponding to OH or C=O absorption but had a weak band at 875 cm.$^{-1}$ indicative of O–O vibration.[14]

Oxidation of 2,5-dimethylfuran. Freshly distilled 2,5-dimethylfuran (9·6 g.), methanol (250 ml.), and hydrogen peroxide (30 ml. of 100 vol. = 3 equiv.) were stirred together in the dark and sodium hypochlorite solution (125 ml., 4 equiv.) was added from a capillary below the surface of the liquid. After 2 hr. a deposit of sodium chloride was removed and the liquid was diluted with water (500 ml.) and extracted with ether and the extract was evaporated at room temperature leaving a syrup (1·4 g.) which eventually solidified (m. p. 40—72°). On treatment with benzene–ether it yielded a small quantity of a white solid, which when crystallised from ether gave needles, m. p. 132—135° (decomp.) of 2-hydroperoxy-5-hydroxy-2,5-dimethylfuran [lit.,[5] m. p. 134—136° (decomp.)] (Found: C, 49·1; H, 7·0. Calc. for $C_{10}H_{10}O_4$: C, 49·3; H, 6·9%. Available O = 96% of theory]. Its infrared spectrum had a shoulder at 3500 cm.$^{-1}$ (OOH) and a peak at 3600 cm.$^{-1}$ (OH) and its n.m.r. spectrum had singlet peaks at 4·2, 4·9, and 8·7 τ with relative areas 2 : 2 : 6. The OH and O·OH groups apparently undergo rapid proton exchange and so only one line appears (at 4·9 τ).

Oxidations of steroids. Ergosterol did not appear to react with H_2O_2–NaOCl but its acetate (1 g.) gave 10% of a product which was shown by thin-layer chromatography to contain at least 3 components. Lumisterol (m. p. 117—119°) on acetylation (Ac_2O in pyridine) gave both crystals of m. p. 95—98° (lit.,[15] m. p. 100°) and of m. p. 109—111° (lit.,[16] 109—111°). Each modification gave a single spot by thin-layer chromatography on alumina. To the pure acetate (1 g.) in ethanol (170 ml.) and hydrogen peroxide (10 ml. of 100 vol.) sodium hypochlorite solution (30 ml.) equivalent to the peroxide was added, in the dark, during 1 hr. The mixture was then diluted with water (600 ml.) and extracted with ether. The resulting solid (1·02 g., m. p. 104—144°) was chromatographed through deactivated alumina. Benzene extracted unchanged lumisteryl acetate (0·52 g.) and ether–benzene (1 : 19) then separated 3-β-acetoxy-5β,3β-epidoxylumista-6,22-diene (0·14 g.) which after crystallisation from acetone had m. p. 158—159°, $\alpha_D = +40°$ (0·67% in $CHCl_3$ at 18°); Bladon[9] gives m. p. 154—159°, α_D +46·3° (Found: C, 76·3; H, 9·8. Calc. for $C_{30}H_{46}O_4$: C, 76·55; H, 9·85%). Its ultraviolet spectrum had no peak at 273 or 281 mμ indicating unsaturation and its infrared spectrum had a peak at 840 cm.$^{-1}$ (–O–O–link) but no absorption indicative of hydroxyl or ketone groups. Further elution with ether gave a yellow solid (0·05 g.) and eventually a resin [0·19 g., m. p. ca. 115° (decomp.)]. From the former, by crystallisation from ether, a small quantity (2·3 mg.) of impure 3-β-hydroxylumista-5,8(9),22-trien-7-one was obtained (m. p. 228—230° with a change of form before melting), $\lambda_{max.}$ 250 mμ, (ε 10,250) 3380 (OH), 1652, 1600 with shoulder at 1605, and 875 cm.$^{-1}$: Bladon[9] records similar m. p. and spectra.

Cholestadiene, made by dehydration of cholesterol on activated alumina, was a mixture of 3,5-, (20%) and 2,4-isomers (80%), as assessed from their characteristic ultraviolet spectra. The mixture (5 g.) in chloroform (100 ml.) was stirred in the dark with hydrogen peroxide (40 ml. of 100 vol.) and 1 equiv. of aqueous sodium hypochlorite was added during 90 min.; the solution warmed appreciably during the reaction. The chloroform solution was then separated, dried, and evaporated to leave a pale yellow syrupy (4·75 g.) whose ultraviolet spectrum indicated that 65% of the 2,4-diene had reacted but only 25% of the 3,5-diene. Chomatography through deactivated alumina removed unchanged cholestadienes (1·8 g.), then a solid (0·47 g.), m. p. 94—105° with an infrared spectrum indicative of a non-conjugated ketone (peaks at 1715 and 1620 cm.$^{-1}$) possibly containing an epoxy-group (sharp peak at 890 cm.$^{-1}$). Following this, the next fractions (0·26 g.) both gave peroxide reactions (KI in HOAc) and on crystallisation from methanol gave 2α : 5α-epidioxycholest-3-ene, m. p. and mixed m. p. 110—112° (Found: C, 80·5; H, 10·6. Calc. for $C_{27}H_{44}O_2$: C, 81·0; H, 11·0%). We thank Dr. P. Gartside for the authentic specimen. Fractions eluted with greater difficulty proved to be hydroxy-ketones; these were not examined.

Use of Bromine and Hydrogen Peroxide.—9-Methyl-10-phenylanthracene (1 g.) was dissolved in a hot mixture of methanol (50 ml.) and ethanol (100 ml.). Potassium hydroxide (5·6 g.) in methanol (100 ml.) was added and to the solution, stirred vigorously in the dark, hydrogen peroxide (35 ml. of 100 vol.) and bromine (5 ml.) were added simul-

[14] G. J. Minkoff, *Discuss. Faraday Soc.*, 1950, **9**, 851.
[15] A. Windaus, K. Dithmar, and E. Fernholtz, *Annalen*, 1932, **493**, 259.
[16] G. A. Fletcher, Ph.D. Thesis (Manchester), 1956, p. 63.

taneously from capillary tubes placed well below the surface of the stirred mixture. The hydrogen peroxide, a 3-fold excess, was added so as to be always in preponderance. Vigorous evolution of oxygen could be seen if the stirring was interrupted. When half the reactants had been added some free bromine could be seen in the liquid and to remove this 120 ml. of N-sodium hydroxide were added; the mixture was then diluted with water (500 ml.) and extracted with ether. The extract yielded a white solid (1·12 g.), m. p. 153—173° (decomp.), which on crystallisation from ethanol gave 9,10-dihydroanthracene 9,10-*endo*-peroxide (0·42 g.), m. p. 178—182° (decomp.) (Found: C, 83·4; H, 5·3%). The mother-liquors, after evaporation and chromatography through alumina, gave further *endo*-peroxide (total yield 0·5 g.), then (a) 9,10-*dihydro*-9,10-*di-methoxy*-9-*methyl*-10-*phenylanthracene* (0·12 g.), m. p. 210—212° (Found: C, 83·0; H, 6·7; OMe, 20·5. C₂₃H₂₂O₂ requires C, 83·4; H, 6·7; OMe, 19·1%) and (b) probably 9,10-*dihydro*-10(or 9)-*perhydroxy*-9(or 10)-*methoxy*-9-*methyl*-10-*phenylanthracene* (0·15 g.), m. p. 150—153° from ether (Found: C, 77·5; H, 6·0. C₂₂H₂₀O₃ requires C, 79·2; H, 6·0%). Substance (a) had no ultraviolet anthracene absorption and no infrared band indicative of OH or CO, but had strong absorptions at 1070, 1090, 1160, and 1250 cm.⁻¹ and weak absorption at 1220 cm.⁻¹ (C–O–C), while substance (b) absorbed at 3500 and 1095 cm.⁻¹ indicative of OH or O·OH; neither contained bromine.

9-Methyl-10-phenylanthracene (7 g.) and potassium hydroxide (80 g.) were dissolved in ethanol (575 ml.), and hydrogen peroxide (245 ml. of 100 vol.) and bromine (35 ml.) were added simultaneously in the dark. The addition took 3 hr. and the mixture remained at 50—60°. Dilution with water and ether extraction yielded a yellow solid (8·3 g.) which was chromatographed through alumina. Elution with light petroleum–benzene (4 : 1) first gave a yellow solid (1 g.) which by further chromatography was separated into the stereoisomeric (or mesomorphic) forms of 10-dihydro-9,10-diethoxy-9-methylanthracene. *Product A* (0·28 g.) had m. p. 150—151·5° from light petroleum (change of form before melting) (Found: C, 82·5; H, 7·2. C₂₅H₂₆O₂ requires C, 83·7; H, 7·3%). It had no anthracene chromophore and infrared bands diagnostic of ether links at 1040, 1080, 1090, and 1162 cm.⁻¹ whilst its n.m.r. spectrum in CDCl₃ had bands at 2·1—2·9 (aromatic protons), 6·65—6·72 (quartet split into doublets owing to CH₂ of OEt), 8·25 (singlet due to Me), and 8·7—9·1 (triplet split into doublets due to CH₃ of OC₂H₅); the relative peak areas were 13 : 4 : 3 : 6.

Product B (0·18 g.) gave large crystals, m. p. 204°, from benzene or light petroleum, which changed to distinct platelets before melting (Found: C, 83·3; H, 7·4%). Its n.m.r. spectrum was identical with that of isomer A and its infrared spectrum had bands at 1070, 1090, and 1260 cm.⁻¹.

Further elution of the product with benzene gave the expected *endo*-peroxide (2·36 g.), m. p. 177—182° (decomp.) and then a soft yellow solid (1·3 g.), m. p. 89—147°, which was a complex mixture.

9,10-Dimethylanthracene. The preparation of this substance gave unexpected difficulty; the reduction of anthracene (15 g.) in tetrahydrofuran (15 g.) with sodium (15 g.)

¹⁷ B. M. Mikhailov and T. K. Koxminskaya, *Doklady Akad. Nauk S.S.S.R.*, 1947, **58**, 811 (*Chem. Abs.*, 1951, **45**, 9522g).
¹⁸ G. M. Badger, F. Goulden, and F. L. Warren, *J. Chem. Soc.*, 1941, 18.

in liquid ammonia and treatment of the resulting solution with methyl iodide (45 g.) gave mainly 9,10-dihydro-9,9,10,10-tetramethylanthracene,¹⁷ m. p. 170—171·5° (8·3 g.), a little 9,10-*dihydro*-9,9,10-*trimethylanthracene*, m. p. 83—84° (1 g.) [Found: C, 91·8; H, 7·9%; *M*, 214. C₁₇H₁₈ requires C, 91·9; H, 8·1%; *M*, 220); its n.m.r. spectrum showed aromatic protons (τ 2·5—2·9), hydrogen (quartet at τ 5·7—6·2), and methyl groups (triplet at τ, 8·4—8·7), in the ratio 8 : 1 : 9] and only a similar amount of 9,10-di-methylanthracene. It was finally prepared by the method of Badger, Goulden, and Warren,¹⁸ in about 10% yield (m. p. 180—181°).

The reaction as above of 9,10-dimethylanthracene (2 g.) in ethanol (150 ml.) containing potassium hydroxide (11·2 g.) with hydrogen peroxide (30 equiv.) and bromine (10 equiv.) gave first 9,10-diethoxy-9,10-dihydro-9,10-dimethylanthracene, m. p. 191° (1·03 g.), and then 9,10-di-*hydro*-9,10-*dimethylanthracene* 9,10-*endo*-*peroxide*, m. p. 221° (0·48 g.) (Found: C, 80·4; H, 5·8. C₁₆H₁₄O₂ requires C, 80·9; H, 5·9%; available O = 97% of theory). The former compound (II; R¹ = R² = Me, R³ = R⁴ = OEt) had infrared absorptions at 1081, 1105, and 1205 cm.⁻¹ with weak bands at 1135 and 1205 cm.⁻¹, and its n.m.r. spectrum had bands at 2·3—2·8 τ (aromatic protons), 6·7—7·2 τ (quartet due to CH₂ of OEt), 8·5 τ (singlet due to protons of C–CH₃), and 8·85—9·1 τ (triplet due to protons of CH₃ in OEt group), the relative peak areas being 8 : 4 : 6 : 6, consistent with the assigned structure (Found: C, 80·9; H, 8·4; OEt, 30·2. C₂₀H₂₄O₂ requires C, 81·2; H, 8·1; OEt, 30·4%). The final material (0·37 g.) to be eluted from this mixture, m. p. 124—128°, was a mixture containing 9% of bromine and its infrared spectrum showed the presence of OH groups (bands at 3400 and 3502 cm.⁻¹) whilst its n.m.r. spectrum was consistent with the presence of one OH and one OEt group. This accords with compound (II; R¹ = R² = Me, R³ = OH, R⁴ = OEt), admixed with about 25% of product brominated in an aromatic nucleus.

9,10-Diphenylanthracene, by similar treatment in ethanol suspension, gave only traces of the *endo*-peroxide.

9-Methylanthracene, when oxidised as for 9-methyl-10-phenylanthracene, gave a mixture the main component of which proved to be 10-hydroxy-10-methylanthrone, m. p. 156·5° (lit.,¹⁹ 154°) identified by its infrared and n.m.r. spectra (Found: C, 80·3; H, 5·4%; *M*, 237. Calc. for C₁₅H₁₂O₂: C, 80·3; H, 5·4%; *M*, 224).

9-Phenylanthracene, by similar treatment, reacted to the extent of only 30%. The main products were 9-*ethoxy*-9-*phenylanthrone*, m. p. 159° (5%) and 9-hydroxy-9-phenyl-anthrone, m. p. 215° (lit.,²⁰ 214°) (10%). Of these the former had the requisite infrared spectrum and a n.m.r. spectrum showing peaks at 1·4—1·7 and 2·4—2·9 (aromatic protons), 6·6—7·1 (quartet from CH₂ of OEt) and 8·55—9 τ (triplet due to CH₃ of OEt) in relative areas of 13 : 2 : 3 (Found: C, 83·8; H, 5·7; OEt, 14·5. C₂₂H₁₈O₂ requires C, 84·1; H, 5·7; OEt, 14·3%). The latter had a n.m.r. spectrum showing aromatic protons at 1·5—1·75 τ and a singlet at 4·6 τ (OH) (Found: C, 83·7, H, 5·3. Calc. for C₂₀H₁₄O₂: C, 83·8; H, 4·9%).

Reaction of 9,10-diphenylanthracene in chlorobenzene. Potassium hydroxide (11·2 g.) was dissolved slowly, with cooling, in 100 vol. hydrogen peroxide (35 ml.) and 9,10-di-phenylanthracene (1 g.) in chlorobenzene (50 ml.) was

¹⁹ P. L. Julien, W. Cole, and G. Diemer, *J. Amer. Chem. Soc.*, 1945, **67**, 1721.
²⁰ E. de B. Barnet and J. W. Cook, *J. Chem. Soc.*, 1923, 2631.

J. Chem. Soc. (B), 1966

floated on this solution. The mixture was stirred gently so that the boundary between the layers was not broken and, in the dark, bromine (5 ml.) in chloroform (5 ml.) was added to the lower liquid through a capillary tube at such a rate that bromine colour did not appear in the chlorobenzene layer; the addition took 1·5 hr. The organic layer was separated, evaporated under reduced pressure, and chromatographed through alumina (100 g.). Light petroleum–benzene separated unchanged diphenylanthracene (0·51 g.), and then benzene separated 9,10-diphenylanthracene 9,10-*endo*-peroxide (0·5 g.), which on heating decomposed completely into diphenylanthracene before melting (decomp. 130—180°) (Found: C, 86·4; H, 5·1. Calc. for $C_{26}H_{18}O_2$: C, 86·2; H, 5·0%).

Tests of the Reaction Mechanism.—After being boiled for 30 min. a 1% solution of the *endo*-peroxide of 9-methyl-10-phenylanthracene in a 5% solution of potassium hydroxide in ethanol, 97% of the peroxide was recovered. A similar recovery was effected after boiling for 1 hr. with a mixture of 100 vol. hydrogen peroxide (35 ml.), ethanol (150 ml.), and potassium hydroxide (5 g.).

9-Methyl-10-phenylanthracene (2 g.) and potassium hydroxide (11·2 g.) in ethanol (150 ml.) were treated with bromine (5 ml.) and after a few min. hydrogen peroxide (35 ml. of 100 vol.) was gradually added with stirring. On working up this mixture yielded a yellow solid (m. p. 95—185°) which, after thorough chromatographic study, was found not to contain any *endo*-peroxide which can be eluted easily with light petroleum–benzene.

9-Methyl-10-phenylanthracene (1 g.) in chloroform was treated dropwise with bromine (3% in chloroform until the colour persisted) (0·6 g. $Br_2 = 1$ mol.). After 15 minutes' stirring, a solution of potassium hydroxide (11·2 g.) in ethanol (100 ml.) was added and then hydrogen peroxide (35 ml. of 100 vol.) was dripped in; on working up the alkaline solution a pale yellow solid (0·68 g.) was obtained which on crystallisation from benzene melted at 132°, then solidified and remelted at 143°. Its ultraviolet spectrum showed $\lambda_{max.}$ at 263 mμ (ε 72,400) and absorptions at 350, 370, and 390 mμ whilst its infrared spectrum had peaks at 892, 1410, and 1785 cm.⁻¹ (C=CH₂), and a very weak peak at 1680 cm.⁻¹. Its n.m.r. spectrum in CDCl₃ had peaks at 1·5—2·9 τ (aromatic protons), 2 (protons separated from the main multiplet), a sharp singlet at 4·5 τ (CH₂:C), a quartet at 6—6·4 τ CH₂' in OEt), and a triplet at 8·55—8·8 τ (CH₃ of OEt), their relative intensities being 13 : 2 : 2 : 3. This identifies the product as 9-*ethoxy*-9,10-*dihydro*-10-*methylene*-9-*phenylanthracene* (Found: C, 88·1; H, 6·4. $C_{23}H_{20}O$ requires C, 88·2; H, 6·4%) which clearly has been formed by the attack of alcoholic potash on (III). No *endo*-peroxide was found in this mixture but minor fractions of it were found to contain bromine.

Use of Per-acids in Alkali.—9-Methyl-10-phenylanthracene (1·5 g.) in ethanol (200 ml.) was added to di-isoperphthalic acid (72% purity, 2 mol.) and to the stirred mixture, in the dark, 4% alcoholic potassium hydroxide was added to make the pH 8—9. The pH was controlled by further addition of alkali during 4 hr. and the mixture was stored in the dark for another 48 hr. by which time the pH had fallen to 6·7. Alkali was then added to take the pH to 10 and the mixture was diluted with water (1 l.) and extracted with ether. The solid thus obtained was chromatographed whereupon it yielded unchanged, fluorescent, 9-methyl-10-phenylanthracene (0·64 g.) and 9,10-dihydro-9-methyl-10-phenylanthracene 9,10-*endo*-peroxide (0·29 g., m. p. 180—183°); yield 21% based on hydrocarbon or 10% on available oxygen.

9-Methyl-10-phenylanthracene (0·8 g.) and *p*-peroxytoluic acid (4 g., 67% pure) were dissolved in ethanol (100 ml.) and 4% alcoholic potassium hydroxide was added to bring the solution to pH 10. The flask was blacked out and hydrogen peroxide (5 ml. of 100 vol.) was added with stirring during 45 min. The pH was then readjusted to 10 and the mixture was stirred for 1 hr. Dilution with water and ether extraction yielded a yellow solid which in chromatography gave unchanged methylphenylanthracene (0·07 g.) and its *endo*-peroxide [0·16 g., m. p. 183° (decomp.)]. This reaction was repeated with peracetic acid and hydrogen peroxide in methanol at pH 10. 70% of unchanged hydrocarbon was recovered together with only 2% of the *endo*-peroxide.

Peracetic acid in a sodium hydrogen carbonate buffer at pH 8·5 left 80% of the hydrocarbon unchanged and no *endo*-peroxide was isolated.

Other Reactions.—*Trautz reaction.* Pyrogallol (3·8 g.), formalin (10 ml. of 37%), 9-methyl-10-phenylanthracene (1 g.), hydrogen peroxide (10 ml. of 100 vol.), and ethanol (150 ml.) were stirred together in the dark and potassium hydroxide (10 g.) in ethanol was added dropwise. Each drop produced a transient amber glow throughout the liquid; eventually the glow became continuous though no oxygen gas was evolved. Finally the mixture, when poured into water, deposited an insoluble brown tar from which unchanged hydrocarbon was recovered quantitatively.

Modified Trautz reaction. 9-Methyl-10-phenylanthracene (1 g.) in ethanol (100 ml.), hydrogen peroxide (20 ml. of 100 vol.), formalin (10 ml. of 37%), and resorcinol (5 g.) were stirred together at 55° and, in the dark, potassium hydroxide in ethanol (30 ml.) was added dropwise. A slight amber glow was discernible and oxygen was slowly evolved. On dilution with water a brown solid was deposited which after chromatography gave 0·85 g. of unchanged hydrocarbon and 50 mg. of its *endo*-peroxide, m. p. 162—173° (decomp.).

Benzyl cyanide (3 ml.) and hydrogen peroxide (10 ml. of 100 vol.) were stirred in ethanol to give a homogeneous solution to which was added 9-methyl-10-phenylanthracene (2 g.). Potassium hydroxide (2 g.) in aqueous ethanol (10 ml. of 1 : 1) was added slowly whereupon the mixture became hot. After 90 min. the mixture was worked up, 95% of the hydrocarbon being recovered together with phenylacetamide.

A similar mixture was brought only to pH 9 with potassium hydrogen carbonate. On working up after 48 hr. the hydrocarbon was recovered unchanged.

Similar experiments with methyl cyanide in methanol also failed to oxidise the hydrocarbon.

We thank Unilever Ltd. for the peracids. One of us (E. McK.) thanks the Directors of Unilever Ltd. for the Unilever Internal Fellowship.

THE DYSON PERRINS LABORATORY,
OXFORD UNIVERSITY. [6/627 *Received, May 25th*, 1966]

26

Reprinted from *J. Amer. Chem. Soc.*, **90**(4), 975–981 (1968)

Chemistry of Singlet Oxygen. IV.[1] Oxygenations with Hypochlorite–Hydrogen Peroxide[2a]

Christopher S. Foote,[2b] Sol Wexler, Wataru Ando, and Raymond Higgins

Contribution No. 2120 from the Department of Chemistry, University of California, Los Angeles, California 90024. Received September 29, 1967

Abstract: Good acceptors for the dye-sensitized photooxygenation can be oxygenated efficiently by metal hypochlorites and hydrogen peroxide to give products identical with those of photooxygenation. Oxygenations of 2,3-dimethyl-2-butene, $\Delta^{9,10}$-octalin, 2,5-dimethylfuran, tetraphenylcyclopentadienone, 1,3-cyclohexadiene, and anthracene are described. The reactive intermediate is probably $^1\Delta_g$ molecular oxygen. The yield of singlet oxygen depends on solvent and other factors, for reasons which are not yet understood.

Dye-sensitized photooxygenations of organic compounds have been studied extensively.[3] Among the acceptors studied, two types have received particular attention. Conjugated dienoids (cyclic and a few other *s-cis* dienes, polycyclic aromatics, and some heterocycles) undergo addition of oxygen to give (at least as the primary products) 1,4-*endo*-peroxides, in a reaction analogous to Diels–Alder addition (reaction 1). Many olefins with allylic hydrogens produce allylic hydroperoxides with an attendant specific shift of the double bond to the allylic position (reaction 2), in a reaction analogous to the "ene" reaction.[4]

(1) Part III: C. S. Foote, S. Wexler, and W. Ando, *Tetrahedron Letters*, 4111 (1965).

(2) (a) Supported by National Science Foundation Grants No. G-25086, GP-3358, and GP-5835, and by a grant from the Upjohn Company; taken in part from S. Wexler, Ph.D. Thesis, UCLA, 1966; (b) Alfred P. Sloan Research Fellow, 1965–1967.

(3) For leading references, see: (a) G. O. Schenck, *Angew. Chem.*, **69**, 579 (1957); (b) K. Gollnick and G. O. Schenck, *Pure Appl. Chem.*, **9**, 507 (1964); (c) E. J. Bowen, *Advan. Photochem.*, **1**, 23 (1963); (d) A. Nickon, N. Schwartz, J. B. DiGiorgio, and D. A. Widdowson, *J. Org. Chem.*, **30**, 1711 (1965); (e) Yu. A. Arbuzov, *Russ. Chem. Rev.*, **34**, 558 (1965); (f) K. Gollnick and G. O. Schenck in "1,4-Cycloaddition

Reactions," J. Hamer, Ed., Academic Press Inc., New York, N. Y., 1967, p 255.

(4) K. Alder and H. von Brachel, *Ann.*, **651**, 1411 (1962), and earlier papers.

$$\text{(structure)} \xrightarrow[\text{O}_2]{h\nu/\text{dye}} \text{(structure)} \qquad (1)$$

$$X = CH_2, \overset{CH_2}{\underset{CH_2}{|}}, O, \text{ etc.}$$

$$\text{(structure)} \xrightarrow[\text{O}_2]{h\nu/\text{dye}} \text{(structure)} \qquad (2)$$

Several mechanisms have been suggested for the dye-sensitized photooxygenations, but only two are consistent with the kinetic evidence which has been presented.[3] Both mechanisms require the intermediacy of the sensitizer (Sens) in its triplet state.

$$Sens \xrightarrow{h\nu} {}^1Sens$$
$${}^1Sens \longrightarrow {}^3Sens$$

The first mechanism postulates the reaction of ^{3}Sens with oxygen to give a complex of sensitizer and oxygen (Sens $\cdot \cdot$ O$_2$), which transfers oxygen to the acceptor (A) to produce the product (AO$_2$).[3] This mechanism was originally suggested by Schönberg,[5] and was widely accepted until recently.[3]

$${}^3Sens + {}^3O_2 \longrightarrow Sens \cdot \cdot O_2$$
$$Sens \cdot \cdot O_2 + A \longrightarrow AO_2 + Sens$$

Another mechanism (originally suggested by Kautsky[6]) involves energy transfer from ^{3}Sens to oxygen to give singlet molecular oxygen (${}^1O_2$), which reacts with acceptor to form the product peroxide.

$${}^3Sens + {}^3O_2 \longrightarrow Sens + {}^1O_2$$
$${}^1O_2 + A \longrightarrow AO_2$$

Singlet oxygen can be produced by radiofrequency discharge in gaseous oxygen[7] and by the reaction of positive-halogen compounds with hydrogen peroxide[7] among other processes.[8]

Both the ${}^1\Sigma_g^+$ and ${}^1\Delta_g$ states of oxygen are usually produced, and have been identified by their characteristic (although somewhat complex) emission;[7] in addition, the ${}^1\Delta_g$ state gives a characteristic esr absorption in the vapor phase.[9]

The two metastable singlets differ in the electronic configuration of their degenerate highest occupied (antibonding) orbitals. The ${}^1\Delta_g$ state (22 kcal) has both electrons in one orbital, and the other vacant; the ${}^1\Sigma_g^+$ (37 kcal) has one electron in each orbital.[10]

State	Energy (kcal) rel to ground state	Configuration of highest occupied orbitals
${}^1\Sigma_g^+$	37	↑ ↑
${}^1\Delta_g$	22	↑↓ —
${}^3\Sigma_g^-$	Ground state	↑ ↑

Oxygen in the ${}^1\Delta_g$ state is produced in at least 10% yield in the reaction of chlorine with H$_2$O$_2$.[11] This state of oxygen (which resembles ethylene electronically) might be expected to react in two-electron, concerted processes. The ${}^1\Sigma_g^+$ state should resemble the ground state, and would be expected to undergo radical-like reactions. The ${}^1\Delta_g$ state is extremely long-lived (it is known to survive more than 10^8 collisions in the vapor).[12] The ${}^1\Sigma_g^+$ state is shorter lived, and is rapidly quenched by water vapor.[7,13]

We previously reported in preliminary form that an oxidizing species (probably ${}^1\Delta_g$ oxygen) produced in the reaction of sodium hypochlorite and hydrogen peroxide oxygenates many compounds to give products identical with those of the dye-sensitized photooxygenation.[14] Similar observations were made by Corey and Taylor, using singlet oxygen produced by radiofrequency discharge in gaseous oxygen.[15] Since these reports, several investigators have described oxygenations using these methods or similar ones.[8,12b,16] In this paper, the chemical oxygenation of a number of organic substrates which yield a single product on photooxygenation is described from a mainly preparative standpoint, with attention to the effect of certain reaction variables on product yield. A series of papers in preparation will present detailed comparisons of the photooxygenation and hypochlorite–hydrogen peroxide oxygenation in regard to stereoselectivity, substrate reactivity, substituent effects, and mechanism.

Results

Oxygenation of 2,3-Dimethyl-2-butene. Photooxygenation of 2,3-dimethyl-2-butene (tetramethylethylene (I), hereafter referred to as TME) has been reported to produce 3-hydroperoxy-2,3-dimethyl-1-butene (II).[17] The reaction was repeated with rose bengal as sensitizer, in methanol. The properties of the product (isolated by distillation) are in good agreement with those reported, and the infrared and nmr spectra support the assigned[17] structure. Reduction of II with a wide variety of reducing agents produces 3-hydroxy-2,3-dimethyl-1-butene (III). When the crude photooxygenation mixture is reduced, no volatile products other than III can be detected by gas chromatography. The amount of product formed (measured gas chromatographically, using an internal standard) is in excellent agreement with that calculated from the oxygen uptake, which is quantitative when the reaction is carried to completion.

The oxygenation of TME with hypochlorite–hydrogen peroxide takes a very similar course. A methanolic solution of TME containing excess hydrogen peroxide was oxidized by adding dilute aqueous sodium hypochlorite. Somewhat more than 1 equiv of hypochlorite/equiv of olefin was used (the reasons for this are discussed

(5) A. Schönberg, *Ann.*, **518**, 299 (1935).
(6) H. Kautsky, *Biochem. Z.*, **291**, 271 (1937), and earlier papers.
(7) See J. S. Arnold, R. J. Browne, and E. A. Ogryzlo, *Photochem. Photobiol.*, **4**, 963 (1965), and references therein.
(8) E. McKeown and W. A. Waters, *Nature*, **203**, 1063 (1963); *J. Chem. Soc., Sect. B*, 1040 (1966); and H. H. Wasserman and J. R. Scheffer, *J. Amer. Chem. Soc.*, **89**, 3073 (1967), describe other sources of singlet oxygen.
(9) A. M. Falick, B. H. Mahan, and R. J. Myers, *J. Chem. Phys.*, **42**, 1837 (1965).
(10) G. Herzberg, "Molecular Spectra and Molecular Structure. I. Spectra of Diatomic Molecules," 2nd ed, D. Van Nostrand Co., New York, N. Y., 1950, p 560.

(11) R. J. Browne and E. A. Ogryzlo, *Can. J. Chem.*, **43**, 2915 (1965).
(12) (a) R. M. Badger, A. C. Wright, and R. F. Whitlock, *J. Chem. Phys.*, **43**, 4345 (1965); (b) A. M. Winer and K. D. Bayes, *J. Phys. Chem.* **70**, 302 (1966); (c) A. Vallance-Jones and A. W. Harrison, *J. Atmospheric Terrest. Phys.*, **13**, 45 (1958).
(13) L. W. Bader and E. A. Ogryzlo, *Discussions Faraday Soc.*, **37**, 46 (1964).
(14) C. S. Foote and S. Wexler, *J. Amer. Chem. Soc.*, **86**, 3879 (1964).
(15) E. J. Corey and W. C. Taylor, *ibid.*, **86**, 3881 (1964).
(16) (a) J. A. Marshall and A. R. Hochstetler, *J. Org. Chem.*, **31**, 1020 (1966); (b) T. Wilson, *J. Amer. Chem. Soc.*, **88**, 2898 (1966); (c) H. H. Wasserman and M. B. Floyd, *Tetrahedron Suppl.*, **7**, 441 (1966).
(17) G. O. Schenck and K.-H. Schulte-Elte, *Ann.*, **618**, 185 (1958).

in a subsequent section). After extraction with ether and removal of the solvent, the product was distilled *in vacuo* to give a 63% yield of the peroxide II, identical with the product of photooxygenation. Gas chromatographic analysis of crude oxygenation mixtures, after reduction, showed that, in addition to III, no other volatile products were present (except under conditions discussed in a subsequent section).

Oxygenation of 1,3-Cyclohexadiene. Photooxygenation of 1,3-cyclohexadiene (IV) is reported to yield 5,6-dioxabicyclo[2.2.2]octene-2 ("norascaridol" (V)) in 21% yield.[18] The photooxygenation was repeated and produced a 35% yield of impure V; the nmr and infrared spectra support the previously assigned structure. Isolation of completely pure V was not attempted as the compound appears to undergo ready polymerization.

Oxygenation of 1,3-cyclohexadiene in methanol with hypochlorite–hydrogen peroxide also produced V, identical with the product of photooxygenation. The yield was 20% after distillation.

Oxygenation of Tetraphenylcyclopentadienone. Tetraphenylcyclopentadienone (VI) is reported to give *cis*-dibenzoylstilbene (VII) in 65% yield on photooxygenation.[19] This reaction is believed to proceed by loss of carbon monoxide from an initial 1,4-*endo*-peroxide (VIII). Oxygenation with hypochlorite–hydrogen peroxide was carried out in dioxane (subsequently found to be a very poor solvent for the reaction). The reaction was followed by loss of the intense color of tetraphenylcyclopentadienone. A 50-fold excess of hypochlorite was required to cause nearly complete loss of color. The yield of pure VII was 50% (after recrystallization).

Oxygenation of Anthracene. Anthracene (IX) was photooxygenated in chloroform (because of low solubility in methanol) using methylene blue as sensitizer. The product had physical and chemical properties in accord with the reported structure (X).[20]

Hypochlorite–hydrogen peroxide oxygenation of anthracene was carried out in dioxane. Although a 20-

fold excess of hypochlorite was used, only a small fraction of the anthracene reacted. The peroxide X was separated from unreacted anthracene by fractional crystallization, in low yield. Subsequent investigations (reported below) showed that the yield of singlet oxygen is very low in dioxane; however, solubility considerations precluded the use of other, more favorable, solvents. McKeown and Waters have recently described a two-phase system using the aqueous phase reaction of bromine with alkaline hydrogen peroxide to oxidize anthracene derivatives in chlorobenzene, with improved results.[8]

Oxygenation of 2,5-Dimethylfuran. The photooxygenation of 2,5-dimethylfuran (XI) in methanol gives high yields of 2-methoxy-5-hydroperoxy-2,5-dimethyl-dihydrofuran (XII).[21] The reaction apparently proceeds *via* solvolytic ring opening of an initially formed *endo*-peroxide XIII. Oxygenation of this excellent acceptor with hypochlorite–hydrogen peroxide in methanol proceeds very smoothly to give an excellent yield of XII.

Oxygenation of Δ⁹,¹⁰-Octalin.[22] The photooxygenation of Δ⁹,¹⁰-octalin (XIV) is reported to produce 10-hydroperoxy-Δ¹,⁹-octalin (XV).[17] Repetition of the photooxygenation yielded a product with melting point similar to that reported; the infrared and nmr spectra are in agreement with the previously assigned structure. Oxygenation of XIV with hypochlorite–hydrogen peroxide also yielded XV, identified by melting point and infrared spectrum.

Effect of Reaction Variables on Yield. In none of the oxygenations described above was any attempt made to optimize product yields. In order to establish the synthetic utility of the reaction, one of the acceptors was subjected to more detailed study. TME was chosen because it is very reactive; in addition, the oxygenation product II can be quantitatively reduced to the alcohol III which can be determined accurately by gas chromatography. In these studies, a known small amount of hypochlorite was added to a solution of excess TME and excess H₂O₂. The amounts were chosen so that the change in TME or H₂O₂ concentration during the course of the reaction was small. The amount of product formed was measured by gas chro-

(18) G. O. Schenck and W. Willmund, reported by R. Criegee in Houben-Weyl, "Methoden der organischen Chemie," Vol. VIII, 4th ed, E. Müller Ed., Georg Thieme Verlag, Stuttgart, 1952, p 16.

(19) (a) C. F. Wilcox, Jr., and M. P. Stevens, *J. Amer. Chem. Soc.*, **84**, 1258 (1962); (b) G. O. Schenck, *Z. Elektrochem.*, **56**, 855 (1952).

(20) C. Dufraisse and M. Gérard, *Bull. Soc. Chim. France*, **4**, 2052 (1937).

(21) C. S. Foote, M. T. Wuesthoff, S. Wexler, I. G. Burstain, R. Denny, G. O. Schenck, and K.-H. Schulte-Elte, *Tetrahedron*, **23**, 2583 (1967).

(22) These reactions were carried out by Miss Elaine Holstein, whose assistance is gratefully acknowledged.

matography of the reduced reaction mixture after addition of an internal standard, using a calibrated flame-ionization detector.

Acceptor Concentration. Kinetic studies of the photooxygenation have shown that the reactive species (RS, either singlet oxygen or sensitizer–oxygen complex) may either react with acceptor (A) to give peroxide (AO_2, reaction 1) or decay to ground-state oxygen (reaction 2).[3b,c] In this scheme, singlet oxygen and a sensitizer–oxygen complex are kinetically equivalent.

$$A + RS \xrightarrow{k_1} AO_2 \qquad (1)$$

$$RS \xrightarrow{k_2} {}^3O_2 \qquad (2)$$

Application of steady-state kinetics to the reaction gives the result[3a,b,23]

$$\Phi_{AO_2} = \Phi_{RS}\frac{k_1[A]}{k_1[A] + k_2}$$

where Φ_{AO_2} is the quantum yield of product formation and Φ_{RS} is the quantum yield of formation of reactive species, which does not depend on acceptor. The fraction of RS which gives product is

$$\frac{[A]}{\beta + [A]}$$

where $\beta = k_2/k_1$; β is the concentration of A at which half of RS is trapped to give product. Kinetic studies show that the reactive species in the hypochlorite–hydrogen peroxide oxygenation has behavior identical with that of the photooxygenation intermediate, and that in both reactions, β is approximately 0.003 M for TME.[3b,24] Thus at TME concentrations over 0.1 M, more than 95% of the reactive species is trapped, and the yield of oxidation product provides a good measure of the yield of the reactive species. Tetramethylethylene is one of the most reactive olefins studied; most compounds are less reactive, and correspondingly less efficient at trapping the reactive species.[3b,24,25] The studies reported below were all carried out at TME concentrations above 0.1 M. The maximum yield of III under any conditions so far studied was about 80%, based on hypochlorite. This yield was obtained by adding 1.5 M aqueous sodium hypochlorite to a methanol solution containing 0.18 M hydrogen peroxide, 0.012 M NaOH, and 0.21 M TME at −20°. The amount of product formed was proportional to the amount of hypochlorite added, and did not depend on the concentration of hypochlorite added. The yield was constant between −50 and 0°, but dropped to 40% at 30°.

Effect of pH. In the absence of added NaOH (except for that introduced with the sodium hypochlorite), the product yield dropped to about 70%. If the pH was below 7, almost no III was formed, and the major product was the methoxychloro adduct XVI (which is the main product of reaction of hypochlorite with TME in methanol in the absence of H_2O_2). All of the following experiments were carried out without added NaOH, except for that introduced with the hypochlorite.

(23) R. Livingston and V. Subba Rao, *J. Phys. Chem.*, **63**, 794 (1959), and references therein cited.
(24) R. Higgins, C S.. Foote, and H. Cheng, Advances in Chemistry Series, American Chemical Society, Washington, D. C., in press.
(25) K. R. Kopecky and H. Reich, *Can. J. Chem.*, **43**, 2265 (1965).

XVI

Effect of Solvent. The solvent chosen was found to have a drastic effect on product yield for reasons which are not yet understood. Methanol (pure, aqueous, or saturated with sodium chloride), ethanol, and 1:1 (v/v) methanol–*t*-butyl alcohol as solvent gave comparable high yields of III (over 60%). In isopropyl alcohol, the yield dropped to 40%. In pure *t*-butyl alcohol, 50% aqueous *t*-butyl alcohol, tetrahydrofuran, dioxane, and acetonitrile, yields were below 10%. No new products were formed; the only apparent effect was a marked increase in oxygen evolution from the solution.

Other Effects. Considerable difficulty in reproducing yields at high and low peroxide concentrations was observed, and drastically lowered yields were often observed below 0.05 M or above 0.3 M H_2O_2. Part of this effect seems to be associated with the rate of addition of the hypochlorite (rapid addition decreases yields) and may involve local depletion of reactants, but further study is necessary.

Oxygen Evolution. Oxygen is produced by the reaction between sodium hypochlorite and hydrogen peroxide. In the absence of an acceptor, the yield of oxygen is nearly stoichiometric to the amount of hypochlorite added, if hydrogen peroxide is in excess. Some difficulty was encountered in obtaining completely reproducible measurements of oxygen evolution, probably because of some reaction of hypochlorite with solvent; in addition, blank values varied somewhat from solvent to solvent. However, in a variety of solvents and with a variety of acceptors, the sum of the amount of oxygen evolved and the amount of product formed is equal to the amount of hypochlorite added. Under conditions where product yield is high, the amount of oxygen evolved is low, and vice versa. Oxygen evolution can thus be used to follow the course of the oxygenation. The amount of oxygen evolved during the course of the addition of a known amount of hypochlorite is measured and subtracted from the theoretical evolution calculated from blank experiments; the difference corresponds to the amount of acceptor oxidized. If oxidation is less than complete, additional reagent is added.

Discussion

Synthetic Utility of Chemical Oxygenation. The results of the studies reported above show that the hypochlorite–hydrogen peroxide reaction can produce synthetically useful yields of oxygenation products under certain conditions. In order to trap a useful fraction of the reactive intermediate, the substrate must be present in a concentration not much less than its $\beta(k_2/k_1)$ value. Values of β for chemical oxygenation can be calculated from relative reactivity data,[24] and are listed in Table I for several typical substrates.

In general, in chemical oxygenation,[24] as in photooxygenation,[3b,25] reactivity decreases in the series:

Table I. Values of $\beta(k_2/k_1)$ for Various Substrates[a]

Substrate	β, M	Substrate	β, M
[structure: furan]	0.001	[structure: branched alkene]	(0.14)
[structure: tetramethylethylene]	0.0030	[structure: methylcyclohexene]	0.55
[structure: cyclopentene]	0.003	[structure: acyclic alkene]	13
[structure: cyclohexene]	0.01	[structure: acyclic alkene]	70

[a] Calculated from data in ref 24, based on an average value of β for chemical and photochemical oxygenation of 0.14 M for 2-methyl-2-pentene. Values given are for oxygenation with H_2O_2–NaOCl, but are identical within experimental error with the values for photosensitized oxygenation.

tetraalkylated > trialkylated > dialkylated olefins, and compounds with double bonds in a six-membered ring are less reactive than acyclic compounds. This reactivity variation implies that tetra- and trialkylated acyclic olefins, furans, etc. can be oxygenated practically, but difficulty in obtaining sufficient conversion is obtained with methylcyclohexene derivatives.[1] In addition, substituted anthracenes,[23] (but not anthracene itself), cyclohexadienes, and cyclopentadienes are reactive enough to be good acceptors.

For the less reactive acceptors, or those which are limited to low concentrations by solubility, large excesses of hypochlorite may be required to obtain appreciable conversion. Side reactions (probably free radical in nature) also compete with the oxygenation when unreactive acceptors are used; the side reactions can be at least partially suppressed by the use of free-radical inhibitors such as 2,6-di-t-butylphenol.[1] As reagents, aqueous sodium hypochlorite or solutions of calcium hypochlorite (which is less basic) are equally suitable; a wide variety of other positive-halogen sources can also be used.[26]

As solvents, so far only methanol, ethanol, and methanol–t-butyl alcohol have been found to be very practical. Of these, 1:1 (v/v) methanol–t-butyl alcohol offers solubility advantages when aqueous hypochlorite is used. Further investigation of the solvent effect, particularly with other reagents, is under way.[26] One method of avoiding the solubility and solvent limitations described above is the two-phase system described by McKeown and Waters, in which the acceptor is dissolved in chlorobenzene and the hypohalite–hydrogen peroxide reaction carried out in an aqueous layer; a large excess of reagents is required to effect appreciable conversions with this method.[8]

Nature of the Oxidizing Species. On the basis of quenching experiments, Rosenberg and Humphries have suggested that both the dye-photosensitized and hypochlorite–peroxide oxygenations may involve the intermediacy of vibrationally excited ground-state (triplet) oxygen, rather than singlet oxygen.[27] However, the following arguments strongly favor the intermediacy of singlet oxygen. (1) Singlet ($^1\Delta_g$) oxygen is known to be produced by the reaction of chlorine and

H_2O_2 in a yield of at least 10%;[11] high yields (up to 80%) of the oxidizing species are observed in the chemical oxygenation. (2) Energy transfer from triplet sensitizer to triplet oxygen to give singlet oxygen and singlet sensitizer is a spin-allowed process, and could therefore be very efficient (as is the production of the excited species in the photoreaction[3b] and the reaction of oxygen with triplet sensitizers).[3] (3) Kinetically and stereochemically, the reactive intermediates in the photooxygenation and the chemical oxygenation behave identically, and would appear to be the same.[1,24] Thus arguments applying to one reaction should apply also to the other. (4) The observed chemistry in both reactions seems to be a concerted or nearly concerted cycloaddition of oxygen to acceptors.[1,3] Vibrationally excited triplet oxygen should resemble a free radical in its reactions, as should $^1\Sigma_g{}^+$ oxygen. The presence of free-radical inhibitors does not inhibit the formation of oxygenated products.[1] Molecular oxygen in the $^1\Delta_g$ stage, in contrast, has both electrons paired in the same orbital, and should be capable of concerted two-electron reactions. The chemistry would be predicted to be that of a reactive dienophile, as observed.[28] (5) A minimum lifetime can be calculated for the reactive species by simple considerations.[29] The most reactive substrate yet studied is 1,3-diphenylisobenzofuran,[16b] which has a $\beta(k_2/k_1)$ value which can be calculated to be 10^{-4} M from the published data.[3b,16b,24] The rate constant (k_1) for reaction of the reactive species with acceptors cannot exceed the diffusion-controlled rate, about 10^{10} M^{-1} sec^{-1}. Therefore, the decay rate (k_2) of the reactive species must be no larger than 10^6 sec^{-1} in solution. In fact, since most Diels–Alder reactions have a large negative entropy of activation, the maximum reaction rate is probably less than diffusion controlled, and the decay rate is therefore probably less than 10^6 sec^{-1}.[30] Since collision rates in solution are of the order of 10^{13} sec^{-1},[31] the reactive species must survive roughly 10^7 solvent collisions. Vibrationally excited oxygen in the gas phase requires about 10^3 collisions with small molecules per vibrational quantum for deactivation;[27,32] however, the quenching efficiency of methanol should be greater than this. Thus it seems very unlikely that vibrationally excited oxygen is the reactive species. Similarly, $^1\Sigma_g{}^+$ oxygen is rapidly quenched by collisions with water (and thus presumably also with methanol).[7,13] On the other hand, $^1\Delta_g$ oxygen survives more than 10^8 collisions in the vapor, and thus has a lifetime long enough to be consistent with the lifetime of the intermediate.[12] Thus from both the chemical and the longevity criteria, the reactive species would seem to be $^1\Delta_g$ O$_2$. Gaffron showed that in at least one case photooxidation could be effected with light of insufficient energy to produce $^1\Sigma_g{}^+$ O$_2$.[33] Curiously,

(26) C. S. Foote and J. Barnett, unpublished results.

(27) J. L. Rosenberg and F. S. Humphries, *Photochem. Photobiol.*, **4**, 1185 (1965).

(28) Very recent results suggest that O$_2$ ($^1\Sigma_g{}^+$) does indeed behave as a free radical: D. R. Kearns, R. A. Hollins, A. U. Khan, R. W. Chambers, and P. Radlick, *J. Amer. Chem. Soc.*, **89**, 5455 (1967); D. R. Kearns, R. A. Hollins, A. U. Khan, and P. Radlick, *ibid.*, **89**, 5457 (1967); we are grateful to Professor Kearns for a prepublication copy of his manuscripts.

(29) G. O. Schenck and E. Koch, *Z. Elektrochem.*, **64**, 170 (1960).

(30) The enthalpy of activation is near zero with reactive acceptors.[29] However, no determination of absolute rates has been made.

(31) J. E. Leffler and E. Grunwald, "Rates and Equilibria of Organic Reactions," John Wiley and Sons, Inc., New York, N. Y., 1963.

(32) (a) N. Basco and R. G. W. Norrish, *Discussions Faraday Soc.*, **33**, 99 (1962); (b) H. Knotzel and L. Knotzel, *Ann. Physik*, **2**, 383 (1948).

(33) H. Gaffron, *Biochem. Z.*, **287**, 130 (1936).

this experiment was used as evidence against the intermediacy of *any* form of singlet oxygen, although Kautsky pointed out that ample energy was available to excite the $^1\Delta_g$ state.[6]

Evidence against the intermediacy of a sensitizer–oxygen complex has already been presented,[1,34] and will be supplemented in subsequent papers.[24]

Experimental Section[35]

Materials. H_2O_2 was J. T. Baker reagent grade 30% solution. NaOCl was General Chemical reagent grade ("5%") or Purex 14 (up to 17%). $Ca(OCl)_2$ solutions were prepared from J. T. Baker reagent grade solid (30–35% "active Cl") mixed with approximately twice its weight of water and filtered. All peroxide and hypochlorite solutions were standardized frequently by reaction with KI solution and titration of I_2 formed with thiosulfate.[36]

Photosensitized Oxygenations. A water-cooled immersion irradiation apparatus similar to the one described by Gollnick and Schenck was used.[3f] O_2 was recirculated by a diaphragm pump (Cole-Palmer Dyna-Vac Model 3). The solutions were irradiated with a Sylvania "Sungun" Type DWY 625-W tungsten–iodine lamp; oxygen uptake was measured by a gas buret. The maximum rate of oxygen uptake in this apparatus was about 200 cc/min, which corresponds to about 1.5×10^{-4} mol/sec.

Hypochlorite–Hydrogen Peroxide Oxygenations. In all runs, the acceptor was dissolved in the solvent, the solutions were chilled, and H_2O_2 was added. The amount of H_2O_2 in the solution was always in excess of the amount of hypochlorite to be added. Hypochlorite was added dropwise to the vigorously stirred solution.

The volume of O_2 evolved from the solution was measured with a gas buret in several experiments. Blank experiments with no olefin present showed that, in methanol, somewhat less than the theoretical amount of O_2 was evolved, presumably because some of the NaOCl was consumed by reaction with methanol. The volumes were therefore always corrected for this consumption, which limits the accuracy of the calculations, since it is not entirely reproducible.

Control experiments showed that under the reaction conditions used, neither H_2O_2 nor the products remaining in solution after reaction of NaOCl and H_2O_2 cause significant oxidation of olefins; in addition, reaction of hypochlorite with olefin is negligible when H_2O_2 is present in excess.

In the absence of H_2O_2, however, or at pH below 7, hypochlorite reacts with olefins. In the case of tetramethylethylene, the product is apparently the methoxy chloro adduct XVI. The compound was isolated gas chromatographically and had ir bands (CCl_4) at 3.40, 6.85, 7.30, 7.37, 8.65, 9.00, 9.40, 11.25, and 14.1 μ; in the nmr, there were sharp singlets at τ 8.70, 8.47, and 6.78, relative areas 2:2:1. The compound was not further characterized.

Photooxygenation of 2,3-Dimethyl-2-butene.[17] A solution of 2.72 g (32.3 mmol) of TME (I) (Columbia Organic Chemicals) and 30 mg of rose bengal in 250 ml of CH_3OH was irradiated in the immersion apparatus; 776 cc (32 mmol) of O_2 was absorbed in 3.75 min. The solution was decolorized and most of the solvent removed by distillation (Vigreux) at atmospheric pressure. The residue, approximately 20 ml, was distilled (6-in. concentric-tube column) to give 2.02 g (17.4 mmol, 54%) of hydroperoxide II, bp 53–55° (12 mm) (lit.[17] bp 55° (12 mm)). The ir spectrum (neat) had principal bands at 2.83, 6.04, 6.86, 7.25, 7.32, 7.68, 8.27, 8.51, 8.69, 9.83, 11.04, and 11.94 μ. The nmr spectrum (CCl_4) had peaks at τ 1.62 (broad, OOH), 5.08 (multiplet, CH_2), 8.22 (doublet, $CH_2C{=}C$), and 8.70 (singlet, CH_3CO); excluding the hydroperoxyl proton, the peaks were in the ratio 2:3:6.

Hypochlorite–Hydrogen Peroxide Oxygenation of 2,3-Dimethyl-2-butene. A solution of 5.05 g (60 mmol) I in 300 ml of methanol

was chilled to 10° and 19.2 ml of 9.36 M (180 mmol) H_2O_2 was added. To this solution 145.5 ml (150 mmol) of 1.03 M NaOCl solution was added dropwise with continual cooling and stirring during 90 min. The reaction mixture was diluted with water and extracted with ether. The dried ether extracts were concentrated (Vigreux) at atmospheric pressure, and the residue was distilled. The fraction with bp 51–54° (9 mm) was 4.43 g (38.3 mmol, 64%) of hydroperoxide II, with nmr and ir spectra identical with those of the photoproduct and $n^{24.5}$D 1.4408 (lit.[17] n^{20}D 1.4428).

Reduction of Hydroperoxide II. A solution from photooxygenation of 7.1 g (0.085 mol) of TME (O_2 uptake 1660 cc, 0.074 mol) was reduced by the addition of excess sodium borohydride. The crude product was purified by vpc to give alcohol III, with ir (CCl_4) bands at 2.75, 2.86, 3.24, 6.10, 6.92, 7.30, 7.37, 7.63, 8.62, 10.42, 10.70, and 11.10 μ. The nmr spectrum had bands at τ 8.70 (singlet, 6 H), 8.22 (singlet, 3 H), 6.35 (singlet, 1 H), 5.32 (singlet, 1 H), and 5.05 (singlet, 1 H); mass spectrum (m/e of molecular ion), calcd for $C_6H_{12}O$: 100.08881; found: 100.08881.

Photosensitized Oxygenation of 1,3-Cyclohexadiene.[18] A solution of 3.206 g of cyclohexadiene IV (Aldrich, 40 mmol) and 30 mg of rose bengal in 250 ml of methanol was irradiated; 913 cc of oxygen (40 mmol) was absorbed in 8 min. The irradiated solution was evaporated below room temperature and the resulting 4.00 g of syrup taken up in ether. A reddish precipitate was filtered off and the yellow ethereal solution dried. After removal of solvent, 3.47 g of semisolid material was obtained. A portion (0.528 g) of this semisolid was distilled (bulb to bulb at 1 μ) yielding 0.226 g of liquid containing waxy solid. Both phases were investigated by nmr and ir and found to be identical in spectral characteristics. The yield of peroxide V was 35%. No attempt was made to obtain pure crystalline material. The principal ir bands (CCl_4) were 7.28, 8.25, 8.60, 9.50, 10.40, 10.80, 13.80, and 14.60 μ. The nmr spectrum (CCl_4) was at τ 3.43 (overlapping doublets), 5.47 (broad), and 8.26 (multiplet) in the ratio 1:1:2, assigned to vinylic, bridgehead, and methylene protons, respectively.

Hypochlorite–Hydrogen Peroxide Oxygenation of 1,3-Cyclohexadiene. A solution of 3.21 g (40 mmol) of IV in 250 ml of methanol was chilled to −5° and treated with 25.6 cc (240 mmol) of 9.36 M H_2O_2, followed by 168 ml (150 mmol) of 0.89 M NaOCl solution at −5 to −10°. Evolved O_2 amounted to 2640 cc at 23° (109 mmol). The reaction mixture was diluted with water and extracted with ether. The solvent was removed from the dried ether extracts on a rotary evaporator and the residue (1.18 g) purified by bulb-to-bulb distillation at ∼1 μ. The product, *endo*-peroxide V, weighed 0.88 g (7.85 mmol, 19.6%) and had nmr and ir spectra identical with that produced by photooxygenation.

Oxygenation of Tetraphenylcyclopentadienone with Hypochlorite–Hydrogen Peroxide. Tetraphenylcyclopentadienone[37] (VI) (mp 219–220°), 0.17 g (0.44 mmol), was dissolved in 125 ml of dioxane, cooled in ice water, and treated with 5.0 ml of 8.8 M H_2O_2 (44 mmol). To this solution 32.8 ml of 0.67 M NaOCl solution (22 mol) was added dropwise with stirring. Only a small amount of color remained. The solution was extracted with benzene and the solvent evaporated. The resulting crystals were recrystallized from ethanol and yielded 0.085 g of *cis*-dibenzoylstilbene (VII) (0.22 mmol, 50%), mp 215.9–216.3° (lit.[19] 215–216°). The ir spectrum (in $CHCl_3$) was in good agreement with that reported[19] (a Nujol mull spectrum).

Photooxygenation of Anthracene. Anthracene (Eastman, blue-violet fluorescence), 0.535 g (3 mmol), and 50 mg of methylene blue were dissolved in 250 ml of freshly prepared, purified[38] $CHCl_3$ and irradiated. The lamp was operated at 60 V to remove any small amount of uv light and to keep the solution temperature low. In 10 min, 58 cc (2.4 mmol) of O_2 was absorbed, and the rate of absorption had decreased appreciably. The solution was filtered through silica gel to remove the dye. The colorless solution was evaporated to dryness and the residue recrystallized from purified $CHCl_3$. The first crop of crystals (needles, mp 139–143°, 0.19 g, 0.91 mmol, 30% yield) was almost pure anthracene peroxide (X) (lit.[20] mp 120° dec).

Anal. Calcd for $C_{14}H_{10}O_2$: C, 79.98; H, 4.79. Found: C, 79.89; H, 4.71.

The principal ir bands (Nujol mull) were at 7.57, 7.75, 8.12, 8.33, 8.56 m, 8.71, 9.00, 9.86, 10.10, 10.25, 10.55, 11.14, 11.26, 11.42, 11.51 m, 11.63 m, 11.82, 12.12 m, 12.25, 12.42 m, 13.02 s,

(34) C. S. Foote and S. Wexler, *J. Amer. Chem. Soc.*, **86**, 3880 (1964).

(35) Infrared (ir) spectra were recorded on a Perkin-Elmer Model 137 spectrophotometer. Nuclear magnetic resonance (nmr) spectra were recorded on a Varian A-60 spectrometer. Molecular weights were determined on a Mechrolab vapor pressure osmometer using benzene as the solvent and benzil as the standard. Melting points were determined on a Fisher-Johns melting block and are corrected except where otherwise indicated. Gas chromatography was carried out on a Perkin-Elmer Model 800, a Varian Aerograph HI-FI III (flame ionization), or an Aerograph A90-P (thermal conductivity) instrument. Elemental analyses were performed by Miss H. King (UCLA), and methoxyl determinations were done by Elek Laboratories, Los Angeles, Calif.

(36) I. M. Kolthoff and R. Belcher, "Volumetric Analysis," Vol. 3, Interscience Publishers, Inc., New York, N. Y., 1957, p 283.

(37) E. C. Horning, Ed., "Organic Syntheses," Coll. Vol. III, John Wiley and Sons, Inc., New York, N. Y., 1955, p 806.

(38) L. F. Fieser, "Experiments in Organic Chemistry," D. C. Heath and Co., Boston, Mass., 1957, p 283.

and 13.42 s μ. All the above bands are weak except those marked m (medium) and s (strong). The nmr spectrum ($CDCl_3$) had bands at τ 2.64 (multiplet) and 3.98 (singlet, bridgehead protons) in the ratio 4:1.

Oxygenation of Anthracene with Hypochlorite–Hydrogen Peroxide. Anthracene, 0.535 g (3 mmol), was dissolved in 100 ml of *p*-dioxane and treated with 6.4 ml (60 mmol) of 9.38 *M* H_2O_2 at 10–15°. Over a period of 30 min, 79.5 ml of 0.32 *M* (50 mequiv) $Ca(OCl)_2$ solution was added at 5–10°, with vigorous stirring. The O_2 evolved was roughly 1350 cc (55 mmol). The reaction mixture was diluted with water and extracted with chloroform. The dried chloroform extracts were evaporated. The residue (0.535 g) was taken up in hot chloroform and successive crops of crystals were separated by filtration. The first crops were mainly anthracene. The third and fourth crops (0.5 g) were combined and extracted with carbon tetrachloride. The undissolved residue, 32 mg (5.1%), of yellowish powder, had ir and nmr spectra identical with that of anthracene peroxide (X) produced by photooxygenation.

Hypochlorite–Hydrogen Peroxide Oxygenation of 2,5-Dimethylfuran. A solution of 3.01 g (0.031 mol) 2,5-dimethylfuran in 250 ml of methanol was chilled to 5° and 3.0 ml of 9.75 *M* H_2O_2 (0.0292 mol) added. The solution was stirred and 21 ml of 0.67 *M* NaOCl solution (0.0141 mol) was added dropwise in 10 min. The reaction mixture was evaporated under reduced pressure and below 0° to about 20 ml, and ice was added. The mixture was extracted with three 30-ml portions of cold ether. The residue after evaporation of the ether was recrystallized from ether to give 1.89 g (0.0118 mol) of the methoxyhydroperoxide XII (84% yield based on NaOCl used), mp 75–77°. The nmr and ir spectra were identical with those of the product of photooxygenation, mp 75–76°.[21]

Photooxygenation of $\Delta^{9,10}$-Octalin.[22] A solution of 1.0 g of $\Delta^{9,10}$-octalin[39] (7.3 mmol) and 45 mg of rose bengal in CH_3OH was irradiated. Within 5 min, 162 cc (7.2 mmol) of O_2 had been taken up, and the reaction became very slow. The solvent was removed, and the product taken up in ether, washed with water, and dried. The product was recrystallized from hexane to give crystals, mp 55–58° (lit.[17] mp 60°). The ir spectrum had principal bands at 2.72, 2.85, 3.35, 3.45, 6.90, 7.83, 8.43, 8.52, 8.71, 10.00, 10.49, 11.09, 11.69, and 11.85 μ.

Hypochlorite–Hydrogen Peroxide Oxygenation of $\Delta^{9,10}$-Octalin. To a stirred solution of 1.0 g of $\Delta^{9,10}$-octalin (7.3 mmol) and 2.4 ml of 30% aqueous H_2O_2 (22.2 mmol) in 45 ml of ethanol at 10° was added 20.8 ml of 0.89 *M* NaOCl (18.5 mmol) during 1 hr. At the end of this time, an additional 22.2 mmol of H_2O_2 and then 18.5 mmol of NaOCl were added. The solution was extracted with ether, the ether was dried and evaporated, and the product was recrystallized from hexane to give 0.069 g (6%) of the hydroperoxide, mp 54–57°, with infrared spectrum identical with that of the photooxygenation product. Substantial work-up and crystallization losses were encountered, and the actual product yield was much higher. A second oxidation with 2-propanol as solvent gave a 12% yield of peroxide, with work-up losses again being substantial.

Quantitative Analysis of Oxygenation Products. A. Photooxygenation. A weighed quantity of TME was dissolved in the solvent; an aliquot of a stock solution of rose bengal was added and the solution diluted to a known volume. After irradiation, the hydroperoxide was reduced with a tenfold excess of trimethyl phosphite.[40]

The solution was left overnight and analyzed directly with a flame-ionization gas chromatograph using isoamyl alcohol, dioxane, or *n*-decane as internal standard. The detector was calibrated by injection of solutions containing weighed amounts of the internal standard and the pure product alcohol.

B. Chemical Oxygenation. A weighed quantity of TME was dissolved in the solvent and an aliquot of 30% H_2O_2 added; the solution was then diluted to the desired volume and transferred to a three-necked flask equipped with a magnetic stirrer, a septum, and, if required, a gas buret. The flask was immersed in a bath at the required temperature and allowed to reach equilibrium. Aqueous hypochlorite was then added with an all-glass syringe through a capillary reaching below the surface. The solution was then reduced, using enough excess reductant to consume any remaining H_2O_2, and analyzed in the same manner as with the photooxygenation experiments. In some cases, sodium borohydride was used as the reductant; jn these cases, a small amount of the saturated alcohol (2,3-dimethyl-2-butanol) was also produced, apparently by hydroboration of the olefin. Control experiments showed that this alcohol could be produced (in low yield) by the addition of borohydride to solutions containing only olefin and hydrogen peroxide.

Dependence of Product on NaOCl. To a 0.39 *M* solution of TME in methanol at $-20°$ containing 0.18 *M* H_2O_2 were added various amounts of NaOCl solution. The yield of product is summarized in Table II.

Table II

Amount of NaOCl added, mmol	Concn of NaOCl added, *M*	Yield of product based on NaOCl, %
1.13	1.51	64
1.51	1.51	69
3.02	1.51	70
4.53	1.51	64
6.04	1.51	65
1.51	0.30	69

Solvent Effect. To 100 ml of the chosen solvent containing H_2O_2 and *ca.* 0.15 *M* tetramethylethylene was added a measured amount of 1 *M* NaOCl. Yields of product determined by the usual technique are shown in Table III.

Table III

Solvent	Temp, °C	$[H_2O_2]$, *M*	Product yield, %
Methanol	-20	0.18	61
Methanol	25	0.18	40
Ethanol	-20	0.18	69
Isopropyl alcohol	-20	0.18	38.4
t-Butyl alcohol	25	0.18	7.5
Methanol–*t*-butyl alcohol 1:1 (v/v)	25	0.18	35
Tetrahydrofuran	25	0.45	5.5
Dioxane	25	0.45	2.3
Acetonitrile	25	0.45	1
Methanol	3	0.50	28
Methanol	3	0.90	4.6

(39) R. A. Benkeser and E. M. Kaiser, *J. Org. Chem.*, **29**, 955 (1964).
(40) D. B. Denny, W. F. Goodyear, and B. Goldstein, *J. Amer. Chem. Soc.*, **82**, 1393 (1960); M. S. Kharasch, R. A. Mosher, and I. S. Bengelsdorf, *J. Org. Chem.*, **25**, 1000 (1960). Yields determined by oxygen absorption agree well with product yields measured by this technique, but are less accurate.

Editor's Comments
on Papers 27 and 28

27   FONER and HUDSON
Metastable Oxygen Molecules Produced by Electrical Discharges

28   COREY and TAYLOR
A Study of the Peroxidation of Organic Compounds by Externally Generated Singlet Oxygen Molecules

PRODUCTION OF SINGLET OXYGEN BY ELECTRIC DISCHARGE

Foner and Hudson (Paper 27) demonstrated in 1956 that $^1\Delta_g$ molecular oxygen is produced by passing a stream of oxygen through an electric discharge. It was estimated that 10–20 percent of the oxygen molecules are in the $^1\Delta_g$ state.

Corey and Taylor (Paper 28) were the first to recognize that this source of singlet oxygen might be used to oxygenate organic substrates. Oxygen was passed through an electric discharge and bubbled into a solution containing the acceptors to yield products that were identical to those obtained from sensitized photooxygenation of the same acceptors.

The observations of Corey and Taylor and of Foote and Wexler showed that singlet oxygen was capable of oxidizing organic substrates, as suggested by Kautsky. However, unambiguous evidence for the intermediacy of $^1O_2$ in photooxygenations was to be presented later (Papers 38–40, 45, 46).

27

Reprinted from *J. Chem. Phys.*, 25(3), 601–602 (1956)

Metastable Oxygen Molecules Produced by Electrical Discharges*

S. N. FONER AND R. L. HUDSON

*Applied Physics Laboratory, The Johns Hopkins University,
Silver Spring, Maryland*

(Received July 12, 1956)

WHEN pure oxygen was sent through an electrical discharge and analyzed with a mass spectrometer, it was noticed that the O_2^+ ion intensity at low electron energy was higher than expected. This suggested that metastable oxygen molecules might be present in the gas, in addition to oxygen atoms and normal oxygen molecules. An analysis of the data subsequently obtained has confirmed the presence of a surprisingly high percentage of O_2 in the $^1\Delta_g$ state.[1]

Although the radiative lifetime of O_2 in the $^1\Delta_g$ state is very long because decay to the $^3\Sigma_g^-$ state represents a magnetic dipole transition and is a singlet-triplet intercombination, this molecule has not been observed previously by mass spectrometry, probably because in a conventional instrument an excited molecule undergoes many wall collisions which could bring it to the ground state. There is, however, some indirect evidence that metastable oxygen molecules may be produced in a mass spectrometer by reaction on a heated filament.[2]

The mass spectrometer used here and the essentially collision-free molecular beam sampling system have been described.[3,4] The experimental arrangement was similar to that in Fig. 1 of reference 4 with the reactor at room temperature and no gas flowing in the central tube. High purity tank oxygen at a pressure of about 4 mm was sent through a conventional Wood's discharge tube. The distance from the discharge tube to the sampling pinhole was 30 cm corresponding to a gas transit time of about 0.05 sec. Figure 1 shows the appearance potential curves for normal oxygen and for oxygen subjected to discharge. The 30% decrease in the O_2^+ intensity at high energy with "Discharge On" can be accounted for by the amount of atomic oxygen observed. The increase in ion

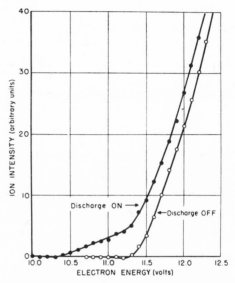

FIG. 2. "Deboltzmannized" appearance potential curves for O_2^+ ions from oxygen. For comparing the curves, a minor slope adjustment has been made. The slight undershooting of the curves on the axis is expected and results from the least squares treatment of the data. The Discharge On curve is clearly the sum of two components having different ionization potentials. The voltage scale is uncorrected.

intensity at low energy with Discharge On is clearly due to the presence of O_2 with a lower ionization potential than O_2 in the ground state. Although the results reported here were obtained with a 60-cycle discharge, we have also observed metastable O_2 with an electrodeless 5-mc discharge and with a 2450-mc microwave discharge.

An analysis of the data was made by the "deboltzmannizing" procedure of Foner, Kossiakoff, and McClure[5,6] to resolve the structure of the ionization curves. The results of these calculations are shown in Fig. 2. The "Discharge On" curve is the superposition of a normal O_2 ionization curve and another curve starting 0.93 ± 0.1 ev below that of O_2. A comparison of this energy separation with the spectroscopic separation of the $^1\Delta_g$ and $^3\Sigma_g^-$ states of O_2,[7] 7882.39 cm^{-1}=0.9772 ev, strongly suggests that the excited species of O_2 are $^1\Delta_g$ molecules.

A rough estimate of the $O_2(^1\Delta_g)$ concentration can be made by comparing initial slopes of the two portions of the oxygen curve in Fig. 2. Assuming that the ionization cross sections for the metastable and ground states are the same, it appears that between 10 and 20% of the oxygen molecules are in the $^1\Delta_g$ state.

* This work was supported by the Bureau of Ordnance, Department of the Navy, under NOrd 7386.

[1] For a discussion of the metastable states of the oxygen molecule, see G. Herzberg, *Spectra of Diatomic Molecules* (D. Van Nostrand Company, Inc., New York, 1950).
[2] H. Hagstrum and J. T. Tate, Phys. Rev. 59, 509 (1941).
[3] S. N. Foner and R. L. Hudson, J. Chem. Phys. 21, 1374 (1953).
[4] S. N. Foner and R. L. Hudson, J. Chem. Phys. 23, 1974 (1955).
[5] Foner, Kossiakoff, and McClure, Phys. Rev. 74, 1222 (1948).
[6] *Physical Measurements in Gas Dynamics and Combustion* (Princeton University Press, Princeton, 1954), p. 450.
[7] L. Herzberg and G. Herzberg, Astrophys. J. 105, 353 (1947).

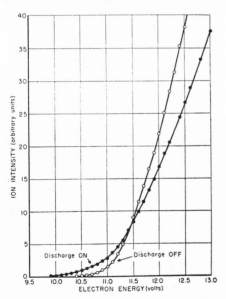

FIG. 1. Appearance potential curves for O_2^+ ions from oxygen. The curve "Discharge Off" is the normal oxygen curve. The curve "Discharge On" shows the change produced by the electrical discharge. The voltage scale is uncorrected.

117

28

Copyright © 1964 by the American Chemical Society

Reprinted from *J. Amer. Chem. Soc.*, **86**, 3881–3882 (1964)

A Study of the Peroxidation of Organic Compounds by Externally Generated Singlet Oxygen Molecules

Sir:

The existence of two low-lying metastable singlet states of diatomic oxygen ($^1\Delta_g$ and $^1\Sigma_g$ which are 0.977 and 1.63 e.v. above the ground state)[1] and the "forbiddenness" of radiative return to the ground state suggested the possibility that singlet O_2 might be a useful and selective reagent in organic chemistry. We were further interested in this case as a result of our previous experience with another highly reactive small molecule, diimide[2] (N_2H_2), and because of the possible importance of metastable O_2 in photosensitized peroxidations.[3,4] It seemed to us that the O_2 molecule, raised above its ground state by *ca.* 22.5 or 37.5 kcal./mole and converted to a singlet state (removing spin-conservation difficulties in forming singlet products from reactants of singlet multiplicity), could reasonably be expected to afford the same reactions as observed in sensitized photooxidation.[3]

Our initial studies have been conducted using gaseous oxygen subjected to electrodeless discharge at 6.7 Mc., a method for producing singlet O_2 first described by Foner and Hudson.[5] The radiofrequency unit was a T21/ARC-5 U. S. surplus aircraft transmitter (modified to permit resonant coupling to the discharge tube) with the output leads attached (by platinum wire) to two aluminum foil bands (2 cm. apart) fitted around quartz tubing (12-mm. o.d.). Oxygen was passed through the quartz tube at *ca.* 20 mm. with the discharge operating to produce a definite glow in the tube between the aluminum terminals, and the emerging gas was bubbled into a solution of the reactant (usually stirred). The reactor was located downstream from the discharge zone by 20–30 cm.; the exact distance did not appear to be a critical variable, however. A water aspirator was employed at the end of the system to pull gas through. The vessel containing the organic reactant was shielded from light by a covering of aluminum foil.

Using this apparatus and bromobenzene as solvent at 0°, anthracene, 9,10-diphenylanthracene, and 9,10-dimethylanthracene were cleanly converted to the corresponding 9,10-endoperoxides, identical with authentic samples prepared by the photooxidation route. *No other product could be detected.* In control experiments in the same apparatus, under the same conditions *but with the radiofrequency unit and discharge off*, no more than trace quantities of peroxide were formed (analysis by chromatography). Therefore, the zero level of the $^3\Sigma_g$ ground state of O_2 is not responsible for endoperoxide formation. It also seems improbable that vibrationally excited $^3\Sigma_g$ molecules could persist long enough to effect oxidation in solution. Ozone and monatomic oxygen can also be excluded since these would lead to other types of products.[6-8] The species

(1) See G. Herzberg, "Molecular Spectra and Molecular Structure. I. Spectra of Diatomic Molecules," 2nd Ed., D. Van Nostrand Co., New York, N. Y., p. 560, for example.

(2) E. J. Corey and W. L. Mock, *J. Am. Chem. Soc.*, **84**, 865 (1962), and previous papers; see also S. Hunig, H. Muller, and W. Thier, *Tetrahedron Letters*, No. **11**, 353 (1961); E. E. van Tamelen, R. S. Dewey, and R. J. Timmons, *J. Am. Chem. Soc.*, **83**, 3725 (1961).

(3) For a recent review see G. O. Schenck, *Angew. Chem.*, **69**, 579 (1957).

(4) This possibility appears to have been suggested first by H. Kautsky and H. deBruijn, *Naturwiss.*, **19**, 1043 (1931).

(5) S. N. Foner and R. L. Hudson, *J. Chem. Phys.*, **25**, 601 (1956); **23**, 1974 (1955).

(6) R. J. Cvetanovic, "Advances in Photochemistry," Vol. 1, Interscience Publishers, Inc., New York, N. Y., 1963, p. 115.

(7) R. E. Erickson, P. S. Bailey, and J. C. Davis, Jr., *Tetrahedron*, **18**, 389 (1962).

(8) P. S. Bailey, *Chem. Rev.*, **58**, 926 (1958); P. S. Bailey, P. Kolsaker, B. Sinha, J. B. Ashton, F. Dobinson, and J. E. Batterbee, *J. Org. Chem.*, **29**, 1400 (1964).

which can account most reasonably for the observed results are $^1\Delta_g$ or $^1\Sigma_g$ forms of O_2. Both of these are known to be produced by the technique used in the present work,[5,9,10] the former in greater amount by at least 10^2.

The relative rates of peroxidation of the three anthracenes studied were 9,10-dimethylanthracene > 9,10-diphenylanthracene > anthracene. With the particular apparatus used in the present work, 19 hr. was required for complete conversion of 100 mg. of 9,10-dimethylanthracene to the 9,10-peroxide. Even in the case of this relatively reactive substrate there does not seem to be a highly efficient capture of reactive oxygen, based on the supposition that 10% of the oxygen molecules are converted to the $^1\Delta_g$ state[5] and the estimated flow rate. The considerable difference in reactivity of 9,10-dimethylanthracene and anthracene (roughly 100-fold) indicates that this reactive O_2 is rather selective.

An interesting effect of solvent on the rate of peroxidation of 9,10-diphenylanthracene was noted. Chlorobenzene, bromobenzene, and nitrobenzene gave faster reaction rates than anisole, dimethyl sulfoxide, or iodobenzene (which were all about the same), and these in turn led to faster oxidation than p-cymene or decalin. The rate factor from the most effective solvent chlorobenzene to decalin, the poorest, was over twenty (at $0°$).

Reactive 1,3-dienes are also susceptible to endoperoxidation. For example, and in analogy with the classical case of photooxidation,[3] exposure of α-terpinene to singlet oxygen afforded ascaridol. 2,5-Diphenyl-3,4-isobenzofuran afforded 1,2-dibenzoylbenzene[3] in high yield. On the other hand, attempts to convert olefins to allylic hydroperoxides have so far not succeeded in the case of α-pinene, 1-phenylcyclohexene, tetramethylethylene (gas phase), or cholest-4-en-3β-ol. Work is in progress to ascertain the significance of these preliminary results.

(9) L. Elias, E. A. Ogryzlo, and H. I. Schiff, *Can. J. Chem.*, **37**, 1680 (1959).

(10) M. A. A. Clyne, B. A. Thrush, and R. P. Wayne, *Nature*, **199**, 1057 (1963).

Although the instrumentation and techniques used in the present work are quite simple (comparable to those conventionally employed for ozonolysis), they are satisfactory only on a limited scale; with quantities greater than several millimoles, inconveniently long reaction times are required for complete conversion. Improvements in the metastable oxygen generator are clearly desirable.

We have followed with interest several recent publications on the chemical generation of singlet oxygen by the oxidation of hydrogen peroxide,[11-14] but we have done no work on the application of these systems to peroxidation of organic substrates. We have been informed by Professor Christopher Foote[15] that he has succeeded in effecting a number of peroxidation reactions by the oxidation of hydrogen peroxide in the presence of various substrates.

It is possible that singlet O_2 might also be a product of other types of chemical processes, *e.g.*, the reaction of hydrogen peroxide with oxalyl chloride,[16] which yields both molecular oxygen and luminescence, and the reactions of ozone with phosphines, phosphite esters, sulfides, etc.[17]

Work on the chemistry of externally generated metastable oxygen is being continued along the lines suggested by the results described above.

Acknowledgment.—We thank the National Science Foundation for financial support under Grant GP-221 and Professors William Klemperer and Dudley Hershbach for helpful discussions.

(11) A. U. Khan and M. Kaska, *J. Chem. Phys.*, **39**, 2105 (1963).

(12) S. J. Arnold, E. A. Ogryzlo, and H. Witzke, *ibid.*, **40**, 1769 (1964).

(13) E. J. Bowen and R. A. Lloyd, *Proc. Chem. Soc.*, 305 (1963).

(14) R. J. Browne and E. A. Ogryzlo, *ibid.*, 117 (1964).

(15) C. S. Foote, personal communication, June 22, 1964; see C. S. Foote and S. Wexler, *J. Am. Chem. Soc.*, **86**, 3879, 3880 (1964). Professor Foote has also indicated that Professor K. Bayes of U.C.L.A is also engaged in the study of singlet O_2.

(16) E. A. Chandross, *Tetrahedron Letters*, **No. 12**, 761 (1963).

(17) See Q. E. Thompson, *J. Am. Chem. Soc.*, **83**, 845 (1961), and references cited therein.

DEPARTMENT OF CHEMISTRY   E. J. COREY
HARVARD UNIVERSITY   WALTER C. TAYLOR
CAMBRIDGE, MASSACHUSETTS 02138

RECEIVED JUNE 25, 1964

Editor's Comments
on Papers 29 Through 36

ADDITIONAL SUPPORT FOR THE KAUTSKY
PHOTOOXYGENATION MECHANISM

Paper 29 describes some of the elegant experimental work from the laboratory of G. O. Schenck on dye-sensitized photooxygenation. Gollnick and Schenck report a method for determining the quantum yield of triplet sensitizer formation by measuring the quantum yield of the "photosensitized oxygen transfer reaction." This procedure depends on quenching all the triplets with O_2 and subsequently trapping all the excited sensitizer–oxygen adduct (or all the singlet oxygen) with a highly reactive acceptor. Under these conditions, one observes reaction rates that are independent of acceptor concentration. By using this method, it was determined that the quantum yield for Rose Bengal (tetraiodo-tetrachloro-fluorescein) triplet formation is 0.76. The stereochemistry of the "ene" reaction is also discussed. Photooxygenation of (+)-limonene and (+)-carvomenthene yields several hydroperoxides that are optically active. In contrast, free radical autoxidation of these acceptors gives racemic products. These results indicate that the photooxygenation reaction, whether mediated by a sensitizer–oxygen complex or singlet oxygen, does not involve free-radical intermediates. A mechanism involving a concerted oxygen transfer from the sensitizer–oxygen complex to the acceptor via a cyclic transition state is proposed.

In Paper 30, Foote, Wexler, and Ando report that similar product distributions are obtained from oxygenation of (+)-limonene with sodium hypochlorite and hydrogen peroxide or by dye-sensitized photooxygenation. More importantly, it was demonstrated (in contrast to the results reported in Paper 29) that the oxygenation of a-pinene by hypochlorite–hydrogen peroxide is not primarily a free-radical autoxidation, although in some cases there may be a free-radical component.

Evidence for the intermediacy of singlet oxygen in photooxygenations was also provided by the kinetic investigations described in Papers 31–35. Kopecky and Reich report that the relative rate of photooxygenation of several acceptors is independent of the sensitizer used. They conclude that a common reactive intermediate ($^1O_2$) is involved in the photooxygenations with the various sensitizers. The results are in contrast to what one might expect if a sensitizer–oxygen complex was the reactive intermediate. Wilson similarily observed that the competitive photooxygenation of pairs of acceptors gives the same results whether the reaction is sensitized by a dye or by the aromatic substrate itself (Paper 32). In Paper 33, Foote and coworkers report that the relative reactivities of several substrates are the same toward dye-sensitized photooxygenation and reaction with chemically generated singlet oxygen.

In Papers 34 and 35, Stevens and Algar describe kinetic and spectroscopic results that are consistent with the production of $^1\Delta_g$ oxygen via the spin-allowed energy transfer from aromatic hydrocarbons in the triplet state.

In 1968, Schnuriger and Bourdon (Paper 36) observed that dye-sensitized photooxygenations can occur across an oxygen-permeable membrane. Like the original Kautsky experiments (Papers 7–9), these results demonstrate that a diffusible intermediate such as singlet molecular oxygen is involved in the photooxygenation reaction.

29

Reprinted from *Pure Appl. Chem.*, **9**, 507–525 (1964)

MECHANISM AND STEREOSELECTIVITY OF PHOTOSENSITIZED OXYGEN TRANSFER REACTIONS

K. Gollnick and G. O. Schenck

*Max-Planck-Institut für Kohlenforschung, Abt. Strahlenchemie,
Mülheim-Ruhr, Germany*

INTRODUCTION

Reactions of molecular oxygen with substrates, A, in solution forming compounds of the composition AO_2 can be sensitized by the presence of a sensitizer, S, and light according to the overall reaction

$$A + O_2 \xrightarrow[\text{Solution}]{\text{Sensitizer}/h\nu} AO_2 \qquad (1)$$

These reactions can occur *via* different reaction mechanisms.

Two types of chemical relay mechanisms of sensitization are fairly well understood at the present time[1, 2]. In both cases S in its ground state S_0 is electronically photo-excited by unpairing two electrons, thus forming photo-biradicals, $\cdot S \cdot$ (^1S, ^3S). Type 1 and Type 2 mechanisms differ in the propagation and termination reactions.

Type 1: Here, monoradicals are involved in the propagation and termination reactions. This is due to the primary chemical reaction of $\cdot S \cdot$ with hydrogen donors, AH, *e.g.* isopropanol, to give rise to pairs of mono-radicals $\cdot$SH and $\cdot$A. In further propagation reactions hydroperoxides AOOH, *e.g.* isopropanol hydroperoxide[3, 4], are formed by $\cdot A + O_2 \rightarrow AOO \cdot$, $AOO \cdot + AH \rightarrow AOOH + \cdot A$. In the termination step S_0 is regenerated by $AOO \cdot + \cdot SH \rightarrow AOOH + S_0$. Type 1 sensitization of autoxidation may be called "Bäckström-type photosensitized autoxidation" or "primary dehydrogenation photosensitized reaction with oxygen".

Type 2: In reactions which proceed by the Type 2 mechanism only biradicals take part. Here, $\cdot S \cdot$ adds oxygen to give a short-lived sensitizer-oxygen adduct $\cdot SOO \cdot$, which transfers its oxygen to unsaturated substrates A to AO_2 and the sensitizer in the ground state. This reaction type is called "photosensitized oxygen transfer".

Both these mechanisms are also valid for photosensitized reactions in which compounds other than oxygen are involved; *e.g.*, the photosensitized AH-addition of isopropanol and maleic acid to form terebic acid proceeds by Type 1 sensitization[5], and the photosensitized cyclo-addition of maleic anhydride to benzene or toluene occurs by Type 2 sensitization[6–8].

Only photosensitized oxygen transfer reactions will be discussed here.

The most convenient and thoroughly studied oxygen acceptors are the acenes[9-25], cyclohexadiene derivatives[9, 26-35], olefins with isolated double bonds and allylic hydrogen atoms[36-50], furans[51-55], sulphides[56], and sulphoxides[57]. These compounds are oxidized to endoperoxides, allylic hydroperoxides, ozonides of the yet unknown cyclobutadienes (which in polar solvents such as alcohols react spontaneously to alkoxy-hydroperoxides)[55], sulphoxides, and sulphones, respectively.

$$-\underset{\underset{H}{|}}{C_1}=\underset{|}{C_2}-\underset{|}{C_3}- \quad \longrightarrow \quad -\underset{|}{C_1}-\underset{\underset{OOH}{|}}{C_2}=\underset{|}{C_3}-$$

$$2 \ R—S—R \quad \longrightarrow \quad 2 \ R—SO—R$$

$$2 \ R—SO—R \quad \longrightarrow \quad 2 \ R—SO_2—R$$

Acenes are able to act as sensitizers as well as substrates. Therefore, in most experiments with these compounds no extra-sensitizers were added. Consequently these reactions are, strictly speaking, unsensitized. Nevertheless, as the mechanism of these "photo-oxidations" requires two acene molecules to take part in the reaction, one acting as a light absorber and oxygen carrier, and the other acting as the substrate, these reactions belong to the class following the Type 2 mechanism. In order to distinguish those cases in which the sensitizer and the substrate are identical molecules, and those in which they are different, we may refer to these reactions as occurring *via* "eigen relay mechanisms" and "foreign relay mechanisms", respectively.

In the photosensitized oxygen transfer reactions at least three short-lived intermediates take part which give rise to a number of very fast monomolecular and bimolecular reactions. These short-lived intermediates are the sensitizer in the first excited singlet state and in the lowest excited triplet state, which may be considered from a chemical point of view as a biradical[6, 58, 59] and an oxygen-containing species which in our opinion[60] is the excited sensitizer-oxygen adduct with a biradical structure $\cdot SOO \cdot$ rather than some excited oxygen molecule, *e.g.*, in its lowest excited singlet state[61–65].

Independently of the problem of the nature of the third intermediate we are able to investigate the effectiveness of fluorescein and fluorescein derivatives as sensitizers of the photosensitized oxygen transfer reaction, and to investigate the intrinsic mechanism of the oxygen transfer to olefins which contain isolated double bonds and allylic hydrogen atoms. The nature of the oxygen transferring species will also be discussed.

EFFECTIVENESS OF FLUORESCEIN AND DERIVATIVES AS SENSITIZERS

In *Figure 1* all the individual reaction steps are shown which lead to the formation of AO_2 and which decrease the quantum yield of the formation of AO_2 by side reactions.

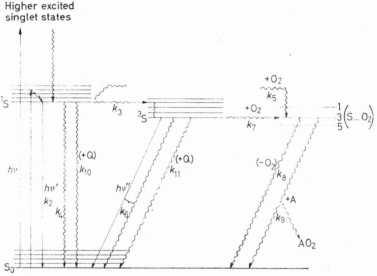

Figure 1. Term scheme; S_0 = sensitizer in the singlet ground state; ^1S = sensitizer in the first excited singlet state; ^3S = sensitizer in the lowest excited triplet state; $\frac{1}{3}(S \ldots O_2)$ = sensitizer-oxygen adduct (described as a Mulliken charge transfer complex in either the singlet-, triplet-, or quintuplet state); A = substrate; Q = quencher; straight arrows = reactions taking place under absorption (*hv*) or emission of light (*hv'* = fluorescence, *hv''* = phosphorescence); curly arrows = radiationless mono- and bimolecular reactions

The absorption of polychromatic light by the sensitizer S_0 leads to higher excited singlet states and to higher vibrational levels of the first excited singlet state. The radiationless transitions from all these levels to the zero vibrational level of the first excited singlet state, $^1S(0)$, take place in a time which is short compared with the lifetime of $^1S(0)$[66, 67]. Therefore practically all the processes such as fluorescence, internal conversion to the ground state, intersystem crossing to the triplet state, and bimolecular reactions, occur from the $^1S(0)$ level. For example, it was shown[68] that with fluorescein and tetrabromo-fluorescein the excess vibrational energy of ^1S is dissipated within 10^{-10} to 10^{-11} sec so that the fluorescence occurs only from the $^1S(0)$ level. Consequently, the quantum yield of fluorescence, γ_f, as well as the quantum yield of the photosensitized oxygen transfer, γ_{AO_2}, is independent of the exciting wavelength, at least with fluorescein dyes as sensitizers.

The intersystem crossing from $^1S(0)$ leads to a triplet excited molecule in a higher vibrational level. From this level two paths for transformation exist: (*i*) intersystem crossing back to $^1S(0)$, (*ii*) transition to the zero level of the triplet state, $^3S(0)$, by dissipation of the excess vibrational energy. By absorption of thermal energy the $^3S(0)$ molecule can reach the crossing point and by path (*i*) the $^1S(0)$ state; delayed fluorescence can then occur. This transition is not shown in *Figure 1* because the probability of path (*i*) compared with path (*ii*) is ten thousand times smaller in the case of tetra-bromofluorescein in ethanol[69].

As well as $^1S(0)$, the triplet excited sensitizer in the $^3S(0)$ state can undergo a radiationless transition to S_0 and/or a transition to the ground state by emission of a photon (phosphorescence). As both transitions are accompanied by a change of multiplicity these transitions are quasi-forbidden, and consequently the life time of $^3S(0)$ is much longer than the life time of $^1S(0)$. Therefore the probability of bimolecular reactions with $^3S(0)$ is some orders of magnitude greater than those with $^1S(0)$. Nevertheless, if the concentration of the reactant such as O_2 or Q is sufficiently high, reactions with $^1S(0)$ have to be taken into account.

In our opinion reactions of $^1S(0)$ and $^3S(0)$ with O_2 form some sensitizer-oxygen adducts, $\cdot SOO \cdot$, rather than produce excited oxygen molecules by mere physical energy transfer[61-65]. This adduct may be interpreted as a Mulliken charge transfer complex in either the excited singlet-, triplet-, or quintuplet state[70] which can undergo a transition to the ground state according to $\frac{1}{3}(S\ldots O_2) \rightarrow {}^3(S_0\ldots O_2) \rightleftharpoons S_0 + O_2$, or may react with the substrate A to give $AO_2 + S_0$†.

For the mechanism shown in *Figure 1* the quantum yield of the oxygen transfer reaction is given by

$$\gamma_{AO_2} = \frac{[O_2]}{k_2 + k_3 + k_4 + k_5\,[O_2] + k_{10}\,[Q]}$$

$$\left(k_5 + k_3\,\frac{k_7}{k_6 + k_7\,[O_2] + k_{11}\,[Q]}\right)\frac{k_9\,[A]}{k_8 + k_9\,[A]} \qquad (2)$$

We investigated the dianions of fluorescein, tetrabromo-, tetraiodo-, and tetraiodo-tetrachloro-fluorescein as sensitizers of the photosensitized oxygen transfer to 2,5-dimethylfuran. For these sensitizers in methanol at 20° the absolute rate constants have been determined in the following way‡:

† It may be assumed that Q also reacts with $^1S(0)$ or $^3S(0)$ to form similar adducts which then decompose to Q and S_0[71, 72].

‡ Experimental details are given in references 73, 74.

The natural lifetimes, τ_e, of the sensitizers in the $^1S(0)$ state were calculated from the absorption spectra and the O-O-transitions, which for all the dyes are shifted to longer wavelengths by about 400 cm^{-1}, using the expression (3).

$$\frac{1}{\tau_e} = k_2 = \frac{8\pi n^2 (\ln 10) c}{N'} F_{\tilde{\nu}} \ (\text{sec}^{-1})^{75} \tag{3}$$

where

$n = 1{,}3288$ (refractive index of methanol at 20°)
$c = 3 \times 10^{10}$ cm/sec (velocity of light)

$$F_{\tilde{\nu}} = \int_0^\infty \frac{(2\tilde{\nu}_0 - \tilde{\nu}_a)^3}{\tilde{\nu}_a} \times \epsilon(\tilde{\nu}_a) \ d\tilde{\nu} \ \text{cm}^{-1}/\text{mmole (area of the absorption band at the long wavelength side)}$$

$\tilde{\nu}_0 =$ wavenumber of the O-O-transition, cm^{-1}
$\tilde{\nu}_a =$ wavenumber (absorption spectrum), cm^{-1}
$\epsilon(\tilde{\nu}_a) =$ molar extinction coefficient (absorption spectrum,) cm^2/mmole.

The quantum yields of fluorescence, γ_f, in methanol were measured by comparing the intensities of fluorescence in methanol with those in aqueous alkaline[76-80] or ethanolic [69, 80] solutions for which the quantum yields are known.

If O_2 and Q are absent γ_f is given by

$$\gamma_f = \frac{k_2}{k_2 + k_3 + k_4} \tag{4}$$

As k_2 is known the sum of $k_3 + k_4$ can be calculated. In the presence of Q we get

$$\gamma_f^Q = \frac{k_2}{k_2 + k_3 + k_4 + k_{10} [Q]} \tag{5}$$

Only with fluorescein and the tetrabromo derivative could an influence of Q (= Cyclo-octatetraene, COT) on γ_f be obtained. Although the concentration of Q was as high as 1 mole/l. no influence on γ_f was observed with either of the tetraiodo compounds.

The lifetimes of the triplet excited sensitizers, $\tau_{3S} = 1/k_6$, and the rate constant k_{11} of the reaction $^3S(0) + Q \rightarrow S_0 + Q$ (with Q = COT) were determined by means of flash photolysis techniques[81-83].

In the absence of Q, and with dimethylfuran as the substrate, A, the quantum yield of the oxygen transfer reaction, γ_{AO_2}, was independent of the concentration of A with $[A] > 2 \times 10^{-3}$ mole/l.; i.e. $k_9 [A] \gg k_8$ and $k_9[A]/(k_8 + k_9[A]) = 1$.
Consequently

$$\gamma_{AO_2} = \frac{[O_2]}{k_2 + k_3 + k_4 + k_5[O_2]} \left(k_5 + k_3 \frac{k_7}{k_6 + k_7[O_2]} \right) \tag{6}$$

Furthermore, γ_{AO_2} was independent of the oxygen concentration with

$[O_2] > 2 \times 10^{-3}$ mole/l. in the case of tetraiodo- and tetraiodo-tetrachloro-fluorescein as sensitizers. At such high oxygen concentrations $k_6 \ll k_7[O_2]$ because generally k_7 is found to be of the order of 10^8 to 10^{10} l./mole sec[21, 23, 84–87]. Therefore equation (6) is reduced to

$$\gamma_{AO_2} = \frac{k_3 + k_5[O_2]}{k_2 + k_3 + k_4 + k_5[O_2]} \tag{7}$$

which allows two interpretations of the fact that there is no oxygen concentration dependence of γ_{AO_2}:

either $k_5[O_2] \gg k_2 + k_3 + k_4$ and $\gamma_{AO_2} = 1$
or $k_5[O_2] \ll k_3$ and $\gamma_{AO_2} < 1$.

If the first alternative were true, k_5 has to be of the order of 10^{12} to 10^{13} l./mole sec (because $k_2 + k_3 + k_4 = 2 \cdot 3 \times 10^9$ sec^{-1}), which is much greater than the maximum value of about 10^{11} l./mole sec. Furthermore, γ_{AO_2} is in every case smaller than unity. We are, therefore, forced to assume that $k_5[O_2] \ll k_3$. This assumption is supported by the fact that γ_f was always found to be independent of $[O_2]$[75, 88, 89]. Consequently, we arrive at the expression

$$\gamma_{AO_2} = \frac{k_3}{k_2 + k_3 + k_4} = \gamma_{3S} \tag{8}$$

By this method it is possible to determine the quantum yield of the triplet formation, γ_{3S}, by measuring the quantum yield of the photosensitized oxygen transfer reaction. The sum of $k_3 + k_4$ being known, the absolute rate constants k_3 and k_4 can be calculated†.

In the case of fluorescein and tetrabromo-fluorescein there was a slight dependence of γ_{AO_2} on the oxygen concentration. By equation (7) k_5 was determined, though the limits of error were great.

In the presence of COT as the quencher Q, the quantum yield γ_{AO_2} depends on both the quencher and oxygen concentration. With the quencher concentrations used (from 1×10^{-2} to 6×10^{-2} mole/l.) the quantum yields of fluorescence, γ_f, were not affected. $\gamma_{AO_2}^Q$ for the inhibited reaction is given by

$$\gamma_{AO_2}^Q = \gamma_{3S} \frac{1}{1 + k_{11}[Q]/k_7[O_2]} \tag{9}$$

which is correct in the case of the two tetraiodo-fluoresceins and nearly so for the tetrabromo-fluorescein. As required by equation (10)

$$\gamma_{AO_2}/\gamma_{AO_2}^Q = 1 + \frac{k_{11}}{k_7[O_2]}[Q] \tag{10}$$

† According to Noyes[90] there exists no experimental evidence for the assumption of a radiationless transition $^1S(0) \to S_0$. With the sensitizers used in our work the sums of $\gamma_f + \gamma_{3S}$ are found to be from 0·7 to nearly 1 (*Table 1*). On account of the difficulties involved in quantum yield determinations there is some doubt whether the $^1S(0) \to S_0$ radiationless transitions really exist. Nevertheless, our preliminary results with rhodamin 3B as a sensitizer gave $\gamma_{AO2} = 0\cdot02$, $\gamma_{3S} < 0,02$, and $\gamma_f = 0,65$ in methanol at 20°C which seems to support the existence of some path for the radiationless transition $^1S(O) \to S_0$.

straight lines are observed for each oxygen concentration by plotting the ratio of the γ's against [Q]. From the slope of the lines $k_{11}/k_7[O_2]$ is calculated. As k_{11} and $[O_2]$ are known k_7 can be determined for each sensitizer.

The results of our measurements and calculations are shown in *Table 1*.

Table 1. Rate constants, γ_t, γ_{3s}, and γ_{AO_2}; sensitizers: fluorescein, tetrabromo-, tetraiodo-, and tetraiodo-tetrachloro-fluorescein

	Fluorescein X = H Y = H	Tetrabromo- X = Br Y = H	Tetraiodo- X = I Y = H	Tetraiodo-tetrachloro- X = I Y = Cl
k_2	2·2	2·2	1·8	$1·8 \times 10^8$ sec^{-1}
k_3	0·08	0·88	14·1	$17·3 \times 10^8$ sec^{-1}
k_4	0·09	0·42	6·8	$3·6 \times 10^8$ sec^{-1}
γ_t	0·93	0·63	0·08	0·08
γ_{3s}	0·03	0·3	0·6	0·76
k_6		2·2	5·8	$6·5 \times 10^3$ sec^{-1}
k_5	~1	~5		$\times 10^9$ l./mole sec
k_7		1·2	1·2	$1·2 \times 10^9$ l./mole sec
k_{10}^{COT}	~4·5	~3·5		$\times 10^8$ l./mole sec
k_{11}^{COT}		1·4	1·1	$0·5 \times 10^9$ l./mole sec
γ_{AO_2}	0·1	0·4	0·6	0·76

Schenck and Koch[2] determined the lifetime of the triplet state of tetraiodo-tetrachloro-fluorescein in oxygen saturated methanol at 20° to be less than 2×10^{-7} sec. With $k_7 = 1·2 \times 10^9$ l./mole sec and $[O_2] = 1·057 \times 10^{-2}$ mole/l., this lifetime should be $1/k_7[O_2] = 0·9 \times 10^{-7}$ sec.

As was already mentioned γ_{AO_2} is independent of A when $[A] > 2 \times 10^{-3}$ mole/l. (A = dimethyl furan). This permits us to estimate the minimum lifetime of the sensitizer-oxygen-adduct. Assuming k_9 to be a diffusion-controlled rate constant $k_9 = 8RT/3000\eta \approx 10^{10}$ l./mole sec ($\eta = 0·584 \times 10^{-2}$ erg sec/cm^3 for methanol at 20°). Because of $k_8 \ll k_9[A]$ we get $1/k_8 \gg 5 \times 10^{-8}$ sec. Measurements of the lifetime of the sensitizer-oxygen-adduct with the sensitizers tetrabromo- and tetraiodo-tetrachloro-fluorescein led to lifetimes $1/k_8 < 13 \times 10^{-8}$ and $< 20 \times 10^{-8}$ sec, respectively[2].

The dependence of γ_{AO_2} on the oxygen concentration is given by

$$\gamma_{AO_2} = \gamma_{3S} \frac{k_7[O_2]}{k_6 + k_7[O_2]} \tag{11}$$

for tetraiodo- and tetraiodo-tetrachloro-fluorescein. With the rate constants of *Table 1* γ_{AO_2} is found to be $\frac{1}{2}\gamma_{3S}$ at $[O_2] = 6 \times 10^{-6}$ mole/l. Experiments on the determination of this "half-value concentration of oxygen" were carried out very recently by Franken in our laboratory. His preliminary results show this particular oxygen concentration to be between 5×10^{-6} and 2×10^{-5} mole/l.

With the assumption that triplet excited tetrabromo-fluorescein in ethanol at $-196°$ decays exclusively by emission of phosphorescence light quanta Parker and Hatchard[69] calculated $\gamma_{3S} = 0.025$ at $25°$ in ethanol. With the same sensitizer in methanol and ethanol at $20°$ we found (by the method discussed) $\gamma_{3S} = 0.3$. In our opinion this discrepancy indicates that even at $-196°$ the radiationless transition $^3S(0) \to S_0$ is very effective.

Adelman and Oster[78] investigated a series of fluorescein dyes in aqueous solution. They found that all the γ_{3S}'s had approximately the same values while the γ_f's decreased in the same order as they do in our experiments. The authors concluded that only the radiationless transitions $^1S(0) \to S_0$ are enhanced by halogenation of the fluorescein molecule. Similar results were obtained by Forster and Dudley[91]. This is in contrast to the behaviour of these sensitizers in alcoholic solutions.

As is shown in *Table 1* there is nearly no influence of halogenation on the k_2-values (or natural lifetimes $1/k_2$) of the different sensitizers. On the other hand substitution of $X = H$ by $X = Br$ and $X = I$ enhances the internal conversion, $^1S(0) \to S_0$, as well as (somewhat more strongly) the inter-system crossing, $^1S(0) \to {}^3S(0)$. Consequently, γ_{3S} is increased by the introduction of "heavy" halogen atoms into the fluorescein molecule. This effect of heavy halogen atoms is due to enhanced spin-orbit coupling which causes some mixing of singlet and triplet states[92-95].

As a consequence of the very short lifetime of $^1S(0)$ the oxygen transferring intermediate is exclusively (or nearly exclusively) formed by the reaction of $^3S(0)$ with oxygen.

MECHANISM OF OXYGEN TRANSFER TO OLEFINS CONTAINING ISOLATED DOUBLE BONDS

In considering the termination stage of the photosensitized oxygen transfer reactions we restrict our discussion to the reactions of the oxygen trans-ferring species with olefins which contain isolated double bonds and allylic hydrogen atoms.

Since the discovery of this particular reaction[36-38] no exception from the rule given in *Figure 2* has been found:

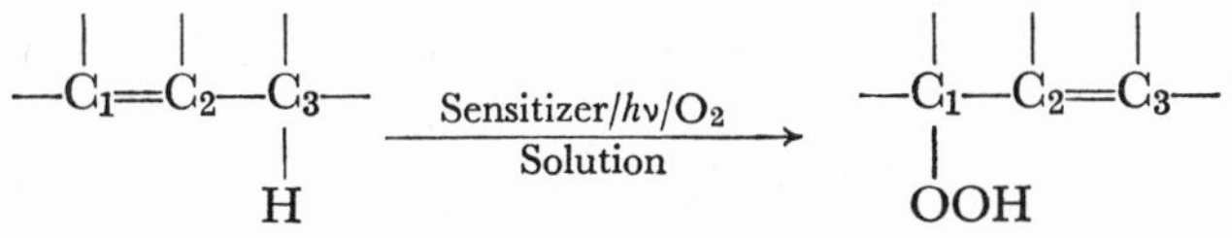

Figure 2

i.e., when oxygen in a photosensitized reaction is transferred to olefins the oxygen is added to one carbon atom of the double bond (in this case to C–1), the double bond is shifted into the allyl position (with formation of the double bond $C_2=C_3$), and the allylic hydrogen atom at C–3 is moved to the oxygen atom which is not attached to a carbon atom. As will be shown later this reaction takes place as a concerted reaction[48, 73, 74, 96].

This reaction type differs considerably from that of the Type 1 autoxidation which is initiated, *e.g.*, by the decomposition of peroxides or by irradiation. In the Type 1 autoxidation reaction[97–99] the allylic hydrogen atom at C–3 is abstracted by an initiator radical to give a mesomeric monoradical

$$-\overset{|}{C_1}=\overset{|}{C_2}-\overset{|}{\underset{\bullet}{C_3}}- \quad \longleftrightarrow \quad -\overset{|}{\underset{\bullet}{C_1}}-\overset{|}{C_2}=\overset{|}{C_3}-$$

O_2 is then attached at both radical sites, C_1 and C_3, and the peroxy radicals thus formed abstract hydrogen from the C–3 position of the olefin to give hydroperoxides and a new mesomeric monoradical. In contrast to the photosensitized oxygen transfer reaction a chain propagation can occur.

With (+)-limonene (Ia) and (+)-carvomenthene (Ib) (*Figure 3*), both these reactions were studied in detail[46, 100–102].

In addition to other products, *cis* and *trans* carveyl hydroperoxides and carvotanacetyl hydroperoxides respectively are formed in both oxidation

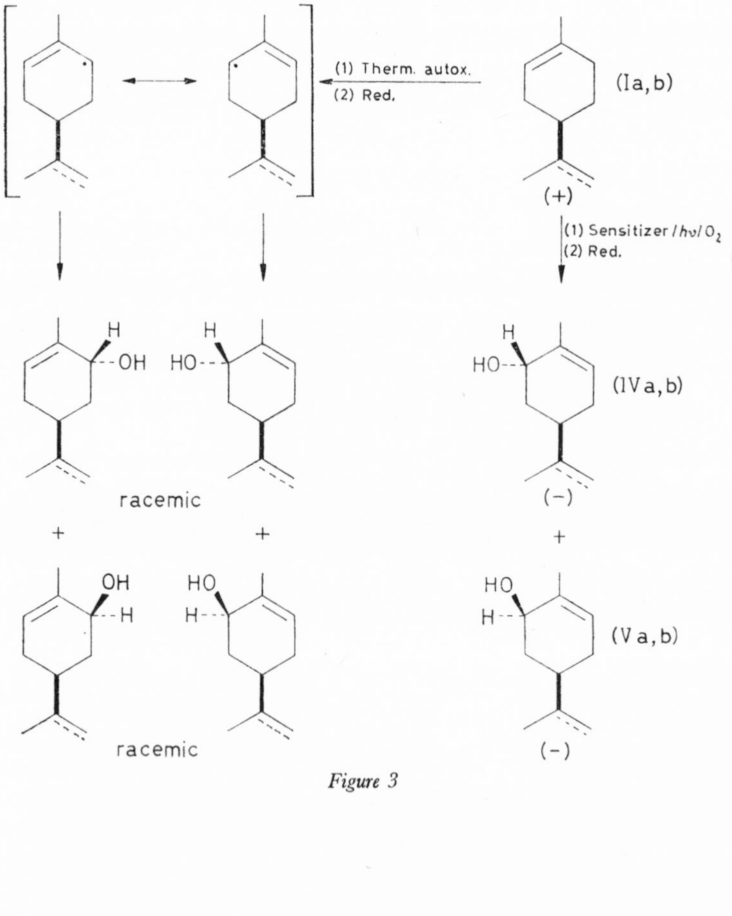

Figure 3

reactions. After reduction of the hydroperoxides with retention of configuration, the corresponding carveols and carvotanacetols respectively show a remarkable difference: those obtained from the photosensitized oxygen transfer reaction are optically active, those obtained from the thermal autoxidation reaction are optically inactive (racemic) (*Figure 3*).

Furthermore, the exclusive formation of (−)-*cis* and (−)-*trans*-carveols and of (−)-*cis*- and (−)-*trans*-carvotanacetols respectively shows that, in the photosensitized oxygen transfer reaction, there is no insertion of oxygen between the carbon atom α to the double bond and the allylic hydrogen atom; if there were, the (+)-carveols and the (+)-carvotanacetols respectively would be formed.

According to *Figure 2* four other products of the photosensitized transfer of oxygen to (+)-limonene and (+) -carvomenthene are formed in addition to the carveols and carvotanacetols respectively†.

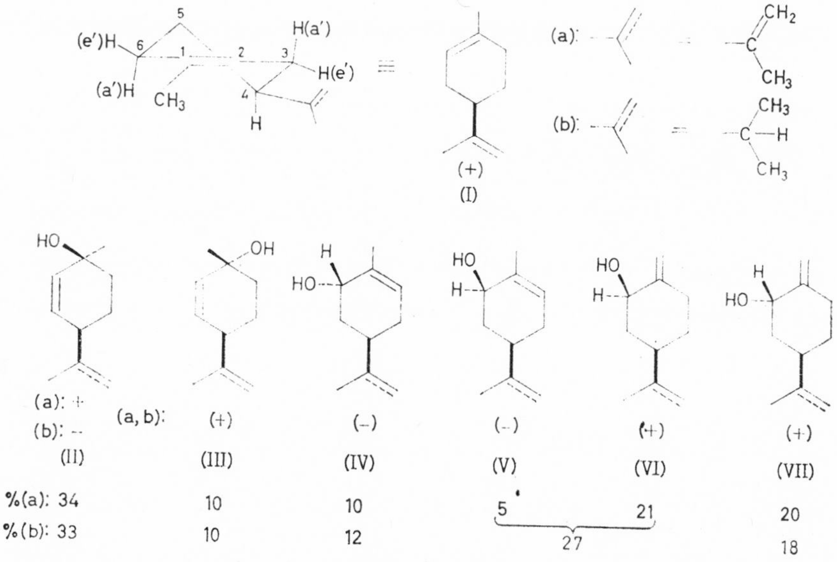

Figure 4

As can be seen from *Figure 4* the products are not formed in a statistical manner. If this were the case there should be 25 per cent of both the tertiary alcohols and 12·5 per cent of each of the four secondary alcohols. The deviations from the statistical product distribution can be explained by conformational analysis. According to these results[103] the most probable conformations of limonene and carvomenthene at room temperature are those in which the largest substituents are in an equatorial position as shown in *Figure 4*. In these conformations the allylic hydrogen atoms at C–3 and C–6 are in quasi-axial (a') and quasi-equatorial (e') positions. In

† The Δ⁸-double bond of limonene is not attacked under the conditions used. This is in accord with the rule that oxygen is transferred much faster to tri- and tetra-substituted ethylenes than to disubstituted ethylenes[10, 51, 55, 73, 74].

contrast to this the allylic hydrogen atoms of the CH_3-group are equally present on both sides of the molecule because of the free rotation of this group. If we further take into account the results of Nickon and Bagli[47, 48], who found that with cyclohexene derivatives only the allylic hydrogen atom is used which is *cis* to the oxygen attack, we are able to correlate the product distribution with the availability of allylic hydrogen atoms.

With the product pairs (II/III) and (IV/V) it is easily seen that (II) and (IV) are formed in a reaction in which quasi-axial allylic hydrogen atoms at C–3 and C–6, respectively, are used. On the other hand (III) and (V) which are formed in smaller amounts, arise from reactions in which quasi-equatorial allylic hydrogen atoms at C–3 and C–6, respectively, shonld have been used. In agreement with the theory (VI) and (VII) are formed in equal amounts.

The fact that (VI) is formed in the same amount as (VII), and that the yield of (II) is more than three times that of (III), excludes a steric effect of the substrates. Otherwise the yields of (II) and (VI), formed by an oxygen attack *cis* to the side-chain at C–4, would have been smaller than those of (III) and (VII), respectively.

We therefore conclude that the stereoselectivity of the photosensitized oxygen transfer to limonene and carvomenthene arises from the different availability of allylic hydrogen atoms.

This effect was also shown in the aliphatic series[55].

(VIIIa): $R_1 = H$ $R_2 = H$ (IXa): 46%   (Xa): 54%
   (b): $R_1 = H$ $R_2 =$ Alkyl   (b): 45%   (b): 55%
   (c): $R_1 = R_2 = CH_3$ (c): 95%   (c): 5%

While in (VIIIa) and (VIIIb) there is always an "axial" allylic hydrogen atom of the freely rotating $C_3HR_1R_2$ group available, this is no longer the case with (VIIIc). An inspection of Stuart-Briegleb models shows that in (VIIIc) the free rotation of the $CH(CH_3)_2$ group is strongly hindered. Furthermore, the most stable conformation of this molecule is that in which the H atom lies in the plane of the double bond ("equatorial"). As a consequence of the "equatorial" position of this allylic hydrogen atom (IXc) is found in 95 per cent yield although C_2 is more shielded against an oxygen attack than C_1.

The stereoselectivity of photosensitized oxygen transfer is also influenced by the steric effect of bulky groups in the substrate molecule. This is demonstrated in *Figure 5* with (+)–Δ^3–carene (XI) as a substrate[44, 104].

The equilibrium of the two conformations of (+)-Δ^3-carene, (XIa) and (XIb), at room temperature is not known. From the ease of interconversion of Dreiding models of (XIa) and (XIb), the equilibrium constant is probably near unity. The allylic hydrogen atoms at C–2 and C–5 possess "axial" positions when they are *cis* to the dimethyl cyclopropane ring in (XIa), and when

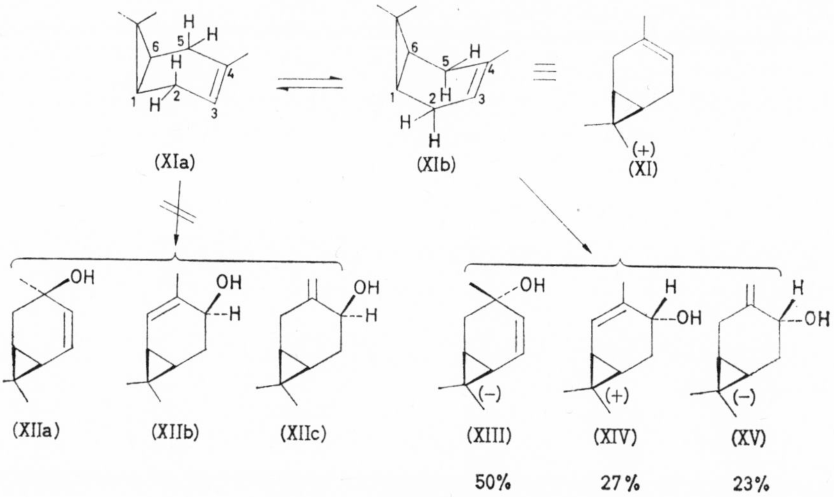

Figure 5

they are *trans* to this ring in (XIb). Therefore, all these hydrogen atoms should be as available as those of the freely rotating CH_3 group. This should give rise to a product containing equal amounts of tertiary and secondary alcohols; *i.e.* 25 per cent of each of the tertiary, and 12·5 per cent of each of the secondary alcohols. In fact only the *trans* alcohols (XIII–XV) are formed. Nevertheless, our assumption on the equal availability of the allylic hydrogen atoms seems to be correct as is seen from the 2 : 1 : 1 ratio of the products formed. Consequently, the absence of the formation of *cis* alcohols must be due to the complete inhibition of a *cis* oxygen attack due to the bulky dimethyl cyclopropane ring[45].

A factor which influences the ease of photosensitized oxygen transfer to ethylene compounds as well as to cyclohexadiene derivatives, is the electron density of the double bonds[73, 74]. In the cyclohexadiene series, for example, the formation of an endoperoxide occurs only with (XVI). No reaction takes place with (XVII)[51].

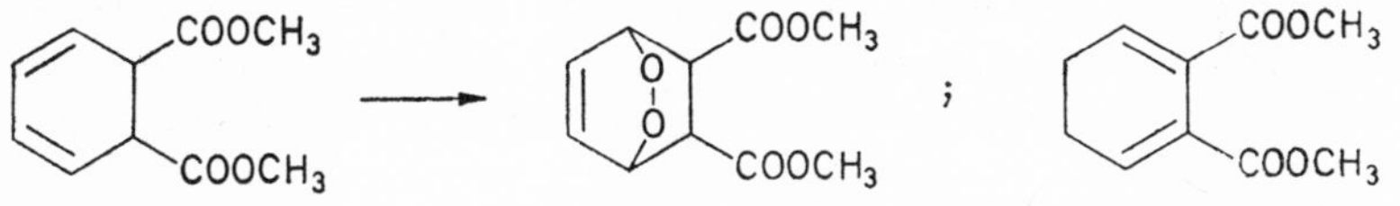

(XVI)   (XVII)

With olefins an increase in the quantum yield is observed in the order 1,2-dimethylethylene < trimethylethylene < tetramethylethylene, and cyclohexene < 1-methylcyclohexene < 1,2-dimethylcyclohexene[40, 55, 105]. This increase is due to the enhanced electron densities caused by the electron-donating methyl groups[73, 74].

Cis and *trans* dimethylstilbene (XVIII and XIX, respectively) give the same oxidation product (XX)[55], but the rate of oxygen transfer, r_1, with (XVIII) is considerably larger than that, r_2, with (XIX).

The higher oxidation rate of (XVIII) seems again to be due to the enhanced electron density in the *cis* compound as compared with the *trans*.

It is most interesting that the rate r_1 is not changed to r_2, nor is the rate r_2 changed to r_1, during the reactions of (XVIII) or (XIX), respectively. These observations, together with the fact that the oxygen transfering species is not able to isomerize *cis* stilbene into the *trans* compound and *vice*

(XVIII)   (XX)   (XIX)

versa[55], provide strong support for our view that the photosensitized oxygen transfer reactions occur as concerted reactions *via* a cyclic transition state[48, 73, 74, 96]. With olefins the transition state consists of the two carbon atoms of the double bond, the carbon atom α to this double bond, the allylic hydrogen atom at C–α, and the oxygen molecule.

NATURE OF THE OXYGEN TRANSFERRING SPECIES

In this part we are concerned with the problem of the nature of the third intermediate in the photosensitized oxygen transfer reactions.

Two possible mechanisms were soon excluded for spectroscopic, kinetic, energetic, and chemical reasons[58, 60–65, 106, 107]: the "acceptor activation mechanism" proposed by Gaffron[14, 15], in which the electronic energy of the excited sensitizer is transferred to the substrate, and the mechanism proposed by Weiss[108–110], in which an electron is transferred from the excited sensitizer to the oxygen molecule. So we are left with the question: is the oxygen transferring species an excited oxygen molecule in its lowest excited singlet state as was proposed by Kautsky[61–65], or is it an excited oxygen-sensitizer adduct (complex) as was independently proposed by Schönberg[12, 13] Terenin[106, 107], and Schenck[58]?

As the very effective quenching of the fluorescent state, $^1S(0)$, of poly-cyclic hydrocarbons such as anthracene cannot be brought about by the energy transfer reaction $^1S(0) + {}^3O_2 \rightarrow S_0 + {}^1O_2$, on account of the violation of the spin conservation rule, Terenin[106] proposed a "paramagnetic" quenching of oxygen of the type $^1S(0) + {}^3O_2 \rightarrow {}^3S(0) + {}^1O_2$: "The main feature of such a "paramagnetic" quenching is that not only an active O_2 molecule is formed during this process, but that the second partner, *i.e.* the aromatic molecule, has also acquired chemical reactivity, having been trans-formed into the biradical $^3S(0)$†. We expect therefore, that "paramagnetic"

† The electronic states such as the triplet state *etc.* are given in this and other citations in the notation of the present authors.

quenching will be accompanied by association or dimerization processes, induced by the biradical $^3S(0)$. In the first place, the latter can easily react with O_2 before they separate, according to the mechanism: $^1S(0) + {}^3O_2 \rightarrow {}^3(S..O_2)$, with the formation of a more-or-less stable peroxide, which, certainly, will be produced in one step, and is already in the transition complex formed by a suitable encounter of O_2 with the excited molecule". The existence of such an unstable addition product was proved by an extensive study of the photoconductivity of solid dye films[106].

In a thorough investigation of intramolecular and intermolecular energy conversion involving change of multiplicity Porter and Wright[111] concluded that, in the reaction of a triplet state, or biradical state, molecule with a paramagnetic quencher such as oxygen, a collision complex is formed which should have a stability of several kcal. Spin–spin interaction will therefore be strong and the complex may have a considerable life. These authors further concluded: "It is to be expected that the radiationless transition probability is increased in the presence of an efficient paramagnetic quencher only by an amount corresponding to the difference between a spin-forbidden and a spin-allowed transition, i.e. by a factor of about 10^4. Now the lifetime of triplet anthracene in n-hexane in the absence of quenchers is 10^{-3} sec so that its lifetime when the spin restriction is removed should be about 10^{-7} sec. The average lifetime of the collision complex between triplet anthracene and oxygen, nitric oxide, or a second triplet state, should therefore also be about 10^{-7} sec which is much longer than the duration of an encounter not involving chemical interaction and is in accordance with kinetic studies of anthracene photosensitized oxidation".

It may be stated that our own results fit very well with the last statement: with tetraiodo-tetrachloro-fluorescein we found

$$\tau_{3S} = 1 \cdot 5 \times 10^{-4} \text{ sec and } 5 \times 10^{-8} \text{ sec} < \tau^{\frac{1}{3}}_{(S \cdots O_2)} < 2 \times 10^{-7} \text{ sec}$$

Linschitz and Pekkarinen[112], as well as Tsubomura and Mulliken[70], explain the quenching of excited singlet and triplet states by molecules like oxygen as due to the formation of short-lived charge transfer adducts. In the theory of the latter authors there exists in addition to the $^1_3(S...O_2)$ state given in *Figure 1* a triplet charge transfer state, ^3CT, with a great probability of interaction of both these states. Mulliken describes the reaction of excited molecules with oxygen with respect to Kautsky's theory as follows: "the energy difference between the singlet and triplet states of the donor is dissipated as thermal energy (vibrational first), and no excitation of the oxygen molecule to its metastable states is necessarily involved". Although the Mulliken theory has one feature in common with the Weiss theory in relating the quenching action of oxygen to charge transfer "such a complete electron transfer as he (Weiss) proposes is very unlikely in non-polar solutions, where the present theory of the mechanism of quenching seems to be much more adequate". Similar theories were also proposed by Murrel[113] and Hoijtink[114].

Taking very seriously the theory that triplet and singlet excited molecules have more or less biradical character (compare e.g. Porter[115]) Schenck[1, 2, 6, 58–60, 71, 72, 116] proposed that electronically excited molecules

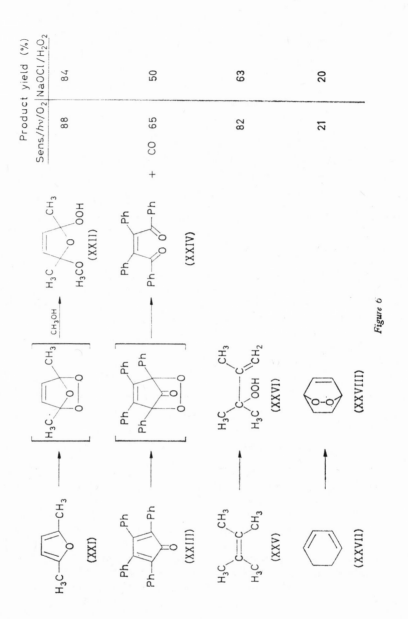

Figure 6

136

should behave chemically in the manner of free radicals. Therefore, reactions with oxygen should not only be very effective in quenching $^1S(0)$ and $^3S(0)$, but should also lead to new biradical molecules in which the oxygen is attached to a former radical site by a normal chemical bond. The effect of steric hindrance to oxygen attack, caused by bulky groups in substrates where these groups are somewhat removed from the reaction centres, led us again to the assumption that the oxygen must have the big sensitizer molecule in close proximity when it reacts with the substrate.

However, the problem of the oxygen transferring species is not yet settled. Very recently the assumption that a sensitizer-oxygen adduct is the oxygen transferring species was severely criticized by Foote and Wexler[117]†. In the reaction of hypochlorite with hydrogen peroxide excited singlet oxygen in its $^1\Delta g$ state is formed, at least as an intermediate in the formation of the light emitting species $O_4^{*[118, 119]}$. Foote made some experiments with substrates such as dimethylfuran (XXI), tetraphenylcyclopentadienone (XXIII), tetramethylethylene (XXV), and cyclohexadiene (XXVII) in methanol containing hydrogen peroxide (*Figure 6*). By dropwise addition of hypochlorite to the solution at $10°$ he succeeded in isolating 2,5-dimethyl-2-hydroperoxy-5-methoxydihydrofuran (XXII), *cis*-dibenzoyl stilbene (XXIV), 2,3-dimethyl-3-hydroperoxybutene-(1) (XXVI), and norascaridol (XXVIII), respectively.

These results are very interesting, and in the case of the cyclic diene compounds no other explanation than that given by Foote seems to be possible for the moment in order to account for the formation of the products isolated. But with the tetramethylethylene (XXV) a normal autoxidation reaction induced by hydrogen abstraction might have taken place.

In order to check Foote's assumption in the case of olefins we carried out some experiments with (+)-α-pinene (XXIX) under the same conditions (in the presence of air) as were used in the experiments cited above, as well as in the presence of nitrogen and oxygen, respectively. The results of these experiments were compared with those obtained by photosensitized oxygen transfer to (XXIX) at $20°^{[38, 120]}$, and by benzoyl peroxide induced autoxidation at $80–120°^{[121]}$.

As can be seen from *Figure 7* the reactions of α-pinene with hydrogen peroxide/sodium hypochlorite are very similar to the thermal autoxidation process in which chain reactions are induced mainly by hydrogen abstraction from C–4, resulting in the formation of the products (XXXIII–XXXVI), and to a smaller degree by hydrogen abstraction from C–7, resulting in the formation of the products (XXX–XXXII).

In our opinion this is a striking proof that in the reactions of olefins with excited singlet oxygen this reactive oxygen has enhanced dehydrogenating properties, as was proposed by Dupont[122], rather than the properties of the oxygen-transferring species in the photosensitized oxygen transfer reactions.

References

[1] a. G. O. Schenck. *Ind. Eng. Chem.* **55**, 40 (1963).
   b. G. O. Shenk, O.-A. Neumüller, and R. Koch. Strahlentherapie, **114**, 321 (1961).
[2] G. O. Schenck and E. Koch. *Z. Elektrochem.* **64**, 170 (1960).
[3] G. O. Schenck and H.-D. Becker. *Angew. Chem.* **70**, 504 (1958).

† We thank Dr Foote for making his results available to us prior to publication.

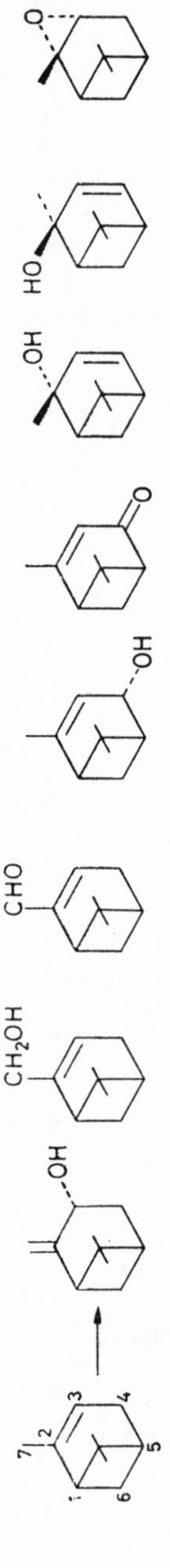

(XXIX)	(XXX)	(XXXI) (XXXII)	(XXXIII)	(XXXIV)	(XXXV) (XXXVI) (XXXVII)
Sens./$h\nu$/O_2	94				
Autoxidation	14	13	33	9	31
NaOCl/H_2O_2 in the presence of air	9	10	21	10	50
N_2	12	10	23	9	46
O_2	9	11	22	6	52

Figure 7

[4] G. O. Schenck, H.-D. Becker, K.-H. Schulte-Elte, and C. H. Krauch. *Chem. Ber.* **96**, 509 (1963).

[5] G. O. Schenck, G. Koltzenburg, and H. Grossmann. *Angew. Chem.* **69**, 177 (1957).

[6] G. O. Schenck. *Z. Elektrochem.* **64**, 997 (1960).

[7] G. O. Schenck and R. Steinmetz. *Tetrahedron Letters* No. 21, 1 (1960).

[8] G. O. Schenck and R. Steinmetz. *Bull. Soc. Chim. Belges* **71**, 781 (1962).

[9] W. Bergmann and M. J. McLean. *Chem. Rev.* **28**, 367 (1941).

[10] Ch. Dufraisse. *Bull. Soc. Chim. France* (4) **53**, 789 (1933).

[11] A. Etienne, "Photo-oxydes d'acènes," *Traité de chimie organique* de V. Grignard, Vol. 17, 1299 (1949).

[12] A. Schönberg. *Ber.* **67**, 633 (1934).

[13] A. Schönberg. *Ann.* **518**, 299 (1935).

[14] H. Gaffron. *Biochem. Z.* **264**, 251 (1933).

[15] H. Gaffron. *Z. Physik. Chem.* **B37**, 437 (1937).

[16] W. Koblitz and H.-J. Schumacher. *Z. Physik. Chem.* **B35**, 11 (1937).

[17] H.-J. Schumacher. *Z. Physik. Chem.* **B37**, 462 (1937).

[18] E. J. Bowen and F. Steadman. *J. Chem. Soc.* **1934**, 1098.

[19] E. J. Bowen and D. W. Tanner. *Trans. Faraday Soc.* **51**, 475 (1955).

[20] R. Livingston, in L. J. Heidt, R. S. Livingston, E. Rabinowitch, and F. Daniels. *Photochemistry in the Liquid and Solid States.* Symp. Endicott House, Dedham, Mass., USA, 3–7 Sept. (1957), p. 76, J. Wiley and Sons, Inc., New York (**1960**).

[21] R. Livingston. *J. Chim. Phys.* **55**, 887 (1958).

[22] R. Livingston and D. W. Tanner. *Trans. Faraday Soc.* **54**, 765 (1958).

[23] G. Porter and M. W. Windsor. *Discussions, Faraday Soc.* **17**, 178 (1954).

[24] G. O. Schenck, W. Müller, and H. Pfennig. *Naturwiss.* **41**, 374 (1954).

[25] G. O. Schenck. *Naturwiss.* **40**, 212 (1953).

[26] A. Windaus and J. Brunken. *Ann.* **460**, 225 (1928).

[27] G. O. Schenck and K. Ziegler. *Naturwiss.* **32**, 157 (1944).

[28] G. O. Schenck. *Angew. Chem.* **61**, 434 (1949).

[29] G. O. Schenck and H. Ziegler. *Naturwiss.* **38**, 356 (1951).

[30] G. O. Schenck, K. G. Kinkel, and H.-J. Mertens. *Ann.* **584**, 125 (1953).

[31] G. O. Schenck and D. E. Dunlap. *Angew. Chem.* **68**, 248 (1956).

[32] G. O. Schenck and K. Ziegler. *Synthesen zweier cyclischer Naturstoffe mit Brückensauerstoffatomen, Festschrift.* A. Stoll, p. 620, Birkhäuser Verlag, Basel (1956).

[33] R. N. Moore and R. V. Lawrence. *J. Am. Chem. Soc.* **80**, 1438 (1958).

[34] R. N. Moore and R. V. Lawrence. *J. Am. Chem. Soc.* **81**, 458 (1959).

[35] W. H. Schuller, R. N. Moore and R. V. Lawrence. *J. Am. Chem. Soc.* **82**, 1734 (1960).

[36] G. O. Schenck. *Dtsch. Bundes-Pat.* 933925 from 24 Dec. (1943). (*C.Z.* **1956**, 3998).

[37] G. O. Schenck. *Fiat-Review of German Science* 1939–1946, *Preparative organic chemistry*, Part II, Senior Author K. Ziegler; XII, Reaktionen mit molekularem Sauerstoff, p. 167 (1948).

[38] G. O. Schenck, H. Eggert, and W. Denk. *Ann.* **584**, 177 (1953).

[39] G. O. Schenck, K. Gollnick, and O.-A. Neumüller. *Ann.* **603**, 46 (1957).

[40] G. O. Schenck and K.-H. Schulte-Elte. *Ann.* **618**, 185 (1958).

[41] G. O. Schenck and O.-A. Neumüller. *Ann.* **618**, 194 (1958).

[42] G. Ohloff, E. Klein and G. O. Schenck. *Angew. Chem.* **73**, 578 (1961).

[43] G. O. Schenck, E. Koerner von Gustorf, and H. Köller. *Angew. Chem.* **73**, 707 (1961).

[44] G. O. Schenck, S. Schroeter, and G. Ohloff. *Chem. Ind.* (*London*) **1962**, 459.

[45] G. O. Schenck, K. Gollnick, G. Buchwald, G. Ohloff, G. Schade, and S. Schroeter. *Angew. Chem.* **76**, 582 (1964).

[46] G. O. Schenck, K. Gollnick, G. Buchwald, S. Schroeter, and G. Ohloff. *Ann.* **674**, 93 (1964).

[47] A. Nickon and F. Bagli. *J. Am. Chem. Soc.* **81**, 6330 (1959).

[48] A. Nickon and F. Bagli. *J. Am. Chem. Soc.* **83**, 1498 (1961).

[49] R. L. Kenney and G. S. Fisher. *J. Org. Chem.* **28**, 3509 (1963).

[50] W. H. Schuller and R. V. Lawrence. *J. Am. Chem. Soc.* **83**, 2563 (1961).

[51] G. O. Schenck. *Angew. Chem.* **64**, 12 (1952).

[52] G. O. Schenck. *Ann.* **584**, 156 (1953).

[53] G. O. Schenck and Ch. Foote. *Angew. Chem.* **70**, 505 (1958).

[54] G. O. Schenck. *Angew. Chem.* **69**, 579 (1957).

[55] K.-H. Schulte-Elte. *Diss.*, Göttingen (1961).

[56] G. O. Schenck and C. H. Krauch. *Angew. Chem.* **74**, 510 (1962).

[57] G. O. Schenck and C. H. Krauch. *Chem. Ber.* **96**, 517 (1963).

[58] G. O. Schenck. *Naturwiss.* **35**, 28 (1948).

[59] G. O. Schenck. *Z. Naturforsch.* **3b**, 59 (1948).

[60] G. O. Schenck. *Naturwiss.* **40**, 205, 229 (1953).

[61] H. Kautsky and A. Hirsch. *Ber.* **64**, 2677 (1931).

[62] H. Kautsky, A. Hirsch, and F. Davidshöfer. *Ber.* **65**, 1762 (1932).
[63] H. Kautsky, H. de Bruin, R. Neuwirth, and W. Baumeister. *Ber.* **66**, 1588 (1933).
[64] H. Kautsky and A. Hirsch. *Biochem. Z.* **274**, 423 (1934).
[65] H. Kautsky, A. Hirsch, and W. Flesch. *Ber.* **68**, 152 (1935).
[66] B. Sveshnikov. *Dokl. Acad. Nauk SSSR* **58**, 49 (1947).
[67] G. A. Mokeeva and B. Ya. Sveshnikov. *Opt. Spectr. (USSR)* **10**, 41 (1961).
[68] G. Weber and F. W. J. Teale. *Trans. Faraday Soc.* **54**, 640 (1958).
[69] C. A. Parker and C. G. Hatchard. *Trans. Faraday Soc.* **57**, 1894 (1961).
[70] H. Tsubomura and R. S. Mulliken. *J. Am. Chem. Soc.* **82**, 5966 (1960).
[71] G. O. Schenck and R. Wolgast. *Naturwiss.* **48**, 737 (1961).
[72] G. O. Schenck and R. Wolgast. *Naturwiss.* **49**, 36 (1962)
[73] K. Gollnick. *Diss.* Göttingen (1962).
[74] G. O. Schenck and K. Gollnick. *Forschungsber. Land Nordrhein-Westfalen, No. 1256,* Westdeutscher Verlag, Köln and Opladen (1963).
[75] Th. Förster. *Fluoreszenz organischer Verbindungen,* Vandenhoeck and Ruprecht, Göttingen (1951).
[76] M. Imamura. *Bull. Chem. Soc. Japan* **31**, 62 (1958).
[77] L. S. Forster and R. Livingston. *J. Chem. Phys.* **20**, 1315 (1952).
[78] A. H. Adelman and G. Oster. *J. Am. Chem. Soc.* **78**, 3977 (1956).
[79] G. Weber and F. W. J. Teale. *Trans. Faraday Soc.* **53**, 646 (1957).
[80] V. Zanker and H. Rammensee. *Z. Phys. Chem. (Frankfurt)* **26**, 168 (1960).
[81] G. Porter. *Proc. Roy. Soc. (London)* **A200**, 284 (1950).
[82] G. Porter. *Radiation Res. Suppl.* **1**, 479 (1959).
[83] L. Lindqvist. *Arkiv Kemi* **16**, 79 (1960).
[84] G. Porter and M. R. Wright. *J. Chim. Phys.* **55**, 705 (1958).
[85] R. Livingston and K. E. Owens. *J. Am. Chem. Soc.* **78**, 3301 (1956).
[86] E. Fujimori and R. Livingston. *Nature* **180**, 1036 (1957).
[87] L. I. Grossweiner and E. F. Zwicker. *J. Chem. Phys.* **34**, 1411 (1961).
[88] E. Schneider. *Z. Phys. Chem.* **B28**, 311 (1935).
[89] E. W. Schpolsky and G. D. Sheremetiev. *Acta Physicochim. USSR* **5**, 575 (1936).
[90] W. A. Noyes and I. Unger *Pure Appl. Chem.* **9**, 461 (1964).
[91] L. S. Forster and D. Dudley. *J. Phys. Chem.* **66**, 838 (1962).
[92] M. Kasha. *J. Chem. Phys.* **20**, 71 (1952).
[93] M. Kasha. *Radiation Res. Suppl.* **2**, 243 (1960)
[94] E. H. Gilmore, G. E. Gibson, and D. S. McClure. *J. Chem. Phys.* **20**, 829 (1952).
[95] D. S. McClure, N. W. Blake, and Ph. L. Hanst. *J. Chem. Phys.* **22**, 255 (1954).
[96] G. O. Schenck. Preprint, Fifth Int. Conf. Free Radicals, Uppsala, 1961; *Arbeitsgem. Forschung, Land Nordrhein-Westfalen, No. 120,* Westdeutscher Verlag, Köln und Opladen, "Mehrzentren-Termination".
[97] E. H. Farmer. *Trans. Faraday Soc.* **38**, 340 (1942).
[98] E. H. Farmer. *Trans. Faraday Soc.* **42**, 228 (1946).
[99] O. Lundberg. *Autoxidation and Antioxidants,* J. Wiley and Sons, New York (1962).
[100] J. P. Bain. *B.P.* 761 686, from 21 April (1954).
[101] J. P. Bain, A. B. Booth, and E. A. Klein. *U.S. P.* 2 863 882, from 9 December (1958).
[102] G. O. Schenck, O.-A. Neumüller, G. Ohloff, and S. Schroeter. To be published.
[103] W. Klyne. *Progress in Stereochemistry,* Vol. 1, 81, Butterworths, London (1954).
[104] K. Gollnick, S. Schroeter, G. Ohloff, G. Schade, and G. O. Schenck. To be published.
[105] W. Müller. *Diss.* Göttingen (1956).
[106] A. N. Terenin. *Acta Physiocochim. USSR* **18**, 210 (1943).
[107] A. N. Terenin. *Photochemistry of Dyes and Related Organic Compounds,* Chapt. 7 (translated by "Kresge-Hooker Scientific Library"), Academy of Sciences Press, Moscow and Leningrad (1947).
[108] J. Weiss. *Naturwiss.* **23**, 610 (1935).
[109] J. Weiss. *Trans. Faraday Soc.* **35**, 48 (1939).
[110] J. Weiss. *Trans. Faraday Soc.* **42**, 133 (1946).
[111] G. Porter and M. R. Wright. *Discussions Faraday Soc.* **27**, 18 (1959).
[112] H. L. Linschitz and L. Pekkarinen. *J. Am. Chem. Soc.* **82**, 2411 (1960).
[113] J. N. Murrel. *Mol. Phys.* **3**, 319 (1960).
[114] G. J. Hoijtink. *Mol. Phys.* **3**, 67 (1960).
[115] G. Porter. *Proc. Chem. Soc.* **1959**, 291.
[116] G. O. Schenck and K. H. Ritter. *Naturwiss.* **41**, 374 (1954).
[117] C. S. Foote and S. Wexler. *J. Am. Chem. Soc.* **86**, 3897 (1964).
[118] A. U. Khan and M. Kasha. *J. Chem. Phys.* **39**, 2105 (1963).
[119] S. J. Arnold, E. A. Ogryzlo, and H. Witzke. *J. Chem. Phys.* **40**, 1769 (1964).
[120] G. Helms. *Diss.* Göttingen (1961).
[121] K. Gollnick. To be published.
[122] G. Dupont. *Bull. Soc. Chim. France* **15**, 838 (1948).

30

Reprinted from *Tetrahedron Lett.*, No. 46, 4111–4118 (1965)

CHEMISTRY OF SINGLET OXYGEN

III. PRODUCT SELECTIVITY[1]

Christopher S. Foote[2], S. Wexler, and Wataru Ando
Contribution No. 1846 from the
Department of Chemistry
University of California
Los Angeles, California 90024

(Received 27 August 1965)

The well-known dye-photosensitized autoxidations of olefins and dienoid compounds have been postulated to proceed through an intermediate adduct (·Sens-O-O·) of excited sensitizer and oxygen which reacts with acceptor (A), to give the product peroxide (AO_2).[3] Excited singlet molecular oxygen (produced by reaction of H_2O_2 and sodium hypochlorite[1a] or by radio-frequency discharge[4]) gives products which are very similar to those of the photooxidations. Energy transfer from triplet sensitizer to oxygen to produce singlet molecular oxygen as the reaction intermediate (originally suggested by Kautsky[5] and, more recently, by Sharp[6]) was shown to be an alternate mechanism for the photooxidations which is consistent with the available evidence.[1a]

$$^3\text{Sens} + {}^3O_2 \longrightarrow \text{Sens} + {}^1O_2$$

$$^1O_2 + A \longrightarrow AO_2$$

In this study we report a detailed comparison of the product distributions from oxidation of several olefins both by the photochemical reaction and with reagents which produce singlet oxygen, and show that the two reactions give product distributions which are indistinguishable. The photooxidations were carried out

141

in a water-cooled immersion irradiation apparatus using a Sylvania DXY incandescent

lamp, with Rose Bengal as sensitizer. The singlet oxygen oxidations were carried

out by dropwise addition of aqueous hypochlorite solution to a solution containing

the olefin to be oxidized and excess H_2O_2. Inhibitors were added as indicated.

In the workup of both types of reaction, peroxides in the crude reaction

mixture were reduced by adding excess $NaBH_4$[7]; after reduction was complete, water

was added and the products were extracted. The dried solutions were analyzed

by gas chromatography. All major products were collected from the gas chromato-

graph and characterized by infrared and nuclear magnetic resonance spectroscopy;

the spectra were consistent in all cases with the structures previously assigned

to the photooxidation products.

The product distributions from 2-methyl-2-butene and limonene are summarized

in Tables 1 and 2. The product distributions from the two oxidations are identical

within experimental error, and agree well with those reported for the photo-

oxidations.[8,9]

TABLE 1

Products of Oxidation of 2-Methyl-2-butene[a]

Products	Per Cent in Reaction Mixture[b]	
	Photosensitized Autoxidation[c,d]	Singlet Oxygen $(Ca(OCl)_2 + H_2O_2)$[d]
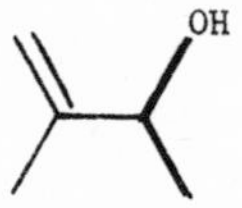	51	48
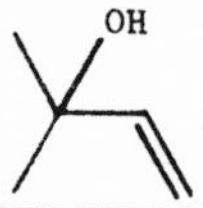	49	52

a) In methanol.
b) After reduction.
c) Reported distribution 46 and 54%, respectively.[8]
d) Products with unshifted (trisubstituted) double bonds were absent.

TABLE 2

Products from Oxidation of (+)- Limonene[a] ( )

Product	Per Cent in Reaction Mixture[b]	
	Photosensitized Autoxidation[c,d]	Singlet Oxygen (NaOCl + H_2O_2)[d,e]
	31	34
	11	9
	21	18
	10	9
	3	7
	25	24

a) In 1:1 t-butanol methanol.
b) After reduction.
c) A very similar product distribution was reported from photosensitized autoxidation; free radical oxidation gives a drastically different product mixture.[9]
d) Products with unshifted double bonds were not found.
e) $Ca(OCl)_2$ + H_2O_2 with added 2,6-di-t-butyl phenol gave a virtually identical product distribution.

A further sensitive characterization of the stereoselectivity of singlet

oxygen is provided by the optical activity of the trans-carveol (I), one of the

alcohols which is produced by reduction of the hydroperoxide mixture formed on

oxidation of limonene (see Table 2). If the reaction proceeded by initial hydrogen

abstraction from limonene to give the allylic free radical (II), I would be

I   II

racemic.[8,9] Trans-carveol (I) was isolated gas-chromatographically from the

$Ca(OCl)_2/H_2O_2$ oxidation of (+)- limonene,[10] and found to have $[\alpha]_D^{24}$ = -131°

$(CHCl_3$, c = 0.04).[11] For comparison, I, isolated gas chromatographically from

the photooxidation, had $[\alpha]_D^{24}$ = -141° $(CHCl_3$, c = 0.08).[13] Again, singlet oxygen

displays a stereospecificity nearly identical with that of the reactive inter-

mediate in the photosensitized autoxidations.

Recently a study of the oxidation of α-pinene reported that the product

distribution from $NaOCl/H_2O_2$ oxidation was completely different from that of

photosensitized oxidation, and closely resembled that of free radical oxidation.[8]

α-Pinene is an extremely unreactive acceptor for both the photooxidation and the

hypochlorite-H_2O_2 oxidation as shown by the fact that both photochemical quantum

yield and singlet oxygen utilization are less than 0.5%, and it seemed possible

that side reactions can compete with the singlet oxygen reaction in this case.

We have reinvestigated the oxidation of this substrate, and this suggestion is

dramatically confirmed. In the presence of free radical inhibitors, such as

2,6-di-t-butyl phenol, the product distribution becomes very similar to that

from photooxidation; the products from various conditions are summarized in

Table 3.

It is apparent from these results that serious side reactions (apparently

free radical in nature) can compete with the H_2O_2/hypochlorite reaction when the

TABLE 3

Products from Oxidation of α-pinene

Conditions	Per Cent in Reaction Mixture[a]	
	OH	Other Products[b]
Photosensitized Autoxidation	93[c]	7
Free Radical Oxidation[8]	14	86
NaOCl + H_2O_2 Ref. 8	9 - 12	88 - 91
Present Study[d]	35	65
$Ca(OCl)_2$ + H_2O_2[d]	43	57
$Ca(OCl)_2$ + H_2O_2[d,e] + 2,6-di-t-butyl phenol	85	15

a) After reduction.
b) Not all identified.
c) Ref. 8 reports 94%.
d) At -20°; the study in Ref. 8 was presumably done at room temperature.
e) Addition of other free radical inhibitors gave similar results.

olefin acceptor is unreactive. It is interesting to note that the reactivity
of acceptors is extraordinarily sensitive to structure in both photosensitized
autoxidation[8,14] and hypochlorite/H_2O_2 oxidation[15], trialkylated olefins being
in general around 100 times less reactive than tetraalkylated and dialkylated
olefins being still less reactive. It is for this reason that the disubstituted
double bond in limonene is unreactive in both reactions. This unreactivity of
many olefins which are less than tetraalkylated sets an important limitation
on the synthetic utility of the hypochlorite/H_2O_2 oxidation, since very large
excesses of reagents are required for only modest conversions of unreactive
olefins: most of the singlet oxygen decays to ground-state oxygen and escapes
from the solution unless a reactive acceptor is present. Corresponding limitations
are not present in the photochemical reaction, because even if the quantum yield
is low, conversions can be made high by increasing the length of irradiation.

It is apparent from the results of this and previous[1a,4] work that singlet
oxygen displays a chemistry identical to that of the intermediate in the photosensitized autoxidations. Gollnick and Schenck have recently reformulated the
hypothetical ·Sens-O-O· intermediate as a charge-transfer complex between excited
sensitizer and oxygen.[8] We cannot rule out the possibility rigorously that a
sensitizer-oxygen complex is the reactive intermediate, but our results require
that <u>sensitizer</u> <u>exert</u> <u>no</u> <u>steric</u> <u>influence</u> <u>whatever</u> <u>on</u> <u>the</u> <u>oxygen</u>, since otherwise
one should have expected to see differences in product distribution and stereo-
chemistry; there appears to be no convincing evidence which requires <u>any</u> participation of sensitizer in the transition state for oxygen transfer.[1a]

The photosensitized autoxidation of olefins has been shown by extensive
stereochemical studies to proceed by a cycloaddition mechanism, as shown below
with[8,9,16] or without[16] the oxygen complexed with sensitizer. As complexing appears
unlikely, the reactions of singlet oxygen can be visualized as those of a reactive
dienophile, which undergoes the "ene" reaction[17] with suitable olefins, and the

Diels-Alder reaction with dienes.

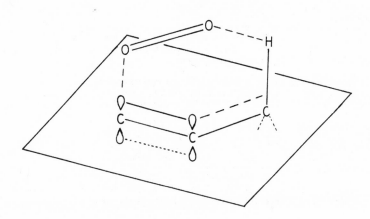

REFERENCES

1. (a) Parts I and II: C. S. Foote and S. Wexler, J. Am. Chem. Soc., 86, 3879, 3880 (1964). (b) Portions of this work were presented at the 149th meeting of the American Chemical Society, Detroit, Michigan, April, 1965. This work was supported by N.S.F. grants G-25086 and GP-3358, and by a grant from the Upjohn Company.

2. Alfred P. Sloan Fellow.

3. (a) G. O. Schenck, Naturwiss., 35, 28 (1948); (b) ibid., 40, 205, 229 (1953); (c) G. O. Schenck and E. Koch, Z. Elektrochem., 64, 170 (1960); (d) E. J. Bowen, in "Advances in Photochemistry", Vol. 1, W. A. Noyes, G. S. Hammond, and J. N. Pitts, Jr., Ed., Interscience Publishers, Inc., New York, N. Y., 1963, p. 23, and references therein cited.

4. E. J. Corey and W. C. Taylor, J. Am. Chem. Soc., 86, 3881 (1964).

5. H. Kautsky, Biochem. Z., 291, 271 (1937), and earlier papers.

6. D. B. Sharp, Abstracts, 138th Meeting of the American Chemical Society, New York, N. Y., Sept. 1960, p. 79P.

7. Occasionally, addition of dry $NaBH_4$ to the reaction mixtures caused fires or small vapor explosions; addition in aqueous solution proceeds without incident.

8. K. Gollnick and G. O. Schenck, Pure and Appl. Chem., 9, 507 (1964).

9. G. O. Schenck, K. Gollnick, G. Buchwald, S. Schroeter, and G. Ohloff, Ann., 674, 93 (1964).

10. The limonene used had $[\alpha]_D^{26}$ = +122.7° (that used in ref. 9 had $[\alpha]_D^{20}$ = +122.4°). In this reaction, 2,6-di-t-butyl phenol was added; the overall product distribution was essentially identical to that reported in Table 2 for a run using NaOCl + H_2O_2 with no inhibitor, but uninhibited runs gave products with lower rotations. The epimeric cis-carveol could not be isolated in sufficient quantity to permit measurement of its rotation.

11. Rotations were measured in $CHCl_3$ on milligram quantities using a Cary 60 spectropolarimeter. The reported rotation for pure I as a neat liquid is $[\alpha]_D^{22}$ = -213.8°[9]; $[\alpha]_D^{22}$ = +213°, for the enantiomer.[12]

12. R. G. Johnston and J. Read, J. Chem. Soc., 233 (1934).

13. This product was reported to be optically active,[9] but rotations were measured on derivatives which had been purified by repeated recrystallization, so that optical purity is uncertain.

14. K. Kopecky and H. Reich, Can. J. Chem., 43, 2265 (1965). We are grateful to Dr. Kopecky for a prepublication copy of this manuscript.

15. C. S. Foote and H. Cheng, unpublished results.

16. (a) A. Nickon and J. F. Bagli, J. Am. Chem. Soc., 83, 1498 (1961); (b) A. Nickon and W. L. Mendelson, Can. J. Chem., 43, 1419 (1965); (c) A. Nickon, N. Schwartz, J. B. DiGiorgio, and D. A. Widdowson, J. Org. Chem., 30, 1711 (1965).

17. K. Alder and H. von Brachel, Ann., 651, 141 (1962) and earlier papers.

31

Reprinted from *Can. J. Chem.*, **43**, 2265–2270 (1965)

REACTIVITIES IN PHOTOSENSITIZED OLEFIN OXIDATIONS

KARL R. KOPECKY AND HANS J. REICH
University of Alberta, Edmonton, Alberta
Received January 15, 1965

ABSTRACT

The rate of the methylene blue photosensitized oxidation of olefins increases in the order: 1-nonene < 4-methylcyclohexene < cyclohexene < 2,3-dimethyl-1-butene < 2-hexene < cyclopentene < 1-methylcyclohexene < 2,3-dimethylcyclohexene < 1-methylcyclopentene <1,2-dimethylcyclohexene < 2,3-dimethyl-2-butene. The last compound is oxidized 5 500 times as fast as cyclohexene. Comparison of this reactivity sequence with those obtained in other reactions confirms that the reactive intermediate in the photosensitized oxidations is electrophilic. The relative rates (k_A/k_B) of photo-oxidation of 1-methylcyclopentene (A) and of 1-methylcyclohexene (B) are 10 and do not vary significantly with five different sensitizers. Thus the same reactive intermediate, presumably singlet oxygen, is formed using each of the five sensitizers.

INTRODUCTION

The dye-photosensitized oxidation of olefins to allylic hydroperoxides has been studied extensively by Schenck and his co-workers (1). A shift of the double bond was reported always to take place. The reaction is stereospecific in cyclic systems (2) and has recently been used to advantage in several synthetic studies (3–5).

$$-\overset{|}{C^1}=\overset{|}{C^2}-\overset{|}{C^3}- \quad \overset{O_2,\ h\nu}{\underset{dye}{\longrightarrow}} \quad -\overset{|}{C^1}-\overset{|}{C^2}=\overset{|}{C^3}-$$
$$\underset{H}{|} \qquad\qquad\qquad \underset{O-O-H}{|}$$

Two mechanisms for this transformation are currently popular.

Schenck, adapting a mechanism suggested by Schönberg (6) has proposed the following mechanism for the photosensitized oxidation of mono-olefins (1).

$$\text{Sens} \overset{h\nu}{\to} {}^1\text{Sens}$$

$$^1\text{Sens} \to {}^3\text{Sens}$$

$$^3\text{Sens} + {}^3O_2 \rightleftarrows \cdot\text{Sens}-O-O\cdot$$

$$\cdot\text{Sens}-O-O\cdot + -\overset{|}{C}=\overset{|}{C}-\overset{|}{C}- \to -\overset{|}{C}-\overset{|}{C}-\overset{|}{C}- + \text{Sens}$$

In this mechanism a sensitizer molecule (Sens) is excited by radiation (light absorbed only by the dye molecule is sufficient for reaction) to an excited singlet state which undergoes internal conversion to give the triplet (7). This reacts with oxygen to give a "diradical" adduct (·Sens—O—O·). The adduct reacts with an olefinic double bond, regenerating the sensitizer and forming a "diradical" adduct of oxygen and olefin. Hydrogen transfer then yields the allylic hydroperoxide.

149

Kautsky (8), and, more recently, Sharp (9) and Foote (10, 11) have suggested that the metastable $^1\Delta_g$ state of the oxygen molecule is the reactive intermediate involved in photosensitized oxidations.

$$^3Sens + {}^3O_2 \rightarrow Sens + {}^1O_2$$
$$^1O_2 + olefin \rightarrow hydroperoxide$$

Energy transfer from triplet sensitizer to triplet oxygen yields the ground state sensitizer and singlet oxygen, because of conservation of electron spin angular momentum during energy transfer (12, 13). Singlet oxygen then reacts with the olefin to produce the hydroperoxide.

In support of the second mechanism Foote (10) has shown that products identical to those formed in dye photosensitized photo-oxidations are produced when sodium hypochlorite is added to solutions containing hydrogen peroxide and an oxidizable substrate. Singlet oxygen has been observed spectroscopically when sodium hypochlorite was added to hydrogen peroxide (14–17). In addition, singlet oxygen generated by electrodeless discharge oxidizes anthracenes to the corresponding endoperoxides (18).

The rate of disappearance of the reactive intermediate in the sensitized oxidation of α-terpinene to ascaridole is independent of sensitizer used (19a). It has been suggested that a common intermediate is thus involved when different sensitizers are used (11). However, the activation energy for this reaction is only ca. 0.5 kcal/mole (19a), and the only conclusion one may safely make regarding any intermediates in these reactions is that they are all very reactive.

Schenck (19b) and Sharp (9) have measured the rate of sensitized oxidation of several olefins and found that increasing substitution of the double bond greatly increased reactivity. We report here similar reactivity studies but find that the response of the reactive intermediate to variation in olefin structure is much greater than that reported by Sharp but similar to that found by Schenck. We also compare the relative rates of photo-oxidation of olefins with the relative rates of their reactions with a number of free radicals and electrophiles. In addition, it is shown that the relative rates of photo-oxidation of two mono-olefins, a reaction which must proceed with activation energy (19a), does not vary with several sensitizers.

RESULTS AND DISCUSSION

Schenck's adduct would be expected to show radical reactivity while singlet oxygen

$$(:\ddot{O}=O: \leftrightarrow :\overset{\oplus}{\ddot{O}}-\overset{\ominus}{\ddot{O}}:)$$

(9) may be electrophilic. A number of studies have been made in which the relative reactivities of a number of olefins toward a variety of reagents have been measured. A comparison of the relative reactivities of olefins of different structural types toward photosensitized oxidation and towards the other reagents can thus be made.

We have measured the relative rates of the photo-oxidation, sensitized by methylene blue, of a number of olefins. These were measured by oxidizing olefins in pairs, thus allowing them to compete for the reactive intermediate, and measuring the amounts of olefin remaining at various intervals by g.l.c. analysis. The relative rates are summarized in Table I along with the relative rates of the reactions of a variety of olefins with peracetic acid (20), phthaloyl peroxide (21), bromine (22), methyl radical (23), trifluoromethyl radical (24), and trichloromethyl radical (25, 26). The photo-oxidation reaction is seen to be extremely sensitive to structure. Indeed, the reactive intermediate involved is the most discriminating reagent of those listed in Table I. There is a close similarity in the

TABLE I

Relative reactivity of olefins towards various reagents

Olefin	Photo-oxidation	Peracetic acid[a]	Phthaloyl peroxide[b]	Bromine[c]	Methyl radical[d]	Trifluoromethyl radical[e]	Trichloromethyl radical[f]
Ethylene	0.1[g]	0.032[i]	0.021[e]	1	1	1	4.2[n]
Alkylethylene	1.7[h]				0.65[l]	1.4	
2-Methylpropene		0.71		2.03[i]	1.1	3.8	
cis-2-Butene		0.71[j]		5.5	0.1	1.3	
trans-2-Butene					0.2	1.3	
2-Hexene	9	0.76					
4-Methylcyclohexene	0.67						
Cyclohexene	1	1	1				1
Cyclopentene	16	1.51					3.3
1-Methylcyclohexene	45						3.8
2-Methyl-2-butene		9.61	9.3	10.4			
1-Methylcyclopentene	390	17.2	31.3				
1,2-Dimethylcyclohexene	29 00						
2,3-Dimethyl-2-butene	55 00	3 000[k]	100	14		1.2[m]	

[a]Reference 20.
[b]Reference 21.
[c]Reference 22.
[d]Reference 23.
[e]Reference 24.
[f]References 25, 26.
[g]1-Nonene.
[h]2,3-Dimethyl-1-butene.
[i]Propylene.
[j]Isomer not indicated.
[k]Estimated by Greene (27) from data collected by Swern (20).
[l]1-Decene.
[m]Calculated from data of ref. 28.
[n]1-Octene.

variation of rate with structure between the photosensitized oxidation of olefins and their reaction with peracetic acid, bromine, and phthaloyl peroxide. No such similarity exists at all between the rate of the photosensitized oxidation of olefins and the rate of addition of any of the radical reagents listed in Table I to olefins. Even the electrophilic trifluoro-methyl radical (24) does not respond to structural changes in the manner that the reactive intermediate in the photosensitized oxidation reaction does. This intermediate thus has strong electrophilic character and does not exhibit any radical properties. It is quite unlikely that the intermediate has the structural features of the diradical described by Schenck (29). The reactivity is best explained by the suggestion (9–11) that singlet oxygen is the reactive intermediate.

Product formation is best described by the suggestion of Sharp (9) that an intermediate perepoxide is formed which rearranges with a concurrent proton shift to yield the allylic hydroperoxide (Chart I). However, our data do not provide evidence against a concerted cycloaddition as proposed by Nickon (30).

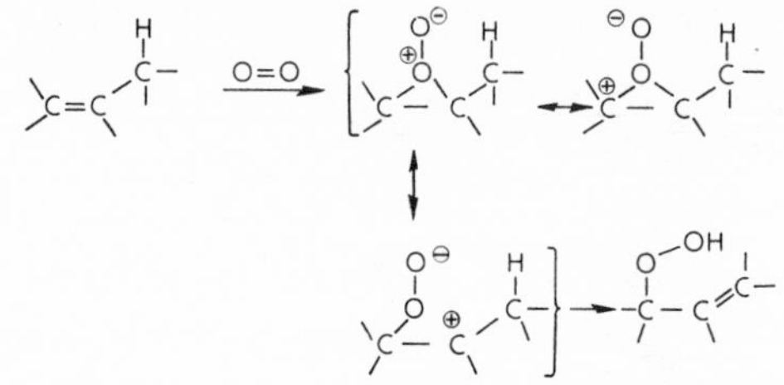

CHART I.

Sharp reports that the relative rates of tetraphenylporphine sensitized photo-oxidation of 1-hexene, 2-methyl-2-butene, and 2,3-dimethyl-2-butene are 1, 94, and 100, respectively. We find a much greater difference in reactivity with comparably substituted olefins, as does Schenck (19b). The relative rates of photo-oxidation of 1-nonene, 1-methylcyclo-hexene, 1-methylcyclopentene, and 2,3-dimethyl-2-butene are, from Table I, 1, 450, 3 900, and 55 000, respectively. The reason for this discrepancy is not clear. It is probably not due to a difference in sensitizer. The relative rates of photo-oxidation of 1-methylcyclo-hexene and 1-methylcyclopentene with a variety of sensitizers are listed in Table II and do not vary with sensitizer. It may be that Sharp's relative rates were measured indirectly. If so, the relative errors may have accumulated to yield the data reported.[1]

TABLE II

Effect of sensitizer on relative reactivities

Sensitizer	k_A/k_B*
Methylene blue	8.6 ± 0.9
Eosin Y	10.8 ± 1.4
Rose bengal	11.5 ± 1.9
Erythrosin B	8.8 ± 1.6
Hematoporphyrin	9.4 ± 1.8

*A = 1-methylcyclopentene, B = 1-methylcyclo-hexene.

The oxidation reactions reported in this paper must have an activation energy signifi-cantly different from zero. The data of Table II therefore provide the first evidence that a common reactive intermediate is formed in oxidations involving different sensitizers.

[1]A referee has pointed out that reactivity depends upon concentration, and that the descrepancies may be due to this factor.

152

The activation energies of the rose bengal sensitized photo-oxidation of dicyclohexylidene and β-pinene are reported to be 1.3 and 4.5 kcal, respectively (19a). Dicyclohexylidene is thus about 700 times as reactive as β-pinene at 25°. This is in good agreement with the reactivity difference to be expected on the basis of the data in Table I.

EXPERIMENTAL

Preparation of Olefins

Cyclohexene (Eastman Kodak White Label), 2-hexene (Phillips 66 Pure Grade) and 4-methylcyclohexene (Eastman Kodak White Label) were distilled from lithium aluminium hydride. Pyrolysis of 1-nonylacetate yielded 1-nonene. The remaining olefins were obtained from the phosphoric acid catalyzed dehydration of the appropriate alcohols. The physical properties of those olefins obtained in the pure state were (boiling points in this study were obtained at 700 mm unless otherwise indicated and are uncorrected; the boiling points cited in the literature references are all for 760 mm unless otherwise indicated): cyclohexene b.p. 79–80°, n_D^{25} 1.4441 (reported (31) b.p. 82.9°, n_D^{25} 1.4438); cyclopentene b.p. 43°, n_D^{25} 1.4199 (reported b.p. 44° (32), n_D^{25} 1.4194 (33)); 1-methylcyclopentene b.p. 72–74°, n_D^{25} 1.4293 (reported (34) b.p. 75.8°, n_D^{20} 1.4330); 1-methylcyclohexene b.p. 106–107°, n_D^{25} 1.4478 (reported (31) b.p. 110°, n_D^{25} 1.4478); 4-methylcyclohexene b.p. 98–99°, n_D^{25} 1.4392 (reported (31) b.p. 102.7°, n_D^{25} 1.4389); 2-hexene b.p. 65–66°, n_D^{25} 1.3941 (reported for *cis*-2-hexene (35) b.p. 68.8°, n_D^{25} 1.3948; for *trans*-2-hexene (35) b.p. 67.8°, n_D^{25} 1.3907); 2,3-dimethyl-2-butene b.p. 70–71°, n_D^{25} 1.4098 (reported (35) b.p. 73°, n_D^{25} 1.4094); 1-nonene b.p. 61–62° at 48 mm, n_D^{25} 1.4136 (reported b.p. 33.5° at 11 mm (36), n_D^{25} 1.4133 (37)). These olefins were shown to be more than 97% pure by g.l.c. analysis.

Dehydration of 1,2-dimethylcylohexanol produced a mixture, shown by g.l.c. analysis to consist of 58% 1,2-dimethylcyclohexene, 37% 2,3-dimethylcyclohexene, and 5% of an unidentified compound. This mixture was used as such in the rate studies. The small amount of 2,3-dimethyl-1-butene, which was formed along with 2,3-dimethyl-2-butene upon dehydration of 2,3-dimethyl-2-butanol and which was used in the rate studies, was shown by g.l.c. analysis to contain 16% ether and 9% 2,3-dimethyl-2-butene as the only detectable impurities. The results obtained from these impure compounds fit well into the general pattern and are considered accurate enough for the present purpose.

Determination of Relative Reactivities of Olefins

Pairs of olefins were chosen for the competition reactions on the basis of suitable reactivity. Approximately equimolar amounts of the two olefins (0.5 to 2.0 ml), 1.0 ml of an internal standard (benzene, toluene, or ethylbenzene) and 30 mg of methylene blue were dissolved in 200 ml of redistilled methanol. The resulting solution was poured into an immersion photochemical reactor (38) just filling the annular space. A slow stream of oxygen was passed through the solution which was kept at 15 °C while it was being irradiated. The light source was a 200 W Hanovia mercury vapor lamp, type S, No. 654A. A filter sleeve, Hanovia No. 516-27-116 was used to cut off all radiation below 3 600 °A. Samples were withdrawn from the solution initially and at intervals, after mixing, during the irradiation until one component had decreased to about 10% of its original concentration. The samples were analyzed by g.l.c. using the internal standard method of Keulemans (39). A Wilkins Aerograph 202 gas chromatographic instrument was used with a 5 ft column packed with Apiezon M on 60–80 Chromosorb W which gave good separation between methanol and all the olefins. From the relative amounts of olefins remaining, the relative rate constants were calculated by the method of Ingold and Shaw (40). The results are summarized in Table III. The relative rates given represent the average of several determinations and the average errors are indicated. Control experiments showed that no olefin was consumed in the absence of light or in the absence of dye when the lamp was on for the time required for 90% oxidation of the more reactive component except in the case where 1-nonene was a

TABLE III

Relative rates of photosensitized oxidation of olefins

Expt. No.	Olefin A	Olefin B	k_B/k_A
1	1-Nonene	Cyclohexene	10*
2	4-Methylcyclohexene	Cyclohexene	1.5±0.2
3	Cyclohexene	2-Hexene	8.5±0.2
4	2-Hexene	Cyclopentene	1.8±0.4
5	Cyclopentene	2,3-Dimethylcyclohexene	3.0±0.1
6	1-Methylcyclohexene	2,3-Dimethylcyclohexene	1.07±0.03
7	1-Methylcyclohexene	1-Methylcyclopentene	8.6±0.9
8	1-Methylcyclopentene	1,2-Dimethylcyclohexene	7.5±0.7
9	1,2-Dimethylcyclohexene	2,3-Dimethyl-2-butene	1.9±0.2
10	Cyclohexene	2,3-Dimethyl-1-butene	1.7±0.4

*Estimated value. Oxidation of 1-nonene proceeded so slowly that evaporation losses became significant.

competitor. In another control experiment a solution containing 0.4 g of 2,3-dimethyl-3-hydro-peroxy-1 butene (41) and 1.0 g of 2,3-dimethyl-2-butene were photolyzed in the above apparatus. After 15 min 5% of the olefin had been consumed. Thirty mg methylene blue was added and the photolysis was continued. After 15 min 90% of the olefin had been consumed. The relative reactivities reported in Table II were determined in the manner described above using solutions made up from 2.0 ml ethylbenzene, 1.0 ml each of 1-methylcyclopentene and 1-methylcyclohexene, 30 mg of sensitizer, and 200 ml of distilled methanol.

ACKNOWLEDGMENT

This research was supported in part by a grant from the National Research Council.

REFERENCES

1. G. O. SCHENCK. Angew. Chem. **69**, 579 (1957).
2. A. NICKON and J. F. BAGLI. J. Am. Chem. Soc. **83**, 1498 (1961).
3. R. A. BELL and R. E. IRELAND. Tetrahedron Letters, No. 4, 269 (1963).
4. S. MASAMUNE. J. Am. Chem. Soc. **86**, 290 (1964).
5. A. NICKON and W. L. MENDELSON. J. Am. Chem. Soc. **85**, 1894 (1963).
6. A. SCHÖNBERG. Ann. **518**, 299 (1935).
7. G. N. LEWIS and M. KASHA. J. Am. Chem. Soc. **66**, 2100 (1944).
8. H. KAUTSKY and H. DE BRUIJN. Naturwiss. **19**, 1943 (1931).
9. D. B. SHARP. Abstracts, 138th National Meeting of the American Chemical Society, New York, N.Y. September, 1960. p. 79 P.
10. C. S. FOOTE and S. WEXLER. J. Am. Chem. Soc. **86**, 3879 (1964).
11. C. S. FOOTE and S. WEXLER. J. Am. Chem. Soc. **86**, 3880 (1964).
12. G. PORTER and M. R. WRIGHT. J. Chem. Phys. **55**, 705 (1958).
13. G. PORTER and M. R. WRIGHT. Discussions Faraday Soc. **27**, 18 (1959).
14. E. J. BOWEN and R. A. LLOYD. Proc. Chem. Soc. 305 (1963).
15. R. J. BROWNE and E. A. OGRYZLO. Proc. Chem. Soc. 117 (1964).
16. A. U. KHAN and M. KASHA. J. Chim. Phys. **39**, 2105 (1963).
17. S. J. ARNOLD, E. A. OGRYZLO, and H. WITZKE. J. Chem. Phys. **40**, 1769 (1964).
18. E. J. COREY and W. C. TAYLOR. J. Am. Chem. Soc. **86**, 3881 (1964).
19. (a) G. O. SCHENCK and E. KOCH. Z. Electrochem. **64**, 170 (1960). (b) G. O. SCHENCK, H. MERTENS, W. MÜLLER, E. KOCH, and O. P. SCHMIEMENZ. Angew. Chem. **68**, 303 (1956).
20. D. SWERN. J. Am. Chem. Soc. **69**, 1692 (1947).
21. F. D. GREENE and W. W. REES. J. Am. Chem. Soc. **80**, 3432 (1958).
22. S. V. ANANTAKRISHNANA and R. VENKATARAMAN. Chem. Rev. **33**, 27 (1943).
23. R. P. BUCKLEY and M. SZWARC. Proc. Roy. Soc. London, Ser. A, **240**, 396 (1957).
24. A. P. STEFANIC, L. H. HERK, and M. SZWARC. J. Am. Chem. Soc. **83**, 4732 (1961).
25. P. S. SKELL and A. Y. GARNER. J. Am. Chem. Soc. **78**, 5430 (1956).
26. C. WALLING. Free radicals in solution. John Wiley and Sons, Inc., New York, N.Y. 1957. p. 254.
27. F. W. GREENE and W. ADAM. J. Org. Chem. **29**, 136 (1964).
28. H. KOMAZAWA, A. P. STEFANIC, and M. SZWARC. J. Am. Chem. Soc. **85**, 2043 (1963).
29. G. O. SCHENCK, O.-A. NEUMULLER, and R. KOCH. Strahlentherapie, **114**, 321 (1961).
30. A. NICKON and J. F. BAGLI. J. Am. Chem. Soc. **83**, 1498 (1961).
31. F. D. ROSSINI et al. Selected values of chemical thermodynamic properties. U.S. Government Printing Office, Washington, D.C. 1955. p. 65.
32. A. I. VOGEL. J. Chem. Soc. 1323 (1938).
33. Reference 31, p. 64.
34. C. E. BOORD, A. L. HENNE, K. W. GREENLEE, W. L. PERILSTEIN, and J. M. DERFER. Ind. Eng. Chem. **41**, 609 (1949).
35. Reference 31, p. 55.
36. A. W. SCHMIDT, V. SCHOELLER, and K. EBERLEIN. Ber. **74**, 1314 (1941).
37. Reference 31, p. 52.
38. K. R. KOPECKY, G. S. HAMMOND, and P. A. LEERMAKERS. J. Am. Chem. Soc. **84**, 1015 (1962).
39. A. I. M. KEULEMANS. Gas chromatography. Reinhold Publishing Corp., New York, N.Y. 1957. p. 32.
40. C. K. INGOLD and F. R. SHAW. J. Chem. Soc. 2918 (1927).
41. G. O. SCHENCK and K. H. SCHULTE-ETTE. Ann. **618**, 185 (1958).

Reprinted from *J. Amer. Chem. Soc.*, **88**(13), 2898–2902 (1966)

Excited Singlet Molecular Oxygen in Photooxidation

Thérèse Wilson

*Contribution from Converse Memorial Laboratory of Harvard University,
Cambridge, Massachusetts 02138. Received February 18, 1966*

Abstract: Competitive photooxidation of three pairs of unsaturated compounds yields a single set of relative reactivities, whether sensitized by a dye (methylene blue or rose Bengale) or by the aromatic substrates themselves. This result seems, except in case of a remarkable coincidence, uniquely consistent with excited singlet oxygen as the reactive species and inconsistent with a series of biradical-like "moloxides" which would be different for each sensitizer. In pyridine, the relative reactivities with singlet oxygen are as follows: tetramethylethylene, 1.0; 9,10-dimethylanthracene, 1.0; 9,10-diphenylanthracene, ~0.2; rubrene, 2.25; and 1,3-diphenylisobenzofuran, 38.5. Direct addition to the triplet state of these acceptors is undetectable, with the possible exception of the last one where its contribution would be small. Energy considerations would indicate that the active molecules are in the $^1\Delta_g$ state.

The purpose of this note is to present additional experimental evidence for a mechanism *via* excited singlet oxygen molecules for the photooxidation in solution of organic compounds (acceptors) such as the polyacenes, in which a transannular peroxide is formed. This type of reaction has attracted much attention.[1] It is generally agreed that it does not pro-

ceed by direct addition of a ground-state oxygen molecule to a singlet excited molecule of the acceptor during the fluorescence quenching step. On the basis of the complex kinetic data previously accumulated, three

(1) For recent discussions of photooxidation, see E. J. Bowen, *Advan. Photochem.*, **1**, 23 (1963); and C. S. Foote and S. Wexler, *J. Am. Chem. Soc.*, **86**, 3880 (1964).

main mechanisms have been proposed, differing in the reactive intermediates assumed: (a) the triplet state of the acceptor reacting with ground-state oxygen, (b) a hypothetical transient "moloxide" of the acceptor or of the sensitizer that later transfers its oxygen to the acceptor, or (c) "active" oxygen reacting with ground-state acceptor. A strong argument for the latter can be inferred, by analogy, from the recent work of Corey[2] and Foote[3] on the reactivity of electrically or chemically excited molecular oxygen in a singlet state (here $^1O_2*$).

In the present work, the competitive photooxidation of a mixture of two acceptors has been investigated. It shows that the hypothesis (a) above of a reaction of the triplet state of the acceptor can be ruled out in such cases as the photooxidations of rubrene, 9,10-dimethylanthracene, or 9,10-diphenylanthracene, which all give transannular peroxides. The results are consistent with a singlet oxygen mechanism. They will be presented at first exclusively on that basis, in anticipation of the Discussion where it will be shown that a Schenck-like interpretation by a moloxide-type of reaction (b) meets with difficulties. A method is outlined by which relative rate constants of the reactions of singlet oxygen with different acceptors can be obtained.

It is obvious that if the energies required to excite the first triplet as well as the first singlet excited states of an acceptor are greater than the energy absorbed by the sensitizer, the efficient participation in the oxidation of any excited states of the acceptor can be excluded. Such an acceptor, if photosensitized, acts unambiguously as a captor of singlet oxygen. Tetramethylethylene (TME) fits the energy and reactivity requirements for this role of $^1O_2*$ captor. It reacts with chemically produced singlet oxygen, mainly to give 2,3-dimethyl-3-hydroperoxybutene-1 (TMEO$_2$) identical with the photooxidation product.[3] Its first triplet state is about 83 kcal above ground state,[4] yet it has been found here that its photooxidation proceeds readily when sensitized by methylene blue at 640 mμ, corresponding to quanta of 44.6 kcal only.

Methylene blue as sensitizer (S) must transfer its excitation energy to an oxygen molecule, which in turn reacts with TME to form its hydroperoxide.[5]

$$S + h\nu \longrightarrow S* \tag{1}$$

$$S* \longrightarrow S + h\nu' \tag{2}$$

$$S* + {}^3O_2 \longrightarrow S + {}^1O_2* \tag{3}$$

$$^1O_2* \longrightarrow {}^3O_2 + h\nu'' \tag{4}$$

$$^1O_2* + TME \longrightarrow TMEO_2 \tag{5}$$

If tetramethylethylene acts as an inhibitor of the sensitized photooxidation of another acceptor (A), it is evidence that the photooxidation of the latter also proceeds *via* singlet oxygen, by the same sequence of reactions with an additional step (6).

$$^1O_2* + A \longrightarrow AO_2 \tag{6}$$

(2) E. J. Corey and W. O. Taylor, *J. Am. Chem. Soc.,* **86,** 3881 (1964).

(3) (a) C. S. Foote and S. Wexler, *ibid.,* **86,** 3879 (1964); (b) C. S. Foote, S. Wexler, and W. Ando, *Tetrahedron Letters,* No. 46, 4111 (1965).

(4) Its exact value is not known but has been estimated at 3.6 ev, *i.e*, 3 ev below the first excited singlet state of TME: see C. A. Coulson and E. T. Stewart in "The Chemistry of Alkenes," E. Patai, Ed., John Wiley and Sons, Inc., New York, N. Y., 1964, pp 137, 144.

(5) The electronic states involved are left unspecified here, but will be considered in the Discussion. The nature of the deactivating processes (eq 2 and 4) does not need to be specified here; radiative transitions are written for simplicity.

Total inhibition at a high enough ratio of concentration of the captor TME to that of the acceptor A rules out the contribution to the oxidation of A of a side path involving, for example, the reaction of triplet $^3A*$ with $^3O_2$ (reaction 7, with the competitive radiative or rationless deactivation of $^3A*$).

$$^3A* + {}^3O_2 \longrightarrow AO_2 \tag{7}$$

$$^3A* \longrightarrow A + h\nu \tag{8}$$

Assuming steady states for S* and $^1O_2*$ and assuming that reaction 4 is negligible[6] compared to reactions 5 and 6, one arrives at the following linear relation between the reciprocal of the rate of oxidation of the acceptor and the concentration of TME

$$\frac{1}{\dfrac{d[AO_2]}{dt}} = \frac{k_5}{k k_6 [A]}[TME] + \frac{1}{k} \tag{9}$$

with $k = (k_1 k_3 [S][^3O_2])/(k_2 + k_3 [^3O_2])$ for a series of runs at constant light intensity and constant initial concentrations of oxygen, acceptor, and sensitizer, the values of which determine k.

A study of the direct photooxidation where the acceptor acts as its own sensitizer provides a way of checking the validity of the proposed mechanism (1 to 6). With the following reaction sequence now replacing reactions 1, 2, and 3, relation 9 should still be observed, with a

$$A + h\nu \longrightarrow A* \tag{1'}$$

$$A* \longrightarrow A + h\nu' \tag{2'}$$

$$A* + {}^3O_2 \longrightarrow A + {}^1O_2* \tag{3'}$$

different value of k.

Experimental Section

Pyridine was used as solvent when not otherwise specified. Bromobenzene was the solvent in one series of runs with 9,10-dimethylanthracene.

Oxygen at pressure of 180 mm was admitted to an outgassed stirred solution of the acceptors. The course of the reaction induced by monochromatic light, from a Bausch and Lomb 33.86.-26.07 grating monochromator, at about 10° was checked by the absorption of oxygen; plots of the pressure against time were nearly linear during the reaction times. The rates of oxidation were determined photometrically with a Beckman DU2 from the initial and final concentrations of the acceptors after appropriate dilution. Final concentrations of TME were not measured.

Solvents and reagents were of high commercial grade, used without further purification; their absorption spectra and extinction coefficients agree with the literature values. The samples of 9,10-dimethylanthracene and 1,3-diphenylisobenzofuran were kindly provided by Miss B. Kaski. Control tests were performed to make sure that dark reactions were negligible, and that in the absence of oxygen no photochemical reaction took place.

Results

1. Rubrene (R). Tetramethylethylene acts indeed as an inhibitor of the photooxidation of rubrene[7] sensitized by methylene blue at 640 mμ, as well as of its direct photooxidation at 540 mμ (Figure 1). The inhibition is total when TME is present at an initial concentration about 100 times that of rubrene, the oxygen being then absorbed by the solution exclusively to oxidize TME.

Figure 2 shows that the results fit the linear relation 9. It gives for the ratio k_5/k_{6R} the value of 0.42 for the

(6) This follows from the observation that the rate of oxidation is independent of the acceptor concentration in the range studied.

(7) 9,10,11,12-Tetraphenylnaphthacene.

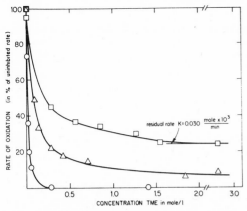

Figure 1. Effect of tetramethylethylene (TME) on photooxidation: O, rubrene at 540 mµ (initial concentration 3.4×10^{-3} mole/l.); □, diphenylisobenzofuran (DPBF) at 425 mµ (initial concentration 2.6×10^{-3} mole/l.); △, diphenylisobenzofuran (DPBF) at 575 mµ (initial concentration 2.6×10^{-3} mole/l.).

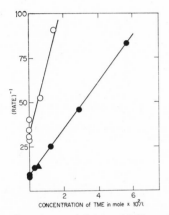

Figure 2. Effect of tetramethylethylene (TME) on the photo-oxidation of rubrene: ●, rubrene at 540 mµ (initial concentration 3.4×10^{-3} mole/l.); O, sensitized rubrene at 640 mµ (initial concentration 3.8×10^{-3} mole/l.); ▲, with 9,10-dimethylanthracene (DMA instead of TME (same units of concentration).

sensitized reaction, and 0.47 for the direct oxidation. Thus, the photooxidation of rubrene proceeds entirely *via* $^1O^*_2$. It reacts with $^1O_2^*$ about twice as fast as TME.

2. 9,10-Dimethylanthracene (DMA). TME inhibits the photooxidation of DMA sensitized by methylene blue at 640 mµ in pyridine, as well as its direct photooxidation at 403 mµ in bromobenzene. Figure 3 shows that the results fit relation 9, with $k_5/k_{6_{DMA}} = 1.0$, both at 640 and at 403 mµ.

Changing the solvent from pyridine to bromobenzene, which could have been expected to have a heavy-atom influence, is shown to affect neither the rate of photooxidation of pure DMA, nor the ratio $k_5/k_{6_{DMA}}$.

As in rubrene, the photooxidation of 9,10-dimethylanthracene proceeds entirely *via* $^1O_2^*$; it reacts with $^1O_2^*$ at about the same rate as TME.

This suggests that the effect of DMA on the photooxidation of rubrene should be identical with that of

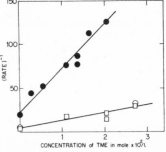

Figure 3. Effect of tetramethylethylene (TME) on the photooxidation of 9,10-dimethylanthracene (DMA): ●, in pyridine at 640 mµ (initial concentration 4.4×10^{-3} mole/l.); O, in pyridine at 403 mµ (initial concentration 5.3×10^{-3} mole/l.); □, in bromobenzene at 403 mµ (initial concentration 5.0×10^{-3} mole/l.).

TME. One run with a mixture of rubrene and DMA confirms this prediction, thereby supporting a simple mechanism *via* $^1O_2^*$ for these acceptors (Figure 2).

3. 9,10-Diphenylanthracene (DPA). The direct photooxidation of DPA at 400 mµ is much slower than that of DMA. Its inhibition by TME is already total when TME is present at half the concentration of DPA; this creates less satisfactory experimental conditions because of the rapid change with time of the small concentrations of TME needed.

Four runs at an initial concentration of DPA of 1.4×10^{-2} mole/l. gave a value of $k_5k_{6_{DPA}}$ of the order of 5.6 (±1.3).

4. 1,3-Diphenylisobenzofuran (DPBF). The situation is somewhat different in the case of the direct photooxidation of DPBF at 425 mµ.[8] The addition of TME has an inhibiting effect on the rate of oxidation, but higher concentrations of TME are necessary, and the inhibition is never total (Figure 1) no matter how much TME is present. More explicitly, it is found that the final concentration of DPBF is always smaller than the initial concentration, as it would be if some of the DPBF were oxidized in spite of the TME. However, as only DPBF concentrations were measured in these runs instead of the final amount of the oxidized products, it is impossible without further experiments to ascertain the significance of this residual rate of change K of DPBF (Figure 1, $K = 0.030 \times 10^{-3}$ mole/min).[9] It could be the effect of the solvent cage surrounding the DPBF and the singlet oxygen just formed, if the probability of reaction of $^1O_2^*$ with DPBF were comparable to the probability of diffusion of oxygen out of the cage.[10] It could also, however, result from a second concurrent path of oxidation, not proceeding *via* excited oxygen molecules and therefore

(8) The sensitized photooxidation of DPBF is believed to yield first a transannular, ozonide-like peroxide (DPBFO₂), which then rearranges into *o*-dibenzoylbenzene: see G. O. Schenck, *Ann.*, **584**, 156 (1953); A. Le Berre and R. Ratsinbazafy, *Bull. Soc. Chim. France*, 229 (1963); and references therein cited. The peroxidation of DPBF by externally generated singlet oxygen is very fast, according to Corey.[2]

(9) There is no significant autoxidation of DPBF in the dark and no reaction of DPBF with TME in the dark or at 425 mµ, in the time scale and at the temperature of these runs. There does seem to be some photodecomposition of pure DPBF at 425 mµ in pyridine, but not enough to account for more than about one-third of the residual rate observed.

(10) The author is grateful to one of the referees for suggesting this interpretation.

insensitive to the presence of TME, such as a reaction of DPBF in a triplet state with $^3O_2$ (reactions 7 and 8 above). In any case, the rate of photooxidation as a function of TME can be expressed as the sum of two terms, one being this constant rate K, the other resulting from the usual sequences ($1'$ to 6) *via* singlet oxygen. By analogy with (9), and substituting for the experimental value of K

$$\frac{1}{\frac{d[DPBFO_2]}{dt} - 0.03} = \frac{k_5}{k'k_{6_{DPBF}}}[TME] + \frac{1}{k'} \quad (10)$$

Figure 4 gives $k_5/k_{6_{DPBF}} = 0.026$. Thus DPBF reacts about 20 times faster with $^1O_2^*$ than rubrene does, and at least 75 % of its photooxidation proceeds *via* $^1O_2^*$.

On the basis of the cage effect mentioned above, any sensitizer could be expected to suppress the residual rate. Such seems to be the situation at 575 mμ with rose Bengale as a sensitizer. The inhibition by tetramethylethylene is now almost total (Figure 1) (nearly within the limits of errors of these measurements), yet the results agree well with the value of $k_5/k_{6_{DPBF}}$ obtained above at 425 mμ (Figure 4). However, this sensitizer effect could also be interpreted on the basis of a direct reaction of DPBF in its triplet state. One could indeed expect the residual rate to be smaller at a longer wavelength, where the incident energy may be below the triplet level of DPBF or inefficiently transferable to it. Further experiments are needed to clarify this point.

5. Mixtures of Rubrene and DPBF. From the ratios of rate constants k_5/k_{6_R} and $k_5/k_{6_{DPBF}}$ one predicts $k_{6_{DPBF}}/k_{6_R} = 17.3$. Therefore, rubrene and DPBF should inhibit each other's photooxidation by competing for the available singlet oxygen. Light of 540 mμ that excites rubrene was used to induce both the direct oxidation of rubrene and the rubrene-sensitized oxidation of DPBF. As the initial and final concentration of both rubrene and DPBF can be measured, the ratio $k_{6_{DPBF}}/k_{6_R}$ can be calculated from eq 11, based on reaction 6 for R and for DPBF and on assuming a steady state for $^1O_2^*$. Three runs gave the following

$$\log\frac{[DPBF]_0}{[DPBF]} = \frac{k_{6_{DPBF}}}{k_{6_R}}\log\frac{[R]_0}{[R]} \quad (11)$$

values for $k_{6_{DPBF}}/k_{6_R}$: 20.0, 20.7, and 25.0; one run in which none of the rubrene was oxidized gave a lower limit of 10. (No attempt was made here to correct for the so-called residual rate.) These results can be considered in adequate agreement with the value of $k_{6_{DPBF}}/k_{6_R} = 17.3$ deduced from the effect of TME on rubrene and on DPBF separately.

6. Experiments with Gaseous Singlet Oxygen. In order to check further the $^1O_2^*$ mechanism for the photooxidations, a few rough runs were carried out on the competitive oxidation of rubrene and DPBF by metastable oxygen externally generated by Corey's electrodeless discharge; they gave rate ratios similar to the photochemical ones in order of magnitude. Two runs with a pyridine solution of rubrene and DPBF shielded from light gave $k_{6_{DPBF}}/k_{6_R}$ between 9 and 73, by eq 9. This is compatible with a value of 20 obtained above for the photochemical reactions. In another pair of runs, no oxidation of rubrene was detectable after 20 min in a solution containing TME, whereas

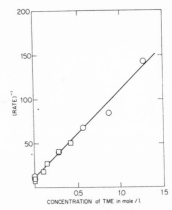

Figure 4. Effect of tetramethylethylene (TME) on the photooxidation of 1,3-diphenylisobenzofuran (DPBF) (initial concentration 2.6×10^{-3} mole/l.): O, at 425 mμ (Y axis represents reciprocal of corrected rates (rate $- 0.03)^{-1}$); □, sensitized at 575 mμ.

rubrene was 90 % oxidized in a run without TME. Such a total inhibition was observed in the photooxidation when the same concentration of TME was present (3.10^{-1} mole/l.). This experiment also illustrates the sensitivity of rubrene as a photometric detector of oxidation *via* $^1O_2^*$.

Discussion

All the data presented here on the competitive photooxidation of rubrene, dimethyl- and diphenylanthracenes, and tetramethylethylene are entirely consistent with a mechanism *via* singlet oxygen, thus ruling out any significant role of the triplet excited state of these acceptors in the peroxide-forming step. Anthracene itself has not been investigated here, because of its even slower rate of oxidation and its tendency to dimerize. There is no reason to suspect that it would not behave like the substituted anthracene as far as the mechanism of its photooxidation is concerned. Although the case of DPBF is more complex, $^1O_2^*$ is the main reactive intermediate.

The kinetic data, thus presented, do not completely rule out the alternative but unnecessary hypothesis of a moloxide of the sensitizer or of the acceptor (SOO or AOO).[11] Reactions 5 and 6 could conceivably be replaced by reactions 12 and 13 in the photosensitized oxidation

$$SOO + TME \longrightarrow TMEO_2 + S \quad (12)$$
$$SOO + A \longrightarrow AO_2 + S \quad (13)$$

or by reactions 14 and 15 in the direct photooxidation of the same acceptor.

$$AOO + TME \longrightarrow TMEO_2 + A \quad (14)$$
$$AOO + A \longrightarrow AO_2 + A \quad (15)$$

However, no moloxide has been isolated yet or physically characterized in any way. Besides, the relative rates for two different acceptors were found to be independent of the sensitizer used (*i.e.*, $k_{12}/k_{13} = k_{14}/k_{15}$). Thus, the

(11) For a recent discussion of the moloxide hypothesis, see K. Gollnick and G. O. Schenck, *Pure Appl. Chem.*, **9**, 507 (1964). See also the arguments of Foote.[3b]

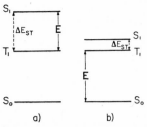

Figure 5. Energy diagrams illustrating the energy (E) available for excitation of oxygen through a triplet–triplet transfer (a) or through a triplet–triplet annihilation (b). Case a is typical of the configuration of energy levels in a polyacene, case b in a dye.

tendency of a given sensitizer–oxygen moloxide to give away its oxygen to one acceptor rather than another must depend solely on the nature of the acceptors involved, and not at all on the relationship between sensitizer and acceptor. In the case of a direct photooxidation where the acceptor is also the sensitizer, one does not quite see why a transfer of oxygen should be necessary at all, and why the moloxide could not be internally rearranged in the peroxide form at the occasion of collisions with the solvent molecules (hence $k_{14}/k_{15} < k_{12}/k_{13}$), for instance.[12] Moreover, no adduct between oxygen and sensitizer can be invoked in the "dark" experiments carried out with gaseous $^1O_2*$; yet they behave very much like the photochemical ones.

Returning to the singlet oxygen mechanism, the energy transfer reactions 3 and 3′ appear quite remarkable in their efficiency, no matter what the amount of energy present in the excited sensitizer. The electronic states in which the participating molecules are, is of course not directly known. Because of spin conservation, only a triplet–triplet transfer (eq 16) or a triplet–triplet annihilation (eq 17) could be responsible for the excitation of oxygen (the donor D represents either the sensitizer S or the acceptor A above). Both

$$^1D* + \,^3O_2 \longrightarrow \,^3D* + \,^1O_2* \qquad (16)$$

$$^3D* + \,^3O_2 \longrightarrow \,^1D + \,^1O_2* \qquad (17)$$

should be frequent events in the unusual situation existing here due to the triplet ground state of oxygen. Therefore, it may very well be that both play a part in the excitation of oxygen, with relative efficiencies depending on the energy levels of the donor present (Figure 5).

In the polyacenes, the singlet–triplet splitting (ΔE_{ST}) is of the order of 30 kcal,[13] which is sufficient to excite the first singlet state of oxygen ($^1\Delta_g$ at 22.5 kcal above ground state) *via* triplet–triplet transfer (eq 16), but too low to excite the next singlet state ($^1\Sigma_g^+$ at 37.5 kcal above ground state). In dyes, the singlet–triplet splitting is

much smaller. Unfortunately, no information on the energies of the triplet states of methylene blue or rose Bengale seems available in the literature. But if their ΔE_{ST} is smaller than 10 kcal as is the case with eosin, crystal violet, and various fluorescein derivatives,[14] then a triplet–triplet transfer could clearly not be responsible for the excitation of even the lowest $^1\Delta_g$ state of $^1O_2*$. This could only be achieved by a triplet–triplet annihilation (reaction 17) in which the whole of the triplet energy of the sensitizer is available. With rose Bengale as sensitizer ($E_S = 50$ kcal) there would probably be sufficient energy to excite also the $^1\Sigma_g^+$ state.

Such a triplet–triplet annihilation could explain as well the autosensitization by such polyacenes as the anthracene derivatives studied here and perhaps rubrene, although the height of its triplet state (as yet unknown) can be expected to be very close to the $^1\Delta_g$ state of $^1O_2*$. However, if ΔE_{ST} remains roughly constant along the linear polyacenes series, a triplet–triplet annihilation (reaction 17) could no longer result in excitation of singlet oxygen in pentacene[15] and diphenyl-6,13-pentacene;[16] for example, their triplet level is likely to be well below 20 kcal, yet they are reported to undergo direct photooxidation. A triplet–triplet transfer remains therefore a more generally valid interpretation for the photooxidation of the polyacenes.

Finally, a very indirect clue on the spectroscopic state of the excited oxygen molecules involved may be gathered from the unsensitized photooxidation of rubrene. By either eq 16 or 17 it can only produce singlet oxygen in the $^1\Delta_g$ state; yet the relative rates of reactions of rubrene, TME, and DPBF with that "pure" $^1\Delta_g$ oxygen are not distinguishable from their relative rates of reactions with other singlet oxygen possibly made of $^1\Sigma_g^+$ oxygen. Either the oxidizing properties of $^1\Delta_g$ or $^1\Sigma_g^+$ singlet oxygen are the same towards the acceptors studied here, or else by the time singlet oxygen reacts it is always mostly in the lowest $^1\Delta_g^+$ state. This latter interpretation seems to agree with the findings of Ogryzlo[17] on the predominance of the $^1\Delta_g$ state in chemically excited $^1O_2*$ in solution as well as in electrically excited $^1O_2*$ in the gas, from an analysis of their luminescences.

Acknowledgment. The author is grateful to Professor P. D. Bartlett for calling her attention to the work on excited molecular oxygen and for offering her hospitality in his laboratory. She wishes to thank him and Professor E. J. Corey for very helpful discussions, Miss Barbara Kaski for her help with the externally generated singlet oxygen experiments, and Mr. Paul Engel for his comments on the manuscript. The author's association with the Radcliffe Institute for Independent Study is gratefully acknowledged.

(12) Since these results were obtained, K. R. Kopecky and H. J. Reich, *Can. J. Chem.*, **43**, 2265 (1965), have published data demonstrating that the rates of dye-sensitized photooxidation of olefins are independent of the dye employed, and have similarly used this as an argument in favor of the singlet oxygen hypothesis.

(13) See ref 1 and N. J. Turro, "Molecular Photochemistry," W. A. Benjamin, Inc., New York, N. Y., 1695, pp 86, 132; G. Porter and N. M. Windsor, *Proc. Roy. Soc.* (London), **A245**, 238 (1958).

(14) See N. J. Turro, ref 13, p 132; P. Pringsheim, "Fluorescence and Phosphorescence," Interscience Publishers, Inc., New York, N. Y., 1949; L. S. Foster and D. Dudley, *J. Phys. Chem.*, **66**, 838 (1962).

(15) E. Clar and F. John, *Ber.*, **63**, 2967 (1930).

(16) C. F. H. Allen and A. Bell, *J. Am. Chem. Soc.*, **64**, 1253 (1942).

(17) R. J. Browne and E. A. Ogryzlo, *Proc. Chem. Soc.*, 117 (1964); S. J. Arnold, E. A. Ogryzlo, and H. Witzke, *J. Chem. Phys.*, **40**, 1769 (1964).

33

Reprinted from *Advan. Chem. Ser.*, **77**, 102–117 (1968)

Chemistry of Singlet Oxygen

V. Reactivity and Kinetic Characterization

RAYMOND HIGGINS, CHRISTOPHER S. FOOTE, and HELEN CHENG

University of California, Los Angeles, Calif. 90024

Reactivities of olefinic and other acceptors are determined for oxygenation, both dye-photosensitized and with singlet oxygen (generated by reaction of metal hypochlorites with hydrogen peroxide). The ratio of the rate constant for reaction with acceptor to that for decay is the same for the intermediate in both reactions. Relative reactivities of various acceptors (from competition experiments) are the same in both reactions and indicate that the intermediate is electrophilic and quite discriminating. In addition, product distributions in the two reactions are identical. The reactive intermediate in the dye-sensitized photo-oxygenation of these substrates cannot be distinguished by any criterion yet devised from chemically generated $^1\Delta_g$ oxygen.

The dye-sensitized photo-oxygenation of olefins and dienoid compounds has been the subject of much recent interest (*3, 13, 14, 15, 16, 32, 34*). Other papers in this volume provide a good introduction to the background and scope of the reaction (*17, 23, 27, 33*). Previous work has shown that singlet oxygen (*1, 24*), produced by the reaction of sodium hypochlorite and hydrogen peroxide (*8, 9, 10, 29*) or by electrodeless discharge (*6*), reacts with typical photo-oxygenation acceptors to give products which are indistinguishable from those of photosensitized oxygenation. In particular, the stereoselectivities of the reactive species in the photo-oxygenation and of chemically produced singlet oxygen are identical (*9*). In this paper, the reactivity of the photochemical intermediate is compared with that of chemically produced singlet oxygen.

The kinetics of the dye-sensitized photo-oxygenation have been studied by several groups (*3, 4, 5, 13, 14, 15, 16, 28, 32, 34, 35*). With

Rose Bengal and many other sensitizers, the reaction proceeds by way of the triplet dye (^{3}Sens); at oxygen concentrations above $2 \times 10^{-5}M$ (exceeded for methanol solutions in equilibrium with air or oxygen), all of the triplet dye reacts with oxygen to give a reactive species (RS) which is the actual oxygenating agent (*13, 14, 15, 34*). This reactive species has often been assumed to be a complex of sensitizer and oxygen (Sens-O_2), but singlet oxygen ($^1O_2$), formed by energy transfer from the dye triplet to oxygen, is kinetically equivalent (*11, 19*). The reactive species either decays, regenerating ground state oxygen (Reaction 1) or reacts with acceptor (A) to give the product (AO$_2$, Reaction 2).

$$^3\text{Sens} + {}^3\text{O}_2 \rightarrow \text{RS (Sens-O}_2 \text{ or } {}^1\text{O}_2)$$

$$\text{RS} \xrightarrow{k_1} {}^3\text{O}_2 \qquad (1)$$

$$\text{RS} + \text{A} \xrightarrow{k_2} \text{AO}_2 \qquad (2)$$

The instantaneous quantum yield for product formation (Φ_{AO_2}) is given by the following expression.

$$\Phi_{AO_2} = \Phi_{3\,\text{sens}} \frac{k_2[\text{A}]}{k_1 + k_2[\text{A}]}$$

Thus, the limiting quantum yield of product formation at high acceptor concentration ($k_2[\text{A}] \gg k_1$) is the quantum yield of triplet dye formation (Φ_{3_sens}); at limiting low acceptor concentration ($k_2[\text{A}] \ll k_1$), Φ_{AO_2} is proportional to [A]. The constant β is defined as k_1/k_2 and is characteristic for a given acceptor. The numerical value of β gives the acceptor concentration at which half the reactive species gives product. Values of β for a few compounds have been reported (*4, 5, 13, 14, 16, 34, 36, 37*).

The instantaneous yield of product (Y_{AO_2}) in the hypochlorite–H_2O_2 oxygenations would be expected to be given by an expression very similar to that for Φ_{AO_2} in the photo-oxygenation:

$$Y_{AO_2} = Y_{{}^1O_2} \frac{k_2[\text{A}]}{k_1 + k_2[\text{A}]}$$

In this expression, $Y_{{}^1O_2}$ is the yield of singlet oxygen produced from a given amount of hypochlorite, and the other terms have the same meaning as stated previously.

For both the photochemical and chemical oxygenation, a plot of $1/\Phi_{AO_2}$ or $1/Y_{AO_2}$ against $1/[\text{A}]$ should be a straight line and can be used to determine β for a given acceptor.

$$\frac{1}{\Phi_{AO_2}} = \frac{1}{\Phi_{3_{Sens}}} \left(1 + \frac{\beta}{[A]} \right)$$

and

$$\frac{1}{Y_{AO_2}} = \frac{1}{Y_{^1O_2}} \left(1 + \frac{\beta}{[A]} \right)$$

The ratio of slope to intercept is β in both expressions.

These relationships are valid only if [A] does not change during the reaction (negligible conversions). This point has been overlooked at times. In practice, instead of Φ_{AO_2}, the amount of product formed in a given time of irradiation will be used since it is proportional to Φ_{AO_2} if the light flux is constant within a series of reactions.

Reaction 3 is an important mode of decomposition of $^1O_2$ in the vapor phase (1, 24).

$$^1O_2 + {}^1O_2 \xrightarrow{k_3} 2\ {}^3O_2 \tag{3}$$

The expressions derived above assume all decay of the reactive species is first order. If Reaction 3 is significant under these conditions, the expression for the chemical oxygenation would have to be modified as follows:

$$Y_{AO_2} = Y_{^1O_2} \frac{k_2[A]}{k_1 + k_2[A] + 2\ k_3[{}^1O_2]}$$

and a similar correction would be necessary for the photochemical process. The fraction of intermediate giving product would depend on the concentration of intermediate. Even though in photochemical experiments the rate of production of intermediate is constant (for a given apparatus), its steady state concentration would depend on acceptor concentration, and curved plots should result. In the chemical reaction, nothing resembling a constant rate of $^1O_2$ formation is obtained in our experiments because hypochlorite is not introduced at a steady rate. Since excellent linear plots are observed (see Results), $2k_3[{}^1O_2] \ll k_1$ in solution.

From the ratio of $\beta(k_1/k_2)$ values for different acceptors, the relative rate constant (k_2) for the reaction of each with the reactive species can be obtained. However, it is much easier to use a competition method. This method has been used by Kopecky and Reich (26) and by Wilson (39) for the photo-oxygenation, using different sensitizers and acceptors. It has been shown that the relative value of k_2 does not depend on sensitizer (26, 39), and a ratio of k_2 values obtained using $^1O_2$ generated by electrodeless discharge was found to be compatible with that obtained in photosensitized oxygenation (39). These results were used to support the suggestion (11, 19) that singlet oxygen is intermediate in the photosensitized oxygenations (26, 39).

For the photochemical reaction, the expression shown below governs competition reactions with two acceptors (A and B).

$$\frac{-d[A]}{-d[B]} = \frac{d[AO_2]}{d[BO_2]} = \frac{\Phi_{AO_2}}{\Phi_{BO_2}} = \frac{\dfrac{k_2{}^A[A]}{k_1 + k_2{}^A[A] + k_2{}^B[B]}}{\dfrac{k_2{}^B[B]}{k_1 + k_2{}^A[A] + k_2{}^B[B]}} = \frac{k_2{}^A[A]}{k_2{}^B[B]}$$

This differential expression can be used directly for small conversions. Since the denominators cancel, the expression can also be integrated easily to give the equations necessary for higher conversions:

$$\frac{k_2{}^A}{k_2{}^B} = \frac{\log(1 - [AO_2]/[A]_o)}{\log(1 - [BO_2]/[B]_o)} = \frac{\log([A]_f/[A]_o)}{\log([B]_f/[B]_o)}$$

These expressions are for measurement of product (AO_2, BO_2) appearance or acceptor disappearance, respectively. Initial and final concentrations are indicated by the subscript o or f, respectively. This result is equivalent to the method of Ingold and Shaw (*18*), derived for the less complex case of aromatic nitration, where no decay term (k_1) for the intermediate interferes, and was used by Kopecky and Reich for the photo-oxidation, apparently without derivation (*26*). Previous authors followed acceptor disappearance, which is very difficult to measure accurately, particularly for unreactive acceptors, and is subject to severe errors if starting material is consumed by side reactions (*26, 39*). However, the technique has the advantage that gas-chromatograph detector calibration is not required since only concentration ratios are measured. We have used this technique also, but we found it far more accurate to measure product appearance since highly characteristic products are formed, which in many cases distinguish the desired reaction from all side reactions. However, this method presents the difficulty that the product peroxides must be quantitatively reduced for gas chromatography, and all products must be isolated and characterized and detector response carefully calibrated.

Experimental

Materials. *cis*-4-Methyl-2-pentene, 2-methyl-2-butene, 1-methylcyclohexene and 2,3-dimethyl-2-butene were Phillips Petroleum Co. 99% grade. Sodium hypochlorite was Purex 14 (Purex Corp.). Other reagents were from standard sources.

Since the olefins used for product appearance studies always contained sufficient oxygen-containing impurities to interfere with analyses, they were purified immediately before use by repeated passage through silica gel until shown to be sufficiently pure.

Preparative Photo-Oxygenations. The apparatus used has been described (*8, 16*). Peroxides were reduced with a fourfold excess of sodium borohydride, with cooling. After workup by partial solvent removal and ether–water partition, the crude alcohol mixtures were separated by preparative GLC. Structures were unambiguously determined by infrared, nuclear magnetic resonance, and high resolution mass spectra, which will be reported in more detail later.

Kinetic Measurements. PHOTO-OXYGENATIONS. The reaction vessel was a small immersion irradiation apparatus similar to that used for preparative work (8, 16). It was immersed in a constant temperature bath (30°C.) and cooled internally by circulating 0.2M Cu_2Cl_2 solution (also held at 30°C.) through the water jacket; the coolant also served as an infrared filter. To ensure stability, the light source (Sylvania DWY) was operated from a constant voltage transformer; to avoid overheating and to ensure convenient reaction times (5.00 min.), the voltage was reduced to 50 volts. The temperature in the solution rose less than 2°C. when the lamp was in operation.

Stock solutions of Rose Bengal were prepared in the appropriate solvent, filtered, and stored in the dark. Weighed quantities of 2-methyl-2-pentene were dissolved in the solvent, an aliquot of sensitizer solution was added, and the solution diluted volumetrically (solutions were handled in subdued light after addition of sensitizer). After reaction, the solutions were immediately reduced by adding excess trimethylphosphite (7, 25). They were left overnight, and an aliquot of isoamyl alcohol internal standard was added. The solutions were analyzed on a Perkin-Elmer Model 800 flame-ionization gas chromatograph (N_2 carrier) using an 8-foot × 1/8-inch copper column with 10% UCON WS on 80/100 mesh Anakrom ABS at 80°C. Peak areas were evaluated by triangulation and by the peak height-half height width method. Absolute yields were calculated by calibrating the detector by injecting weighed mixtures of pure product alcohols and the internal standard in the appropriate solvent. The data are summarized in Tables I and II.

Table I.   Effect of Acceptor Concentration on Product Yield for Photo-Oxygenation of 2-Methyl-2-pentene (A) in *tert*-Butyl Alcohol

[A], M	[AO$_2$],[a] mM	Conversion, %
0.0109	0.88	8.1
0.0131	1.01	7.7
0.0199	1.43	7.2
0.0313	1.92	6.1
0.0504	2.40	4.8
0.1003	3.22	3.2
0.1989	4.69	2.4

[a] Sum of yields of two products; mean of three or more analyses.

CHEMICAL OXYGENATIONS. Weighed amounts of 2-methyl-2-pentene were dissolved in 50:50 (by volume) methanol–*tert*-butyl alcohol. Hydrogen peroxide (1 ml. of 9.08M) was added, and the solution was diluted to 100 ml. To 50.0 ml. of this solution was added 0.50 ml. of 1.05M aqueous NaOCl over 10 min. with stirring at 25°C. NaOCl was

Table II.   Effect of Acceptor Concentration on Product Yield for the
Photo-Oxygenation of 2-Methyl-2-pentene (A) in
Methanol–*tert*-butyl Alcohol

[A], M	[AO$_2$],[a] mM	Conversion, %
0.0098	0.422	4.3
0.0118	0.479	4.1
0.0141	0.616	4.3
0.0203	0.840	4.1
0.0270	0.995	3.7
0.0325	1.250	3.8
0.0401	1.513	3.8
0.0518	1.680	3.2
0.0988	2.520	2.5
0.2067	3.510	1.7

[a] Sum of yields of two products; mean of three or more analyses.

Table III.   Effect of Acceptor Concentration on Product Yield
for the Chemical Oxygenation of 2-Methyl-2-pentene (A)

[A], M	[AO$_2$],[a] mM	Conversion, %
0.0121	0.310	2.6
0.0220	0.552	2.5
0.0284	0.737	2.6
0.0490	0.972	2.0
0.0933	1.680	1.8
0.1995	2.280	1.1

[a] Sum of yields of two products; mean of three or more analyses.

introduced through a capillary below the solution surface. The capillary
was fed with a syringe driven by a thumbscrew. After addition, the
peroxides were reduced overnight with excess trimethyl phosphite, and
the solutions were analyzed by gas chromatography, as above. The results
are summarized in Table III.

COMPETITION REACTIONS: PRODUCT ANALYSIS METHOD. The procedures used were identical to those described above, except that pairs
of olefins were present in the solutions. In some cases trimethyl phosphite
had a retention time near that of a product, and triphenyl phosphite was
used as the reducing agent. In chemical oxygenation of the less reactive
olefins, products other than those found in photo-oxygenation reactions
were detected. These side products could be suppressed by adding free-
radical inhibitors such as 2,6-di-*tert*-butylphenol and conducting the
reaction near 0°C. The products from these olefins are listed in Table
IV. The results are listed in Table V.

COMPETITION REACTIONS; ACCEPTOR DISAPPEARANCE METHOD. *Photo-
Oxygenation.* Weighed quantities of the two olefins (such that the final
solution was approximately 0.24*M* in each) and an internal standard

Table IV. Products of Olefin Oxygenations[a]

Olefin	Alcohol A	Alcohol B	Photo-Oxygenation %A	%B	OCl⁻/H₂O₂ %A	%B
(structure)	*(structure)*	*(structure)*	100		100	
(structure)	*(structure)*	*(structure)*	49	51	51	49
(structure)	*(structure + structure)*	*(structure)*	48	52	49	51
(structure)	*(structure)*	*(structure)*	51	49	52	48
(structure)	*(structure)*	*(structure)*	96	4	94	6
	(structure)	*(structure)*	100		[b]	

[a] After reduction of the peroxides to the corresponding alcohols: mean of four or more analyses.
[b] Free radical products could not be suppressed sufficiently to permit analysis.

Table V. Relative Reactivities of Olefins by Product Analysis[a]

Olefin A	Olefin B	Photo-Oxygenation[b] k_2^A/k_2^B	OCl⁻/H₂O₂[c] k_2^A/k_2^B
2,3-Dimethyl-2-butene	2-Methyl-2-butene[d]	41	35
2-Methyl-2-butene	2-Methyl-2-pentene[d]	1.32	1.28
2-Methyl-2-pentene	1-Methylcyclohexene[d]	4.5[e]	3.9[e,f]
cis-4-Methyl-2-pentene	Cyclohexene[g]	5.4[e]	[h]
cis-4-Methyl-2-pentene	2-Methyl-2-pentene[i]	0.014	0.011[j]
trans-4-Methyl-2-pentene	2-Methyl-2-pentene[i]	0.0025	0.0020[j]

[a] Mean of two or more experiments in methanol–*tert*-butyl alcohol (50:50); reduced with trimethyl phosphite unless otherwise stated.
[b] Sensitized by Rose Bengal, 30°C.
[c] At 25°C. unless otherwise stated; solution volume 50 ml.
[d] Internal standard, isoamyl alcohol.
[e] Reduced with triphenyl phosphite.
[f] 0.024M in 2,6-di-*tert*-butylphenol; reaction temperature 3–4°C.
[g] Internal standard, n-amyl alcohol.
[h] Free radical products could not be suppressed sufficiently to permit analysis.
[i] Internal standard, isobutyl alcohol.
[j] 0.097M in 2,6-di-*tert*-butylphenol; reaction temperature 3–4°C.

were dissolved in methanol–*tert*-butyl alcohol (50:50 by volume) containing Rose Bengal, and a small quantity of this solution was removed for analysis. The remaining solution was then irradiated under oxygen, and the reaction was followed by oxygen consumption. Irradiation was continued until the extent of reaction was approximately equal that used in the chemical oxygenation of the same pair of olefins (*see below*). Analyses were performed on a Perkin-Elmer Model 154 gas chromatograph. Peak heights of the two olefins relative to the internal standard in the final solution were compared with the same quantities in the unreacted solution to calculate the fractions of each of the olefins converted. The columns and internal standards used are given in Table VI. In some reactions, a free radical inhibitor was also present to make conditions strictly parallel to those used in the calcium hypochlorite experiments. Addition of this compound had no detectable influence on the results, which are summarized in Table VI.

Chemical Oxygenations. These were carried out using the reaction of either sodium or calcium hypochlorite with hydrogen peroxide. In the sodium hypochlorite experiments, solutions of the two olefins and internal standard (again 0.24M in the three compounds) and hydrogen peroxide (in excess of the quantity of hypochlorite to be added) in methanol–*tert*-butyl alcohol (50:50 by volume) were prepared, a small quantity removed for analysis, and the remaining solution cooled to below −20°C. Aqueous sodium hypochlorite solution was then added dropwise to this solution under stirring, and the progress of the reaction was followed by measuring the oxygen evolved. The initial and final solutions were then analyzed as in the photo-oxygenation. The experiments using calcium hypochlorite were essentially identical, except that the solutions also

Table VI.   Relative Reactivities of Acceptors by
Acceptor Disappearance[a]

Compound A	Compound B	Photo-Oxygenation[b] k_2^A/k_2^B	NaOCl/ H_2O_2 k_2^A/k_2^B	Ca(OCl)$_2$/ H_2O_2 k_2^A/k_2^B
2,5-Dimethylfuran	2,3-Dimethyl-2-butene[c,e]	2.4	5.2	1.5
Cyclopentadiene	2,3-Dimethyl-2-butene[d,f]	1.2	2.0	0.7
2,3-Dimethyl-2-butene	1,3-Cyclohexadiene[d,f]	13[j]		3
	2-Methyl-2-butene[c,g]	22	15	23
	cis-3-Methyl-2-pentene[c,h]	36	25	
2-Methyl-2-butene	cis-3-Methyl-2-pentene[c,h]	1.8		2.5
	trans-3-Methyl-2-pentene[c,h]	1.2		1.4
	1-Methylcyclopentene[c,g]	1.0		1.6
	1-Methylcyclohexene[d,i]	23		20
trans-3-Methyl-2-pentene	1-Methylcyclohexene[d,i]	22		7

[a] Mean of two or more determinations in methanol–*tert*-butyl alcohol (50:50).
[b] Sensitized by Rose Bengal.
[c] Analyzed on a 6-foot × 1/4-inch aluminum column packed with 20% UCON WS on 60/80 Chromosorb W.
[d] Analyzed on a 6-foot × 1/4-inch aluminum column packed with 10% 3,3′-thiodipropionitrile on 70/80 Anakrom ABS.
[e] Internal standard, toluene.
[f] Internal standard, carbon tetrachloride.
[g] Internal standard, hexane.
[h] Internal standard, heptane.
[i] Internal standard, octane.
[j] Result of a single determination.

contained 2,4,6-tri-*tert*-butylphenol (5-10% of the olefin concentration), and the calcium hypochlorite was added as a solid. The results are summarized in Table VI.

Results

Determination of $\beta(k_1/k_2)$. The previous discussion shows that the amount of product formed changes most rapidly with changes in acceptor concentration when the range of [A] brackets β. Since the convenient experimental range of [A] is 0.01-0.2M, 2-methyl-2-pentene, [reported $\beta = 0.18M$ in methanol (*13, 14, 34*)], was selected.

Initial experiments were run in *tert*-butyl alcohol, which dictated a reaction temperature of 30°C., but since methanol–*tert*-butyl alcohol (50:50 by volume) was found to be the best solvent for the chemical oxygenation, the photo-oxygenation experiments were rerun in this solvent.

Solutions of 2-methyl-2-pentene of concentrations in the range 0.01 to 0.2M containing identical quantities of Rose Bengal were prepared and irradiated under identical conditions for 5 minutes. Figure 1 is a plot of 1/[AO$_2$], the amount of product formed in photo-oxygenations in the two solvents against 1/[A]$_{mean}$, the mean concentration of acceptor during a reaction. Since conversions were purposely kept low (in no case exceeding 8%), [A]$_{mean}$ differs only slightly from [A]$_{initial}$, and the equation discussed in the introduction is valid. The least-square line through the points gives a value of β equal to 0.13 ± 0.02M for methanol–*tert*-butyl alcohol, and 0.051 ± 0.003M for *tert*-butyl alcohol. The results of the chemical oxygenation in methanol–*tert*-butyl alcohol are plotted in Figure 2 with the least-square line giving β equal to 0.16 ± 0.04M.

Relative Reactivities. BY PRODUCT ANALYSIS. Solutions of pairs of olefins in methanol–*tert*-butyl alcohol were irradiated using Rose Bengal as sensitizer. The peroxides were reduced to the corresponding alcohols, and the products were analyzed by gas chromatography. Methods similar to those described above were used for the chemical oxygenations. With unreactive olefins, large amounts of reagents had to be used to give sufficient conversions, and side reactions became important in some cases. The side reactions could be suppressed by adding free radical inhibitors and lowering the temperature.

BY ACCEPTOR DISAPPEARANCE. Weighed quantities of two compounds, chosen to be similar in reactivity, were dissolved along with an unreactive internal standard in methanol–*tert*-butyl alcohol containing Rose Bengal. Analyses of the solution before and after irradiation were carried out, and the data were treated as described in the introduction. Similar methods were used for the chemical oxygenations, which were carried out at −20°C. Radical inhibitors were added in some of the runs. Severe

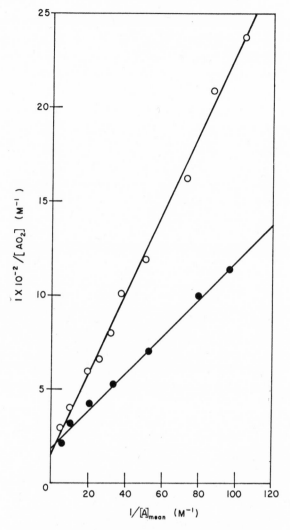

Figure 1.   Photo-oxygenation of 2-methyl-2-pentene at 30°C.

○: *in methanol–tert-butyl alcohol (β = 0.13M ±* 0.02)
●: *in tert-butyl alcohol (β = 0.051M ± 0.003)*

difficulties were encountered in using this technique with unreactive acceptors because large amounts of aqueous hypochlorite had to be added to convert enough acceptor for accurate measurement.

Reactivities of various acceptors (values of k_2 relative to k_2 for 2,3-dimethyl-2-butene), calculated by the method of Ingold and Shaw (*18*) (as discussed in the introduction) are assembled in Table VII. Values determined by previous workers for photo-oxygenation are included for comparison. Those of Gollnick and Schenck (*13, 14, 34*) are calculated from β values obtained from oxygen uptake measurements, whereas those of Kopecky and Reich (*26*) were determined by acceptor disappearance studies.

Discussion

The excellent linear plot (Figure 2) obtained for the chemical oxygenation of 2-methyl-2-pentene demonstrates that a transient reactive intermediate is formed in this reaction and that the fraction of the intermediate which is trapped depends on acceptor concentration in the expected way. [From the intercept of this plot, it can be determined that the yield of $^1O_2$, based on hypochlorite is 43% under these conditions. At lower temperatures, yields as high as 65% can be obtained reproducibly (*8*)].

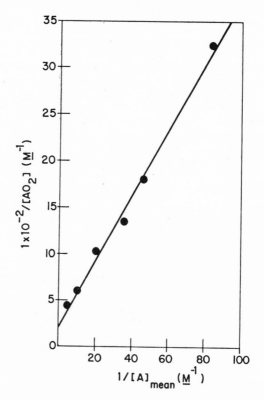

Figure 2.   Chemical oxygenation (NaOCl/ H₂O₂) of 2-methyl-2-pentene in methanol– tert-butyl alcohol ($\beta = 0.16M \pm 0.04$)

The value of β (k_1/k_2) obtained from Figure 2 is 0.16 $\pm$ 0.04M, in excellent agreement with that obtained for the photochemical oxygenation (0.13 $\pm$ 0.02M). These values (for methanol–*tert*-butyl alcohol) are in good agreement with that obtained for 2-methyl-2-pentene in methanol (0.18M) by oxygen absorption methods (*13, 14, 34*). [The small discrepancy is probably caused by a solvent effect, as shown by the fact that the photochemical value in pure *tert*-butyl alcohol is 0.051 $\pm$ 0.003M. No large solvent effect on β is observed; values in solvents as different as methanol and benzene differ by less than a factor of 2 (*12*).] That the values for the two reactions are so similar provides strong support for the suggestion (*11, 19*) that singlet oxygen is the reactive intermediate in both reactions. It would be a remarkable coincidence if the ratio of the two rate constants were the same for Sens-O_2 and for $^1O_2$; complexing with sensitizer would have to affect both the reaction rate of oxygen with acceptor and the decay rate by the same fraction.

Relative reactivities (Table VII) determined in this study are in excellent agreement with previous photo-oxygenation values (*13, 14, 26, 34*), considering that the reactivity spread is 10^5. Results of the product appearance studies are highly reproducible, and the values for photosensitized and chemical oxygenation agree within the experimental error. The results of the acceptor disappearance studies are more erratic; results of chemical oxygenation are reproducible from run to run only within about a factor of two (most of the values in Table VII are the average of many determinations). The lack of reproducibility is probably caused by the fact that low conversions had to be used to avoid solubility problems caused by adding aqueous hypochlorite. In addition, even small amounts of acceptor lost by evaporation or side reactions cause a large error in results obtained by this method, particularly with unreactive acceptors. Nevertheless, the results obtained for the two reactions are in good agreement. The worst agreement is found for 1,3-cyclohexadiene, where the results for the two reactions differ by a factor of 4; only one run was carried out in the photo-oxygenation of this substrate, and the value may well be in error.

The reactivity of cyclohexene determined is lower by a factor of 3–5 than that found in previous studies. Particularly with this very unreactive acceptor, product appearance studies are more reliable since acceptor disappearance methods (*26*) or oxygen uptake studies (*13, 14, 34*) can lead to a deceptively high reactivity for the reasons mentioned above. However, side reactions occurred to such an extent in the chemical oxygenation that the desired product could not be distinguished reliably from side products.

If the reactive intermediate in the photo-oxygenations were a complex of sensitizer and oxygen, the sensitizer should influence the electron

Table VII.   Relative Reactivities of Acceptors

	Product Analysis[a]		Olefin Disappearance[a]			
	Photo-Oxygenation	NaOCl/H_2O_2	Photo-Oxygenation	OCl$^-$/H_2O_2[b]	Photo-Oxygenation[c]	Photo-Oxygenation[d]
2,5-Dimethylfuran			2.4	3.8		
Cyclopentadiene			1.2	1.1		
2,3-Dimethyl-2-butene	(1.00)	(1.00)	(1.00)	(1.00)	(1.00)	(1.00)
1,3-Cyclohexadiene			0.08[e]	0.3		
2-Methyl-2-butene	0.024	0.028	0.05	0.06	0.055	
1-Methylcyclopentene			0.05	0.03		0.070
2-Methyl-2-pentene	0.019	0.022			0.017	
Trans-3-methyl-2-pentene			0.04	0.03		
Cis-3-methyl-2-pentene			0.03	0.03		
1-Methylcyclohexene	0.0041	0.0056	0.002	0.003	0.0025	0.0082
Cyclohexene	0.000048				0.00012	0.00018
Cis-4-methyl-2-pentene	0.00026	0.00024				
Trans-4-methyl-2-pentene	0.000047	0.000044				

[a] Present work.
[b] Combined results from NaOCl and Ca(OCl)$_2$ experiments.
[c] Calculated from β values reported in Refs. 13, 14, 34.
[d] From Ref. 26.
[e] Result of a single determination.

demand of the oxygen (and presumably reduce it). The complex should also be bulkier than singlet oxygen. Hence, it would be expected to differ from singlet oxygen in sensitivity to both electron density and steric effects. However, the relative reactivities of acceptors are the same in both reactions.

As already pointed out (9), the stereoselectivity of the reactive intermediate should be altered if a bulky sensitizer were attached to the oxygen, but no differences in stereoselectivity between photosensitized and chemical oxygenation were observed (9). The results in Table IV provide further strong evidence that no differences in stereoselectivity exist since the product distributions in the two reactions are identical with these acceptors also.

Thus, by the criteria of reactivity, decay rate, and stereoselectivity the chemically generated reactive species is indistinguishable from the reactive intermediate in the photo-oxygenation. Strong arguments based on lifetime and chemistry have been advanced that the intermediate in the chemical oxygenation is $^1\Delta_g$ O_2 (rather than $^1\Sigma_g^+$ or vibrationally excited ground state O_2) (8). The reactive intermediate in the photo-oxygenations is thus probably also $^1\Delta_g$ O_2.

The results of Litt and Nickon (27) show small but reportedly significant differences in isotope effects with different photosensitizers and in chemical oxygenation. Further study is obviously desirable. The formation of a sensitizer–oxygen complex which dissociates before reacting (13, 14, 17, 23, 34) would be entirely consistent with our results.

A recent study indicates that with sensitizers of triplet energy above 37 kcal./mole, $^1\Sigma_g^+$ O_2 is the principal primary product of oxygen quenching of the dye triplet (20-23). Rose Bengal has a triplet energy of 45.4 kcal./mole and should give substantial amounts of $^1\Sigma_g^+$ O_2 (20-23). The results reported here show that $^1\Sigma_g^+$ O_2 either does not react appreciably with these substrates, or else (much less likely) it has an identical ratio of reaction to decay rates and gives an identical product distribution. If this were not the case, curvature of the plots in Figures 1 and 2 would be expected, and the product distribution would depend on acceptor concentration since the lifetime of $^1\Sigma_g^+$ O_2 is less than that of $^1\Delta_g$ O_2 (1, 2, 33). Similar conclusions were reached by Wilson on the basis of more limited evidence (39). Probably $^1\Sigma_g^+$ O_2, if formed, decays to $^1\Delta_g$ O_2 rather than directly to ground state O_2, since the maximum quantum yield for product formation in the photo-oxygenation is equal to the quantum yield of Rose Bengal triplet formation (13, 14, 15, 34) so that no intermediate is lost. The results reported by Ogryzlo (33) show that the lifetime of $^1\Sigma_g^+$ O_2 is so short that the reaction rate of an acceptor must be 10^5–10^8 times faster with $^1\Sigma_g^+$ than with $^1\Delta_g$ O_2 for the $^1\Sigma_g^+$ reaction to be detected.

The factors influencing the reactivity of various substrates in the photo-oxygenation have been discussed by Kopecky and Reich (26) and by Gollnick and Schenck (13, 14, 15, 34). The results reported here confirm that electron-rich olefins are particularly reactive and that cyclohexenes are comparatively unreactive compared with acyclic compounds. Kopecky and Reich (26) rationalized this reactivity sequence by using an extreme resonance form of singlet oxygen, reacting *via* a perepoxide, as originally proposed by Sharp (38).

While it is very difficult to rule out such a mechanism positively, it seems far more likely that the reaction is a nearly concerted cycloaddition, as previously suggested by Gollnick and Schenck (*13, 14, 15, 34*) and by Nickon (*27, 30, 31, 32*) on stereochemical grounds. Singlet oxygen ($^1\Delta_g$) has an electronic structure similar to ethylene (*8*), and the electronegativity of the oxygen atoms is quite adequate to rationalize the electrophilic character of the reaction without recourse to the perepoxide mechanism. Litt and Nickon have presented additional arguments against the perepoxide mechanism (*27*). Singlet oxygen can be visualized as a dienophile, undergoing the Diels-Alder reaction with dienes, and the ene reaction with olefins (*9*). (It should be emphasized that it is a reagent of remarkable selectivity and undergoes no reaction at all with most compounds.)

The concerted nature of the reaction is indicated by the fact that a Markovnikoff-type directing effect is essentially absent. Secondary and tertiary products are formed in nearly equal amounts from unsymmetrically substituted olefins (*9, 13, 14, 15, 17, 27, 34*). In addition, substituents on the phenyl group of trimethylstyrene exert no effect on the product distributions (*12*). Furthermore, β values for 2-methyl-2-pentene show only a very small solvent effect (*12*). All these facts suggest that little polarity is developed at the transition state and are consistent with a concerted reaction.

J. H. Knox has asked whether singlet oxygen might not be implicated in spontaneous initiation processes at elevated temperatures since the activation energy would be fairly low. However, $^1\Delta_g$ O_2, even if spontaneously formed (a fairly slow process, even at 300°C. because of the forbiddenness) would not react appreciably with ethylene or propylene; more highly substituted olefins might in principle react, but the over-all rate would probably be too low. However, little is known about gas-phase reactions of singlet O_2.

Acknowledgments

We thank K. Gollnick for unpublished information and helpful discussions and D. R. Kearns for prepublication copies of his manuscripts. C. S. Foote was an Alfred P. Sloan Research Fellow, 1965-67. Part of the work reported here was taken from the Master's Thesis of Helen Cheng, U.C.L.A., 1966.

Literature Cited

(1) Arnold, J. S., Browne, R. J., Ogryzlo, E. A., *Photochem. Photobiol.* **4**, 963 (1965).
(2) Bader, L. W., Ogryzlo, E. A., *Discussions Faraday Soc.* **37**, 46 (1964).
(3) Bowen, E. J., *Advan. Photochem.* **1**, 23 (1963).

(4) Bowen, E. J., *Discussions Faraday Soc.* **14**, 143 (1953).
(5) Bowen, E. J., Tanner, D. W., *Trans. Faraday Soc.* **51**, 475 (1955).
(6) Corey, E. J., Taylor, W. C., *J. Am. Chem. Soc.* **86**, 3881 (1964).
(7) Denny, D. B., Goodyear, W. F., Goldstein, B., *J. Am. Chem. Soc.* **82**, 1393 (1960).
(8) Foote, C. S., Wexler, S., Ando, W., Higgins, R., *J. Am. Chem. Soc.* **90**, 975 (1968).
(9) Foote, C. S., Wexler, S., Ando, W., *Tetrahedron Letters* **1965**, 4111.
(10) Foote, C. S., Wexler, S., *J. Am. Chem. Soc.* **86**, 3879 (1964).
(11) *Ibid.*, p. 3880.
(12) Foote, C. S., Denny, R., unpublished observations.
(13) Gollnick, K., Ph.D. Dissertation, University of Göttingen, Germany, 1962.
(14) Gollnick, K., *Advan. Photochem.*, in press.
(15) Gollnick, K., Schenck, G. O., *Pure Appl. Chem.* **9**, 507 (1964).
(16) Gollnick, K., Schenck, G. O., "1,4-Cycloaddition Reactions," J. Hamer, Ed., p. 255, Academic Press, New York, 1967.
(17) Gollnick, K., ADVAN. CHEM. SER. **77**, 78 (1968).
(18) Ingold, C. K., Shaw, F. R., *J. Chem. Soc.* **1927**, 2918.
(19) Kautsky, H., *Biochem. Z.* **291**, 271 (1937).
(20) Kawaoka, K., Khan, A. U., Kearns, D. R., *J. Chem. Phys.* **46**, 1842 (1967).
(21) Kearns, D. R., Hollins, R. A., Khan, A. U., Chambers, R. W., Radlick, P., *J. Am. Chem. Soc.* **89**, 5455 (1967).
(22) Kearns, D. R., Hollins, R. A., Khan, A. U., Radlick, P., *J. Am. Chem. Soc.* **89**, 5456 (1967).
(23) Khan, A. U., Kearns, D. R., ADVAN. CHEM. SER. **77**, 143 (1968).
(24) Khan, A. U., Kasha, M., *J. Am. Chem. Soc.* **88**, 1574 (1966).
(25) Kharasch, M. S., Mosher, R. A., Bengelsdorf, I. S., *J. Org. Chem.* **25**, 1000 (1960).
(26) Kopecky, K. R., Reich, H. J., *Can. J. Chem.* **43**, 2265 (1965).
(27) Litt, F. A., Nickon, A., ADVAN. CHEM. SER. **77**, 118 (1968).
(28) Livingston, R., Subba Rao, V., *J. Phys. Chem.* **63**, 794 (1959).
(29) McKeown, E., Waters, W. A., *J. Chem. Soc.* **1966**, B1040.
(30) Nickon, A., Bagli, J. F., *J. Am. Chem. Soc.* **83**, 1498 (1961).
(31) Nickon, A., Mendelson, W. L., *Can. J. Chem.* **43**, 1419 (1965).
(32) Nickon, A., Schwartz, N., DiGorgio, J. B., Widdowson, D. A., *J. Org. Chem.* **30**, 1711 (1965).
(33) Ogryzlo, E. A., ADVAN. CHEM. SER. **77**, 133 (1968).
(34) Schenck, G. O., Gollnick, K., *Forschung. Landes Nordrhein-Westfalen* No. **1256** (1963).
(35) Schenck, G. O., Koch, E., *Z. Elektrochem.* **64**, 170 (1960).
(36) Schenck, G. O., Schulte-Elte, K.-H., *Ann.* **618**, 185 (1958).
(37) Schenck, G. O., *Angew. Chem.* **69**, 579 (1957).
(38) Sharp, D. B., "Abstracts of Papers," 138th Meeting, ACS, Sept. 1960, 79P.
(39) Wilson, T., *J. Am. Chem. Soc.* **88**, 2898 (1966).

RECEIVED October 9, 1967. Supported by NSF Grants GP-3358 and GP-5835.

34

QUANTUM YIELD OF TRIPLET STATE FORMATION
FROM NAPHTHACENE PHOTOPEROXIDATION

B. STEVENS and B. E. ALGAR

Department of Chemistry, The University, Sheffield 10, U.K.

Received 15 March 1967

The photoperoxidation quantum yields and fluorescence decay constants of naphthacene in benzene have been measured as a function of oxygen concentration at 25°C. From an analysis of the data in terms of $O_2\ ^1\Delta g$ participation it is concluded that the quantum yield of naphthacene triplet state formation is $0.63\ ^{+0.05}_{-0.09}$, and that the singlet oxygen molecule is produced solely by oxygen quenching of the naphthacene triplet state.

Recent spectroscopic [1,2], analytical [3-5] and kinetic evidence [6,7] support the suggestion [8,9] that the photoaddition of molecular oxygen to unsaturated compounds involves the $O_2\ ^1\Delta g$ state as intermediate. Since the quantum yield of autoperoxidation of an aromatic hydrocarbon A increases with its concentration, but is independent of absorbed light intensity, the peroxide AO_2 must be formed by process 9 of the following scheme in which the singlet state of oxygen, denoted $^1O_2^*$, may be produced by the spin-allowed exothermic quenching of the singlet state $^1A^*$ (process 5), or the triplet state ^3A (process 7) of the aromatic hydrocarbon, or by both:

$$^1A^* \rightarrow A + h\nu \qquad\qquad 1$$

$$^1A^* \rightarrow {}^3A \qquad\qquad 2$$

$$^1A^* \rightarrow A \qquad\qquad 3$$

$$^1A^* + O_2 \rightarrow {}^3A + O_2 \qquad\qquad 4$$

$$^1A^* + O_2 \rightarrow {}^3A + {}^1O_2^* \qquad\qquad 5$$

$$^3A \rightarrow A \qquad\qquad 6$$

$$^3A + O_2 \rightarrow A + {}^1O_2^* \qquad\qquad 7$$

$$^3A + O_2 \rightarrow A + O_2 \qquad\qquad 8$$

$$A + {}^1O_2^* \rightarrow AO_2 \qquad\qquad 9$$

$$^1O_2^* \rightarrow O_2 \qquad\qquad 10$$

Under photostationary conditions processes 1-10 lead to expression (1) for the quantum yield γAO_2 of peroxide formation

$$\gamma AO_2 = \left\{ \frac{k_9 A}{k_9 A + k_{10}} \right\} \qquad (1)$$

$$\left\{ \frac{k_5 O_2(k_6 + k_7 O_2 + k_8 O_2) + k_7 O_2(k_2 + k_4 O_2 + k_5 O_2)}{(k_6 + k_7 O_2 + k_8 O_2)(k_1 + k_2 + k_3 + k_4 O_2 + k_5 O_2)} \right\}$$

where k_i denotes the rate constant of the ith species and the symbols A and O_2 refer to concentrations of the appropriate species; this accounts for the observed [9] linear dependence of γAO_2 on A^{-1}. Under the same conditions the addition of oxygen reduces the lifetime of the excited singlet state $^1A^*$ from τ_0 to τ where

$$\frac{\tau_0}{\tau} = \frac{k_1 + k_2 + k_3 + k_4 O_2 + k_5 O_2}{k_1 + k_2 + k_3} \qquad (2)$$

and at relatively high concentrations of dissolved oxygen, such that[‡]

$$k_6 \ll (k_7 + k_8)O_2 \qquad (3)$$

expressions (1) and (2) may be rearranged to give

$$\gamma AO_2 \frac{\tau_0}{\tau} = \phi_A \left\{ \frac{k_5 O_2}{k_1 + k_2 + k_3} + \frac{k_7}{k_7 + k_8} \left(\frac{k_2 + k_4 O_2 + k_5 O_2}{k_1 + k_2 + k_3} \right) \right\}$$

where $\phi_A = k_9 A/(k_9 A + k_{10})$. If it is assumed that

---

[‡] From the linear dependence of $\gamma_{AO_2}^{-1}$ on O_2^{-1} at low concentrations of O_2 ($\leqslant 10^{-4}$M) shown in fig. 1, the value $k_6/k_7 = 1.76 \times 10^{-6}$M is provided by eq. (1) with $k_5 = k_8 = 0$ and $k_2 \gg k_4 O_2$ for the system described, i.e. $k_7 O_2 \geqslant k_6 \times 300$ over the concentration range covered in fig. 1.

either process 4 or 5, and either process 7 or 8 are operative, the further reduction of expression (4) depends on the mode of production of $O_2 {}^1\Delta g$, i.e.,

(i) $^1O_2^*$ from quenching of $^1A^*$ only ($k_4 = k_7 = 0$)

$$\gamma_{AO_2} \frac{\tau_0}{\tau} = \phi_A \left\{ \frac{k_5 O_2}{k_1 + k_2 + k_3} \right\} = \phi_A \left\{ \frac{\tau_0}{\tau} - 1 \right\}. \quad (5)$$

(ii) $^1O_2^*$ from quenching of ^3A only ($k_5 = k_8 = 0$)

$$\gamma_{AO_2} \frac{\tau_0}{\tau} = \phi_A \left\{ \frac{k_2 + k_4 O_2}{k_1 + k_2 + k_3} \right\} = \phi_A \left\{ \gamma_{ISC} + \frac{\tau_0}{\tau} - 1 \right\} \quad (6)$$

where $\gamma_{ISC} = k_2 / (k_1 + k_2 + k_3)$ is the quantum yield of triplet state formation;

(iii) $^1O_2^*$ from quenching of both $^1A^*$ and ^3A ($k_4 = k_8 = 0$)

$$\gamma_{AO_2} \frac{\tau_0}{\tau} = \phi_A \left\{ \frac{k_2 + 2k_5 O_2}{k_1 + k_2 + k_3} \right\} = \phi_A \left\{ \gamma_{ISC} + 2\left(\frac{\tau_0}{\tau} - 1 \right) \right\} \quad (7)$$

Quantum yields of photoperoxidation of naphthacene in benzene at 25°C have been estimated as a function of oxygen concentration by monitoring the optical density of naphthacene as a function of exposure time under conditions (5.6×10^{-4}M) where solute photoassociation is negligible; the observation of an isosbestic point in the total absorption spectrum confirmed the absence of side reactions and the data are plotted on a reciprocal basis in fig. 1.

Fluorescence decay constants ($1/\tau$), measured directly under the same conditions using the pulsed flash technique [10], were found to vary with oxygen concentration according to the expression

$$1/\tau = 1.92 \times 10^8 + 2.36 \times 10^{10} O_2 \text{ sec}^{-1}.$$

The experimental quantity $\gamma_{AO_2} \tau_0/\tau$ is plotted as a function of $(\tau_0/\tau - 1)$ in fig. 2; the estimated value

$$\text{intercept/slope} = 0.63 {}^{+0.05}_{-0.09}$$

is inconsistent with eq. (5) which predicts that $\gamma_{AO_2} = 0$ when $\tau_0 = \tau$, and with (7) which would require

$$\gamma_{ISC} = 1.26 {}^{+0.10}_{-0.18} > 1.$$

On the basis of this analysis of these data it is therefore concluded that
a) the quantum yield of naphthacene triplet state production in benzene at 25°C is given by

$$\gamma_{ISC} = \frac{k_2}{k_1 + k_2 + k_3} = 0.63 {}^{+0.05}_{-0.09}$$

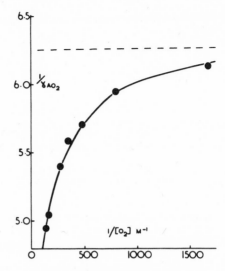

Fig. 1. Plot of $\gamma_{AO_2}^{-1}$ against O_2^{-1} for photoperoxidation of naphthacene in benzene at 25°C. Solid curve drawn in accordance with eq. (6); dashed line denotes extrapolation of data obtained at low concentrations of oxygen ($\leqslant 10^{-4}$M).

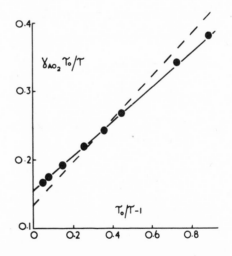

Fig. 2. Plot of $\gamma_{AO_2} \tau_0/\tau$ against $(\tau_0/\tau - 1)$; see eqs. (5) - (7). Dashed line drawn through arithmetic mean of experimental points to give maximum intercept/slope ratio (for $\gamma_{ISC} \leqslant 1 - \gamma_F = 0.84$) from eq. (7).

which, with measured [11] values of 0.16 ± 0.04 for the quantum yield of fluorescence in the same solvent at this temperature, indicates that internal conversion to the ground state is responsible for at least 12% of the overall relaxation of lowest excited singlet states;

b) simultaneous excitation of O_2 $^1\Delta g$ (at ≈ 8000 cm^{-1} above the O_2 ground state) accompanies oxygen-quenching of the triplet state of naphthacene (at 10 250 cm^{-1}) but not that of the lowest excited singlet state (at 21 100 cm^{-1}); this could reflect a preferential external spin-orbit coupling of the lowest excited singlet with a higher triplet state, suggested for intersystem crossing in other molecules [12], which would reduce the energy available for excitation of the quenching species.

The award of an SRC maintenance grant is gratefully acknowledged (BEA).

REFERENCES

[1] S.J.Arnold, E.A.Ogryzlo and H.Witzke, J.Chem. Phys. 40 (1964) 1769.
[2] A.U.Khan and M.Kasha, J.Chem.Phys. 39 (1963) 2105.
[3] C.S.Foote, S.Wexler and W.Ando, Tetrahedron Letters 46 (1965) 4111.
[4] E.J.Corey and W.C.Taylor, J.Amer.Chem.Soc. 86 (1964) 3881.
[5] E.McKeown and W.A.Waters, J.Chem.Soc. B (1966) 1040.
[6] K.R.Kopecky and H.J.Reich, Canad.J.Chem. 43 (1965) 2265.
[7] T.Wilson, J.Amer.Chem.Soc. 88 (1966) 2898.
[8] H.Kautsky and H.de Bruijn, Naturwiss. 19 (1931) 1943.
[9] E.J.Bowen and D.W.Tanner, Trans.Faraday Soc. 51 (1955) 475.
[10] R.G.Bennett, Rev.Sci.Instr. 31 (1960) 1275; J.B.Birks, T.A.King and I.H.Munro, Proc.Phys. Soc. 80 (1962) 355.
[11] B.E.Algar, unpublished; M.W.Windsor, private communication.
[12] R.G.Bennett and P.J.McCartin, J.Chem.Phys. 44 (1966) 1969; B.Stevens, M.F.Thomaz and J.Jones. J.Chem. Phys. 46 (1967) 405.

35

Reprinted from *J. Phys. Chem.*, **72**(10), 3468–3474 (1968)

The Photoperoxidation of Unsaturated Organic Molecules. II.

The Autoperoxidation of Aromatic Hydrocarbons

by B. Stevens[1] and B. E. Algar

Department of Chemistry, Sheffield University, Sheffield, England   (*Received March 25, 1968*)

The quantum yields of autoperoxidation of 9,10-dimethylanthracene, 9,10-dimethyl-1,2-benzanthracene, naphthacene, and rubrene have been measured as a function of the dissolved oxygen concentration down to 2.6×10^{-6} M in benzene at 25°. An analysis of the data in terms of the participation of an $O_2(^1\Delta_g)$ intermediate provides estimates of the triplet-state formation yields of the substrates and leads to the conclusion that $O_2(^1\Delta_g)$ is produced solely by oxygen quenching of the aromatic hydrocarbon triplet state, which is a yield-limiting process at very low oxygen concentrations.

Introduction

The over-all photosensitized peroxidation[2] of an unsaturated organic molecule M in the presence of molecular oxygen may be represented by the process

$$\mathrm{M} \xrightarrow[\mathrm{O_2}]{\mathrm{S}(h\nu)} \mathrm{MO_2}$$

where the sensitizer S absorbs the incident radiation but remains chemically unchanged, and the peroxide or hydroperoxide $\mathrm{MO_2}$ is the sole product. Three types of substrate, M, subject to this process are: (a) those molecules containing an isolated double bond with a β-hydrogen atom,[3] *e.g.*

$$\underset{\mathrm{CH_3}}{\overset{\mathrm{CH_3}}{\mathrm{C}}} = \underset{\mathrm{CH_3}}{\overset{\mathrm{CH_3}}{\mathrm{C}}} \xrightarrow[\mathrm{O_2}]{\mathrm{S}(h\nu)} \underset{\mathrm{CH_3}}{\overset{\mathrm{CH_3}}{\mathrm{C}}} - \underset{\mathrm{O_2H}}{\overset{}{\mathrm{C}}} \Big\backslash\mathrm{CH_2}$$

(b) cyclic conjugated hexadienes,[4] of which α-terpinene is a classic example[5]

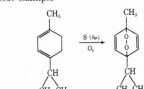

(c) the larger catacondensed aromatic hydrocarbons,

(1) Department of Chemistry, University of South Florida, Tampa, Fla. 33620.

(2) The term peroxidation is used here to distinguish the addition of molecular oxygen from oxidative electron and H atom transfer processes.

(3) G. O. Schenck, *Angew. Chem.*, **69**, 579 (1957).

(4) R. N. Moore and R. V. Lawrence, *J. Amer. Chem. Soc.*, **80**, 1438 (1958).

(5) G. O. Schenck and K. Żiegler, *Naturwissenschaften*, **32**, 157 (1944).

179

e.g., anthracene, naphthacene, 1,2-benzanthracene, and their derivatives

which are capable of acting as the sensitizer in a process of autoperoxidation.[6]

Previous studies of the reaction characteristics have shown that the reciprocal quantum yield of the over-all process is a linear function of the reciprocal substrate concentration[7,8] and that the rate of reaction is directly proportional to the absorbed light intensity,[8] which eliminates processes involving more than one electronically excited species. The simplest general scheme consistent with these observations is

$$S + h\nu \longrightarrow S^*$$

$$S^* \longrightarrow S \quad (+h\nu) \qquad (a)$$

$$S^* + O_2 \longrightarrow T \qquad (b)$$

$$T + M \longrightarrow MO_2 + S \qquad (c)$$

$$T \longrightarrow S + O_2 \qquad (d)$$

which leads to the photostationary expression (eq I) for the over-all quantum yield, γ_{MO_2}, but which raises

$$\frac{1}{\gamma_{MO_2}} = \left\{1 + \frac{k_a}{k_b[O_2]}\right\}\left\{1 + \frac{k_d}{k_e[M]}\right\} \qquad (I)$$

questions concerning the nature of the transferring species, T, and the multiplicity of its electronically excited precursor, S*.

It has been suggested that the intermediate, T, is either a sensitizer–oxygen complex,[9] possibly of significant charge-transfer character,[10] or a singlet excited state of molecular oxygen[11] which may be either O_2($^1\Sigma_g^+$) or O_2($^1\Delta_g$) at 13,300 or 8000 cm^{-1} above the O_2($^3\Sigma_g^-$) ground state.[12] Although to date no spectroscopic evidence is available for either species in a system undergoing photoperoxidation, identical products[13,14] with similar isomeric distributions[15] are obtained from reactions of the substrate with O_2($^1\Delta_g$) produced by an electrodeless discharge in gaseous O_2[16] or by the alkaline H_2O_2–NaOCl reaction[17] in which the presence of O_2($^1\Delta_g$) is established in emission. Kinetic studies[18,19] of the competitive photoperoxidation of two substrates have shown that the relative rates of peroxide formation are independent of the sensitizer used, contrary to expectation for a complex intermediate of varying stability, while the intergranular migration[20] and diffusion of the transferring species in ethylcellulose films,[21] necessary to account for photosensitized peroxidation under conditions where molecular transport is limited, are also inconsistent with the assignment of a sensitizer–oxygen complex to the role of intermediate.

The available evidence strongly supports the suggestion that a singlet oxygen molecule, particularly the O_2($^1\Delta_g$) state, is directly involved in the photosensitize addition of molecular oxygen to an unsaturated su strate. This may be formed in a process of electron energy transfer from either the excited singlet or triple states of the sensitizer in the following exothermic, spi allowed quenching processes

$$^1S^* + {}^3O_2 \longrightarrow {}^3S^* + {}^1O_2^* \qquad (e)$$

$$(\nu^0_F - \nu^0_P > 8000 \text{ cm}^{-1})$$

$$^3S^* + {}^3O_2 \longrightarrow {}^1S + {}^1O_2^* \qquad (f)$$

$$(\nu^0_P > 8000 \text{ cm}^{-1})$$

under the conditions stated parenthetically, whe ν^0_F and ν^0_P denote $0''-0'$ fluorescence and pho phorescence band frequencies, respectively. Evidenc for the multiplicity of the participating electronic sta S* of the sensitizer is based largely on the dependenc of γ_{MO_2} on the concentration of dissolved oxyger Thus Gollnick and Schenck[9] found that for the dy sensitized peroxidation of 2,5-dimethylfuran in met anol dγ_{MO_2}/d[O_2] ≈ 0 at concentrations of dissolve oxygen greater than 2×10^{-3} M, or with $k_b \approx 10$ M^{-1} sec^{-1} (eq I), the actual lifetime τ_{S^*} of the electro ically excited sensitizer is limited by

$$\tau_{S^*} = 1/k_a \gg 1/k_b[O_2] \approx 5 \times 10^{-8} \text{ sec}$$

indicative of a triplet-state precursor of $^1O_2$ (proce f); in these systems process e is in any case i operative, since the necessary condition ($\nu^0_F - \nu > 8000$ cm^{-1}) is not fullfilled for the sensitizers use On the other hand, Livingston and Rao[22] reported

(6) C. Moureau, C. Duffraisse, and P. M. Dean, *Compt. Rend.*, 18 1440, 1584 (1926); *cf.* W. Bergmann, and M. J. McLean, *Chem. Re* 28, 367 (1941).

(7) W. Koblitz and H. J. Schumacher, *Z. Phys. Chem.* (Leipzig B35, 11 (1935); B37, 462 (1937).

(8) E. J. Bowen and D. W. Tanner, *Trans. Faraday Soc.*, 51, 4 (1955).

(9) A. Schonberg, *Ann.*, 518, 299 (1935); *cf.* K. Gollnick and O. Schenck, *Pure Appl. Chem.*, 9, 507 (1964).

(10) H. Tsubomura and R. S. Mulliken, *J. Amer. Chem. Soc.*, 8 5966 (1960).

(11) H. Kautsky and H. de Bruijn, *Naturwissenschaften*, 19, 10 (1931).

(12) *Cf.* G. Herzberg, "Spectra of Diatomic Molecules," D. Va Nostrand Co., Inc., New York, N. Y., 1950.

(13) E. J. Corey and W. C. Taylor, *J. Amer. Chem. Soc.*, 86, 38 (1964).

(14) C. S. Foote and S. Wexler, *ibid.*, 86, 3879 (1964); E. McKeov and W. A. Waters, *J. Chem. Soc., B*, 1040 (1966).

(15) C. S. Foote, S. Wexler, and W. Ando, *Tetrahedron Lett.*, 46, 4 (1965).

(16) J. S. Arnold, E. A. Ogryzlo, and H. Witzke, *J. Chem. Phy* 40, 1769 (1964); J. S. Arnold, R. J. Brown, and E. A. Ogryz *Photochem. Photobiol.*, 4, 963 (1965).

(17) A. H. Kahn and M. Kasha, *J. Chem. Phys.*, 39, 2105 (1963).

(18) K. R. Kopecky and H. J. Reich, *Can. J. Chem.*, 43, 22 (1965).

(19) T. Wilson, *J. Amer. Chem. Soc.*, 88, 2898 (1966).

(20) H. Kautsky, *Trans. Faraday Soc.*, 35, 216 (1939).

(21) J. Bourdon and B. Schnuriger, *Photochem. Photobiol.*, 5, 5 (1966).

dependence of γ_{MO_2} on the dissolved oxygen concentration for the photoperoxidation of 9,10-diphenylanthracene in benzene ($\gamma_F \approx 1$) but not for anthracene in bromobenzene ($\gamma_F \approx 0$) and concluded that the intermediate is formed from the singlet or triplet state of the sensitizer depending on whether this has a high or low fluorescence yield, γ_F.[23]

This article reports an investigation of the autoperoxidation of four aromatic hydrocarbons over a wide range of dissolved oxygen concentrations interpreted in terms of a singlet oxgen intermediate with a view to establishing the multiplicity of the sensitizer electronic state from which it is formed.

Experimental Section

Materials. The sources and methods of purification of the sensitizers, substrates, and solvent are described in part I.[24]

Quantum Yields. The quantum yields of peroxidation are obtained from measurements of light transmission, I_t, of the substrate as a function of time as follows. A parallel beam of filtered radiation at wavelength λ_{ex} from a stabilized 125-W high-pressure mercury arc (Mazda MBL/D) was directed onto a plane face of a cylindrical quartz cell, having the same diameter as the light beam, a volume, V, of 3 ml, and a depth, d, of 1 cm, containing the substrate solution through which an O_2–N_2 mixture of prearranged composition was continuously bubbled from a narrow tube through the neck of the cell; this provided both stirring of the exposed solution and a constant concentration of dissolved oxygen. The transmitted beam was focused onto the slit of a small Hilger grating monochromator set at λ_{ex} (to eliminate solute fluorescence) and was fitted with a Mazda 27M3 photomultiplier at the exit slit; the photocurrent was earthed through a 7.5 kohm load resistance across which the potential difference was fed into a Brown potentiometric recorder (0–10 mV) with a variable chart speed to obtain recordings of I_t at λ_{ex} as a function of exposure time. Following the initial measurement of light transmitted by the cell and solvent only, a backing-off voltage was applied to the potentiometer to allow amplification of the change in I_t. A computer program was written to obtain the quantum yield of substrate consumption γ_M from the slopes of the recorded curves given by

$$\frac{-d \ln (I_t/I_0)}{dt} \simeq \frac{\ln (I_t/I_0) - \ln (I_{t'}/I_0)}{t' - t}$$

over the short time interval $t' - t$ as a function of substrate concentration expressed as

$$[M] = (1/\epsilon d) \ln (I_0/I_t)$$

using the relationship

$$\gamma_M = \gamma_{MO_2} = -\frac{1}{I_0 - I_t} \frac{dn_M}{dt} =$$

$$\frac{VN}{1000\epsilon dI_0[1 - (I_t/I_0)]} \frac{d \ln (I_t/I_0)}{dt}$$

where N is Avogadro's number and ϵ denotes the molar extinction coefficient of the substrate at the monitoring (actinic) wavelength λ_{ex}. The absolute incident light intensity I_0, measured by ferrioxalate actinometry[25] before and after each run, was found to vary by less than 2% over a period of 2 weeks. In all cases λ_{ex} was chosen to avoid absorption by and possible photodecomposition of the peroxide produced, i.e., λ_{ex} 365 mμ for 9,10-dimethylanthracene (DMA) and 9,10-dimethyl-1,2-benzanthracene (DMBA) and λ_{ex} 435.8 mμ for rubrene and naphthacene.

Concentrations of dissolved oxygen were stabilized by flowing O_2–N_2 mixtures through the solutions for a period of at least 1 hr prior to and during exposure and were estimated from the partial pressure of O_2 in the flowstream, which could be varied from 1 to 0.0004 atm by the use of calibrated flowmeters.[24]

Variations in the incident light intensity were effected by placing wire-mesh screens of known transmission characteristics between the light source and the cell.

Results

Photodimerization, photodecomposition, and photochemical reactions of the solute with the solvent (benzene) were shown to be of negligible significance at the solute concentrations employed ($\leq 10^{-3}$ M) by exposure of deoxygenated solutions for periods of time in excess of those required for photoperoxidation to approach completion; in the absence of dissolved oxygen, the optical density of the solute remained unchanged.

Absorption spectra of all substrate solutions in air-saturated cyclohexane, recorded as a function of exposure time, were found to exhibit isosbestic points at the following wavelengths λ_{iso}.

	Rubrene	Naph-thacene	DMA	DMBA
λ_{iso}, mμ	263–264	258–259	233	250

The final spectrum in the case of rubrene corresponded to that reported for rubrene peroxide[26] with peaks at 242 and 290 mμ. This evidence confirms the linear rela-

(22) R. Livingston and V. S. Rao, *J. Phys. Chem.*, **63**, 794 (1959).

(23) Although both groups of investigators[9,22] interpret their data in terms of a sensitizer–oxygen complex intermediate, this does not invalidate the conclusions reached, since the complex is kinetically indistinguishable from singlet oxygen in the approximation of processes a–d.

(24) B. Stevens and B. E. Algar, *J. Phys. Chem.*, **72**, 2582 (1968).

(25) C. G. Hatchard and C. A. Parker, *Proc. Roy. Soc.*, **A235**, 518 (1956).

(26) G. M. Badger, R. S. Pearce, H. J. Rodda, and I. S. Walker, *J. Chem. Soc.*, 3151 (1954).

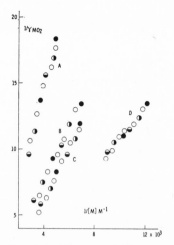

Figure 1. Dependence of the autoperoxidation quantum yield on the substrate concentration at various incident light intensities, I_0: A, naphthacene; B, rubrene; C, DMA; D, DMBA; ○, $I_0 = 1.0I$; ◑, $I_0 = 0.75I$; ◕, $I_0 = 0.50I$; ●, $I_0 = 0.25I$. $[O_2] = 3.65 \times 10^{-3} M$. The solvent is benzene at 25°.

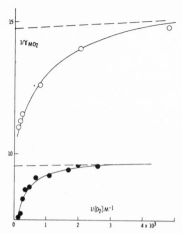

Figure 3. Dependence of the autoperoxidation quantum yield on the concentration of dissolved oxygen in benzene at 25° for $10^{-4} M$ DMBA (○) and $4 \times 10^{-4} M$ naphthacene (●): solid curves, drawn according to eq V with tabulated values of rate parameters; dashed lines, extrapolation of low oxygen concentration data in Figure 4.

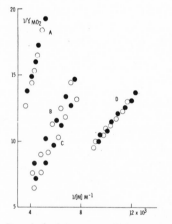

Figure 2. Dependence of the autoperoxidation quantum yield on the substrate concentration in the absence (○) and presence (●) of added peroxide at $5 \times 10^{-4} M$: A, naphthacene; B, rubrene; C, DMA; D, DMBA. $[O_2] = 3.65 \times 10^{-3} M$. The solvent is benzene at 25°.

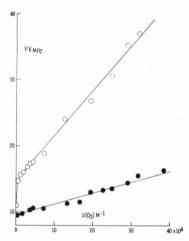

Figure 4. Dependence of the autoperoxidation quantum yield on the concentration of dissolved oxygen in benzene at 25° for $10^{-4} M$ DMBA (○) and $4 \times 10^{-4} M$ naphthacene (●): solid curves, drawn in accordance with eq V.

tionship between the reactant and the product concentration[27] on which the rate of peroxide formation, given by

$$d[MO_2]/dt = -d[M]/dt$$

and the estimation of the quantum yield (γ_{MO_2}) of peroxidation is based.

As shown in Figures 1 and 2, γ_{MO_2} is independent of the incident (or absorbed) light intensity, as previously reported,[8] and of the product (peroxide) concentration, confirming that the peroxide decomposition is unimportant during the observation period. Moreover, in agreement with published data,[7,8] $1/\gamma_{MO_2}$ is a linear function of the reciprocal substrate concentration, as

(27) M. D. Cohen and E. Fischer, *J. Chem. Soc.*, 3044 (1962).

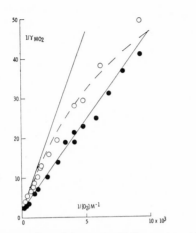

Figure 5. Dependence of the autoperoxidation quantum yield of rubrene (O) and of DMA (●) on dissolved oxygen concentration: solid lines, drawn from eq VIII with $k_2 = 0$; dashed line, drawn according to eq VIII with an intersystem-crossing yield of 0.02 for rubrene.

required by eq I.[28] The data lines $\gamma_{MO_2}^{-1}([M]^{-1})$ obtained for different partial pressures of oxygen in the flowstream are used to obtain the variation of γ_{MO_2} with the concentration of dissolved oxygen at a given substrate concentration, shown in Figures 3–5.

Discussion

The oxygen concentration dependence of the autoperoxidation quantum yield of DMA, DMBA, naphthacene, and rubrene is interpreted in terms of the following kinetic scheme in which oxygen quenching of the substrate singlet and triplet states is described by both energy-transfer processes 5 and 7 and collision-induced intersystem-crossing processes 4 and 8

$$^1M^* \longrightarrow M + h\nu_F \qquad (1)$$

$$^1M^* \longrightarrow {}^3M^* \qquad (2)$$

$$^1M^* \longrightarrow M \qquad (3)$$

$$^1M^* + {}^3O_2 \longrightarrow {}^3M^* + {}^3O_2 \qquad (4)$$

$$^1M^* + {}^3O_2 \longrightarrow {}^3M^* + {}^1O_2^* \qquad (5)$$

$$^3M^* \longrightarrow M \qquad (6)$$

$$^3M^* + {}^3O_2 \longrightarrow M + {}^1O_2^* \qquad (7)$$

$$^3M^* + {}^3O_2 \longrightarrow M + {}^3O_2 \qquad (8)$$

$$M + {}^1O_2^* \longrightarrow MO_2 \qquad (9)$$

$$^1O_2^* \longrightarrow {}^3O_2 \qquad (10)$$

where the asterisk denotes electronic excitation. Under photostationary conditions, processes 1–10 lead to eq II for the quantum yield of autoperoxidation and eq III for the reduction in fluorescence intensity of the

substrate from F_0 to F in the presence of dissolved oxygen at the same concentration.

$$\gamma_{MO_2} = \left\{ \frac{k_9[M]}{k_9[M] + k_{10}} \right\} \times$$

$$\left\{ \frac{\begin{array}{c} k_5[O_2](k_6 + k_7[O_2] + k_8[O_2]) + \\ k_7[O_2](k_2 + k_4[O_2] + k_5[O_2]) \end{array}}{\begin{array}{c} (k_6 + k_7[O_2] + k_8[O_2]) \times \\ (k_1 + k_2 + k_3 + k_4[O_2] + k_5[O_2]) \end{array}} \right\} \quad (II)$$

$$F_0/F = (k_1 + k_2 + k_3 + k_4[O_2] +$$
$$k_5[O_2])/(k_1 + k_2 + k_3) \quad (III)$$

Higher Oxygen Concentrations. At concentrations of dissolved oxygen where fluorescence quenching is experimentally significant, it is assumed that

$$k_6 \ll (k_7 + k_8)[O_2]$$

in which case the product of eq II and III reduces to

$$\gamma_{MO_2}F_0/F \simeq \phi_M\{\beta\gamma_{IS} + (\alpha + \beta)[(F_0/F) - 1]\} \quad (IV)$$

where ϕ_M $(= k_9[M]/(k_9[M] + k_{10}))$ is the $^1O_2^*$-substrate addition efficiency; γ_{IS} $(= k_2/(k_1 + k_2 + k_3))$ defines the substrate triplet-formation efficiency in the absence of dissolved O_2; α $(= k_5/(k_4 + k_5))$ is the $^1O_2^*$ yield from O_2 quenching of substrate fluorescence; and β $(= k_7/(k_7 + k_8))$ is the $^1O_2^*$ yield from O_2 quenching of the substrate triplet state.

The fluorescence-quenching data reported in part I of this series[25] are used to compute the experimental quantity $\gamma_{MO_2}F_0/F$, which is shown to be a linear function of $(F_0/F) - 1$ at the same oxygen concentration for the compounds examined at the concentrations stated in Figures 6 and 7. The intercept:slope ratios of the data lines given by (eq IV)

$$(\gamma_{MO_2}F_0/F)_{[O_2]=0}\, d[(F_0/F) - 1]/d(\gamma_{MO_2}F_0/F) =$$
$$\gamma_{IS}\beta/(\alpha + \beta)$$

are collected in Table I together with values[24] for the corresponding fluorescence yield, γ_F, in the same deoxygenated solvent.

An examination of the experimental findings subject to the condition

$$[\gamma_{IS}\beta/(\alpha + \beta)] + [\gamma_{IS}\alpha/(\alpha + \beta)] + \gamma_{IC} + \gamma_F = 1$$

shows that in the case of (a) DMBA

$$[\gamma_{IS}\alpha/(\alpha + \beta)] + \gamma_{IC} = 0.00 \pm 0.09$$

i.e., (i) the internal conversion efficiency defined as

$$\gamma_{IC} = k_3/(k_1 + k_2 + k_3)$$

is zero within the combined limits of error, and (ii) $\alpha \ll \beta$ or $^1O_2^*$ is produced almost entirely by oxygen

(28) Owing to the scatter of data points over a necessarily limited range of low substrate concentrations and the relatively long extrapolation involved, it was not possible to assign values to the quantum yields at infinite substrate concentration within useful limits of error.

Table I: Rate Parameters from Autoperoxidation

	DMA	DMBA	Naphthacene	Rubrene
ν^0_F, cm^{-1}	25,000	26,100	20,600	18,900
ν^0_P, cm^{-1}	14,000	15,000	10,250	<10,250
$\gamma_F{}^a$	0.90 ± 0.04	0.36 ± 0.03	0.19 ± 0.02	0.98 ± 0.02
$\gamma_{IS}\beta/(\alpha+\beta)$	0.02 ± 0.05	0.66 ± 0.06	0.61 ± 0.07	0.04 ± 0.02
$\phi_M\beta$ ([M])	0.50^b (4 × 10^{-4} M)	0.097 (10^{-4} M)	0.164^b (4 × 10^{-4} M)	0.55^b (4 × 10^{-4} M)
$10^{-7}k_1,{}^a$ sec^{-1}	6.5 ± 0.5	1.48 ± 0.12	3.65 ± 0.38	6.0 ± 0.3
$10^{-7}k_2,{}^b$ sec^{-1}	0.15 ± 0.15	2.70 ± 0.25	11.9 ± 1.2	0.24 ± 0.12
$10^{-7}k_3,{}^b$ sec^{-1}	0.58 ± 0.43	0.00 ± 0.35	3.65 ± 1.54	0.00 ± 0.12
$10^{-10}k_4,{}^c M^{-1}$ sec^{-1}	3.15 ± 0.20	2.76 ± 0.16	2.38 ± 0.18	1.18 ± 0.08
$10^3k_6,{}^d$ sec^{-1}	...	11.0 ± 6.0	4.2 ± 1.4	...

^{a} From part I.[24] ^{b} Assuming that $\alpha \ll \beta$ as in DMBA. ^{c} From part I, with $k_5 = 0$ and $k_4 = k_{O_2}$. ^{d} Based on an encounter quench-ing probability of 0.24 ± 0.06 (see text).

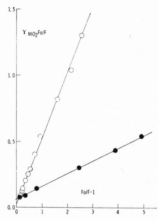

Figure 6. Plot of data for 4 × 10^{-4} M DMA (O) and 10^{-4} M DMBA (●) according to eq IV.

quenching of the triplet state of this molecule; (b) DMA and rubrene

$$\gamma_{IC} \approx \gamma_{IS} \approx 0$$

within the limits of error and

$$\alpha + \beta > 0$$

(c) naphthacene

$$\gamma_{IC} + \gamma_{IS}\alpha/(\alpha+\beta) = 0.20 \pm 0.09$$

providing the alternative interpretations: (i) $\alpha \ll \beta$ with $\gamma_{IC} = 0.20 \pm 0.09$, and (ii) $\gamma_{IC} = 0$, where $\alpha/\beta = 0.33 \pm 0.20$. The assumption of a mechanism common to all four systems requires that $\alpha \ll \beta$ which allows computation of the values tabulated for k_2 and k_3 and provides the quoted data for $\phi_M\beta$ from the slopes of the data lines in Figures 6 and 7 at the substrate concentrations stated.

Low Oxygen Concentrations. With $k_5 \ll k_4$ ($\alpha \ll \beta \leq 1$), eq II reduces to

$$\frac{1}{\gamma_{MO_2}} = \frac{1}{\phi_M\beta}\left\{1 + \frac{k_1+k_3}{k_2+k_4[O_2]}\right\}\left\{1 + \frac{k_6\beta}{k_7[O_2]}\right\} \quad ($$

$$\simeq \frac{1}{\phi_M\beta\gamma_{IS}}\left\{1 + \frac{k_6\beta}{k_7[O_2]}\right\} \quad (V$$

at concentrations of dissolved oxygen such that

$$k_4[O_2] \ll k_2 \quad (VI$$

The data for naphthacene and DMBA at low oxyge concentrations are plotted according to eq VI in Figure which yields the oxygen-quenching constants of t triplet state given by (intercept)/(slope) = k_7/k = $(k_7 + k_8)/k_6 = k_q\tau^3{}_M = (5.7 \pm 1.9) \times 10^5 M^{-1}$ f naphthacene and $(2.2 \pm 0.7) \times 10^5 M^{-1}$ for DMF under the conditions stated. The rate constant k_q = + k_8 has reported values[29,30] of 4 × 10^9 and 3.7 × 1

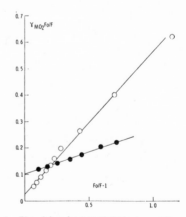

Figure 7. Plot of data for 4 × 10^{-4} M rubrene (●) and 4 × 10^{-4} M naphthacene (O) according to eq IV.

(29) G. Porter and M. R. Wright, *Discussions Faraday Soc.*, **27**, (1959).

(30) G. Jackson, R. Livingston, and A. C. Pugh, *Trans. Farad Soc.*, **56**, 1635 (1960).

M^{-1} sec^{-1} for oxygen quenching of the triplet state of anthracene in hexane, corresponding to an encounter quenching probability $p = 0.24 \pm 0.06$ if k_q is computed from the expression

$$k_q = 8pRT/3000\eta$$

for a solvent of viscosity η. Use of the same encounter probability in this expression leads to a value of $k_q = (2.4 \pm 0.6) \times 10^9 M^{-1}$ sec^{-1} for benzene at 25°, which in combination with the quenching constants $k_q\tau^3{}_M$ reported above yields values of $(2.4 \pm 1.4) \times 10^{-4}$ and $(9.0 \pm 5.3) \times 10^{-5}$ sec for the triplet-state lifetimes, $\tau^3{}_M$, of naphthacene and DMBA, respectively; the former is close to that of $(5.9 \pm 1.1) \times 10^{-4}$ sec measured directly[31] by flash kinetic spectrophotometry for naphthacene in benzene.

The solid lines in Figures 3 and 4 are drawn according to eq V with values of $k_1 - k_4$ and $\phi_M\beta$ tabulated and the triplet-state quenching constants given above.

In the case of rubrene and of DMA where $\gamma_{IS} \approx 0$, condition VII is not established at the lowest concentrations of dissolved oxygen for which reliable measurements of the quantum yield could be obtained. The solid lines in Figure 5 are expressed by the reduced form of eq V under conditions of complete quenching of the triplet state $(k_7[O_2] \gg k_6\beta)$, i.e., by

$$\frac{1}{\gamma_{MO_2}} = \frac{1}{\phi_M\beta}\left\{1 + \frac{k_1 + k_3}{k_2 + k_4[O_2]}\right\} \qquad \text{(VIII)}$$

with $k_2 = \gamma_{IS} = 0$. The divergence of the data points for rubrene at lower oxygen concentrations may reflect a finite intersystem-crossing yield for this molecule, as shown by the dashed curve drawn in accordance with eq VIII with $\gamma_{IS} = 0.02$ $(k_2 = 1.2 \times 10^6$ sec$^{-1})$.

The finding that $O_2(^1\Delta_g)$ is produced by oxygen quenching of the triplet states of naphthacene or DMBA, possibly via the higher state $O_2(^1\Sigma_g{}^+)$, is consistent with the conclusion that this energy-transfer probability exceeds that of collision-induced intersystem crossing reached on theoretical grounds.[32] On the other hand, the absence of energy transfer from the excited singlet state of DMBA (and possibly of naph-

thacene and the other substrates) in the spin-allowed exothermic quenching process (reaction e) was not anticipated and may reflect the preferential external spin-orbit coupling of excited singlet with higher triplet states proposed to account for the different absolute rate constants obtained for fluorescence quenching by oxygen.[24]

Conclusions

An analysis of the oxygen-concentration dependence of the quantum yields of autoperoxidation of four aromatic hydrocarbons in terms of $O_2(^1\Delta_g)$ participation has shown the following.

(a) For substrates (sensitizers) of low fluorescence yield, the $O_2(^1\Delta_g)$ intermediate is produced mainly (naphthacene) if not entirely (DMBA) by oxygen quenching of the sensitizer triplet state which is yield limiting at low oxygen concentrations. At higher concentrations of dissolved oxygen, the nonlinear dependence of $\gamma_{MO_2}{}^{-1}$ on $[O_2]^{-1}$ reflects the competition between intramolecular and oxygen-induced intersystem crossing to the sensitizer triplet state.

(b) In the case of substrates (DMA and rubrene) with high ($\sim$100%) fluorescence yields, it is impossible to ascertain whether the intermediate is produced by oxygen quenching of the singlet state (as previously suggested[22]) or of the triplet state of the sensitizer (substrate). However, the assumption of a common mechanism based on the latter alternative requires that, in the absence of intramolecular intersystem crossing, oxygen quenching of the excited singlet state is necessary to produce the triplet state precursor of the $O_2(^1\Delta_g)$ intermediate. This appears to be the yield-limiting process over the range of oxygen concentrations examined.

Acknowledgments. B. E. A. is grateful to the Science Research Council of Great Britain for a maintenance grant.

(31) A. A. Lamola, W. G. Herkstroeter, J. C. Dalton, and G. S. Hammond, *J. Chem. Phys.*, **42**, 1715 (1965).

(32) K. Kawaoka, A. U. Khan, and D. R. Kearns, *ibid.*, **46**, 1842 (1967).

36

Reprinted from *Photochem. Photobiol.*, 8(5), 361–368 (1968)

PHOTOSENSITIZED OXIDATION THROUGH STEARATE MONOMOLECULAR FILMS*

B. SCHNURIGER and J. BOURDON
with the collaboration of Miss J. BEDU
Centre de Recherches KODAK-PATHE, 94-Vincennes, France

(*Received* 29 *January* 1968; *in revised form* 1 *May* 1968)

Abstract—The photosensitized oxidations of rubrene by methylene blue and of diphenylanthracene by eosin have been studied where the sensitizer and the oxidizable substrate are separated by oxygen permeable layers of barium or cadmium stearate monomolecular films. The rate of the photosensitized oxidation reaction was followed by measurements of the hydrocarbon fluorescence, as a function of the stearate layer thickness. It has been found that the sensitized reaction is observable for thickness up to 500 Å and depends only on the deactivation process of the excited species (singlet oxygen) diffusing through the layers. It has been found also that half of the excited oxygen molecules are deactivated after a diffusion path of 115 Å. The starting hydrocarbon can be regenerated by heating the sample after the photoreaction to approximately 120°C showing the transannular peroxide nature of the oxidation product.

WHEN A dye in the vicinity of an oxidizable substrate and in presence of oxygen is excited by light, a photosensitized oxidation of the substrate may occur, called in biology the 'photodynamic effect'. The mechanism of this photosensitized reaction can present quite different characteristics according to the conditions of the reaction: nature of the dye, D, nature of the oxidizable substrate, RH, concentration of oxygen and state of the medium in which the reaction occurs (solid or liquid).

A general scheme of the photosensitized oxidation has been proposed recently[1] (Fig. 1).

After excitation by light, the dye is converted to the triplet state by intersystem crossing. In this state, which has been shown to be the active state in many of these reactions, the dye can react in a first step either with the substrate RH or with oxygen $^3O_2$. The two reactions are competitive, their relative rate depending in particular upon the relative concentration of the two reactants.

The reaction of oxygen with the dye triplet ^3D leads to different processes also in competition, and, depending upon the conditions, one of the species: $[D \ldots O_2]$, $^1O_2$, O_2^- or O_2^{ribr} should prevail. Various experiments (2–5) have given strong arguments against the sensitizer oxygen complex $[D \ldots O_2]$, which might be, if it really exists, an intermediary (complex of collision) in the formation of the activated form of oxygen (Fig. 1). On the other hand, it seems very unlikely, as Foote[5a] has pointed out, that vibrationally excited oxygen O_2^{ribr} is the reactive species, and no convincing evidence of its occurence have been presented as yet.

Excited singlet oxygen in its various forms, $O_2[^1\Delta_g]$ and $O_2[^1\Sigma_g^+]$, and the anion O_2^- seem at present to be the main species which can be formed in the 'quenching' action of the dye triplet by oxygen. Molecular excited singlet oxygen results from an energy transfer[5]:

$$^3D_1 + {}^3O_2 \rightarrow {}^1O_2({}^1\Delta_g \quad \text{or} \quad {}^1\Sigma_g^+) + {}^1D_0$$

The anion O_2^- results from an electron transfer[6]:

$$^3D_1 + {}^3O_2 \rightarrow D^+ + O_2^-$$

---

*Presented at the Second International Conference on Photosensitization in Solids, Tucson, Arizona, January 29–31, 1968.

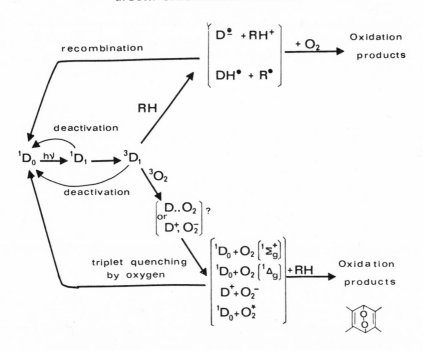

Fig. 1. General scheme of reaction processes in photosensitized oxidations.

The relative rate of each process probably depends upon the dielectric constant of the medium and also upon the nature of the dye. Kashe[7] has noted that, for two closely related dyes, fluorescein and eosin, the rate constant of the electron transfer from the triplet dye to oxygen is markedly higher for the former dye than for the latter.

According to present data in the literature[1, 8] it appears that dyes used in photosensitized oxidations (xanthenes, acridines, thiazines...) are able to produce excited singlet oxygen as well as anionic oxygen and that excited singlet oxygen O_2 $[^1\Delta_u]$ is the active species in these reactions[5].

Molecular excited singlet oxygen O_2 $[^1\Delta_u]$ is known to be longlived and able to survive $3 \cdot 5 \times 10^4$ wall collisions and more than 10^8 collisions in the vapor phase[9]. It reacts as a dienophile with dienes and gives ene-reactions with olefins[5]. These characteristics are quite different from those of anionic oxygen O_2^- and of $O_2[^1\Sigma_g^+]$, the latter being much more collision sensitive.

In order to study each of the different processes represented on Fig. 1, experimenters have usually chosen conditions which favor the reaction of the triplet dye either with RH(absence of oxygen) or with oxygen. For instance, the physical separation of the dye from the substrate favors exclusively the reaction with oxygen.

Indeed, experiments with the aim of demonstrating the occurrence of photosensitized oxidations in which the sensitizer and the oxidizable substrate are not in contact have been performed in the past[2, 4] and it has been shown in particular[4] that the photosensitized oxidation of 4-methoxy-naphthol by erythrosin in ethylcellulose still occurred at an average distance of 80 Å with a quantum yield as high as 0·2–0·3. However these experiments did not give any guaranty on the certainty of the separation and little information on the nature of the oxidizing entity.

In the present study the dye sensitizer was separated from the oxidizable substrate by several monomolecular layers of stearate salts, which has been shown to be an oxygen permeable membrane. Recent work[10] has demonstrated that such films of stearate provide quite convenient spacers of accurately known thickness in a sandwich-like set-up, each external layer being one of the two components under investigation. Such a sandwich allows one to study the rate of the photosensitized oxidation as a function of the distance separating the dye from the oxidizable substrate.

The oxidizable substrates used in this work are aromatic hydrocarbons which are good acceptors of $O_2(^1\Delta_g)$: diphenylanthracene and rubrene[5, 8]. The dyes have been selected among good sensitizers, with their absorption bands located at longer wavelength than the band of the corresponding substrate: eosin for diphenylanthracene and methylene blue for rubrene. This choice precludes the possibility of singlet–singlet energy transfer from the excited dye to the hydrocarbon. Triplet–triplet energy transfer although energetically possible is very unlikely because of the distance. These experimental conditions were selected to favor excited singlet oxygen $O_2(^1\Delta_g)$ as the intermediate oxidizing agent.

EXPERIMENTAL

1. *Sample preparation*

Two types of layered structures have been used in these experiments without noticeable differences between the results (Fig. 2): (a) The dye is adsorbed first onto a hydrophobic glass slide, then several monolayers of stearate salt are deposited, and finally the oxidizable substrate is put on as a mixed layer with stearic acid. (b) The order of the layers in inverted, the mixed layers of oxidizable substrate with stearic acid are transferred first on the glass slide, then the spacer of two or more monolayers of stearate salt and finally the dye is adsorbed.

The technique of Langmuir and Blodgett[11] was used to spread monolayers on the surface of a water subphase contained in a trough. A microscope glass slide which previously had been made hydrophobic by rubbing with chips of ferric stearate was slowly dipped in the trough by mean of an automated device. During this immersion the monolayer of stearate salt existing on the surface of the water was transferred to the glass slide. Then the slide was slowly pulled out of the water, a second monolayer being then transferred over the first. The whole operation can be repeated, each one resulting in the transfer of two additional monolayers on the sample.

All parts of the trough in contact with water were made of polished Plexiglas. The water was distilled twice in quartz and the pH was ajusted between 5·5 and 5·7, which

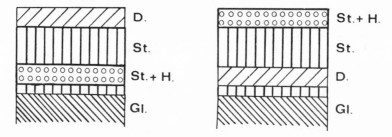

Fig. 2. Disposition of the layers on the samples. D = dye, St = stearate layers, $St + H$ = mixed layers; Gl = glass slide.

corresponds to a domain of best stability for the cadmium stearate spread on water. The cadmium ions (10^{-4}M/l.) were introduced as chloride. In these conditions each film of cadmium stearate is $24 \cdot 8 \pm 1$ Å thick. Stearic acid (99·5 per cent, chromatographically tested, m.p. = 68·6°C) was dropped on the cleaned water surface as a 10^{-4}M/l. solution either in benzene or preferably in n-heptane which is inert towards Plexiglas.

The mixed layers were obtained from solutions containing various proportions of stearic acid and the aromatic hydrocarbon. In the spread film, the stearic acid imparts cohesion to the mixed layer, its concentration varying from 1 to 9 molecules for each molecule of aromatic compound. From area measurements of the spread film on water and from absorption spectra of the samples on the glass slide, it seems that the aromatic hydrocarbons are not dissolved in the stearate, but more probably are distributed in an aggregated state. The dye (eosin or methylene blue) was adsorbed on stearate layers from concentrated aqueous solutions (10^{-3}M/l.). All operations were made in a dust-free box with a laminar flow of filtered air maintaining a slight overpressure.

2. *Irradiation*

The sample was illuminated with a 500 W tungsten lamp at a distance of 15 cm, through an infrared-absorbing filter and a Wratten filter. The latter prevents any excitation in the absorption range of the aromatic hydrocarbons.

3. *Measurements*

The rate of photosensitized oxidation was followed by monitoring the fluorescence intensity of the hydrocarbon as a function of the time of irradiation. The fluorescence of the hydrocarbon was excited through a Wratten filter 18 A ($310 \leqslant \lambda \leqslant 390$ nm) with a mercury discharge lamp. The fluorescence emission, filtered first by a Wratten 2 A filter, was analyzed by a Bausch and Lomb monochromator followed by a RCA 4472 photomultiplier and recorded. The fluorescence intensity of a sample was always very low and to collect the emission more efficiently, the glass slide itself was used as a light pipe to the entrance slit of the monochromator (Fig. 3). With these conditions it was possible to follow the rate of oxidation of two mixed monolayers containing rubrene.

A calibration with unirradiated samples showed a proportionality exists between fluorescence intensity and the amount of hydrocarbon present in the mixed layer. In order to take into account the fluorescence from the base (glass + stearate) the reported values on the graphs are the difference between fluorescence intensity of a sample containing the hydrocarbon, stearate dye and of a blank composed only of a glass slide covered with several monolayers of stearate.

RESULTS

The course of photosensitized oxidation is plotted in the Figs. 4 and 5 as per cent of oxidized compound against the accumulated time of irradiation. The accuracy of these data is estimated to be about 15–20 per cent.

For both couples (eosin-diphenylanthracene and methylene blue-rubrene), the rate of photooxidation decreases when the thickness of the stearate spacers increases. The photosensitized oxidation is the most efficient for a one-layer spacer; it still occurs when 12 layers of stearate (300 Å) separate the two components and in one case it was even observed at distance up to 500 Å (20 layers).

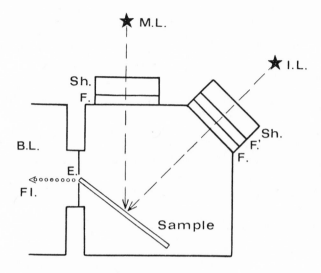

Fig. 3. Equipment for the fluorescence measurement of aromatic hydrocarbons. *I.L.* = irradiation lamp for the photooxidation; *M.L.* = monitoring lamp for fluorescence excitation; *Sh* = shutter; *F.F′* = filters; *E* = entrance slit; *B.L.* = Bausch and Lomb monochromator; *Fl* = fluorescence emission.

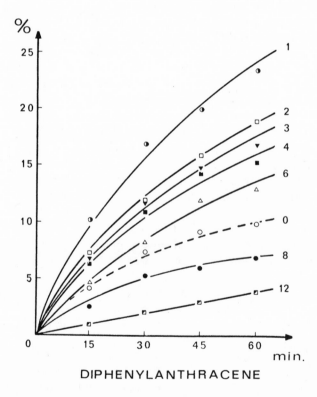

DIPHENYLANTHRACENE

Fig. 4. Percentage of oxidized diphenylanthracene sensitized by eosin against cumulated time of irradiation. Numbers on the curve indicate the numbers of stearate interlayers.

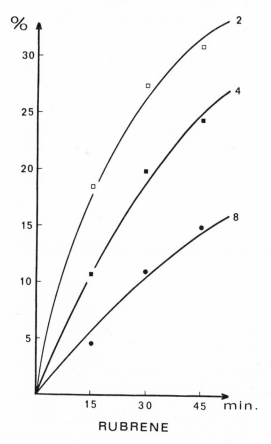

Fig. 5. Percentage of oxidized rubrene sensitized by methylene blue against cumulated time of irradiation. Numbers on the curve indicate the number of stearate interlayers.

An interesting exception is noticeable: the curve which corresponds to the sample without stearate spacer between eosin and diphenylanthracene is anomalous, in that one might expect to find the highest rate of photooxidation when the dye and substrate are in contact.

DISCUSSION

(a) The anomalous behavior of the sample without stearate layers between eosin and diphenylanthracene shows that the contact between the two compounds probably allows others processes to take place. The experiments of Mulder and De Jonge[12] and of Gerischer[13] have indeed shown that an excited dye, when in contact with an aromatic hydrocarbon, can be deactivated by an electron transfer process between both components.

Such a process could compete efficiently with the primary step of the photosensitized oxidation. But as soon as a spacer is present between the two compounds this electron transfer process is made impossible and a strong enhancement of the rate of the photosensitized oxidation is observed (Fig. 4). The contact between the two compounds does not favor the photosensitized oxidation and as a consequence, the anomalous behavior can be taken as an indirect proof of the absence of 'holes' in the stearate layers and of the absence of diffusion of either of the two components through the stearate layers.

(b) It has been noted in the Introduction that excited singlet oxygen $O_2(^1\Delta_g)$ and $O_2(^1\Sigma_{g+})$ are produced by the reaction of the dye triplet with oxygen and that the former is much more resistant to collisions than the latter. Consequently, $O_2(^1\Delta_g)$ appears to be probably the only species able to diffuse at some distance through the stearate films, and, in agreement with the other authors, is held responsible for the oxidation of the aromatic hydrocarbons. If the relative amount of photo-oxidized product, at a given irradiation time, is plotted as a function of the thickness of the spacer on a semi-logarithmic scale (Fig. 6), it appears that the rate of the photosensitized oxidation obeys an exponential decay law and does not depend on the nature of the oxidizable substrate or of the dye sensitizer, both plots being parallel. This decay law can be explained as a first order deactivation of $O_2(^1\Delta_g)$ within the stearate layers, either radiatively: $O_2(^1\Delta_g) \rightarrow {}^3O_2 + h\nu$ or non-radiatively, by collisions: $O_2(^1\Delta_g) \rightarrow {}^3O_2 + $ heat.

The contribution of bimolecular processes such as:

$$^1O_2 + {}^1O_2 \rightarrow [{}^1O_2]_2$$

seems to be excluded in the present experimental conditions.

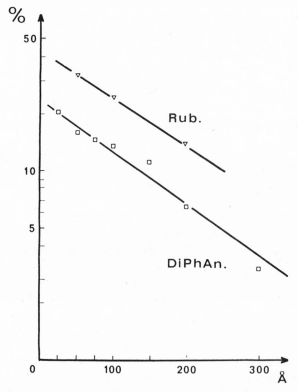

Fig. 6. Percentage of photooxidized product, for 45 min irradiation, as a function of the spacer thickness.

From Fig. 6, is possible to estimate a mean diffusion path for molecular excited oxygen $O_2(^1\Delta_g)$ in the cadmium stearate layers of about 115 ± 20 Å, in which half of the diffusing molecules have been deactivated.

(c) The reaction of singlet oxygen $O_2(^1\Delta_g)$ with aromatic hydrocarbons is known to give transannular peroxides, which are characterized by the disappearance of fluorescence, and by the ability to be decomposed easily by heat, into oxygen and the original hydrocarbon, with restoration of the fluorescence. This has been shown to be the case here with a sample of rubrene sensitized by methylene blue (50 Å apart). 2hr irradiation with a 500 W tungsten lamp (Wratten filter no. 29), produces a decrease of the rubrene fluorescence to 38 per cent of its initial value. Heating the sample for 90 min at 120°C in the dark restored 90 per cent of the characteristic rubrene fluorescence confirming the regeneration of the hydrocarbon and consequently the transannular structure of its oxidation product.

Acknowledgements — We are very grateful to Professor J. Rigaudy, Ecole de Physique et Chimie Industrielle (Paris), for the generous gift of diphenylanthracene and rubrene samples, to Dr. A. Reiser, Kodak Ltd., Research Laboratories (Harrow), for fruitful discussions and to Professor H. Kuhn for his precious advice concerning the technique of monomolecular films.

REFERENCES

1. J. Bourdon and B. Schnuriger, In *Physics and Chemistry of the Organic Solid State* (Edited by D. Fox, M. M. Labes and A. Weissberger), Vol III pp. 59–131. Interscience, New York (1967).
2. H. Kautsky, *Trans. Faraday Soc.* **35**, 216 (1939); J. L. Rosenberg and D. J. Shombert, *J. Am. Chem. Soc.* **82**, 3257 (1960).
3. C. S. Foote and S. Wexler, *J. Am. Chem. Soc.* **86**, 3880 (1964).
4. J. Bourdon and B. Schnuriger, *Photochem. Photobiol.* **5**, 507 (1966).
5. A very extensive literature has appeared on the subject since few years. A good review is to be found in: (a) C. S. Foote. *The Vortex* **27**, 436 (1966): (b) C. S. Foote. S. Wexler. W. Ando and R. Higgins, *J. Am. Chem. Soc.* **90**, 975 (1968): (c) R. Higgins, C. S. Foote and H. Cheng, To be published. Authors are indebted to Professor C. S. Foote for prepublication copies.
6. See for example J. Weiss, *Trans. Faraday Soc.* **42**, 133 (1964); V. Kasche and L. Lindqvist, *J. Phys. Chem.* **68**, 817 (1964).
7. V. Kasche, *Photochem. Photobiol.* **6**, 643 (1967); V. Kasche, AD 602 227 available OTS; from U.S. Govt. *Res. Rept.* **39**, (16) 24 (1964) and CA 62-2396 h.
8. Th. Wilson, *J. Am. Chem. Soc.* **88**, 2898 (1966).
9. A. M. Winer and K. D. Bayes, *J. Phys. Chem.* **70**, 302 (1966).
10. H. Bücher, K. H. Drexhage, M. Fleck, H. Kuhn, D. Möbius, F. P. Schäfer J. Sondermann, W. Sperling, P. Tillmann and J. Wiegand, *Molec. Crystals* **2**, 199 (1967).
11. K. B. Blodgett, *J. Am. Chem. Soc.* **56**, 495 (1934) and **57**, 1007 (1935).
12. B. J. Mulder and J. De Jonge, *Koninkl. Ned. Akad. Wetenschap., Proc.* **66B**, 303 (1963).
13. H. Gerischer, Communication presented at this conference.

Editor's Comments
on Paper 37

37   KOCH
Zur Photosensibilisierten Sauerstoffübertragung: Untersuchung der Terminationsschritte durch Belichtungen bei tiefen Temperaturen

ACTIVATION ENERGIES FOR SENSITIZED PHOTOOXYGENATION OF VARIOUS OLEFINS

In Paper 37, Koch reports a series of investigations on the temperature dependence of dye-sensitized photooxygenation of several olefins and dienes. Singlet oxygen is suggested as the reactive intermediate. The activation energies for the formation of the oxidation product are very low (0.1–5.0 kcal/mol).[1] Photooxygenations can therefore be carried out at low temperatures without a large decrease in the rate of reaction. This is of practical importance as many photooxygenation products, such as 1,2-dioxetanes and endoperoxides, are thermally unstable.

Substituent effects on the reactivity of various acceptors toward singlet oxygen as described in this paper indicate that $^1O_2$ is an electrophilic species.

NOTES AND REFERENCES

1. See also C. A. Long and D. R. Kearns, *J. Amer. Chem. Soc.*, **97**, 2018 (1975)

37

Reprinted from *Tetrahedron*, **24**, 6295–6318 (1968)

ZUR PHOTOSENSIBILISIERTEN SAUERSTOFFÜBERTRAGUNG

UNTERSUCHUNG DER TERMINATIONSSCHRITTE DURCH BELICHTUNGEN BEI TIEFEN TEMPERATUREN

E. KOCH

Max-Planck-Institut für Kohlenforschung, Abt. Strahlenchemie, Mulheim-Ruhr

(*Received in Germany* 20 *May* 1968; *Received in the UK for publication* 29 *May* 1968)

Zusammenfassung—Die Eigenschaften des O_2-haltigen Zwischenproduktes (X) der photosensibilisierten O_2-Übertragung wurden unter Berücksichtigung der vorgelagerten Diffusionsprozesse untersucht. Im Bereich von $+50$ bis $-150°$ folgen die Quantenausbeuten der O_2-Aufnahme für zahlreiche Systeme einem einfachen Zeitgesetz. Aus der Temperaturabhängigkeit der Geschwindigkeitskonstanten wurden die Entropien und Aktivierungsenergien sowohl der Desaktivierung des Zwischenproduktes als auch seiner Reaktion mit dem Akzeptor (A) berechnet.

Die Zerfallskonstante von X ist praktisch vom Sensibilisator ($k_N \sim 5 \times 10^6$ sec^{-1} in Methanol) und von der Temperatur unabhängig. Für die Bildung des Endprodukts ist daher wahrscheinlich elektronisch angeregter Sauerstoff massgeblich, dessen Reaktivität durch benachbarte Sensibilisator- und Lösungsmittelmoleküle nur geringfügig beeinflusst wird.

Die Aktivierungsenergien (E_H) für die AO_2-Bildung nehmen in folgender Reihenfolge ab: Buten-(2), 2-Methyl-propen, 2-Methylbuten-(2), 2·3-Dimethyl-buten-(2), Cyclohexadien-(1·3), Cyclopentadien ($\cong 0$). Allgemein wird E_H durch Methylsubstitution an Doppelbindungen herabgesetzt; der Einfluss voluminöserer Gruppen wird diskutiert.

Die Entropien (ΔS) liegen für 42 der 46 geprüften Akzeptoren zwischen -12 und -18 cal/Mol.Grad.

Abstract—The oxygen containing intermediate X in photosensitized oxygen transfer-reactions as well as the significance of the preceeding diffusion processes have been investigated.

In the range $+50/-150°$ the quantum yield of the O_2 consumption by the acceptor (A) follows a simple rate law. Rate constants, entropies and activation energies have been determined for the deactivation of X as well as for its reaction with A.

The decay constant of X is practically independent of sensitizer ($k_N \approx 5 \times 10^6$ sec^{-1} in methanol) and temperature. Electronically excited O_2 is suggested as the reactive species (X), whose reactivity is only slightly modified by different sensitizers and solvents.

The activation energies (E_H) for the formation of AO_2 decrease in the order: 2-butene, 2-methylpropene, 2-methyl-2-butene, 2,3-dimethyl-2-butene, 1,3-cyclohexadiene, cyclopentadiene ($\cong 0$). In general, E_H is lowered by methyl substitution of the double bonds; the influence of more voluminous groups is discussed.

ΔS is $-12/-18$ cal/deg mole for 42 out of 46 different acceptors.

EINLEITUNG

PHOTOSENSIBILISIERTE Sauerstoffübertragungen auf organische Akzeptoren ($=A$)[1-3] erfolgen in Lösung entweder nach einem durch Dehydrierung von A eingeleiteten monoradikalischen Mechanismus ($=$ Typ I)[3-6] oder durch sensibilisierte Addition des Sauerstoffs ohne Auftreten monoradikalischer Zwischenprodukte ($=$ Typ II).[1,3,5,7-10]

Bei den Reaktionen vom Typ II konnte kinetisch ein oxydierendes Zwischenprodukt nachgewiesen werden, dessen Entstehung auf einer Reaktion zwischen angeregtem Sensibilisator und Sauerstoff im Grundzustand beruht.[11-17] Über dieses

195

Zwischenprodukt bestanden verschiedene Vorstellungen, die bis auf zwei aus spektroskopischen, kinetischen, energetischen und chemischen Gründen ausgeschlossen wurden:[1, 2, 14] Die Forderung nach einem biradikalischen Addukt zwischen angeregtem Sensibilisator und Sauerstoff (Sens.^{rad}.O$_2$) nach Schenck[1, 7, 8, 15–26] und die Annahme eines angeregten Sauerstoffmoleküls im Singlettzustand nach Kautsky.[3, 5, 27–30] Auch bei rein thermischen Prozessen wurde in neuerer Zeit angeregter Singlett-Sauerstoff als Zwischenprodukt diskutiert, so bei der Oxydation von Olefinen durch Wasserstoffperoxyd bei Gegenwart von Hypochlorit[3, 31–36] und durch Ozon-Addukte der Ester der phosphorigen Säure.[37]

Als besonders geeignet zur Untersuchung der Eigenschaften des Zwischenprodukts auf kinetischem Wege erwiesen sich Belichtungen in einem weiten Temperaturbereich. Orientierende Untersuchungen der Temperatur- und Viskositätsabhängigkeit photosensibilisierter Sauerstoff-Übertragungen nach Typ II[38–40] hatten ergeben, dass wegen der minimalen Aktivierungsenergien aller Teilprozesse zahlreiche Akzeptoren noch bei −150° in geeigneten Lösegemischen gut reagieren können. Inzwischen haben wir die zunächst verwendeten Belichtungsapparaturen in ihrer Temperaturkonstanz, Sauerstoffversorgung und Betriebssicherheit verbessert und automatisiert, um eine grosse Anzahl von Versuchen unter Verwendung verschiedenster Akzeptoren, Sensibilisatoren und Lösungsmittel bei möglichst feiner Temperaturabstufung zwischen +20 und −140° rationell durchführen zu können. Das inzwischen sehr reichhaltige Material bestätigt im wesentlichen die wenigen früher ermittelten Zahlenwerte für Aktivierungsenergien und -entropien und erlaubt eine genauere Untersuchung des Einflusses von Akzeptor, Sensibilisator und Lösungsmittel auf die Terminationsschritte.

Kinetik der Tieftemperaturreaktionen

Die Sauerstoff-Aufnahme-Geschwindigkeit $\dot{n}_{O_2}$ einer photosensibilisierten Sauerstoffübertragung ist der Bildungsgeschwindigkeit des Endprodukts AO_2 proportional und gehorcht folgendem Zeitgesetz:

$$\dot{n}_{O_2} = \frac{dAO_2}{dt} = -\frac{dA}{dt} = f_1(O_2)f_2(O_2) \cdot \frac{[A]}{[A] + \beta} \qquad (1a)^{3, 5, 41}$$

$f_1(O_2)$ und $f_2(O_2)$ sind hier Funktionen, die die Konkurrenz der Bildungsreaktion des (Singlett-) angeregten Sensibilisators $^1S_1$ zu den möglichen Folgereaktionen beschreiben und die Reaktionskonstanten von mindestens sieben Teilprozessen enthalten; nämlich Absorption ($^1S_0 \xrightarrow{h\nu} {}^1S_1$), Fluoreszenz ($^1S_1 \rightarrow {}^1S_0 + h\nu'$), "internal conversion" ($^1S_1 \rightarrow {}^1S_0$), "intersystem crossing" ($^1S_1 \rightarrow {}^3S_1$ und $^3S_1 \rightarrow {}^1S_0$) und Bildung des "oxydierenden Zwischenprodukts" aus $^1S_1$ oder $^3S_1$.

β wird auch als "Halbwertskonzentration" bezeichnet und bedeutet diejenige Akzeptorkonzentration, für die die Hälfte des möglichen Maximalwerts der Reaktionsgeschwindigkeit [$= f_1(O_2) \cdot f_2(O_2)$] erreicht wird.

$f_1(O_2)$ und $f_2(O_2)$ sind nur vom Sensibilisator und von der O$_2$-Konzentration abhängig. Wird diese durch ausreichende Gasversorgung konstant gehalten, so vereinfacht sich Gleichung (1a) zu

$$\dot{n}_{O_2} = -\frac{dA}{dt} = \alpha \frac{[A]}{[A] + \beta} \qquad (1b)$$

196

Die allgemeine Gültigkeit dieser Gleichung wurde an zahlreichen Reaktionen zwischen $+50$ und $-140°$ bestätigt[16, 17, 39, 40] (vgl. Abb. 1). β ergibt sich als Quotient der Geschwindigkeitskonstanten zweier Reaktionen des oxydierenden Zwischenprodukts: Der Desaktivierung zu additions-unfähigem molekularem Sauerstoff (ggf. unter Abspaltung eines weiteren Teilchens wie des Sensibilisators) und der Reaktion mit A, die zur Bildung des Endprodukts AO_2 führt. Die β-Werte sind daher ein reziprokes Mass für die Reaktivität der Akzeptoren. Wir bestimmten sie für jeden Versuch durch lineare Interpolation in einem Diagramm, in dem die reziproke Gasaufnahme-Geschwindigkeit als Funktion der reziproken Akzeptorkonzentration, die sich nach Gleichung (1b) aus der Gasaufnahme ergibt, aufgetragen war (Abb. 1). Trägt man β für eine bestimmte Akzeptor-Sensibilisator-Lösungsmittel-

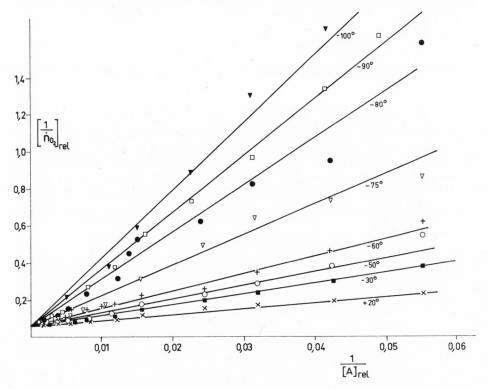

ABB. 1 Beispiel für Bestimmung der Halbwertskonzentration β aus der Geraden

$$\frac{1}{\dot{n}_{O_2}} = \frac{1}{\alpha} + \frac{\beta}{\alpha} \times \frac{1}{[A]}$$

MB/α-Terpinen in Äthanol ($[A]_0 = 0{\cdot}017$ Mol/für alle Versuche; $\frac{1}{\alpha} = 0{\cdot}06$)

Abkürzüngen vgl. S. 6302

Kombination als Funktion von $1/T$ logarithmisch auf, so lassen sich die Systeme nach der Art der Funktion und Grösse der β-Werte in vier Gruppen einteilen (Abb. 2, 3a, 3b):

I Gerade mit minimalem $\beta_{+20°}$ (in Methanol $< 0{\cdot}002$ Mol/l)

II positiv gekrümmte Kurve, die für tiefe Temperaturen asymptotisch in eine Gerade übergeht ($\beta_{+20°}$ = 0·002 bis 0·005 Mol/l.)

III positiv gekrümmte Kurve, die sowohl für tiefe als auch für hohe Temperaturen in je eine Gerade übergeht ($\beta_{+20°}$ = 0·005 bis 0·05 Mol/l.)

IV Gerade im gesamten Bereich ($\beta_{+20°}$ > 0·05 Mol/l.)

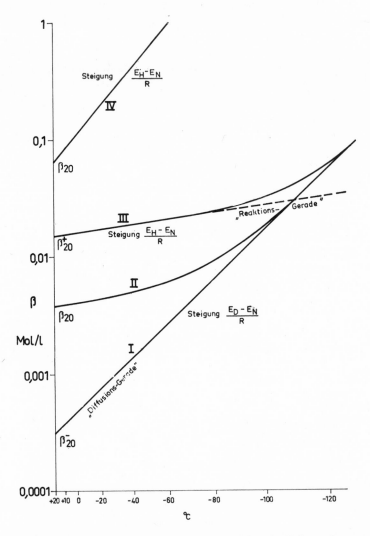

ABB. 2 Verschiedene Typen von Temperaturabhängigkeiten der Halbwertskonzentrationen (in Methanol).
(Angenommen sind Akzeptoren mit gleichen Diffusionskoeffizienten bei gleichbleibendem Sensibilisator.)

Wir nehmen nun an, dass aus dem sauerstoffhaltigen Zwischenprodukt (= X) und dem Akzeptor ein Begegnungskomplex $A \ldots X$ gebildet wird,[11, 12, 24, 39] der entweder in die Ausgangsprodukte zerfällt oder AO_2 bildet (Hauptreaktion in

Schema 1). β lässt sich unter Annahme der Stationaritätsbedingung für die Zwischenprodukte durch die Konstanten der Teilschritte ausdrücken:

$$\beta = k_N \frac{\overleftarrow{k_D} + k_H}{\overrightarrow{k_D} \cdot k_H} \qquad (2)$$

(k_H = Geschwindigkeitskonstante der Hauptreaktion, k_N = Geschwindigkeitskonstante der Nebenreaktion).

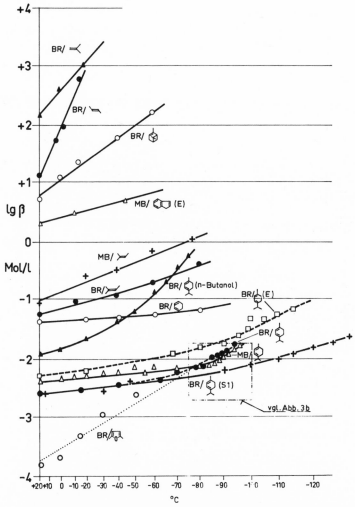

ABB. 3a Beispiele für die Temperaturabhängigkeit der Halbwertskonzentration in Methanol
bzw. anderen Lösungsmitteln (Abkürzungen vgl. S. 6302).

Für die Temperaturabhängigkeit aller Teilprozesse, auch der Diffusionsschritte,[40, 42, 43] gilt näherungsweise

$$k_i = A_i \exp\left(-E_i/RT\right) \qquad (3)$$

199

Durch Einsetzen in Gleichung (2) folgt

$$\ln \beta = \ln \left[\frac{A_N}{A_H} \cdot \frac{1}{K} \exp \frac{E_H - E_N}{RT} + \frac{A_N}{\overrightarrow{A_D}} \exp \frac{E_D - E_N}{RT} \right] \qquad (4)$$

mit

$$K = \frac{\overrightarrow{k_D}}{\overleftarrow{k_D}} \qquad (5)^{44}$$

Unter den durchweg erfüllten Voraussetzungen $E_H > E_N$ und $E_D > E_N$ beschreibt die Gleichung (4) eine Kurve mit positiver Krümmung, die zwei Geraden als Asymptoten hat, wie sie in der Tat für die Akzeptoren der Gruppe III gefunden wurde. Wir

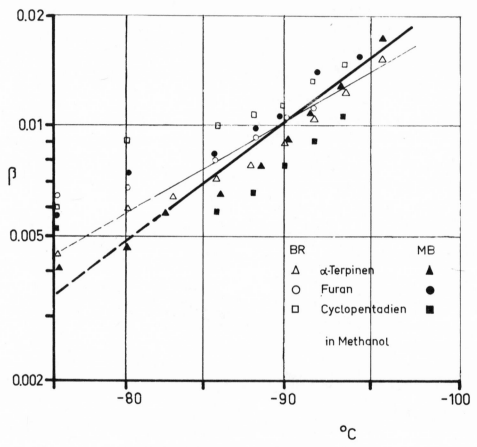

ABB. 3b Ausschnitt aus Abb. 3a; Abkürzungen vgl. S. 6302

wollen die Geraden zur Abkürzung mit "Reaktions"- (bei hohen Temperaturen) und "Diffusions"-Gerade (bei tiefen Temperaturen) bezeichnen. Wird $E_H \ll E_D$, so verschwindet in Gleichung (4) der erste Summand gegen den zweiten, und die Kurve wird zur Diffusions-Geraden. Umgekehrt ist bei langsam reagierenden Akzeptoren der zweite Summand unwesentlich, und wir erhalten die Reaktions-Gerade. Diese beiden Grenzfälle sind in den Gruppen I und IV experimentell realisiert. Die Systeme der Gruppe II unterscheiden sich von denen der Gruppe III dadurch, dass die Reaktions-Gerade sich nicht mehr im zugänglichen Temperaturbereich befindet.

Schema 1

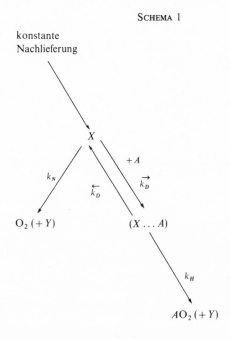

Die Asymptoten haben die Steigungen $(E_D - E_N)/R$ und $(E_H - E_N)/R$, ihre Ordinatenabschnitte β_{20}^+ und β_{20}^- erlauben die Berechnung der Reaktionskonstanten aus den Diffusionskonstanten (vgl. Abb. 2):

$$k_N = \vec{k_D} \cdot \beta_{20}^- \qquad (6)$$

$$k_H = \frac{k_N}{\beta_{20}^+ \cdot \mathrm{K}} \qquad (7)$$

Geschwindigkeitskonstanten für die Desaktivierungsreaktion aus den Ergebnissen mit diffusionsbestimmten Prozessen (Gruppen I–III)

Für alle untersuchten Systeme, die durch geeignete Wahl der Temperatur eine Interpolation auf die beiden Grenz-Geraden der β-Diagramme (Abb. 2) erlauben, erhält man nach dem Vorhergehenden durch Anwendung diffusionskinetischer Vorstellungen Geschwindigkeitskonstanten und Aktivierungsenergien von Haupt- und Nebenreaktion. Da jedoch zur näheren Untersuchung der Hauptreaktion nach Gleichung (7) eine Kenntnis der Desaktivierungs-Geschwindigkeitskonstanten erforderlich ist, haben wir zunächst an Systemen der Gruppen II und III die Lage der Diffusionsgeraden ermittelt. Diese sollte bei gleichbleibendem Sensibilisator und Lösungsmittel vom Akzeptor unabhängig sein, sofern das Modell (Schema 1) richtig ist.

201

Zur Prüfung wählten wir α-Terpinen, Cyclopentadien (Gruppe III) und Furan (Gruppe II) als Akzeptoren, belichteten bei verschiedenen Temperaturen zwischen $-80°$ und $-125°$ und bei kleinen Akzeptor-Anfangskonzentrationen ($< 2 \times 10^{-2}$ Mol/l.) bis zum Ende der O_2-Aufnahme und bestimmten die β-Werte wie oben (Abb. 3a, b). Aus den Steigungen der Temperatur-Zähigkeits-Kurven des Lösungsmittels[40, 42, 43] schätzten wir die Aktivierungsenergien der Diffusion E_D ab (Tabelle 1)

Legende zu Tabellen 1–5:

E_N und E_H in kcal/Mol; k_H [sec^{-1}]; k_N [sec^{-1}]; β[Mol/l] ΔS_H, ΔS_N [cal/Mol. Grad]

	Sensibilisatoren:	
	BR	Rose bengale
	MB	Methylenblau
	TClEO	Tetrachloreosin
	EO	Eosin
	TClF	Tetrachlorfluorescein
	BNT	Binaphthylenthiophen
	HP	Hämatoporphyrin
	PP	Protoporphyrin
	CN	Natrium-Chlorophyllin
	AR	Acridinorange

Lösungsmittelgemische:

	Volumen-Teile				
Lösungsmittel	E	EP	F	S1	K
Methanol	2	2	2	3	
Äthanol					2
n-Propanol	2	2	2	4	
Aceton	1	1		2	
Äther			6	8	4·5
Essigester				4	2
Petroläther <40°			5	4	
Isopentan				4	
Propan		1			

TABELLE 1. MITTLERE "AKTIVIERUNGSENERGIEN DER DIFFUSION" (E_D) IN kcal/Mol

Methanol	3·7	n-Hexanol	5·3	E	2·3
Äthanol	3·9	n-Heptanol	5·8	EP	~2·2
n-Propanol	4·4	n-Oktanol	~5·5	K	1·9
n-Butanol	4·7	n-Decanol	~6	S1	1·9
		Toluol	3·5		

und anhand der weitgehend erprobten Näherungsgleichungen von Wilke und Chang[45, 46] die Diffusionsgeschwindigkeitskonstanten $\overrightarrow{k_D}$.

Diese wie β_{20}^+ und β_{20}^- auf $+20°C$ zu beziehenden Grössen werden für das voluminösere Sens.$^{\text{rad.}}$ O_2 wesentlich geringer als für den Singlett-Sauerstoff. Sie können unter der Annahme kugelförmiger Moleküle über die Atomvolumina ineinander umgerechnet werden.

TABELLE 2. k_N-, E_N- UND A_N-WERTE FÜR CYCLOPENTADIEN (CPD), FURAN (FUR) UND α-TERPINEN (T)

Sens.	Akzeptor	k_N^* 10^6	E_N	A_N^*	ΔS^*
Lösungsmittel: Methanol	(vgl. Abb. 3b)				
BR	T	3·0	−0·2		
	CPD	4·3	0·1	$4·0 \times 10^6$	−30·2
	Fur	3·8	±0		
MB	T	1·3	−0·3		
	CPD	1·8	−0·4	$2·0 \times 10^6$	−31·6
	Fur	1·7	−0·2		
Lösungsmittel: E					
BR	T	12	±0		
	CPD	16	0·2	15×10^6	−27·6
	Fur	11	0·2		
MB	T	13	0·1		
	CPD	11	±0	13×10^6	−27·9
	Fur	—	—		
Fehlergrenze ca.		±25%	±0·2	±25%	±0·4

* Bezogen auf Sens^{rad}.O_2.

Nach der Gleichung (6) ergeben sich innerhalb der Fehlergrenzen gleiche Werte für die Desaktivierungskonstanten k_N (Tabelle 2). Die Verwendung verschiedener Sensibilisatoren hat auf die Werte keinen grossen Einfluss (Tabelle 3). Bei Änderung des Lösungsmittels sind jedoch grössere Unterschiede erkennbar (Tabelle 4): In

TABELLE 3. REAKTIONSGESCHWINDIGKEITSKONSTANTEN DER DESAKTIVIERUNGSREAKTION

Sensibilisator	k_N für Sens^{rad}.O_2 [$\times 10^6$] in MeOH	in E	k_N für $^1O_2$ [$\times 10^6$] in MeOH	in E	E_N	$A_N\ddagger$ [$\times 10^6$] in Methanol	$\Delta S_N\ddagger$
BR	4·0 + 1	12 + 3	7·0	21	±0	4·0	−30·2
TClEO	6·5 + 2		11·0		<0·7	<18·0	−27·3
EO		8 + 3	(13·0)	26		?	?
TClF	6·5 + 2		12·0		<1·0	<30·0	>−26·3
MB	1·7 + 0·5	10 + 4	2·5	20	−0·2	2·0	−31·6
BNT	(3·4 + 0·8)*		(5·4)		+0·3	6·0	−29·5
AR	1·2 + 0·4		2·0			1·4	−32·3
HP	(5 + 2)†						
PP	(4 + 1)†						
CN		11 + 5		25			

* Toluol + 10% Methanol.

† Aceton + 30% Pyridin.

‡ bezogen auf Sens^{rad}.O_2.

polaren Lösungs- oder Mischlösungsmitteln ist die Desaktivierungsgeschwindigkeit deutlich höher. Die Aktivierungsenergien für die Desaktivierungsreaktion liegen bei fast allen getesteten Lösungsmitteln um Null. In einigen Fällen, wie in Äthanol,

TABELLE 4. LÖSUNGSMITTELEINFLUSS

Akzeptor: α-Terpinen *Sensibilisator:* BR

Lösungsmittel		$k_N \times 10^6$	E_N	E_H	k_H	A_N 10^6	A_H 10^9
Methanol	BR	4 ± 1	± 0	0.4 ± 0.2	1.7×10^9	4.0	3.4
Äthanol	BR	9 ± 3	$< +1.0$	<1.0	3.6×10^8	<40.0	< 2.0
	MB	(4 ± 1)		(<0.6)	(1.7×10^8)		
n-Propanol	BR	3.4 ± 1	$< +0.8$	<0.8	3.5×10^8	<14.5	< 1.3
	MB	(1.1 ± 0.3)		<0.6	(1.2×10^8)		
n-Butanol	BR	4 ± 1	$+0.4$	<2.8	2.8×10^8	8	20
n-Hexanol		7 ± 2	$\sim +1.5*$	—	$1.3 \times 10^8†$	21	
n-Heptanol		4 ± 1	$\sim +1.5*$	—	$1.2 \times 10^8†$	55	
n-Oktanol		1 ± 0.2	$\sim +1.0*$	—	$1.4 \times 10^7†$	4	
n-Decanol		0.5 ± 0.2	?	—	$4.0 \times 10^7‡$	1	
Toluol		5 ± 1	~ 0	—	$1.2.0 \times 10^8†$	5	~ 0.1
E		12 ± 2	± 0	0.6 ± 0.2	1.4×10^9	13	~ 1.4
EP		25 ± 5	± 0	$<0.4 \pm 0.2$	1.7×10^9	25	< 1.7
F		25 ± 5	~ 0	0.2 ± 0.2	2.0×10^8	25	~ 0.2
S1		12 ± 3	-0.1	$<0.5 \pm 0.2$	2.4×10^9	12	< 2.4
K		30 ± 7	~ 0	0.3 ± 0.2	3.0×10^8	30	~ 0.3

k_N und A_N liegen um den Faktor 1.7 höher, wenn $^1O_2$ statt $Sens^{rad}O_2$ angenommen wird.
* Nur orientierende Messungen. Fehlergrenze $\sim \pm 0.5$ kcal/Mol.
† Unter der Annahme $E_N = 0$.

ist offenbar auch bei tiefsten Arbeitstemperaturen die "Diffusions"-Gerade noch nicht ganz erreicht, so dass die angegebenen E_N- Werte Maximalwerte sind.

Reaktion des sauerstoffhaltigen Zwischenproduktes mit dem Akzeptor

Für die Akzeptoren der Gruppen III und IV erhält man die Aktivierungsenergien der Hauptreaktion, E_H, aus der Steigung der "Reaktions"-Geraden, während bei den Systemen II nur Maximalwerte nach dem Anlegen einer Tangente an die β-Kurve bei hohen Temperaturen berechnet werden können (Abb. 2 und 3).

Die Berechnung der Geschwindigkeitskonstanten k_H nach Gleichung (7) lässt sich nur orientierend vornehmen. Zwar ist β vor allem im Bereich von ca. 0.01 bis ca. 1 Mol/l recht genau ($\pm 5\%$) zu erhalten. Für die Grössen K und k_N gehen jedoch die Unsicherheiten bei der Abschätzung von Diffusionskoeffizienten ein, weshalb die Fehlergrenze für k_H mit ca. $\pm 50\%$ relativ hoch liegt. Die Werte für k_H erstrecken sich jedoch über fünf Zehnerpotenzen (Tabelle 5a–d) und passen zusammen mit den berechneten Aktivierungsenergien und Entropiewerten recht gut in das Bild, das wir uns heute vom mechanistischen Ablauf der Hauptreaktion machen.[1, 3, 10, 41]

In die Übersicht (Abb. 4) haben wir auch die Akzeptoren der Gruppen I und II sowie einige Verbindungen aufgenommen, deren β-Werte bisher nur bei ca. 20° im Schenckschen Arbeitskreis bestimmt wurden. Für diese nahmen wir plausible

TABELLE 5. AKTIVIERUNGSENERGIEN, REAKTIONSGESCHWINDIGKEITSKONSTANTEN UND ENTROPIEN FÜR VERSCHIEDENE AKZEPTOREN (in Methanol)

(5a) *Carbocyclische konjugierte Diene*

Akzeptor	Sens.	β_{20}^{+}	E_H	k_H	ΔS_H
Gruppe II					
ω-Dimethylfulven	BR	0·006	0·3 ± 0·1	2·5 × 10⁹	−15·0
α-Terpinen	BR	0·0031	0·4 ± 0·2	2·8 × 10⁹	−14·8
	TClEO	0·0033	0·3 ± 0·2	4·4 × 10⁹	−14·3
	TClF	0·006	0·2 ± 0·1	2·2 × 10⁹	−16·2
	MB	0·004	0·4 ± 0·2	1·7 × 10⁹	−15·4
	BNT	0·0014	0·3 ± 0·1	6·0 × 10⁹	−14·0
	AR	0·0013	0·4 ± 0·2	2·0 × 10⁹	−16·3
Gruppe III					
Cyclopentadien	BR	0·0044	0·3 ± 0·1	3·6 × 10⁹	−14·8
	MB	0·0041	0·2 ± 0·1	0·9 × 10⁹	−17·8
α-Phellandren	BR	0·01	1·0 ± 0·1	0·8 × 10⁹	−15·4
Cyclohexadien-(1·3)	BR	0·045	1·2 ± 0·3	3·0 × 10⁸	−17·6
	TClEO	0·025	1·3 ± 0·2	7·8 × 10⁸	−14·4
	TClF	0·04	1·4 ± 0·2	5·0 × 10⁸	−14·8
	MB	0·073	1·2 ± 0·3	1·1 × 10⁸	−9·0
	BNT	0·011	1·4 ± 0·2	1·0 × 10⁸	−17·8

(5b) *Olefine und nicht konjugierte Di- und Triene*

Akzeptor	Sens.	β_{20}^{+}	E_H	k_H	ΔS_H
Gruppe III					
Terpinolen*	BR	∼0·05	0·4 ± 0·2	6·1 × 10⁸	−17·8
2·3-Dimethylbuten-(2)	BR	0·0062	0·5 ± 0·3	2·1 × 10¹⁰	−12·3
Dicyclohexyliden	BR	0·03	1·3 ± 0·3	3·5 × 10⁸	−16·0
2-Methylbuten-(2)	BR	0·6	1·6 ± 0·2	2·3 × 10⁸	−16·1
	TClEO	0·055	1·5 ± 0·2	3·5 × 10⁸	−15·3
	TClF	0·06	1·6 ± 0·1	3·2 × 10⁸	−15·2
	MB	0·1	2·2 ± 0·2	6·0 × 10⁷	−16·8
	BNT	0·03	1·4 ± 0·1	4·0 × 10⁸	−16·4

TABELLE 5—(*Forts.*)

Alloocimen	BR	0·07	2·0 ± 0·2	$1·4 \times 10^8$	−15·4
2-Methylpenten-(2)	BR	0·13	2·0 ± 0·4	$0·7 \times 10^8$	−16·8
d-Limonen*	MB	1·7	2·0 ± 0·4	$2·0 \times 10^7$	−19·6
Gruppe IV					
2-Methyl-4-phenyl-buten-(2)	BR	0·15	2 3 ± 0·3	$1·2 \times 10^8$	−14·4
Inden	BR	1·5	(3·1 ± 0·1)	$4·2 \times 10^6$	−19·4
	MB*	2·0	2·6 ± 0·4	$7·6 \times 10^6$	−19·6
Carvomenthen	MB	∼1·0	3·2 ± 0·2	$1·0 \times 10^7$	−16·8
Nopadien	BR	∼2·0	3·9 ± 0·4	$6·3 \times 10^6$	−15·1
Anethol	BR	∼0·01	4·0 ± 0·6	$1·4 \times 10^9$	− 3·8?
α-Pinen	BR	5·0	4·5 ± 0·2	$2·1 \times 10^6$	−15·3
	TClEO	3·8	4·0 ± 0·2	$4·2 \times 10^6$	−15·4
	TClF	0·5	4·0 ± 0·2	$3·3 \times 10^7$	−11·5?
	MB	7·0	4·7 ± 0·3	$1·0 \times 10^6$	−15·8
	BNT	2·8	5·1 ± 0·2	$3·0 \times 10^6$	−12·2?
β-Pinen	BR	∼1·0	5·0 ± 0·7	$1·1 \times 10^6$	−15·0
Isobutylen	BR	1·60	5·7 ± 0·2	$9·0 \times 10^4$	−16·9
Buten-(2)	BR	12·5	10·0 ± 1·0	$1·1 \times 10^6$	+ 2·7?

(5c) *Heterocyclische Fünfringe*   Sensibilisator : BR

Akzeptor	β_{20}^+	E_H	k_H	ΔS_H
Gruppe I				
2·5-Dimethylfuran	∼0·0002	0·1 ± 0·2	$6·3 \times 10^9$	−14·2
Gruppe II				
Furan	0·0045	0·2 ± 0·1	$4·5 \times 10^9$	−14·1
Gruppe III				
2-Methylfuran	0·0038	0·4 ± 0·2	$6·2 \times 10^9$	−13·2
Furfurylalkohol	0·0033	0·7 ± 0·1	$3·8 \times 10^9$	−12·9
Furfurylamin	0·009	0·7 ± 0·2	$2·5 \times 10^9$	−13·9
Diphenylisobenzofuran	∼0·013	0·7 ± 0·2	$1·0 \times 10^9$	−18·3
2-Vinyl-furan	0·0018	0·9 ± 0·2	$7·5 \times 10^9$	−11·3
Furfuryl-N-methylamin	0·013	1·0 ± 0·3	$1·1 \times 10^9$	−14·7
Tetracyclon	∼0·03	1·0 ± 0·2	$3·2 \times 10^8$	−17·0
Furfurylmethyläther	0·0034	1·2 ± 0·1	$2·3 \times 10^9$	−12·9
2,4-Dimethylfuran	0·002	1·5 ± 0·2	$6·7 \times 10^9$	− 9·8
2,5-Dimethyl-pyrrol	0·16	1·7 ± 0·3	$1·0 \times 10^8$	−14·2
Gruppe IV				
Furan-3,4-dicarbonsäureester	∼0·2	3·0 ± 0·2	$1·1 \times 10^7$	−15·7
Furan-2-aldehyd	0·6	5·0 ± 0·4	$2·3 \times 10^7$	−8·7
Thiophen	>600	(6·0 ± 0·4)	$<10^5$	∼ −14·2

TABELLE 5—(Forts.)

(5d) *Andere Verbindungstypen*

Akzeptor	Sens.	β_{20}^{+}	E_H	k_H	ΔS_H
Gruppe III					
Triphenylphosphin	BR	0·005	1·4 ± 0·2	1·5 × 10⁹	−13·2
	MB	0·006	1·2 ± 0·2	7·5 × 10⁸	−14·9
Diäthanolsulfid	BR	0·025	1·3 ± 0·2	3·0 × 10⁸	−16·4
	TClEO	0·013	0·9 ± 0·2	7·8 × 10⁸	−15·8
	TClF	0·016	1·5 ± 0·3	6·5 × 10⁸	−14·1
	MB	0·022	1·8 ± 0·2	2·1 × 10⁸	−15·2
	BNT	0·01	2·0 ± 0·4	6·0 × 10⁸	−12·6
Dibenzylsulfid	BR	0·012	0·9 ± 0·2	7·0 × 10⁸	−15·7
	TClEO	0·0067	1·3 ± 0·2	1·9 × 10⁹	−13·6
	TClF	0·0055	0·4 ± 0·3	2·3 × 10⁹	−15·3
	MB	0·009	1·7 ± 0·2	6·0 × 10⁸	−18·2
	BNT	0·008	1·6 ± 0·3	9·0 × 10⁸	−17·7
Ascorbinsäure	BR	0·012	1·5 ± 0·3	8·0 × 10⁸	−13·5
Thioharnstoff	MB	0·04	1·5 ± 0·2	2·5 × 10⁸	−16·0
Gruppe IV					
Cyclooctatetraendibromid	BR	∼0·3	2·6 ± 0·7	3·7 × 10⁷	−16·0
Cyclohexylidencyclohexanon	BR	0·35	2·8 ± 0·3	2·3 × 10⁷	−16·4
ω-Dimethyl-fulvenendoperoxid	BR	∼2·5	3·4 ± 0·3	5·1 × 10⁶	−17·2
6,6-Dimethylcyclohexadien-(1·3)-on-(5)	BR	∼1·9	3·6 ± 0·2	4·6 × 10⁶	−16·8
3·6-Endoperoxy-cyclohexen	BR	150	6·2 ± 0·3	7·8 × 10⁴	−16·6

Anmerkung zu Tabelle 5a-d:

   * Lösungsmittel E statt Methanol.

k_H- und ΔS-Werte bezogen auf unassoziiertes $^1O_2$. Die k_H für ein angenommenes Sens.$^{\text{rad.}}O_2$ ergeben sich durch Multiplikation mit einem Faktor von 0·15 bis 0·25, die entsprechenden ΔS-Werte sind um 3–5 cal/Mol Grad negativer (vgl. Gleichüngen (6) und (7)).

Aktionskonstanten durch Vergleich mit konstitutionell ähnlichen bei tiefen Temperaturen untersuchten Akzeptoren an[3,41] und berechneten die E_H-Werte nach Gleichung (3). Alle hier angegebenen E_H-Werte sind jedoch wie auch die Entropien nur Mittelwerte, da wegen des Auftretens mehrerer Reaktionszentren am Akzeptor verschiedenartige O_2-Anlagerungsprozesse erfolgen können. Eine Auftrennung ist nur durch die äusserst zeitraubende präparative Aufarbeitung[47] auf die verschiedenen Endprodukte möglich. Die hier angewandte Messmethodik liefert zu diesem Problem keinen Beitrag, daher sei auf die diesbezügliche Literatur verwiesen.[1,3,41]

    Aufallend sind die minimalen Werte für die Aktivierungsenergien, die für Substanzen wie Furan, 2,5-Dimethylfuran, Cyclopentadien um Null liegen. Bei den strukturell einfachsten Grundtypen, von denen sich alle geprüften olefinischen Substanzen

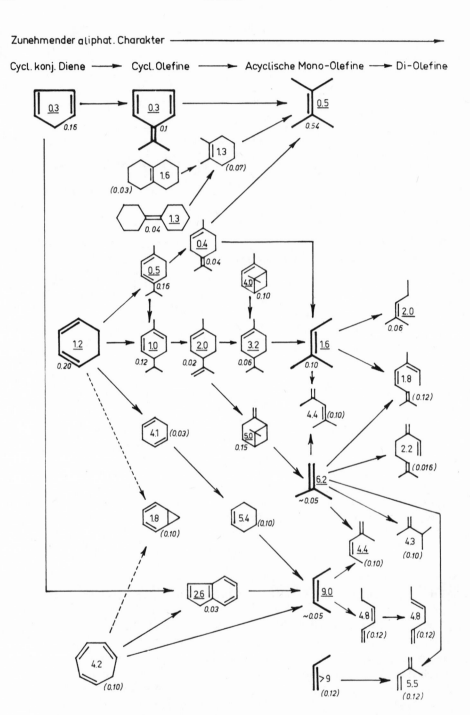

ABB. 4 Strukturabhängigkeiten von Aktivierungsenergien und Aktionskonstanten.

Schräge Ziffern = relative Stossausbeuten.

Gerade Ziffern = Aktivierungsenergien (unterstrichene aus Tieftemperatur-Messung).

ableiten lassen, steigen die Aktivierungsenergien in folgender Reihenfolge an:

Endoperoxid-Bildner:[10, 48, 49] Cyclopentadien, Cyclohexadien
(E_H = 0·3 bzw. 1·2 kcal/Mol)
Hydroperoxid-Bildner:[3, 50—53] 2,3-Dimethyl-buten-(2),
                                 2-Methyl-buten-(2),
                                 2-Methyl-propen-(1),
                                 Propen
(E_H = 0·5 bis ca. 10 kcal/Mol)

Zunehmende Ringgrösse hat ein Zurückgehen der Reaktivität zur Folge. Cyclo-heptatrien, das kein Endoperoxyd mehr bildet,[51, 54] ist aber noch wesentlich reaktiver als das offenkettige Buten-(2), während Cyclooctatetraen gar nicht reagiert, sondern eine starke Inhibitorwirkung für die sensibilisierte Sauerstoff-Übertragung zeigt.[40, 41, 55]

Bei Einführung von Heteroatomen in den Fünfring nimmt die Aktivierungsenergie in folgender Reihe zu:

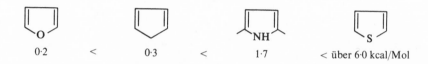

0·2   <   0·3   <   1·7   < über 6·0 kcal/Mol

Methylsubstitution an einer Doppelbindung setzt die Aktivierungsenergie herab.[3, 50—53] Die Verlängerung der C-Ketten bzw. ein Anschluss gesättigter Ringe an die den reaktiven Doppelbindungs-Zentren benachbarten Methylgruppen hat bei gut reagierenden Akzeptoren jedoch den umgekehrten Effekt, wie der Vergleich von Δ^1-Oktalin oder 1.2-Dimethyl-cyclohexen mit dem 2,3-Dimethylbuten-(2) zeigt (Abb. 4).[41, 52, 53]

Den Einfluss von Substituenten untersuchten wir[52, 56—59] u.a. in der Furanreihe (Tabelle 5c). Bei den 2-substituierten Furanen erhöht sich E_H gegenüber dem unsubstituierten Furan:

	H	CH₃	OH	CH₂NH₂	CH=CH₂	CH₂NHCH₃	OCH₃	CHO
E_H	0·2	< 0·4,	< 0·7	⩽ 0·7	⩽ 0·9	< 1·0	< 1·2	< 5·0 kcal/Mol

Die stark elektronegative Formylgruppe setzt die Reaktivität am stärksten herab. Eine Verlängerung der Seitenkette durch Methyl vermindert wiederum die Reaktions-fähigkeit.

Weitere Substituenten-Einflüsse zeigen folgende Beispiele:

(1a) Elektrophile Substituenten in der 5,6-Position des Cyclohexadiens-(1,3)

(zum Problem der Norcaradiencarbonester vgl. Lit.[60, 61]).

E_H 1·2   <   2·8   <   3·3   <   3·6 kcal/Mol

(1b) Elektrophile Substituenten im anellierten gesättigten Ring

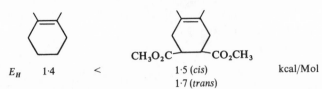

E_H 1·4   <   1·5 (*cis*)   kcal/Mol
1·7 (*trans*)

(2) Konjugation zur Carbonylgruppe

(a)

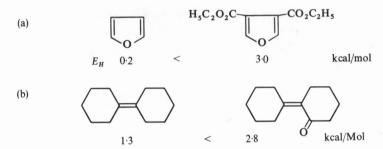

E_H 0·2   <   3·0   kcal/mol

(b)

1·3   <   2·8   kcal/Mol

(3) Konjugation oder Nachbarschaft von Doppelbindungen

(a)

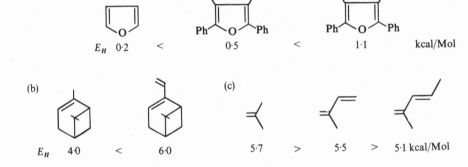

E_H 0·2   <   0·5   <   1·1   kcal/Mol

(b)   (c)

E_H 4·0   <   6·0   5·7   >   5·5   >   5·1 kcal/Mol

(4) Nachbarschaft zur Endoperoxydgruppe

(a)   (b)

5·4   <   5·8   0·5   <   3·4 kcal/Mol

(5) Aliphatische Kettenverlängerung (CH$_3$-Gruppe $\Delta E_H \sim$ 0·4 kcal/Mol) (vergl.[3, 53])

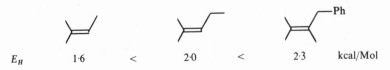

E_H 1·6   <   2·0   <   2·3   kcal/Mol

Aus den Aktivierungsenergien lassen sich über die Gleichung (3) und die Beziehung

$$k_i = (kT/h)\exp(\Delta S_i/R)\exp(-\Delta H_i/RT) \tag{8}$$

auch Aktionskonstanten bzw. Entropiewerte für die Hauptreaktion berechnen. Diese sind, im Gegensatz zu den rein empirischen Aktivierungsenergien, von der Grösse des betrachteten intermediären Moleküls abhängig und daher für Sens.^{rad}. O_2 und $^1O_2$ verschieden. Unter Berücksichtigung des Einflusses der Atomradien auf $\vec{k}_D$ und K (Gleichung 7) berechneten wir je nach Akzeptor Werte zwischen 5×10^8 und 5×10^9 sec^{-1} ($X = $ Sens.^{rad}. O_2) bzw. 10^9 und 10^{10} sec^{-1} ($X = {}^1O_2$) (vgl. Tabelle 4). Daraus ergeben sich Entropiedifferenzen ΔS von -20 bis -15 cal/Mol Grad (Sens.^{rad}. O_2) bzw. -18 bis -13 cal/Mol Grad ($^1O_2$) (Tabelle 5a–d). Die stark unterschiedliche Reaktivität der π-Bindungen—die k_H-Werte liegen zwischen 10^4 und 10^{10} sec^{-1}—ist also viel mehr durch unterschiedliche Aktivierungsenergien als durch unterschiedliche Aktionskonstanten bedingt. Nur bei einigen kaum noch merklich reagierenden Akzeptoren, deren Aktivierungsenergien wesentlich grösser als 5 kcal/Mol sind (z.B. Buten-(2), Furan-2-aldehyd, Anethol), ergeben sich anomal tiefe Entropie-Werte. Vermutlich beginnt in diesen Fällen die Konkurrenz von andersartig ablaufenden Mechanismen, so dass die hier vorausgesetzten kinetischen Gleichungen nicht mehr erfüllt sind. Da die Aktivierungsenergien für die Dehydrierung von Olefinen durch RO_2-Radikale zwischen ca. 5 und 13 kcal/Mol liegen[62–64] könnten hier photoinduzierte Kettenreaktionen nach dem RH-Schema einsetzen.

Die nur schwach ausgeprägte Abhängigkeit der Aktionskonstanten bzw. Entropien von der Struktur des Akzeptors wird deutlich beim Übergang von den cyclischen konjugierten zu den cyclischen nicht konjugierten Dienen. Statt des 1,3-Diensystems beim α-Terpinen und α-Phellandren ist beim, d-Limonen und Terpinolen bei unveränderter Molekülgrösse nur noch eine vergleichbar reaktionsfähige Doppelbindung vorhanden. Die Aktionskonstanten, die den sterischen Faktoren proportional sind, erreichen ihre höchsten Werte bei Furan und Cyclopentadien einerseits und den methyl-substituierten Äthylenen andererseits. Dies ist eine weitere experimentelle Stütze für die Vorstellung, dass die π-Elektronendichte im Molekül den Wirkungsquerschnitt bestimmt.[3, 41]

Die nur mit $\frac{1}{2}$ Mol O_2 reagierenden Sulfide[65 66] und Triphenylphosphin[40] (Tabelle 5d) zeigen mit Entropiewerten von -12 bis -18 cal/Mol Grad ($^1O_2$) und Aktivierungsenergien von 1·0 bis 1·5 kcal/Mol eine ähnliche Reaktivität wie gut reagierende Kohlenstoffverbindungen. Ein einsames Elektronenpaar am Schwefel oder Phosphor kann demnach die Rolle olefinischer π-Elektronen als Reaktionszentrum übernehmen.

Diskussion des Zwischenproduktes

Der Einfluss der verschiedenen Komponenten der untersuchten Systeme auf die Reaktionskonstanten erlaubt weitere Rückschlüsse auf das sauerstoffhaltige Zwischenprodukt X (Tabelle 6).

Die Annahme eines sauerstoffhaltigen Komplexes mit dem Akzeptor[14, 67] scheidet aus, da dann k_N eine starke Abhängigkeit vom Akzeptor zeigen sollte (Tabelle 2). Die Art des Sensibilisators verändert die Reaktionsgeschwindigkeitskonstanten k_H und k_N sowie die Aktivierungsenergien E_H und E_N nur so wenig, dass kein zwingender Grund besteht, für die hier vorgelegten Untersuchungen für X eine feste Bindung zwischen Sensibilisator und Sauerstoff anzunehmen. Immerhin ist aber ein deutlicher

Einfluss vorhanden: Die k_H- und k_N-Werte sind z.B. für Bengalrosa stets höher als für Methylenblau (Tabellen 2, 3 und 5).

Auch die Art des Lösungsmittels wirkt sich auf die beiden konkurrierenden Prozesse gleichartig aus (Tabellen 4 und 6). Da dieses jedoch in mindestens 100 mal so hoher Konzentration wie der Akzeptor vorliegt, ist angesichts der relativ geringen Abhängigkeit (1:60) auch eine feste Bindung von aktivem Sauerstoff an das Lösungsmittel nicht in Betracht zu ziehen. Es ist höchstens an eine assoziative Bindung zwischen Lösungsmittel und Sauerstoff zu denken, wie sie auch zwischen Lösungsmittel und Sensibilisator diskutiert wird.[68]

TABELLE 6.
MAXIMALE BEOBACHTETE ÄNDERUNGEN DER KINETISCHEN KONSTANTEN
(vergl. Tabellen 2–5)

		Änderung von		
	Sensibilisator	Lösungsmittel a	b	Akzeptor
Hauptreaktion k_H	1:4	($\sim$1:700)	1:10	>1:10 000
A_H	1:4	($\sim$1:60)	1:10	1:20
E_H	$\pm$0·4	(<$\pm$1·0)	<$\pm$1·0	<$\pm$5·0
Nebenreaktion k_N	1:4	1:14*	1:4*	1:1·4
A_N	1:4	1:10*	1:2*	1:1·2
E_N	$\pm$0·2	$\pm$0·5	$\pm$0·7	$\pm$0·7

* Bezogen auf eine bimolekulare Reaktion.
a Lösungsmittel der Tabelle 4.
b Lösungsmittel der Tabelle 4 ohne n-Hexanol, n-Heptanol, n-Oktanol und n-Decanol.

Es gäbe auch eine dynamische Deutung. Die Aktionskonstanten der Desaktivierungsreaktionen von X zeigen mit $\sim 10^7$ sec^{-1} für eine monomolekulare Reaktion ($k \approx 10^{10}$–10^{13} sec^{-1}) abnorm tiefe Werte. Dies kann darin begründet sein, dass es sich um einen verbotenen elektronischen Übergang handelt.[69, 70] Nimmt man zum anderen jedoch eine bimolekulare Desaktivierungsreaktion von X mit dem *unangeregten* Sensibilisator an, so ergeben sich bimolekulare Geschwindigkeitskonstanten um 1–2 $\times$ 10^{10} l/Mol sec^{-1}. Ein solcher Vorgang wäre also diffusionsbestimmt, und es müsste praktisch schon beim ersten Stoss mit dem Sensibilisator zur Desaktivierung kommen, während dies erst etwa beim zehntausendsten Stoss mit einem Lösungsmittelmolekül zu erwarten wäre. Wegen der stark unterschiedlichen Konzentrationen müssten beide Prozesse konkurrieren. Versuche mit erhöhten Sensibilisatorkonzentrationen zeigten jedoch, dass auch bei tiefen Temperaturen keine Erhöhung der Desaktivierungsgeschwindigkeit auftritt. Eine bimolekulare Desaktivierungsreaktion mit dem unangeregten Sensibilisator wäre daher nur zu diskutieren, wenn auch die konkurrierende Desaktivierung des Begegnungskomplexes $X \ldots A$ zu AO_2 (Schema 1) erst beim Stoss mit dem Sensibilisator abläuft.

212

Trotzdem stellt sich aufgrund unserer Resultate die Frage, ob das von Schenck formulierte Sens .$^{rad.}$ O_2 eine Lebensdauer aufweist, die über die eines reinen Begegnungskomplexes hinausgeht. Selbst wenn dies der Fall wäre, so würde man für dieses Teilchen andere Eigenschaften als die der Spezies X erwarten, die allein für die Sauerstoff-Übertragung verantwortlich ist. Der Sauerstoff in der Spezies X kann weder an Sensibilisator noch an Lösungsmittel noch an Akzeptor fest gebunden sein. Daher vertragen sich unsere Resultate besser mit der Annahme von elektronisch angeregtem Sauerstoff als Zwischenprodukt,[9, 30, 33, 34, 71–73] dessen Eigenschaften durch assoziierte Sensibilisator- oder Lösungsmittel-Moleküle geringfügig beeinflusst werden.

Der Sauerstoff könnte auch dimer auftreten. So wurde das O_4 zur Erklärung verschiedener Anomalien von UV- und IR-Spektrum, Paramagnetismus, spezifischer Wärme, Dampfdruck u.a. in flüssigem oder gasförmigem Sauerstoff diskutiert.[74–78]

Es scheint jedoch fraglich, ob sich die Verhältnisse in flüssigem Sauerstoff auf gelösten Sauerstoff, der selbst bei tiefstmöglichen Arbeitstemperaturen hier nur etwa in einem Dreihundertstel der Konzentration vorliegt,[40, 41] übertragen lassen. Da die Rückspaltung von O_4 mit steigender Temperatur zunimmt, wäre bei Vorliegen von O_4 als Zwischenprodukt die Temperatur-Unabhängigkeit der Desaktivierungsreaktion nicht zu verstehen. Auch müsste die Quantenausbeute der O_2-Aufnahme bezogen auf die O_2-Konzentration in der Lösung einem Zeitgesetz mit Termen 2. Ordnung folgen. Versuche von Gollnick und Franken, bei denen die O_2-Konzentration elektrochemisch verfolgt wurde, ergaben jedoch hierfür keine Hinweise.[1, 10, 47]

Die Gleichartigkeit der Einflüsse von Sensibilisator und Lösungsmittel auf die Peroxyd-Bildungs- und Desaktivierungsreaktion lässt sich zwanglos durch eine beiden Prozessen gemeinsame Veränderung erklären. Als solche existiert ausser der erwähnten sekundären Assoziation nur der Verlust elektronischer Anregungsenergie. Der Befund, dass bei Gegenwart von Rose bengale mit seinen schweren Halogenatomen beide Reaktionen schneller als bei "leichteren" Molekülen ablaufen, erinnert an die bekannten Schweratom-Einflüsse auf die Geschwindigkeit elektronischer Übergänge.[3, 70, 79–82] Analoge beschleunigende Wirkung haben polare Gruppen bzw. H-Brückenbindungen im Lösungsmittelmolekül. was besonders bei Aceton oder Alkohol enthaltenden Mischlösungsmitteln auffällt (Tabelle 4). Auch die Temperatur-Unabhängigkeit der Desaktivierungsreaktion ist typisch für pseudomonomolekulare Prozesse, die nur in elektronischen Veränderungen im Molekül bestehen.[3, 70]

Für die beobachtete geringe Sensibilisator-Abhängigkeit gäbe es noch eine andere Erklärung. Molekularer Sauerstoff kann unter den Bedingungen der photosensibilisierten O_2-Übertragung ggf. in zwei unterschiedlichen elektronischen Zuständen ($^1\Delta_g$ und $^1\Sigma_g^+$) als Zwischenprodukt auftreten,[3, 33, 34, 83–85] die je nach dem Energiegehalt des angeregten Sensibilisators in verschiedenem Verhältnis gebildet werden und unterschiedliche Lebensdauern und Reaktivitäten aufweisen sollen.[86, 87]

So sollte sich z.B. bei Gegenwart von Acridinorange, dessen Triplettenergie ($E_T \sim 51$ kcal/Mol) zur Bildung des $^1\Sigma_g^+$-Zustandes (37·7 kcal/Mol) ausreicht, für X eine geringere maximale Lebensdauer ergeben als bei Gegenwart von Methylenblau ($E_T \sim 33$) oder Rose bengale ($E_T \sim 44$), da der $^1\Sigma_g^+$-Zustand kurzlebiger ist als der $^1\Delta_g$-Zustand (22·6 kcal/Mol).[69, 78, 85, 87]

Das Gegenteil ist jedoch der Fall: Bei Gegenwart von Acridinorange zeigt X die geringste Desaktivierungsgeschwindigkeit bzgl. der hier geprüften Sensibilisatoren (Tabelle 3). Das spricht für unsere Versuche gegen die Vorstellung, dass für das stationäre Konzentrationsverhältnis beider Zustände nur die Triplett-Energie des Sensibilisators massgebend ist.[88] Der angeregte Sensibilisator, der sich nach der sehr schnellen Thermalisierung zunächst im $^1S_1$-Zustand befindet, kann neben Prozessen wie Fluoreszenz und "intersystem crossing" zum Triplett-Zustand auch direkt bimolekulare Reaktionen, ggf. mit bereits assoziiertem Sauerstoff, eingehen.[3, 40, 41]

Das stationäre Verhältnis beider O_2-Zustände sollte daher nicht nur von der Triplettenergie des Sensibilisators, sondern auch von dem Verhältnis der Geschwindigkeitskonstanten aller weiteren am angeregten Sensibilisator beteiligten Prozesse abhängig sein. Die Frage, ob es sich bei dem oxydierenden Zwischenprodukt stets um den $^1\Delta_g$-Zustand des Sauerstoffs handelt oder ob der $^1\Sigma_g^+$-Zustand statistisch mitbeteiligt ist, kann daher vorerst nicht beantwortet werden, zumal ein unterschiedliches kinetisches Verhalten der beiden Sauerstoff-Arten beim Stoss mit dem Akzeptor nicht unbedingt zu erwarten ist.

Die vorliegenden Untersuchungen mit polychromatischem Licht geben nur über die Temperaturabhängigkeiten der Brutto-Terminationsreaktionen der photosensibilisierten Sauerstoff-Übertragungen Auskunft. Bei zahlreichen andersartigen photosensibilisierten Prozessen in Lösung sollten ähnliche Verhältnisse vorliegen. Treten jedoch zwei parallel reagierende Zwischenprodukte auf, so scheint eine getrennte Untersuchung ihrer Eigenschaften auf dem hier geschilderten Weg nur möglich, wenn zusätzlich mit monochromatischem Licht gearbeitet wird.[89, 90] Durch Licht der Wellenlänge >600 nm kann aus energetischen Gründen nur der $^1\Delta_g$-Sauerstoff entstehen,[29, 33, 91] falls durch Wahl geeigneter Sensibilisatoren überhaupt eine Reaktion erfolgt. Wie die Ergebnisse erster Versuche in dieser Richtung zeigen,[90] ist das zwar prinzipiell ein gangbarer Weg. Es bleibt jedoch abzuwarten, ob bei den technisch bedingten minimalen Umsätzen (vor allem bei tiefsten Temperaturen) eine ausreichende Gas-Messgenauigkeit erreicht werden kann, um die Lage der "Diffusionsgeraden" zu ermitteln und daraus die maximale Lebensdauer des $^1\Delta_g$-Sauerstoffs abzuschätzen.

EXPERIMENTELLER TEIL

Akzeptoren. Handelsüblich "rein" und vor den Versuchsserien destilliert bzw. umkristallisiert. Folgende Verbindungen wurden selbst hergestellt:

2-Vinyl-furan aus Furyl-acrylsäure durch CO_2-Abspaltung $n_D^{20} = 1.4972$ (Lit. 1.4992^{92})

Furfuryl-methyläther aus Furfurylalkohol und Methyljodid bei Gegenwart von KOH $n_D^{20} = 1.4556$ (Lit. $1.4570^{92, 93}$)

2·4-Dimethylfuran aus Mesityloxyd, Acetanhydrid und H_2SO_4 und Zersetzen des entstandenen Sultons mit CaO^{94-96} $n_D^{20} = 1.4472$

Furan-3,4-dicarbonsäure-dimethylester aus dem Addukt von Furan an Acetylencarbonsäure-ester durch Thermolyse.[97]

Die Furane waren z.T. durch Hydrochinon stabilisiert und wurden nach Abdestillieren höchstens kurzzeitig unter Stickstoff im Kühlschrank bei $-25°$ aufbewahrt.

ω-Dimethylfulven aus Cyclopentadien und Aceton bei Gegenwart von methanolischer KOH $n_D^{20} = 1.5055$, Kp. $44.5°/12$ Torr (Lit. $46°/11$ Torr.[98, 99])

Dicyclohexyliden nach[100]

Cyclohexyliden-Cyclohexanon aus Cyclohexanon und konzentrierter HCl durch Zersetzung des entstandenen Chlorcyclohexylcyclohexanons-(2) mit Natriummethylat[53, 100]

ω-Dimethylfulven-endoperoxyd durch photosensibilisierte O_2- Übertragung auf ω-Dimethylfulven bei −80°[98]

3,6-Endoperoxy-cyclohexen analog aus Cyclohexadien-(1,3)[48]

Lösungsmittel. Handelsüblich "rein", destilliert und z.T. getrocknet. Geringe Verunreinigungen haben auf die kinetischen Konstanten meist keinen Einfluss.

Sensibilisatoren. Methylenblau, Eosin, Natrium-Chlorophyllin, Acridinorange, handelsüblich (Merck, Darmstadt). Die Dinatriumsalze von Tetrachlorfluoreszein, Tetrachloreosin, Rose bengale (= Tetrajod-tetrachlorfluoreszein) aus Resorcin und Tetrachlor-phthalsäure-anhydrid, nachfolgende Bromierung oder Jodierung. Reinigung durch Umkristalisation der Diacetate.[47]

Binaphthylenthiophen, hergestellt von G. Helms.[101]

Hochreines chromatographisch geprüftes Rose bengale zeigt praktisch dieselben Reaktionskonstanten wie das sehr unreine Handelsprodukt (Merck, Darmstadt).

Belichtungsapparaturen. Die ersten Belichtungen erfolgten in der bereits beschriebenen[39, 40, 102−104] Tieftemperatur-Tauchlampen-Apparat (Philips-Lampe HPK 125 W in Dewarschacht, Temperatur-konstanz in der Lösung ±0·4° bei Temperaturen bis −150°C, automatische Registrierung des aus Büretten von 150 bis 6000 ml verbrauchten, umgepumpten reinen Sauerstoffs). Reaktionsgefässe von 150 bis 250 ml Inhalt mit Schichtdicken von 6–8 mm wurden verwendet.

Die Versuche mit α-Terpinen, Cyclohexadien-(1,3), Cyclopentadien, α-Phellandren, 2-Methyl-buten-(2), α-Pinen, Isobutylen, Buten-(2), Diäthanol- und Dibenzylsulfid, mit allen Furanen sowie mit Tetrachloreosin, Tetrachlorfluoreszein, Binaphthylenthiophen und Acridinorange wurden in einer stark verbesserten vollautomatischen Apparatur vorgenommen. Ein Teil der ersten Versuche wurde reproduziert.

Lichtquelle: Wassergekühlte Tauchlampe Philips HPK 125 W in einem Solidex- (oder Quartz-) Schacht, der vom 250 ml-Reaktionsgefäss (Solidex-Glas oder verchromtes Messing) umgeben ist. Der Innenschacht ist gegen Farbglasschächte (Wertheim) austauschbar.

Temperierung der Reaktionslösung: Aussenkühlung durch eingespritzten flüssigen Stickstoff. Durch Vorregelung (Temperaturkontrolle des abziehenden Kalt-Stickstoffs) und Hauptregelung (automatische Verstellung des Sollwertes für die Vorregelung durch Widerstandsthermometer in der Reaktionslösung, das über eine Brückenschaltung mit Zeitverzögerung ein Motorpotentiometer ansteuert) wird eine Temperaturkonstanz auf ca. ±0·2° bei Temperaturen von −40 bis −140° erzielt.

Rührung und Gasversorgung: Mit einem Kompressor wird ein starker Gasstrom (bis 7 l O_2/Min.) durch die Lösung gepumpt ("Pari-Optimal-Kompressor" der Fa. P. Ritzau, Starnberg am See. Im Institut umgebaut, um ausreichende Gas-Dichtigkeit zu erreichen).

Messung des Gasverbrauchs: Manometrisch mit einer neuartigen, von F. Schaller[105] erstmalig erprobten und von mir veränderten Anordnung. Eine ventil-zwangsgesteuerte Glas-Dosierpumpe pumpt Öl aus einem thermostatisierten, offenen Gefäss in ein geschlossenes Gasometergefäss (6 l), das auf dem Öl schwimmt. Die Umdrehungen des Pumpen-Motors werden über ein Getriebe (Multurgetriebe der Fa. Halstrup, Kirchzarten, Übersetzung in 10 Stufen bis 1:1000 variierbar) und ein endloses Potentiometer auf einen Kompensationsschreiber (Hartmann und Braun, Frankfurt a.M.) winkelgetreu übertragen. Die Steuerung des Motors erfolgt bei Über- bzw. Unterdruck im Gaskreislauf über ein Kontaktmanometer.[22]

Reproduzierbarkeit der Gasvolumina: besser als ±0·2%.

Gasabgabe- oder -aufnahme-Geschwindigkeit: ∼0·01 bis 200 ml/Min.

Zentrale Steuerung: Nach Einfüllen der zu belichtenden Lösung und Einstellen der Sollwerte für Temperaturregelung und Gasregistrierung kann das gesamte Versuchsprogramm vollautomatisch ablaufen:

1. Abkühlen der Lösung auf die Belichtungs-Temperatur und Spülen mit Sauerstoff
2. Einstellen des Gleichgewichtes zwischen Gas und Lösung
3. Kontrollperiode zur Registrierung eventueller Dunkelreaktionen oder Undichtigkeiten
4. Belichtung und Registrierung von Gasverbrauch und Temperatur.

Bei Störungen (Mangel an flüssigem Stickstoff, Zufrieren der Begasungsfritte, Ausfall der Lampenkühlung, nicht ausreichender O_2-Vorrat im Gasometer u.a.) erfolgt Abschaltung der Apparatur und Warnsignal.

Versuchstechnik.[40, 106] Bei schnell reagierenden Akzeptoren (>5 ml O_2/Min.): Bei anfänglich 0·02–0·1 Mol/l Akzeptor und 5 × 10^{-4} Mol/l Sensibilisator wurde bis zum Aufhören der Gasaufnahme belichtet.

215

Ergab sich bei einer weiteren Belichtung bei gleicher Temperatur, aber anderer Anfangs-Akzeptor-konzentration eine deckungsgleiche O_2-Registrierkurve (was im allgemeinen zutraf), so konnten mehrere Versuche bei verschiedenen Temperaturen mit derselben Lösung und nach jeweils wiederholter Akzeptor-Zugabe durchgeführt werden, ohne dass ein Einfluss des gebildeten Peroxyds auf die Kinetik zu befürchten war. Auswertung wie in Abb. 1.

Bei langsamer reagierenden Akzeptoren (<5 ml/Min.) wurde bei fester Akzeptor-Anfangs-Konzentration und verschiedenen Temperaturen nur kurzzeitig bestrahlt. Dann wurden neue Lösungen mit anderer Anfangs-Konzentration verwendet. So waren die Geschwindigkeiten aus den Registrierkurven zu berechnen, ohne dass sich von einer Arbeitstemperatur zur nächsten die Akzeptor-Konzentration wesentlich änderte. Die Werte ergaben sich nun nach der Formel:

$$\beta = [A]\left(\frac{\alpha}{\dot{n}_{O_2}} - 1\right)$$

Die nur vom Sensibilisator abhängigen α-Werte[16, 17, 39, 40, 106] waren durch Versuche mit schnellen Akzeptoren bereits bekann[e].

Viskositätsmessungen: Durch Entwicklung eines Tieftemperatur-Kleinthermostaten waren die Viskositäten mit einem Rotations-Viskosimeter ("Rotavisko", Fa. Haake, Berlin, Messkopf 50, Messkörper NV, Messbereich ca. 0·2 bis 1000 Centi-Poise) bis $-135°$ zu messen. Mit Literaturangaben über Viskositäten ergab sich gute Übereinstimmung.

Danksagung: Herrn Prof. Dr. G. O. Schenck danke ich für viele anregende Diskussionen und Hinweise.

LITERATUR

[1] K. Gollnick und G. O. Schenck, *Pure appl. Chem.* **9**, 507 (1964).

2 R. Livingston, "Photochemical Butoxidation" in W. O. Lundberg, *Autoxidation and Antioxidants*. Vol. I: S. 249. Interscience, New York (1961).

[3] K. Gollnick "Type II Photo-Oxygenation Reactions in Solution", in W. A. Noyes, Jr., G. S. Hammond and J. N. Pitts, Jr., *Advances in Photochemistry*, Vol. VI. Interscience. New York, in press.

[4] G. O. Schenck, *Angew. Chem.* **61**, 389 (1949).

[5] G. O. Schenck, *Ibid.* **69**, 579 (1957).

[6] G. O. Schenck, H.-D. Becker, K.-H. Schulte-Elte und C. H. Krauch, *Chem. Ber.* **96**, 509 (1963).

[7] G. O. Schenck, *Z. Elektrochem.* **56**, 855 (1952).

[8] C. Dufraisse, *Experientia* **6**, 312 (1950).

[9] Yu. A. Arbuzov, *Russ. Chem. Revs.* **34**, 558 (1965).

[10] K. Gollnick und G. O. Schenck "Oxygen as a Dienophile" in J. Hamer, *1.4-Cycloaddition Reactions: The Diels–Alder Reaction in Heterocyclic Syntheses*. S. 255. Academic Press, New York (1967).

[11] G. O. Schenck, *Naturwissenschaften* **35**, 28 (1948).

[12] G. O. Schenck, *Z. Elektrochem., Ber. Bunsenges. physik. Chem.* **55**, 505 (1951).

[13] G. O. Schenck, *Angew. Chem.* **63**, 286 (1951).

[14] H. Gaffron, *Ber. Dtsch. Chem. Ges.* **68**, 1409 (1935).

[15] A. Schönberg, *Ibid.* **67**, 633 (1934).

[16] H. Mertens, Dissertation, Göttingen (1953).

[17] G. O. Schenck, K. G. Kinkel und H.-J. Mertens, *Liebigs Ann.* **584**, 125 (1953).

[18] A. Schönberg, *Ibid.* **518**, 299 (1935).

[19] A. N. Terenin, *Acta Physicochim. URSS* **18**, 210 (1943).

[20] A. N. Terenin, *Photochemistry of Dyes and Related Organic Compounds*. Chap. 7. Academy of Sciences Press, Moscow and Leningrad (1947).

[21] G. O. Schenck, *Z. Naturforsch.* **3b**, 59 (1948).

[22] G. O. Schenck und K. Kinkel, *Naturwissenschaften* **38**, 355 (1951).

[23] G. O. Schenck, *Ibid.* **40**, 205, 229 (1953).

[24] G. O. Schenck und K. H. Ritter, *Ibid.* **41**, 374 (1954).

[25] G. O. Schenck, *Z. Elektrochem.* **64**, 997 (1960).

[26] E. J. Bowen, in W. A. Noyes, Jr., G. S. Hammond and J. N. Pitts, Jr., *Advances in Photochemistry* Vol. I, S. 23. Interscience, New York (1963).

[27] H. Kautsky und H. de Bruijn, *Naturwissenschaften* **19**, 1043 (1931).

[28] H. Kautsky, H. de Bruijn, R. Neuwirth und W. Baumeister, *Ber. Dtsch. Chem. Ges.* **66**, 1588 (1933).

[29] H. Kautsky, *Biochem. Z.* **291**, 271 (1937).

[30] K. R. Kopecky and H. J. Reich, *Canad. J. Chem.* **43**, 2265 (1965).

[31] A. Nickon und J. F. Bagli, *J. Am. Chem. Soc.* **83**, 1498 (1961).

[32] C. S. Foote und S. Wexler, *Ibid.* **86**, 3879 (1964).

[33] C. S. Foote und S. Wexler, *Ibid.* **86**, 3880 (1964).

[34] E. J. Corey und W. C. Taylor, *Ibid.* **86**, 3881 (1964).

[35] C. S. Foote, S. Wexler und W. Ando, *Tetrahedron Letters* 4111 (1965).

[36] E. McKeown und W. A. Waters, *J. chem. Soc.* B, 1040 (1966).

[37] R. W. Murray und M. L. Kaplan, *J. Am. Chem. Soc.* **90**, 537 (1968).

[38] G. O. Schenck, K. G. Kinkel und E. Koch, *Naturwissenschaften* **41**, 425 (1954).

[39] G. O. Schenck und E. Koch, *Z. Elektrochem.* **64**, 170 (1960).

[40] E. Koch, Dissertation, Göttingen (1957).

[41] G. O. Schenck und K. Gollnick, *Forschungsberichte des Landes Nordrhein-Westfalen*, Nr. 1256 (1963).

[42] W. Jost, *Diffusion in Solids, Liquids, Gases*, Academic Press, New York (1955).

[43] H. Umstätter, *Erdöl und Kohle* **8**, 791 (1955).

[44] M. Eigen, *Z. physik. Chem.* N.F., **1**, 176 (1954).

[45] D. F. Othmer and M. S. Thakar, *Ind. Engng. Chem.* **45**, 589 (1953).

[46] C. R. Wilke und P. Chang, *Amer. Inst. Chem. Engng. J.* **1**, 264 (1955).

[47] T. Franken, noch unveröffentlich[e].

[48] W.-D. Willmund, Diplomarbeit, Göttingen (1951).

[49] G. O. Schenck und D. E. Dunlap, *Angew. Chem.* **68**, 248 (1956).

[50] W. Eisfeld, Dissertation Göttingen (1965).

[51] E. Koerner von Gustorf, Dissertation, Göttingen (1957).

[52] G. O. Schenck und K.-H. Schulte-Elte, *Liebigs Ann.* **618**, 185 (1958).

[53] K.-H. Schulte-Elte, Dissertation, Göttingen (1961).

[54] G. O. Schenck, E. Koerner von Gustorf, B. Kim, G. von Bünau und G. Pfundt, *Angew. Chem.* **74**, 510 (1962).

[55] G. O. Schenck und K. Gollnick, *J. Chim. Phys.* **55**, 892 (1958).

[56] G. O. Schenck, *Liebigs Ann.* **584**, 156 (1953).

[57] E. Koch und G. O. Schenck, *Chem. Ber.* **99**, 1984 (1966): dort weitere Hinweise.

[58] C. S. Foote, M. T. Wuesthoff, S. Wexler, I. G. Burstain, R. Denny, G. O. Schenck und K.-H. Schulte-Elte, *Tetrahedron* **23**, 2583 (1967).

[59] E. Koch, *Chemie-Ing.-Technik* **37**, 1004 (1965).

[60] G. O. Schenck und H. Ziegler, *Liebigs Ann.* **584**, 221 (1953).

[61] G. O. Schenck und A. Ritter, unveröffentlich[e].

[62] J. L. Bolland, *Quart. Rev:* **3**, 1 (1949).

[63] L. Bateman, *Ibid.* **8**, 147 (1954).

[64] F. O. Rice und T. A. Vanderslice, *J. Am. Chem. Soc.* **80**, 291 (1958).

[65] G. O. Schenck und C. H. Krauch, *Angew. Chem.* **74**, 510 (1962).

[66] C. H. Krauch und G. O. Schenck, *Chemie-Ing. Technik* **36**, 978 (1964).

[67] H. Gaffron, *Biochem. Z.* **287**, 130 (1936).

[68] R. Steinmetz, W. Hartmann und G. O. Schenck, *Chem. Ber.* **98**, 3854 (1965).

[69] M. Kasha, *Discuss. Faraday Soc.* **9**, 14 (1950).

[70] S. K. Lower und M. A. El-Sayed, *Chem. Revs.* **66**, 199 (1966).

[71] H. Tsubomura und R. S. Mulliken, *J. Am. Chem. Soc.* **82**, 5966 (1960).

[72] C. S. Foote, *Chem. Engng. News* **43**, No. 16, 41 (1965).

[73] T. Wilson, *J. Am. Chem. Soc.* **88**, 2898 (1966).

[74] W. Finkelnburg und W. Steiner, *Z. Physik* **79**, 69 (1932).

[75] J. W. Ellis und H. O. Kneser, *Ibid.* **86**, 583 (1933).

[76] A. U. Khan und M. Kasha, *J. Chem. Phys.* **39**, 2105 (1963).

[77] S. J. Arnold, E. A. Ogryzlo und H. Witzke, *Ibid.* **40**, 1769 (1964).

[78] J. S. Arnold, R. J. Browne und E. A. Ogryzlo, *Photochem. Photobiol.* **4**, 963 (1965).

[79] M. Kasha, *J. Chem. Phys.* **20**, 71 (1952).

[80] R. Livingston und K. E. Owens, *J. Am. Chem. Soc.* **78**, 3301 (1956).

[81] L. Lindqvist und G. W. Lundeen, *J. Chem. Phys.* **44**, 1711 (1966).

[82] E. Clementi und M. Kasha, *Ibid.* **26**, 956 (1957).

217

[83] D. R. Kearns, R. A. Hollins, A. U. Khan, R. W. Chambers and P. Radlick, *J. Am. Chem. Soc.* **89**, 5455 (1967).

[84] G. Porter und F. Wilkinson, in H. P. Kallmann und G. M. Spruch, *Luminescence of Organic and Inorganic Materials* S. 132. Wiley, New York (1962).

[85] A. M. Winter and K. D. Bayes, *J. Phys. Chem.* **70**, 302 (1966).

[86] C. S. Foote, *Acc. Chem. Res.* **1**, 104 (1968).

[87] D. R. Kearns, R. A. Hollins, A. U. Khan und P. Radlick, *Ibid.* **89**, 5456 (1967).

[88] P. J. Wagner und G. S. Hammond, in W. A. Noyes, Jr., G. S. Hammond und J. N. Pitts, Jr., *Advances in Photochemistry*, Vol. V. S. 21. Interscience, New York (1968).

[89] G. O. Schenck, E. Koch und F. Schaller, *Chemie-Ing.-Technik* **34**, 654 (1962).

[90] E. Koch und F. Schaller, noch unveröffentlicht.

[91] J. Hackikama, M. Imotu und H. F. Rickert, *J. Soc. Chem. Ind. Japan* **45**, 189 (1942).

[92] A. P. Dunlop und F. N. Peters, *The Furans.* S. 230. Reinhold, New York (1953).

[93] T. Morel und P. E. Verkade, *Rec. Trav. Chim.* **67**, 539 (1948).

[94] T. Morel und P. E. Verkade, *Ibid.* **68**, 619 (1949).

[95] T. Morel und P. E. Verkade, *Ibid.* **70**, 35 (1951).

[96] M. Fetizon und J. Guy, *C.R. Acad. Sci. Paris* **247**, 1182 (1958).

[97] K. Alder und H. F. Rickert, *Ber. Dtsch. Chem. Ges.* **70**, 1354 (1937).

[98] J. Hasselmann, Dissertation, Göttingen (1952).

[99] J. Thiele, *Ber. Dtsch. Chem. Ges.* **33**, 666 (1900).

[100] K.-H. Schulte-Elte, Diplomarbeit, Göttingen (1958).

[101] G. Helms, Diplomarbeit, Göttingen (1957).

[102] G. O. Schenck, *Dechema-Monographien* **24**, 105 (1955).

[103] G. O. Schenck in A. Schönberg, *Präparative organische Photochemie* S. 210. Berlin, Springer (1958).

[104] E. Koch, *Tetrahedron* **23**, 1474 (1967).

[105] F. Schaller, Deutsches Bundes-Patent angemeldet.

[106] E. Koch, Diplomarbeit, Göttingen (1955).

Editor's Comments
on Papers 38 Through 40

DIRECT EVIDENCE FOR THE PHOTOSENSITIZED FORMATION OF $^1\Delta_g$ MOLECULAR OXYGEN

Although the evidence for the role of $^1\Delta_g$ oxygen in photooxygenations was relatively strong, there was, nevertheless, no *direct* evidence that singlet oxygen was formed by energy transfer from electronically excited molecules to ground-state oxygen. Support for the Kautsky mechanism came from comparisons of the oxidation products formed by photosensitized oxygenation with those products obtained from oxidation of the same substrates with singlet oxygen generated via NaOCl + H_2O_2 or the electric discharge. The kinetics of photooxygenation reactions were also consistent with the intermediacy of singlet molecular oxygen. It had been shown by many investigators that oxygen quenches excited triplet molecules (phosphorescence quenching); however, there was no evidence that this process was accompanied by $^1\Delta_g$ O_2 formation.

Papers 38–40 report the results of several investigations in which the authors were able to demonstrate that singlet oxygen is formed by energy transfer from electronically excited molecules, as originally proposed by Kautsky. In 1968, Snelling (Paper 38) observed that irradiation of benzene-oxygen mixtures with 2537-Å light resulted in emission at 1.27μm, the characteristic band of the $^1\Delta_g \longrightarrow {}^3\Sigma_g^-$ transition. At low pressures, singlet oxygen is formed exclusively from quenching of triplet benzene. However, at pressures above 1 torr, energy transfer from singlet-excited benzene to yield $^1\Delta_g$ oxygen and triplet benzene becomes important. Kearns (Paper 39) and Wasserman (Paper 40) and coworkers used epr spectroscopy to detect the photosensitized formation of $^1\Delta_g$ oxygen with naphthalene and naphthalene-derivatives sensitization.

219

38

Reprinted from *Chem. Phys. Lett.*, **2**(5), 346–348 (1968)

PRODUCTION OF SINGLET OXYGEN
IN THE BENZENE OXYGEN PHOTOCHEMICAL SYSTEM

D. R. SNELLING

Canadian Armament Research and Development Establishment and
Centre de recherches sur les atomes et les molécules,
P. O. Box 1427, Quebec, P. Q.

Received 19 August 1968

Photochemical production of $O_2(^1\Delta_g)$ in benzene oxygen mixtures has been observed directly and is attributed to quenching of triplet benzene with a rate constant of 5×10^{-11} cm^3 mol^{-1} sec^{-1}. The quenching of singlet benzene by oxygen is also discussed.

1. INTRODUCTION

Photooxygenation reactions have been interpreted [1, 2] by a mechanism involving the excited singlet states of oxygen ($^1\Delta_g$, $^1\Sigma_g^+$) which can be formed by electronic energy exchange from the triplet state of the donor molecule. The quenching of triplet state molecules by oxygen has been examined theoretically [3] and the calculations indicate that quenching by electronic energy transfer is usually 100 - 1000 times faster than quenching by enhanced intersystem crossing in the donor molecule.

The radiative lifetime of the $^1\Delta_g$ and $^1\Sigma_g^+$ states of oxygen (henceforth called $^1\Delta$ and $^1\Sigma$ respectively) are 45 min [4] and 7 sec [5] respectively, and, because of these long lifetimes, these species will decay by non-radiative processes under most laboratory conditions.

2. EXPERIMENTAL

The reaction vessel is a quartz tube 7.5 cm in diameter and 40 cm long surrounded by six 88A45 low pressure Hanovia mercury lamps. The spectrometer and detectors used have been described previously [6].

3. RESULTS AND DISCUSSION

In benzene oxygen mixtures irradiated with 2537 Å radiation we have observed emission at 1.27 μ showing the characteristic P, Q and R branches of the (0, 0) band of the $O_2(^1\Delta_g - {}^1\Sigma_g^-)$ system. The identification has been confirmed by comparison of these spectra with those from $O_2(^1\Delta)$ produced from a microwave discharge in oxygen. This emission is not observed at any oxygen pressure in the absence of benzene so that direct excitation of the $O_2(^1\Delta)$ can be ruled out. We have been unable to detect any emission in the region of the (0, 0) band of the $O_2(^1\Sigma_g^+ - {}^3\Sigma_g^-)$ system at 7619 Å. The variation of the $O_2(^1\Delta)$ emission intensity at 1.27 μ as a function of oxygen pressure, in mixtures of benzene and oxygen containing 20 Torr of benzene, is shown in fig. 1.

3.1. *Fate of singlet oxygen in our system*

We conclude from our observations that the singlet oxygen is formed by quenching of electronically excited benzene molecules. Our results do not preclude the initial formation of $O_2(^1\Sigma)$ rather than $O_2(^1\Delta)$ since there is some indication [7] that

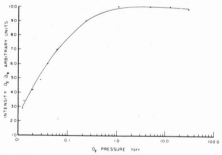

Fig. 1. Intensity of $O_2(^1\Delta)$ emission at 1.27μ versus oxygen pressure for benzene oxygen mixtures containing 20 Torr of benzene.

benzene deactivates $O_2(^1\Sigma)$ to $O_2(^1\Delta)$ and the rates of $O_2(^1\Sigma)$ decay processes [8] are such that deactivation by benzene will probably predominate under all conditions used in this work.

However, the corresponding deactivation reactions (1) - (3) of $O_2(^1\Delta)$ are much slower [8] although k_2 has not been measured.

$$O_2(^1\Delta) + \text{wall} \rightarrow O_2 \qquad (1)$$

$$O_2(^1\Delta) + C_6H_6 \rightarrow O_2 + C_6H_6 \qquad (2)$$

$$O_2(^1\Delta) + O_2 \rightarrow 2\,O_2 \qquad (3)$$

From the value of k_1 and the upper limit to k_3 we conclude that reaction (3) will only be important at oxygen pressures ≥ 5 Torr. Preliminary results at 10 Torr of benzene show no change in the $O_2(^1\Delta)$ intensity of the plateau region (see fig. 1) and, since benzene is optically thick at these pressures, this result suggests that reaction (2) has little effect on the $O_2(^1\Delta)$ steady state concentration. From the rate data we estimate that the $O_2(^1\Delta)$ lifetime is approximately 10^6 that of $O_2(^1\Sigma)$ in our system.

We have observed the emission from discharged oxygen (see also ref. [6]) with the same detection system used in this work. The 0.7619 and $1.27\,\mu$ emissions were observed with similar signal to noise ratios at 2 Torr oxygen and the $1.27\,\mu$ intensity was approximately 50 times that observed in the benzene system. A comparison of these results with the absence of emission at 0.7619μ in the benzene system might be taken to indicate that $O_2(^1\Sigma)$ was not formed in the latter. However, since benzene probably deactivates $O_2(^1\Sigma)$ some 200 times faster than oxygen [8], the rate of decay of $O_2(^1\Sigma)$ in 20 Torr of benzene will be 2000 times faster than in 2 Torr of oxygen. Furthermore, the scattered light from an emission line in the Hanovia mercury lamp falls under one branch of the $O_2(^1\Sigma)$ $0.7619\,\mu$ emission and makes detection of $O_2(^1\Sigma)$ difficult. Thus, our observations indicate either that $O_2(^1\Delta)$ is produced in the quenching of excited benzene or that $O_2(^1\Sigma)$ is formed initially and is rapidly deactivated to $O_2(^1\Delta)$ which then decays more slowly.

3.2. Production of singlet oxygen

The relevant energy levels of benzene and oxygen are shown in fig. 2 and it can be seen that, from energetic consideration alone, the reactions which could produce $O_2(^1\Delta)$ or $O_2(^1\Sigma)$ are:

$$C_6H_6(S_1) + O_2 \rightarrow O_2(^1\Delta \text{ or } ^1\Sigma) + C_6H_6(S_0) \qquad (4)$$

$$C_6H_6(S_1) + O_2 \rightarrow O_2(^1\Delta) + C_6H_6(T_1) \qquad (5)$$

$$C_6H_6(T_1) + O_2 \rightarrow O_2(^1\Delta \text{ or } ^1\Sigma) + C_6H_6(S_0) \qquad (6)$$

The detailed mechanism of $C_6H_6(S_1)$ decay in pure benzene is not completely resolved and has been found to depend on the vibrational level of the S_1 state to which the molecule is excited [9]. At the wavelength we have used (2537 Å from a low pressure mercury arc) the most likely quantum yields [9] of singlet fluorescence and triplet formation by intersystem crossing are 0.18 and 0.63 respectively, leaving 19% of the molecules unaccounted for. However the above triplet quantum yield has been criticised [10] and the data may not be sufficiently accurate [9] to indicate the presence of a third process.

The quenching of $C_6H_6(S_1)$ fluorescence by oxygen in 20 Torr of benzene has been studied [11] over a wide range of oxygen pressures (at oxygen pressures of 1, 11 and 30 Torr the fractions of singlets quenched are 0.32, 0.76 and 0.94 respectively) although the products of the reaction have not been established. We have confirmed these results at several oxygen pressures

However, the results of fig. 1 show that the quenching reaction producing $O_2(^1\Delta)$ is complete at an oxygen pressure of 1 Torr and that no further increase in $O_2(^1\Delta)$ intensity is observed up to an O_2 pressure of 30 Torr. We conclude therefore that the singlet oxygen is produced by quenching of triplet benzene, reaction (6), rather than the singlet benzene quenching reactions (4) and (5). Reactions (4) and (5) may be occurring at higher oxygen pressures and will be discussed later.

Using our data and the triplet lifetime [12] in pure benzene ($2.6 \pm 0.5 \times 10^{-5}$ sec in 20 Torr of benzene) we can deduce a rate constant for the quenching reaction (6). In the plateau region, quenching of $C_6H_6(T_1)$ by O_2 is complete but, as the O_2 pressure is lowered, this quenching reaction competes with other processes which lead to triplet decay. At the O_2 pressure at which the

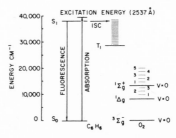

Fig. 2. Energy level diagram for benzene and oxygen.

$O_2(^1\Delta)$ intensity has dropped to 0.5 of its plateau value (0.025 Torr) the rates of these competing reactions are equal. If we assume that the rate of decay of triplets by processes not involving $O_2(^1\Delta)$ production is that observed in pure benzene (we are assuming that the only quenching reaction involving O_2 is that which leads to $O_2(^1\Delta)$ production) then we can equate the rate of quenching by $O_2(k_6(O_2) \text{ sec}^{-1})$ to the triplet decay rate in the absence of O_2 ($1/2.6 \times 10^{-5} \text{ sec}^{-1}$). This gives a value of k_6 of $4.8 \times 10^{-11} \text{ cm}^3 \text{ mol}^{-1} \text{ sec}^{-1}$ which is some 10 times faster than the rate of quenching of triplet naphthalene and anthracene [13] and approximately $\frac{1}{10}$ the hard sphere collision rate. If there is another oxygen quenching reaction not involving $O_2(^1\Delta)$ production this would lead to a lower value of k_6.

Preliminary results at 10 Torr of benzene give a curve similar to fig. 1 but moved to lower oxygen pressures. The half intensity point comes at 0.016 Torr compared to 0.025 Torr in fig. 1. This result suggests that the $C_6H_6(T_1)$ lifetime increases at lower benzene pressure and is a further indication that quenching of $C_6H_6(T_1)$ is the source of $O_2(^1\Delta)$ since the $C_6H_6(S_1)$ fluorescence quantum yield does not change [9, 11] in this range of benzene pressures.

3.3. Quenching of singlet benzene

The major source of $O_2(^1\Delta)$ appears to be due to quenching of $C_6H_6(T_1)$. However at higher oxygen pressures quenching of $C_6H_6(S_1)$ is occurring and we must consider the possible quenching reactions (such as (4) and (5) and the similar processes not involving singlet oxygen production (7) and (8)) and their effect on the $O_2(^1\Delta)$ intensity.

$$C_6H_6(S_1) + O_2 \rightarrow C_6H_6(S_0) + O_2 \qquad (7)$$

$$C_6H_6(S_1) + O_2 \rightarrow C_6H_6(T_1) + O_2 \qquad (8)$$

Reaction (7) can be ruled out since it would quench $C_6H_6(T_1)$ production and lead to a large decrease in $O_2(^1\Delta)$ intensity. Reaction (8) would lead to a small increase in $O_2(^1\Delta)$ production as the $C_6H_6(S_1)$ previously lost by non-triplet forming processes now produce triplets and hence singlet oxygen by reaction (6). Reaction (4) would lead to a similar small increase in $O_2(^1\Delta)$ production since all of the $C_6H_6(S_1)$ formed would now produce singlet oxygen. Finally reaction (5) would lead to a large increase in $O_2(^1\Delta)$ production since it would now be formed in two quenching reactions (5) and (6) and all of the benzene singlets are converted to triplets as before.

Our observations do not support any of these interpretations as there is no change in the $O_2(^1\Delta)$ intensity (within our experimental error of 5%) in the plateau region where $C_6H_6(S_1)$ deactivation increases from 32% to 94%. However, a small rise in the $O_2(^1\Delta)$ production rate might be compensated for by a decrease in the steady state concentration of $O_2(^1\Delta)$ by reaction (3). It seems unlikely that this reaction would fortuitously mask the large increase in $O_2(^1\Delta)$ production expected from reaction (5). Therefore reations (4) or (8) appear to be the most likely for deactivation of $C_6H_6(S_1)$ by O_2. The expected increase in $O_2(^1\Delta)$ intensity at higher oxygen pressures due to reactions (4) or (8) is reduced by a higher $C_6H_6(T_1)$ quantum yield in pure benzene. Therefore, indirectly, our results favour a higher $C_6H_6(T_1)$ quantum yield.

A further study of these reactions over a wider range of benzene and oxygen pressures is in progress.

The author would like to express his gratitude to Professor C. S. Parmenter for helpful discussions.

REFERENCES

[1] D. R. Kearns, R. A. Hollins, A. U. Khan, R. W. Chambers and P. Radlick, J. Am. Chem. Soc. 89 (1967) 5455;
D. R. Kearns, R. A. Hollins, A. U. Khan and P. Radlick, J. Am. Chem. Soc. 89 (1967) 5457.
[2] C. S. Foote, S. Wexler, W. Ando and R. Higgins, J. Am. Chem. Soc. 90 (1968) 975.
[3] K. Kawaoka, A. U. Khan and D. R. Kearns, J. Chem. Phys. 46 (1967) 1842.
[4] R. M. Badger, A. C. Wright and R. F. Whitlock, J. Chem. Phys. 43 (1965) 4345.
[5] W. H. Childs and R. Mecke, Z. Physik 63 (1931) 344.
[6] S. H. Whitlow and F. D. Findlay, Can. J. Chem. 45 (1967) 2087.
[7] E. A. Ogryzlo, private communication.
[8] S. J. Arnold, M. Kubo and E. A. Ogryzlo, Progress in reaction kinetics, to be published.
[9] W. A. Noyes, W. A. Mulac and D. A. Harter, J. Chem. Phys. 44 (1966) 2100.
[10] G. A. Haninger and E. K. C. Lee, J. Phys. Chem. 71 (1967) 3104.
[11] H. Ishikawa and W. A. Noyes, J. Chem. Phys. 37 (1962) 583.
[12] C. S. Parmenter and B. L. Ring, J. Chem. Phys. 46 (1967) 1998.
[13] G. B. Porter and P. West, Proc. Roy. Soc. A279 (1964) 302.

39

Reprinted from *J. Amer. Chem. Soc.*, 91(4), 1039–1040 (1969)

Detection of the Naphthalene-Photosensitized Generation of Singlet (¹Δg) Oxygen by Paramagnetic Resonance Spectroscopy[1]

Sir:

Much of the renewed interest in sensitized photooxygenation reactions was prompted by recent experimental observations which indicate that electronically excited singlet oxygen molecules ($^1\Sigma$ or $^1\Delta$) are involved as reaction intermediates.[2] A crucial assumption in the "singlet oxygen theory" of sensitized photooxygenations is that singlet oxygen molecules are efficiently generated in these reactions by transfer of electronic energy from excited sensitizer to ground triplet-state oxygen molecules.[3] Although the data supporting the involvement of singlet oxygen in sensitizer photooxygenation reactions are quite compelling, evidence for the key energy transfer step is nonetheless circumstantial. Most arguments are based upon comparison of oxygenation products or product distributions obtained in sensitized photooxygenation with those obtained using oxygen which has been excited in a microwave discharge or generated by decomposition of hydrogen peroxide, or they are based upon kinetic studies.[2-4] In this communication we present evidence based on paramagnetic resonance spectroscopy (epr) which conclusively proves that singlet oxygen molecules are generated in a photosensitized process involving an aromatic hydrocarbon sensitizer.

The system which we studied consisted of a cylindrical quartz cell (23-mm i.d.) containing oxygen at pressures ranging from ∼0.1 to 0.3 mm and saturated with naphthalene as the sensitizer (vapor pressure ∼0.5 mm at room temperature). This cell was placed in a large stack cylindrical cavity of a Varian V-4500 X-band epr spectrometer equipped with a Mark II Fieldial, and illuminated by focusing the output of a 1000-W AH-6 mercury lamp on one end of the cell which extended slightly outside of the cavity.

Naphthalene was specifically chosen as the sensitizer in these experiments because it is relatively stable toward photooxidation. Benzene might have been used except that it is known to undergo a photochemical reaction with oxygen to produce long-chain aldehydes.[5] The filtering system (5 cm of water, ∼2 cm of quartz) and the spectral distribution from the lamp ensured that no light of wavelength shorter than about 2200 Å was admitted into the cell.

The epr spectrum of $^1\Delta$ oxygen in the $J = \Lambda = 2$ state is characterized by a nearly symmetrical quartet of lines centered at $|g_J| \sim {}^2/_3$. The spectrum was first observed in the products of a microwave discharge of O_2 by Falick, Mahan, and Myers, who assigned the individual lines to the appropriate $|\Delta M_J| = 1$ transitions.[6] When we irradiated our cell in the epr spectrometer with light under the conditions described above, resonances were observed at exactly the positions expected for $^1\Delta$ oxygen molecules. These results are shown in Figure 1b. For comparison, the $^1\Delta$ oxygen spectrum obtained using a microwave discharge unit to generate $^1\Delta$ oxygen is shown in Figure 1a. The coincidence (within 0.05% of predicted position) of the four lines observed during irradiation with those observed in the microwave discharge and comparison of their relative intensities establishes beyond doubt that the photogenerated resonances reported here are due to $^1\Delta$ oxygen. We further established that, under identical experimental conditions, it was impossible to observe the signals from $^1\Delta$ in the absence of light (Figure 1c), and that no $^1\Delta$ signals were observed when we irradiated O_2 in the absence of naphthalene. This set of observations unequivocally establishes that $^1\Delta$ molecules are generated in a photosensitized process which depends upon excitation of the naphthalene molecules and thus provides a firm foundation for a crucial element of the "singlet oxygen theory" of sensitized photooxygenation reactions. Our results confirm the recent work of Snelling[7] who observed photoinduced luminescence at 12,700 Å from irradiated benzene-oxygen mixtures which he assigned as emission from $^1\Delta$ oxygen.

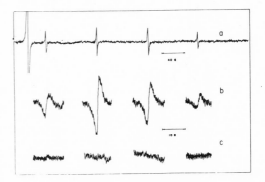

Figure 1. Paramagnetic resonance spectra of $^1\Delta$ oxygen. Magnetic field increases from left to right. (a) $^1\Delta$ oxygen generated by a microwave discharge unit (P_{O_2} = 0.2 mm). The quartet due to $^1\Delta$ is centered at $|g| \sim {}^2/_3$. The intense line to the left is due to $^3\Sigma$ oxygen. (b) $^1\Delta$ oxygen generated by energy transfer from photoexcited naphthalene vapor. Each resonance is the result of 144 passes with a CAT through a 25-G region of field predicted to contain a $^1\Delta$ signal. Note that the photogenerated signals are presented on an expanded gauss scale in comparison with the spectrum above and that the phase of the signal is inverted by the CAT. Some broadening of each line is the result of drift in the microwave cavity frequency during the 1.5-hr run. (c) Paramagnetic resonance signals observed under same conditions as 1b, except without irradiation of the cell.

(1) The support of the U. S. Army Research Office (Durham), Grant No. ARO(D)-31-124-G-804, the National Science Foundation, Grant No. GP-6656, and the National Institutes of Health, Grant No. GM-12327, is most gratefully acknowledged.

(2) (a) C. S. Foote and S. Wexler, *J. Am. Chem. Soc.*, **86**, 3879, 3880 (1964); (b) C. S. Foote, S. Wexler, W. Ando, and R. Higgins, *ibid.*, **90**, 975 (1968); (c) E. J. Corey and W. C. Taylor, *ibid.*, **86**, 3881 (1964); (d) K. Gollnick, *Advan. Photochem.*, in press; (e) D. R. Kearns, R. A. Hollins, A. U. Khan, R. W. Chambers, and P. Radlick, *J. Am. Chem. Soc.*, **89**, 5455, 5456 (1967).
(3) (a) H. Kautsky and H. de Bruijn, *Naturwissenschaften*, **19**, 1043 (1931); (b) H. Kautsky, *Trans. Faraday Soc.*, **35**, 216 (1939).
(4) (a) C. S. Foote, S. Wexler, and W. Ando, *Tetrahedron Letters*, 4111 (1965); (b) K. R. Kopecky and H. J. Reich, *Can. J. Chem.*, **43**, 2265 (1965); (c) R. Higgins, C. S. Foote, and H. Cheng in Advances in Chemistry Series, American Chemical Society, Washington, D. C. in press.
(5) K. Wei, J. C. Mani, and J. N. Pitts, *J. Am. Chem. Soc.*, **89**, 4225 (1967).
(6) A. M. Falick, B. H. Mahan, and R. J. Myers, *J. Chem. Phys*, **42**, 1837 (1965).
(7) D. R. Snelling, *Chem. Phys. Letters*, **2**, 346 (1968).

David R. Kearns et al.

From an estimate of the steady-state concentration of $^1\Delta$ oxygen generated during the irradiation and a measurement of the effective lamp output we find that the quantum yield for the formation of $^1\Delta$ is of the order 0.5 ± 0.3, assuming a $^1\Delta$ lifetime of ~ 1 sec in the cell.[6] While we consider this number uncertain, it indicates that a significant fraction of the excitation energy is utilized to generate $^1\Delta$. Furthermore, if the value of ~ 0.5 could be established, sensitization by the naphthalene triplet would be indicated. This result tends, therefore, to support the theoretical work of Kawaoka, Khan, and Kearns,[8] who predicted that the oxygen quenching of triplet-state molecules proceeds primarily by energy transfer to oxygen. To verify this point, however, a careful study of the pressure dependence of the $^1\Delta$ resonance intensity will be required. Preliminary measurements indicate that over a range of oxygen pressures from 0.1 to 0.3 mm the steady-state concentration of $^1\Delta$ is still increasing with increasing O_2 pressure. A detailed study of this point is in progress and will be presented elsewhere.

(8) K. Kawaoka, A. U. Khan, and D. R. Kearns, *J. Chem. Phys.*, **46**, 1842 (1967).

(9) The support of a Lockheed Summer Research Grant to C. K. Duncan is most gratefully acknowledged.

David R. Kearns, Ahsan U. Khan
Christopher K. Duncan,[9] August H. Maki
Department of Chemistry, University of California
Riverside, California

Received October 21, 1968

Reprinted from *J. Amer. Chem. Soc.*, **91**(4), 1040–1041 (1969)

Electron Paramagnetic Resonance of ¹Δ Oxygen Produced by Gas-Phase Photosensitization with Naphthalene Derivatives

Sir:

We have observed the formation of $^1\Delta\text{-}O_2$ on uv irradiation of vapor-phase mixtures of oxygen with naphthalene and some of its derivatives. Under steady-state conditions as much as 70% of the total oxygen can be converted to the excited form so that the systems could serve as convenient sources of $^1\Delta$ for gas-phase reactions. The excited oxygen was detected by its characteristic epr spectrum, a procedure which had been previously used in demonstrating its formation on decomposition of a triphenyl phosphite–ozone complex.[1] These photosensitization experiments may be regarded as giving strong support to Foote's hypothesis that many photooxidations involve excited oxygen as the reactive intermediate, this upper state being produced from ground-state oxygen by energy transfer from the photosensitizer.[2] To our knowledge, the first direct physical evidence for this transfer has been the very recent report of Snelling who observed the near-infrared emission of $^1\Delta$ oxygen produced by benzene photosensitization.[3] Our experiments, and the similar ones by Kearns, *et al.*, described in the accompanying communication,[4] give additional evidence using other sensitizers and another method of detection which allows direct measurements of concentrations. Also, our results lead to quite high $^1\Delta$ concentrations, some two orders of magnitude larger than the benzene system.[3]

The experiments were performed by including a few crystals or drops of the naphthalene derivative in a quartz tube of ~24 mm o.d. containing from 0.1 to 1.0 mm of oxygen. Irradiation was accomplished with unfiltered medium or high-pressure mercury or xenon lamps of 150–500 W through a Supersil window. The characteristic four-line epr spectrum of $^1\Delta\text{-}O_2$, due to the orbital magnetic moment of the π^* electrons, was readily observed during irradiation between 9–10 kG at X-band. The origin of the spectrum was established by comparison with that observed with a microwave discharge in O_2 as first reported by Falick, *et al.*,[5] and with that observed in the phosphite–ozone decomposition.[1] The concentrations of the $^1\Delta$ and $^3\Sigma$ were established by comparing the intensities of the epr absorptions with those of discharged oxygen where the $^1\Delta$ concentration is 7–10%.[1,5]

Naphthalene (I), 1- (II) and 2-fluoronaphthalene (III), and octafluoronaphthalene (IV) were used as sensitizers. With I–III, steady-state $^1\Delta$ concentrations of 3–6% were readily obtained. Sensitizer, O_2, and light must all be present if $^1\Delta$ is to be observed. With I, and somewhat less readily with II and III, irradiation was also associated with the largely irreversible disappearance of the $^3\Sigma$ oxygen within a period of tens of minutes. While ground state I is supposedly inert to reaction with $^1\Delta\text{-}O_2$, the conditions of our experiment are such that collisions are likely between excited I and excited oxygen, so that reaction may well occur. With IV the irreversible removal of O_2 was slower, and we have concentrated on IV–O_2 mixtures for a more detailed study.

The absolute concentration $^1\Delta\text{-}O_2$ was independent of the original O_2 pressure in the range of 0.2–1.0 mm for a given set of irradiation conditions. The independence implies that effectively all excited IV transferred energy to O_2 before deactivation. Consistent with this conclusion is the observation that even with the more powerful lamps the concentration of the $^1\Delta$ was linearly proportional to the intensity of light irradiating the sample. At the lower O_2 pressures, with the higher proportions of $^1\Delta$, we could easily see the immediate decrease in $^3\Sigma$ associated with rise of $^1\Delta$ on initial illumination. Blocking the light produced a corresponding rise in $^3\Sigma$ and disappearance of $^1\Delta$. In most experiments, the $^3\Sigma$ and $^1\Delta$ changes balanced one another, although in some the loss of $^3\Sigma$ on illumination was greater than the amount of $^1\Delta$ produced. The oxygen balance in the majority of experiments indicates that a substantial amount of O_2 is not present in some metastable form other than $^1\Delta$. These results do not imply that the $^1\Delta$ must be formed immediately on sensitization. The $^1\Sigma$ might be formed first but would be expected to deactivate rapidly to the $^1\Delta$ on collision.[6]

On continued irradiation the $^3\Sigma$ concentration decreases with time, presumably due to irreversible reactions of O_2 with the sensitizer. The $^1\Delta$ also decreases with the total pressure but more slowly. It is possible that at the lower pressures the excited naphthalenes are not completely scavenged by the $^3\Sigma$ to produce $^1\Delta$. At quite low O_2 pressures, ~0.01 mm, the $^1\Delta$ concentrations can be higher than that of the $^3\Sigma$. We have obtained O_2–IV mixtures in which 70% of the total O_2 is converted to $^1\Delta$. Such an inverted population can occur because we are not producing the $^1\Delta$ by direct absorption of light by the $^3\Sigma$ but rather through an intermediate excited aromatic.

One significant point is that the energy transfer we are observing in the above experiments is essentially that postulated by Khan, *et al.*, to explain how significant concentrations of $^1\Delta$ could arise in an atmosphere which contained some aromatic compounds.[7] Direct excitation from the $^3\Sigma$ would not be feasible because of the low transition probabilities. The apparently high efficiencies of the transfer even at low O_2 pressures make the process particularly likely. It would thus appear that substantial amounts of $^1\Delta$ might be available in a polluted atmosphere and could serve as a reactant in subsequent oxidations.

The following compounds have also been found to be sensitizers. The figures given are percentages of $^3\Sigma$ converted to $^1\Delta$: quinoxaline, 3.5%; 1-naphthaldehyde, 2.5; biphenyl, 2.4; 1-acetonaphthone, 0.9; 1-chloronaphthalene, 0.4; phenanthrene, 0.4; anthracene, 0.3; 2-acetonaphthone, 0.2; dibenzfuran, 0.2; anthraquinone, 0.1; phenanthridine, 0.1.

(1) E. Wasserman, R. W. Murray, M. L. Kaplan, and W. A. Yager, *J. Am. Chem. Soc.*, **90**, 4160 (1968).
(2) C. S. Foote, S. Wexler, W. Ando, and R. Higgins, *ibid.*, **90**, 975 (1968), and references cited therein.
(3) D. R. Snelling, *Chem. Phys. Letters*, **2**, 346 (1968).
(4) D. R. Kearns, A. U. Khan, C. K. Duncan, and A. H. Maki, *J. Am. Chem. Soc.*, **91**, 1039 (1969). The simultaneous publication of the two communications is possible through the gracious withholding, by Drs. Maki and Kearns and their coworkers, of the r previously submitted manuscript.
(5) A. M. Falick, B. H. Mahan, and R. J. Meyers, *J. Chem. Phys.*, **42**, 1837 (1965).
(6) S. J. Arnold, M. Kubo, and E. A. Ogryzlo, *Progr. Reaction Kinetics*, in press.
(7) A. U. Khan, J. N. Pitts, Jr., and E. B. Smith, *Environ. Sci. Technol.*, **1**, 656 (1967). We are indebted to Drs. R. W. Murray and P. R. Story for bringing this reference to our attention.
(8) Bell Telephone Laboratories, Inc.
(9) Rutgers, The State University.

E. Wasserman,[8,9] V. J. Kuck,[8] W. M. Delavan,[8] W. A. Yager[8]

Bell Telephone Laboratories, Inc.
Murray Hill, New Jersey 07974
School of Chemistry, Rutgers, The State University
New Brunswick, New Jersey 08900
Received December 7, 1968

Editor's Comments
on Papers 41 and 42

41 EVANS
   Oxidation by Photochemically Produced Singlet States of Oxygen

42 MATHESON and LEE
   Reaction of Chemical Acceptors with Singlet Oxygen Produced by Direct Laser Excitation

OXYGENATION OF ORGANIC ACCEPTORS WITH $^1\Delta_g O_2$ PRODUCED BY LASER IRRADIATION OF MOLECULAR OXYGEN

Papers 41 and 42 describe experiments in which $^1\Delta_g O_2$ is produced by laser irradiation of either the double-molecule transition of $O_2 \cdot O_2$ ($2\ ^3\Sigma_g^- \rightarrow 2\ ^1\Delta_g$) or the single-molecule transitions ($^3\Sigma_g^- \rightarrow ^1\Delta_g$; $^3\Sigma_g^- \rightarrow ^1\Sigma_g^+$). These methods offer the advantage of producing $^1O_2$ cleanly without any extraneous reactive species. These results further demonstrate that no sensitizer–oxygen complex is required for the photooxygenation reaction.

Evans observed that irradiation of a solution of 9,10-dimethylanthracene (DMA) or 1,3-diphenylisobenzofuran (DPBF) under 2000 psi of oxygen with a (He–Ne laser (632.8 nm) led to conversion of the acceptors to the normal singlet oxygen products. The wavelength of the laser coincides almost exactly with the double-molecule transition of oxygen. He also reports that irradiation of the $^3\Sigma_g^- \rightarrow ^1\Sigma_g^+$ and $^3\Sigma_g^- \rightarrow ^1\Delta_g$ transitions at 760 and 1270 nm, respectively, gave rates of oxidation of 9,10-dimethylanthracene which paralleled the intensity of the corresponding absorption band. These results suggested that either $^1\Sigma_g^+$ and $^1\Delta_g$ O_2 react at comparable rates with the acceptor (DMA) or the $^1\Sigma_g^+$ molecules undergo spin-allowed deactivation to $^1\Delta_g$ quantitatively rather than deactivation to the ground state. In view of the results of Ogryzlo on the rapid quenching of $^1\Sigma_g^+$ in solution (Paper 43), Evans suggests that the latter explanation is correct.

Matheson and Lee (Paper 42) demonstrated that irradiation of a solution of oxygen in Freon-113 in the $^3\Sigma_g^- \rightarrow ^1\Delta_g + 1$ eV absorption band with the 1.065-μm output from a Nd–YAG laser produces singlet oxygen which may be trapped by the various acceptors.

41

Oxidation by Photochemically Produced Singlet States of Oxygen

By D. F. EVANS

(*Department of Chemistry, Imperial College of Science and Technology, London S.W.*7)

THERE is now considerable evidence that many photo-oxidations involve the low-lying singlet states of oxygen ($^1\Delta_g$ and $^1\Sigma_g{}^+$) as intermediates.[1] Similar reactions have been carried out using singlet oxygen produced either chemically or in an electric discharge.[1] I describe here oxidations caused by direct photochemical excitation of oxygen at high pressures dissolved in an inert organic solvent.

The Figure (a) shows the absorption spectrum (470—1300 nm.) of oxygen at 2000 lb./sq. in. pressure dissolved in 1,1,2-trichloro-1,2,2-trifluoroethane (a good and safe solvent for oxygen under pressure.) Measurable absorption leading to the $^1\Delta_g$ and $^1\Sigma_g{}^+$ states, and also to the "double" states $(^1\Delta_g)_2$ and $(^1\Delta_g{}^1\Sigma_g{}^+)$ is observed. By a fortunate coincidence the normal wavelength of a He–Ne laser, 632·8 nm., almost exactly coincides with that of the $(^1\Delta_g)_2 \leftarrow (^3\Sigma_g{}^-)_2$ transition. Irradiation with a He–Ne laser of a cell containing oxygen at 2000 lb./sq. in. dissolved in a dilute solution of 9,10-dimethylanthracene in 1,1,2-trichloro-1,2,2-trifluoroethane gave 9,10-dimethylanthracene peroxide. No detectable reaction occurred in the dark. For a $1·8 \times 10^{-4}$M-solution the quantum yield (calculated from the rate of disappearance of the 9,10-dimethylanthracene) was *ca.* 0·13, and for a $4·4 \times 10^{-5}$M-solution *ca.* 0·04₉.† For a 9×10^{-6}M-solution of the very reactive compound 1,3-diphenylisobenzofuran under similar conditions, the quantum yield was *ca.* 0·6₅. The Figure (b) shows the relative rates of photo-oxidation of a 4×10^{-5}M-solution of 9,10-dimethylanthracene saturated with oxygen at 2000 lb./sq. in., with approximately monochromatic radiation of various wavelengths. It is clear that photochemical excitation of the single $^1\Delta_g$ and $^1\Sigma_g{}^+$ states can also cause oxidation of the organic substrate. This

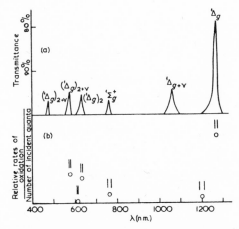

FIGURE. (a) *The absorption spectrum of oxygen at 2000 lb./sq. in. dissolved in 1,1,2-trichloro-1,2,2-trifluoroethane.*

(b) *The relative rates of photo-oxidation of a 4×10^{-5}M-solution of 9,10-dimethylanthracene saturated with oxygen at 2000 lb./sq. in., divided by the number of quanta incident upon the cell. The vertical lines indicate the approximate widths at half-intensity of the interference filters used.*

technique for studying oxidation by singlet states of oxygen has the advantage of high selectivity.

ERRATUM

In the figure, the first peak at 480 mm should read: $^1\Delta_g + \Sigma_g$.

[1] C. S. Foote, *Accounts Chem. Res.*, 1968, **1**, 104.
[2] D. F. Evans, *J. Chem. Soc.*, 1957, 1351.

(*Received, February 6th*, 1969; *Com.* 163.)

† The oxygen-induced singlet–triplet bands[2] of 9,10-dimethylanthracene will also lie in this region, but for the very dilute solutions studied will be very weak.

42

Reprinted from *Chem. Phys. Lett.*, 7(4), 475–476 (1970)

REACTION OF CHEMICAL ACCEPTORS WITH SINGLET OXYGEN
PRODUCED BY DIRECT LASER EXCITATION

I. B. C. MATHESON and J. LEE

Department of Biochemistry, University of Georgia, Athens, Georgia 30601, USA

Received 24 September 1970

$^1\Delta$ oxygen has been produced in 1,1,2-trichlorotrifluoroethane solution by direct laser excitation of the $^1\Delta_{g+v} \leftarrow {}^3\Sigma_g^-$ oxygen transition. The $^1\Delta$ has been characterized by its reaction with known chemical acceptors, and semi-quantitative estimates of the absolute rate constants for the reaction of $^1\Delta$ with the various acceptors have been made.

The output of the Nd-YAG laser at 1065 nm lies within the $^1\Delta_{g+v} \leftarrow {}^3\Sigma_g^-$ absorption band of the oxygen molecule. We wish to report here direct generation of $^1\Delta$ oxygen in solution by excitation of dissolved oxygen with the 35 Watt multimode C.W. output from a Holobeam Type 250 laser. High oxygen pressures, up to 140 atm, are required since the absorption intensity is proportional to the square of the oxygen concentration [1]. Given sufficient oxygen absorption the high laser power available generates enough $^1\Delta$ to react with added chemical acceptors at a high rate. Acceptor decay was detected by measurement of the acceptor absorbance during the course of laser irradiation. Laser scatter interference with the absorbance measurements was eliminated by the use of suitable filters.

In the table are given the pseudo-first order decay constants for five known singlet oxygen acceptors. All except 2,5-dimethyl-furan have visible or near UV absorptions (λ_A) and give colourless oxidation products so that acceptor decay may be measured as absorbance loss. The decay rate constant for 2,5-dimethylfuran was estimated by its inhibition of the oxidation of rubrene.

For the reactions

$$h\nu + 2O_2 \xrightarrow{\phi} {}^1\Delta + O_2 ,$$

$$^1\Delta + A \xrightarrow{k_A} AO_2 ,$$

$$^1\Delta + Q \xrightarrow{k_Q} O_2 + Q ,$$

it may be shown

$$\frac{d[A]}{dT} = \frac{\phi k_A[A]}{k_A[A] + k_Q[Q]} .$$

where A = chemical acceptor and Q = generalized quencher, may be oxygen or solvent.

The observed decay k_1 was strictly first order for most acceptors, i.e. $k_1 = \phi k_A/k_Q[Q]$. In the case of 1,3-diphenylisobenzofuran a zeroth order component was also observed; i.e. for this compound $k_A[A]$ must be competitive with $k_Q[Q]$. The rate constant for 2,5-dimethyl-furan (D) was measured by competition of $k_D[D]$ with $k_Q[Q]$; i.e. a high concentration of D, about 10^{-2} mole litre^{-1} was used with a low concentration of rubrene. This gave an initial slow rubrene decay rate changing after a few seconds to a more rapid decay characteristic of rubrene. The limiting $t = 0$ first order rate constant is given by

$$k_1 = \phi k_{\text{rubrene}}/(k_D[D] + k_Q[Q]) ,$$

from which $k_D/k_Q[Q]$ was extracted by comparison with the rubrene alone decay constant.

120 atm of oxygen in equilibrium with the solvent was used for all the measurements. This gave 2% absorption at 1065 nm. The 35 watt laser output at 1065 nm corresponds to 3×10^{-4} Einsteins sec^{-1}, and thus ϕ is 2% of this, i.e. $^1\Delta$ production rate is 6 μmole sec^{-1}.

The overall uncertainty in the rate constants is estimated to be $\pm 30\%$. The order of reactivity is seen to be comparable with that reported for photosensitized oxidation [2].

The absolute k_A's may be obtained roughly by making an estimate of $k_Q[Q]$. The equilibrium concentration of oxygen in solution may be esti-

Table 1
Reaction rates with $^1\Delta$ oxygen in 1,1,2-trichlorotrifluoroethane

Acceptor	λ_A (nm)	k_1 (sec^{-1})	Relative reactivity	
			Laser ($k_A/k_Q[Q]$)	Photosensitized [a]
9,10-diphenylanthracene	392	7×10^{-4}	1	1
tetraphenylcyclopentadienone	500	4×10^{-3}	7	–
rubrene	522	5×10^{-3}	8	11
2,5-dimethylfuran	522 [b]	2×10^{-4} [b]	20	12 [c]
1,3-diphenylisobenzofuran	410	7×10^{-2}	100	190

a) In pyridine from ref. [2].
b) Measured as inhibition of rubrene oxidation.
c) By comparison with tetramethylethylene in ref. [3].

mated at ≥ 3 mole litre^{-1} by comparison of its absorption with that of the gas at the same pressure. The solvent concentration is 9 mole litre^{-1} so that if the solvent quenching rate is low relative to that of O$_2$ then oxygen quenching will be the dominant process. The $^1\Delta$ quenching rate by *gaseous* oxygen at low pressures is reported as 1.3×10^3 litre mole^{-1} sec^{-1} [4] and so we may estimate a *minimum* value for $k_Q[Q]$ of 4×10^3 sec^{-1}. This would give k_A's ranging from 4×10^5 litre mole^{-1} sec^{-1} for 9,10-diphenyl-anthracene to 4×10^7 litre mole^{-1} sec^{-1} for 1,3-diphenylisobenzofuran.

The absolute rate constant (quenching plus reaction) for 2,5-dimethylfuran in the gas phase has recently been reported as 1.4×10^6 litre mole^{-1} sec^{-1} [5]. From the present work a minimum total quenching rate of 8×10^6 litre mole^{-1} sec^{-1} is calculated for solution which appears to be somewhat higher than the gas phase value.

Evans [6] has previously reported direct photophysical generation of $^1\Delta$ by He-Ne laser excitation (632.8 nm) of the oxygen dimol absorption band. We find that the Nd-YAG laser confers two important advantages:

1) The laser power is at least one thousand times higher than that available from the He-Ne laser, and thus a higher $^1\Delta$ production rate is possible;

2) The possibility exists that some acceptors may have weak absorptions in the visible red so that the He-Ne laser may cause some direct acceptor excitation. The risk of this is much reduced with the 1065 nm output of the Nd-YAG laser.

This work was supported in part by a grant from the Research Corporation.

REFERENCES

[1] I.B.C. Matheson and J. Lee, unpublished data.
[2] T. Wilson, J. Am. Chem. Soc. 88 (1966) 2898.
[3] R. Higgins, C.S. Foote and H. Cheng, in: Oxidation of organic compounds, Vol. 3 (American Chemical Society, Washington, 1968) p. 102.
[4] R.P. Wayne, in: Advances in photochemistry, Vol. 7, eds. J.N. Pitts, G.S. Hammond and W.A. Noyes Jr. (Interscience, New York, 1969) p. 311.
[5] W.S. Gleason, A.D. Broadbent, E. Whittle and J.N. Pitts, J. Am. Chem. Soc. 92 (1970) 2068.
[6] D.F. Evans, Chem. Commun. (1969) 367.

Editor's Comments
on Papers 43 Through 46

$^1\Sigma_g^+$ MOLECULAR OXYGEN

Although considerable evidence had been obtained for the intermediacy of the lower excited state of oxygen ($^1\Delta_g$) in photooxygenations, the possible importance of the $^1\Sigma_g^+$ state in chemical reactions was as yet unclear. Ogryzlo and coworkers (Paper 43) determined that the $^1\Sigma_g^+$ state is very short-lived and is rapidly quenched to $^1\Delta$ (a spin-allowed transition) in about 1 collision in 100. They estimated that the $^1\Delta_g$ state is deactivated to ground state in about 1 collision in 10^9–10^{10}. For $^1\Sigma_g^+$ to react competitively with $^1\Delta_g$, the former species would have to react about 10^8 times faster than the latter. It therefore seems improbable that $^1\Sigma_g^+$ is an important species in chemical reactions of singlet oxygen. Their results give estimated lifetimes in solution for $^1\Delta_g$ and $^1\Sigma_g^+$ of 10^{-5} and 10^{-12} s, respectively.

Gollnick and coworkers (Paper 44) investigated the effect (or lack of it) of the triplet energy of sensitizers on the product distributions obtained from the photosensitized oxygenation of several acceptors. They could find no substantial support for the participation of $^1\Sigma_g^+$ in photooxygenations in solution.

Papers 45 and 46 provide direct evidence for the formation of $^1\Sigma_g^+$ O_2 by photosensitization. This species is apparently formed by energy transfer from a sensitizer with at least 37.5-kcal/mol excitation energy but does not survive sufficiently long to contribute significantly to the chemical reactions of singlet oxygen in solution.

In 1967, Kearns et al. described experiments in which the product ratio from the photosensitized oxygenation of cholest-4-en-3β-ol was a function of the sensitizer used.[1] They cited these results as evidence for the intermediacy of $^1\Sigma_g^+ O_2$ as well as $^1\Delta_g$ in photooxygenation. However, it was subsequently shown that the change in the product distribution in this reaction was attributable to sensitizer interaction.[2]

NOTES AND REFERENCES

1. D. R. Kearns, R. A. Hollins, A. U. Khan, R. W. Chambers, and P. Radlick, *J. Amer. Chem. Soc.*, **89**, 5455 (1967).
2. C. S. Foote and S.-Y. Wang, ACS Division Petroleum Chemistry, *Petrol. Preprints*, **14**(2), A93 (1969).

43

Reprinted from *Advan. Chem. Ser.*, 77, 133–142 (1968)

Relaxation and Reactivity of Singlet Oxygen

S. J. ARNOLD, M. KUBO, and E. A. OGRYZLO

University of British Columbia, Vancouver 8, B. C., Canada

Measurements of some energy transfer, physical quenching, and chemical reaction processes of singlet oxygen are presented. The results of these measurements and those obtained previously are analyzed in an attempt to assess the fate of $O_2(^1\Delta_g)$ and $O_2(^1\Sigma_g^-)$ in oxidation systems.

Two electronically excited singlet states of oxygen are located 22.5 and 37.5 kcal. above the triplet ground state. The one at 22.5 kcal. is commonly designated $a^1\Delta_g$ and will be abbreviated $^1\Delta$. The higher one has the term symbol $b^1\Sigma_g^+$ and will be referred to as $^1\Sigma$. The ground state $X^3\Sigma_g^-$ will be abbreviated $^3\Sigma$. Because their relative importance in chemical reactions has not yet been determined, these two excited states are referred to as singlet oxygen. This paper considers the various physical and chemical processes which the two species can undergo, and an attempt is made to assess their relative importance in oxidation processes.

Experimental

A typical flow system used to prepare $^1\Delta$ molecules for kinetic studies is shown in Figure 1. Oxygen at a pressure between 1 and 10 torr is passed through a microwave discharge. The atoms are removed with a mercuric oxide ring immediately after the discharge. The concentration is measured at one point in the tube by the heat liberated when the molecules are deactivated on a cobalt wire. Relative concentrations of excited molecules along the observation tube can be measured with a movable interference filter and photomultiplier. The details of these methods have been described (1, 3, 4). Tank oxygen is normally selected for low nitrogen content and used without further purification. Quenching gases were treated only to remove higher boiling impurities, especially water.

Radiative Relaxation

In the absence of any external perturbation both $^1\Delta$ and $^1\Sigma$ oxygen do not emit any measurable electric dipole radiation. However, with a lifetime of 7 sec., $^1\Sigma$ can give rise to magnetic dipole radiation at 7619 A., and $^1\Delta$ can give rise to magnetic dipole radiation at 12,683 A. with a lifetime of 45 minutes. In a collision with another molecule electric dipole transitions between these states are made more probable, and the radiative lifetime can be shortened. The exact radiative lifetime of singlet oxygen in a collision complex is difficult to estimate because the duration of a collision is uncertain. However, with a reasonable estimate of about 10^{-13} sec. for this collision time, the following radiative lifetimes for a number of collision complexes can be calculated from the integrated absorption coefficients.

$$^1\Delta^3\Sigma \xrightarrow[\text{12,700 A. and 15,800 A.}]{\tau = 4\ \text{sec.}} \tag{1}$$

$$^1\Delta^1\Delta \xrightarrow[\text{6,340 A. and 7,030 A.}]{\tau = 1.5\ \text{sec.}} \tag{2}$$

$$^1\Sigma^3\Sigma \xrightarrow[\text{7,620 A.}]{\tau = 15\ \text{sec.}} \quad {}^3\Sigma^3\Sigma \tag{3}$$

$$^1\Sigma^1\Sigma \xrightarrow[\text{3,808 A. and 3,612 A.}]{\tau = 0.3\ \text{sec.}} \tag{4}$$

$$^1\Delta^1\Sigma \xrightarrow[\text{4,773 A.}]{\tau = 1.7\ \text{sec.}} \tag{5}$$

When the collision complex is made up of two excited molecules, a novel energy pooling process occurs in which the energy of two molecules appears in a single photon. The above list shows that the probability of such a cooperative event can be comparable with that for a one-electron transition. However, because the fraction of molecules in a state of collision is small, radiative relaxation is not responsible for the decay of a significant number of excited molecules under the usual experimental conditions. For example, the strongest induced radiation for $^1\Delta$ occurs at 6340 and 7030 A. If this were the only mode of decay, the observed lifetime at 1 atm. would be 10^3 sec., whereas the observed lifetime is much less than 1 sec. The 6340-A. band is, however, a convenient and sensitive emission for monitoring the singlet delta concentration since the emission intensity is proportional to the square of its concentration.

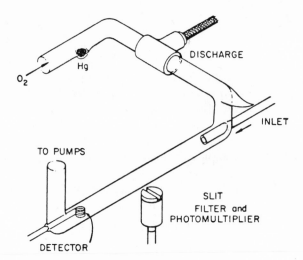

Figure 1.   Core of flow system used for quenching studies

Energy Transfer

For efficient transfer of electronic excitation to another molecule, the acceptor must possess an excited electronic state at or not too far below that of the donor. Not many molecules can meet this requirement when the donor is singlet oxygen. We have observed energy transfer to the following species, (a) another $^1\Delta$ molecule, (b) violanthrone (dibenzanthrone), (c) nitrogen dioxide, and (d) iodine atoms. The mechanism of energy transfer to (b) is not well understood and will be described elsewhere (7). Since (c) and (d) are not directly related to the subject of hydrocarbon oxidation, they will not be discussed in any detail. The transfer to iodine atoms is undoubtedly the most efficient process which we have observed (2), and this can be attributed to the fact that the $^2P_{1/2}$ state of iodine lies 22 kcal. above the $^2P_{3/2}$ ground state—in almost perfect resonance with $^1\Delta$ oxygen. The transfer to nitrogen dioxide is much less efficient and appears to involve energy transfer from $^1\Sigma$ and $^1\Delta$ to raise the acceptor to a radiating state which lies about 60 kcal. above the ground state (2).

$^1\Delta$–$^1\Delta$ **Transfer.** Since the $^1\Sigma$ state lies 15 kcal. above the $^1\Delta$, the latter can act as an acceptor as well as a donor in an energy disproportionation process:

$$^1\Delta + ^1\Delta \xrightarrow{k_D} {}^1\Sigma + {}^3\Sigma \tag{6}$$

Two laboratories have reported rate constants for this reaction. A value of 1.8×10^7 liters/mole/sec. was reported by Young and Black (8), and a value of 1.3×10^3 liters/mole/sec. was more recently reported by Arnold and Ogryzlo (3). Because of the large discrepancy between these two values we have attempted a third determination by measuring the rate of $^1\Delta$ removal directly. The results of these measurements are shown in Figure 2. The $^1\Delta$ concentration was varied by changing the power fed into the discharge. Assuming that in addition to Reaction 6, which is

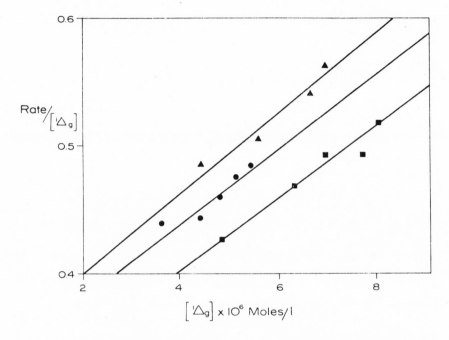

Figure 2. Rate of singlet delta decay ($R = d[^1\Delta]$ dt) divided by singlet delta concentration as a function of the singlet delta concentration

second order in $^1\Delta$, we can have wall and gas-phase quenching that is first order in $^1\Delta$, the rate equation becomes:

$$R = \frac{d[^1\Delta]}{dt} = k_Q[^1\Delta] + k_D[^1\Delta]^2$$

$$R/[^1\Delta] = k_Q + k_D[^1\Delta_g]$$

The slopes of the lines in Figure 2 are then equal to k_D. The average value we obtain is 3×10^4 liters/mole/sec. This value can be compared with 1.3×10^3 liters/mole/sec., previously reported by Arnold and Ogryzlo (3) for Reaction 6. The technique used in the earlier measurement is quite difficult, and possibly the value is somewhat low. However, it is also conceivable that the process we have measured is not the slower, spin-forbidden Reaction 6, but the more rapid spin-allowed process, Reaction 7.

$$^1\Delta + {}^1\Delta \rightarrow {}^3\Sigma + {}^3\Sigma \tag{7}$$

A decision between these possibilities must await further measurements. We can only conclude that the value of k_6 lies between 1.3×10^3 and 3×10^4 liters/mole/sec.

An important consequence of the occurrence of Reaction 6 is that $^1\Sigma$ is constantly being formed in any system which contains $^1\Delta$. We shall see later that the reverse is probably also true.

Radiationless Non-Resonance Relaxation

When an external perturbation such as that caused by a colliding molecule is sufficiently great, the electronic excitation may be degraded into nuclear motion within the collision complex. However, very little information is available about the efficiency of such processes, and consequently no complete theoretical model exists which we could use to predict quenching rates.

The experimental determination of $^1\Delta$ quenching is difficult because of its great stability. In most flow systems the decay is largely on the walls of the vessel where it can suffer about 2×10^5 collisions before deactivation. Collisions with most other molecules are even less effective. At the moment it can only be said that more than 10^8 collision with $O_2(^3\Sigma)$ are necessary to deactivate $^1\Delta$. Other, nonreactive gases cannot be tested simply because so much must be added that it radically affects the flow system and discharge, making the measurements difficult to interpret.

The quenching of $^1\Sigma$ oxygen is somewhat easier to study because it is more easily deactivated. In the absence of any quenching gas, a steady-state concentration of $^1\Sigma$ is maintained in the flow system by the following reactions.

$$^1\Delta + {}^1\Delta \xrightarrow{k_D} {}^1\Sigma + {}^3\Sigma \tag{8}$$

$$^1\Sigma + \text{wall} \xrightarrow{k_w} \text{products} \tag{9}$$

hence,

$$^1\Sigma = \frac{k_D}{k_w} [^1\Delta]^2 \tag{10}$$

and the emission intensity from $^1\Sigma$ is given by

$$I = k[^1\Sigma] = k \frac{k_D}{k_w} [^1\Delta]^2$$

In the presence of a quenching gas (Q) we must add Reaction 11:

$$^1\Sigma + Q \xrightarrow{k_Q} products \tag{11}$$

and therefore the steady state concentration is given by

$$[^1\Sigma] = \frac{k_D[^1\Delta]^2}{k_w + k_Q[Q]} \tag{12}$$

and the emission intensity in the presence of Q:

$$I_Q = \frac{kk_D[^1\Delta]^2}{k_w + k_Q[Q]} \tag{13}$$

The ratio of the emission with and without the quencher is therefore given by:

$$\frac{I_o}{I_Q} = 1 + \frac{k_Q}{k_w}[Q] \tag{14}$$

A plot of I_o/I_Q against Q should yield a straight line with a slope of k_Q/k_w. Such a plot is given in Figure 3 for 15 different gases.

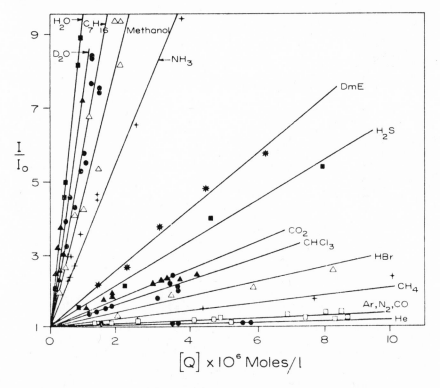

Figure 3. Stern-Volmer plot of 7619-A. emission intensity for a series of quenching gases

Table I. k_Q/k_w from I_0/I_Q vs. Q

| | $k_Q/k_w \times 10^{-6}$ | $k_Q \times 10^{-7}$ ℓ mole^{-1} s^{-1} | |
		if $k_w = 1{,}300$	if $k_w = 65$
He	0.01	1.5	0.07
N_2, Ar, CO	0.02	3	0.15
CH_4	0.11	16	0.8
HBr	0.21	30	1.5
$CHCl_3$	0.31	45	2.2
CO_2	0.39	56	2.8
H_2S	0.56	81	4.0
DME	0.78	110	5.5
NH_3	2.3	330	16
Methanol	3.7	530	26
Heptane	5.0	720	36
D_2O	6.8	980	49
H_2O	8.5	1200	60

The values of k_Q/k_w obtained from the slopes of these lines are given in Table I. To obtain k_Q, we require k_w. This can be obtained from measurements of the steady-state $^1\Sigma$ and $^1\Delta$ concentrations in the absence of any quencher. From Equation 10,

$$\frac{k_D}{k_w} = \frac{^1\Sigma}{[^1\Delta]^2}$$

Values of $[^1\Delta]$ and $[^1\Sigma]$ are listed in Table II together with values of k_D/k_w. From the value of $k_D = 3 \times 10^4$ calculated earlier we obtain $k_w = 1.3 \times 10^3$ on a clean borosilicate glass surface. Combining this value of k_w with the values of k_Q/k_w in Table I, we obtained the values of k_Q listed in the second column of the same table. The third column gives the values of k_Q calculated with the previously determined (3) value of $k_w = 65$.

Table II. Values of $[^1\Delta]$ and $[^1\Sigma]$

P, torr	$[O_2(^1\Sigma_g)] \times 10^9$, Moles Liter	$[O_2(^1\Delta_g)] \times 10^6$, Moles Liter	k_D k_w, Liters Mole
2.4	1.75	9.0	20.5
3.1	2.54	11.0	21.5
3.8	4.20	14.2	20.8
5.1	6.24	17.5	20.3

In this process $^1\Sigma$ may be relaxed into either the $^1\Delta$ or $^3\Sigma$ state. Since the transition to the $^3\Sigma$ state requires the conversion of more electronic energy into nuclear motion and also requires a "spin flip," we would expect the transition to the $^1\Delta$ state to be much more probable. This is confirmed by the observation that the quenching by paramagnetic O_2 is, if anything, less effective than species like Ar and N_2. None of the molecules included in this study display any special resonance effect. There is, however, a rough correlation between quenching efficiency and boiling point. This is a reasonable correlation since one might expect the probability of such an induced transition to be related to the magnitude and duration of the perturbation. We will not attempt a detailed analysis of this correlation here and will simply observe that boiling points reflect both these quantities in a somewhat indirect manner and can therefore be used to estimate quenching rates.

Physical Quenching Processes

From Equation 16 it follows that in the presence of a quenching species Q the ratio of $^1\Sigma$ to $^1\Delta$ concentrations is given by

$$\frac{[^1\Sigma]}{[^1\Delta]} = \frac{k_D[^1\Delta]}{k_Q[Q]}$$

When Q is water or a hydrocarbon with a similar boiling point, the equation becomes:

$$\frac{[^1\Sigma]}{[^1\Delta]} = 2 \times 10^{-6} \frac{[^1\Delta]}{[Q]}$$

In most chemical and photochemical oxidation systems the ratio of $[^1\Delta]$ to $[Q]$ is extremely small, and therefore the $^1\Sigma/^1\Delta$ ratio is very much smaller than 10^{-6}. The only such system in which this ratio might be approached is the Cl_2–H_2O_2 reaction where the partial pressure of $^1\Delta$ probably exceeds 70 torr (6). However, we must also consider the quenching of $^1\Delta$. The relevant processes are then the following in the presence of quenchers such as water:

$$^1\Sigma \; \begin{matrix} \xrightarrow{\;k \simeq 10^9\;} \\ [Q] \\ k = 1.3 \times 10^3 \\ \xleftarrow{\qquad} \\ [^1\Delta] \end{matrix} \; ^1\Delta \; \xrightarrow[\;[Q]\;]{k \approx 10^1 - 10^2} \; ^3\Sigma$$

(Since we are most interested in the possibility that $^1\Sigma$ contributes to the reactivity of singlet oxygen, we have used the lower quenching constants for $^1\Sigma$ calculated from Arnold and Ogryzlo's value of k_w (3).) In about one collision in 100 ($\sim 10^{-9}$ sec. in solution) $^1\Sigma$ is relaxed to $^1\Delta$ by Q. Only if $^1\Delta$ is $\approx 0.1\%$ of Q is $^1\Sigma$ efficiently reformed from $^1\Delta$. Otherwise, it is relaxed to $^3\Sigma$ in about one collision in 10^9–10^{10} (1-10 msec. in solution) with Q. Under these conditions we obtain a $^1\Delta/^1\Sigma$ ratio of about 10^8. Consequently, in such systems when a steady state is established, the rate constant for reaction with $^1\Sigma$ would have to be about 10^8 faster than that for $^1\Delta$ to be competitive.

Chemical Reactions

No technique for measuring the absolute values of rate constants for singlet oxygen reactions in solution has yet been reported. However, such a measurement is possible in the gas phase with the technique described here, provided the species is volatile and highly reactive. We have studied the reaction of singlet oxygen with tetramethylethylene (TME), whose reaction with singlet oxygen in the gas phase was first described by Bayes and Winer (5). The reaction was followed by monitoring the $^1\Delta$ and $^1\Sigma$ concentrations under various conditions. We have found, however, that kinetically the process is not as simple as the preliminary studies suggested. It is possible that this may be a characteristic of exothermic association processes in low density systems where there are an insufficient number of collisions which unreactive molecules to prevent chain reactions from developing. In contrast to the situation in condensed media, it is highly probable that the energy-rich product of the initial reaction will collide with another energetic molecule rather than with an inert species which could relax it to a stable product. We are attempting to study the reaction under conditions which are more comparable with those in the solution reaction, with the hope that the kinetics will become somewhat simpler. Ignoring the complexity of the system we can make a preliminary estimate of 10^8 liters/mole/sec. for this rate constant (TME-$^1\Delta$) from the initial slope of the decay curve.

We have no evidence for a direct reaction between TME and $^1\Sigma$. The effect of TME on the steady-state $^1\Sigma$ concentration is consistent with its boiling point—*i.e.*, it quenches (or reacts with) it with a rate constant somewhat smaller than 10^9 liters/mole/sec. In the last section we concluded that in order to make a significant contribution to the reactions of singlet oxygen in systems where steady-state concentrations of $^1\Sigma$ and $^1\Delta$ are established, the rate constant for the $^1\Sigma$ reaction would have to be between 10^5 and 10^6 times faster than that for $^1\Delta$. This clearly cannot be the case for the TME since collision frequency would be exceeded. However, in "non-steady-state" systems where the initial concentration of $^1\Sigma$ is comparable with $^1\Delta$, the direct importance of the former species depends on the ratio of quenching species to reactive species. However, $^1\Sigma$ is probably relaxed to $^1\Delta$, and even if it does not react directly can indirectly lead to oxidation.

Acknowledgment

The research for this paper was supported by the Defence Research Board of Canada, Grant number 9530-31 and partly by the United States Air Force AFOSR, Grant number 158-65.

Literature Cited

(1) Arnold, S. J., Browne, R. J., Ogryzlo, E. A., *Photochem. Photobiol.* **4**, 963 (1965).
(2) Arnold, S. T., Finlayson, N., Ogryzlo, E. A., *J. Chem. Phys.* **44**, 2529 (1966).
(3) Arnold, S. T., Ogryzlo, E. A., *Can. J. Phys.* **45**, 2053 (1967).
(4) Bader, L. W., Ogryzlo, E. A., *Discussions Faraday Soc.* **37**, 46 (1964).
(5) Bayes, K. D., Winer, A. M., *J. Phys. Chem.* **70**, 302 (1966).
(6) Browne, R. J., Ogryzlo, E. A., *Can. J. Chem.* **43**, 2915 (1965).
(7) Ogryzlo, E. A., Pearson, A. E., *J. Phys. Chem.* **72**, 2913 (1968).
(8) Young, R. A., Black, G., *J. Chem. Phys.* **42**, 3740 (1965).

RECEIVED October 9, 1967.

Reprinted from *Ann. N. Y. Acad. Sci.*, **171**(1), 89–107 (1970)

PHOTOSENSITIZED OXYGENATION AS A FUNCTION OF THE TRIPLET ENERGY OF SENSITIZERS*

Klaus Gollnick, Ph.D.,†‡§ Theodor Franken, Ph.D.,‡¶
Gerhard Schade,‡ and Günther Dörhöfer, Ph.D.†

† Department of Chemistry, The University of Arizona, Tucson, Arizona; ‡ Abteilung Strahlenchemie, Max Planck Institut für Kohlenforschung, Mülheim an der Ruhr, Germany; and ¶ Kernforschungsanlage Jülich, Jülich, Germany

Introduction

Evidence has accumulated over the past years that singlet oxygen, presumably in its $^1\Delta_g$ state, is the responsible intermediate in certain photooxygenation reactions of the two thoroughly studied classes of organic substrates: (1) substrates that contain the structural element of *cis*-1,3-dienes such as cyclic 1,3-dienes, aromatics such as anthracene, and several heterocyclic compounds such as furans, and (2) olefins containing allylic hydrogen atoms. Class 1 substrates give rise to cyclic peroxides in 1,4-cycloaddition reactions analogous to the adducts produced in Diels-Alder reactions, whereas Class 2 compounds form allylic hydroperoxides in which the original double bond has shifted into the allylic position, i.e., the hydroperoxide formation occurs analogous to the "ene" reaction.[1-4]

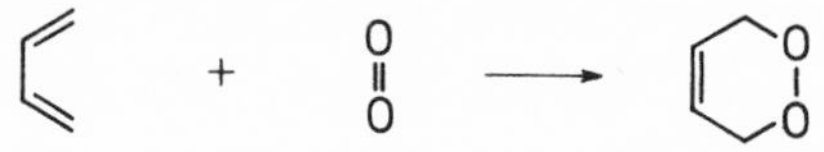

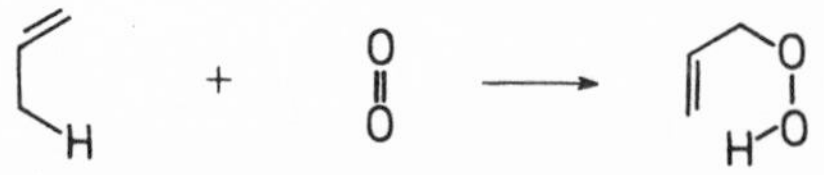

Both these oxygenation reactions may proceed in a highly stereoselective fashion depending upon the stereoelectronic effects exerted by the substrates on the attacking oxygenating species in the assumed one-step concerted mechanism reactions.

In photosensitized oxygenation reactions several short-lived intermediates of potential oxygenating capabilities may be formed according to the following reaction sequence and the energy level diagram for oxygen and a sensitizer-oxygen contact pair (FIGURE 1).[1,2,5]

$$^1S_0 + h\nu \xrightarrow{I_a} {}^1S_1 \qquad \text{absorption} \qquad (1)$$

$$^1S_1 \xrightarrow{k_2} {}^1S_0 + h\nu' \qquad \text{fluorescence} \qquad (2)$$

* The support of the National Science Foundation Science Development Program, granted to the University of Arizona, is gratefully acknowledged. The grant made possible a Visiting Associate Professorship held by Dr. K. Gollnick, and a Postdoctoral Fellowship held by Dr. G. Dörhöfer in 1966–67.

§ All reprint requests should be addressed to Dr. K. Gollnick at the Max Planck Institut.

$$^1S_1 \xrightarrow{k_3} {}^1S_0 \qquad\qquad\qquad \text{internal conversion} \qquad (3)$$

$$^1S_1 \xrightarrow{k_4} {}^3S_1 \qquad\qquad\qquad \text{intersystem crossing} \qquad (4)$$

$$^3S_1 \xrightarrow{k_5} {}^1S_0 \qquad\qquad\qquad \text{intersystem crossing} \qquad (5)$$

$$\left.\begin{array}{l}{}^1S_1 + {}^3O_2 \xrightarrow{k_6} \rightarrow {}^1S_0 + {}^1O_2 \\[4pt] {}^3S_1 + {}^3O_2 \xrightarrow{k_7} {}^1S_0 + {}^1O_2 \end{array}\right\} \quad \begin{array}{l}\text{energy transfer; singlet} \qquad (6)\\[4pt] \text{oxygen formation} \qquad\qquad (7)\end{array}$$

$$^1O_2 \xrightarrow{k_8} {}^3O_2 \qquad\qquad\qquad \text{deactivation of singlet oxygen} \quad (8)$$

$$^1O_2 + {}^1A_0 \xrightarrow{k_9} {}^1(AO_2)_0 \qquad\qquad \text{product formation} \qquad (9)$$

with $^1S_0$, $^1S_1$, and $^3S_1$ = singlet ground-state, excited singlet, and triplet state of sensitizer S, respectively; $^1A_0$ and $^1(AO_2)_0$ = singlet ground-state of organic substrate A and product AO_2, respectively; $^3O_2$ and $^1O_2$ = triplet ground-state and excited singlet oxygen, $^3\Sigma^-$, $^1\Sigma_g^+$ and/or $^1\Delta_g$, respectively. The quantum yield of peroxide formation is thus given by

$$\Phi_{AO_2} = \left(\frac{[O_2]}{k_2 + k_3 + k_4 + k_6[O_2]}\right)\left(k_6 + \frac{k_4 k_7}{k_5 + k_7[O_2]}\right)\left(\frac{k_9[A]}{k_8 + k_9[A]}\right) \qquad (10)$$

or

$$\Phi_{AO_2} = \Phi_{^1O_2}\left(\frac{k_9[A]}{k_8 + k_9[A]}\right) \qquad\qquad\qquad\qquad (11)$$

with

$$\Phi_{^1O_2} = \Phi_{^1O_2}^S + \Phi_{^1O_2}^T \qquad\qquad\qquad\qquad\qquad (12)$$

and

$$\Phi_{^1O_2}^S = \frac{k_6[O_2]}{k_2 + k_3 + k_4 + k_6[O_2]}, \qquad\qquad\qquad\qquad (13)$$

the quantum yield of singlet oxygen formation from energy transfer from excited singlet sensitizer to triplet ground-state oxygen and

$$\Phi_{^1O_2}^T = \left(\frac{k_4}{k_2 + k_3 + k_4 + k_6[O_2]}\right)\left(\frac{k_7[O_2]}{k_5 + k_7[O_2]}\right), \qquad\qquad (14)$$

the quantum yield of singlet oxygen formation from energy transfer form excited triplet sensitizer to triplet gound-state oxygen.

Depending on the lifetimes of the sensitizers in the excited states, energy transfer may occur from singlet and/or from triplet excited sensitizers, and, with regard to the mechanism of the energy transfer process, intermediate formation of sensitizer-oxygen contact pairs is believed to be involved (FIGURE 1). The lowest excited state of the contact pair is the ^{1,3,5}F-state, which may have oxygen-transferring capabilities[6] and from which transitions into the three repulsive states, $^3G_0$, $^1X_1$, and $^1X_2$, should occur, thus giving rise to triplet ground-state oxygen, and the two singlet oxygen species, $^1O_2$ ($^1\Delta_g$) and $^1O_2$ ($^1\Sigma_g^+$), respectively.

$^1O_2$ ($^1\Sigma_g^+$), possessing 37.7 kcal/mole excitation energy, may be produced only by sensitizers having more than 37.7 kcal/mole of triplet energy, whereas $^1O_2$ ($^1\Delta_g$) may be formed by energy transfer from all sensitizers having triplet energies above 22.6 kcal/mole. Singlet-Δ-oxygen is produced in hypochlorite/H_2O_2 systems[4] and by electrodeless discharge[7] and was shown to react with various organic acceptors to products and product distributions which were formerly obtained in oxygenation reactions photosensitized by low triplet-energy sensitizers.[8] In these photooxygenation reactions the final, product-forming step occurs in a one-step concerted

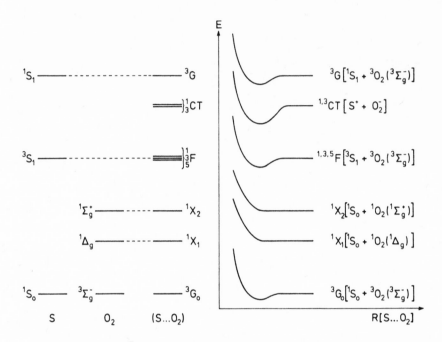

FIGURE 1. Energy level and potential energy curve diagram for a sensitizer, oxygen, and a sensitizer-oxygen contact pair.

mechanism reaction[6,9,10] in agreement with the expected reaction properties of $^1O_2$ ($^1\Delta_g$), but in contrast to those of $^1O_2$ ($^1\Sigma_g^+$), which is expected to undergo one-electron free-radical reactions,[4] for example, hydrogen abstraction.[11]

Hydrogen abstraction, however, occurs readily with ^{3}(n, π^*) sensitizers[12] such as benzophenone. Therefore, a study of the photosensitized oxygenation as a function of the triplet energy as well as of the nature of the triplet states of sensitizers seemed desirable.

Experimental Methods

Anhydrous methanol was used as solvent. Commercially available sensitizers were redistilled or recrystallized. Xanthene dyes were synthesized, purified via their diacetates and the purity controlled by thin layer chromatography as described elsewhere.[13] Tetramethylethylene, trimethylethylene, 2,5-dimethylfuran, (+)-limonene ($[\alpha]_D^{20} = +122°$), (+)-isosylvestrene (synthesized,[14] $[\alpha]_D^{25} = +115.6°$), (−)-caryophyllene and (−)-isocaryophyllene ($\alpha_D^{24} = -9.9°$ and $[\alpha]_D^{24} = -27.4°$, respectively), and clovane were distilled under N_2 until gas-chromatographic purity was established.

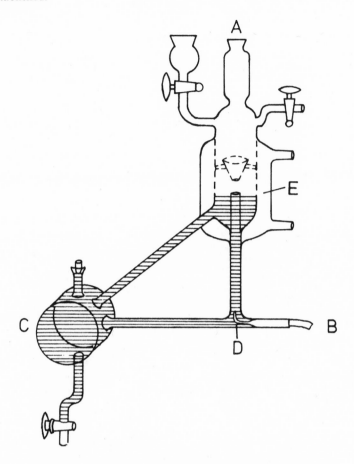

FIGURE 2. Irradiation unit for quantum-yield determinations.

Photooxygenations were carried out at 20°C in immersion type irradiation units described earlier,[3] using high vapor pressure mercury lamps, Philips HPK, 125 W. Product formation was checked in each case: (1) oxygenation products from tetramethylethylene,[15] trimethylethylene,[16] and 2,5-dimethylfuran[17] isolated and identified by melting points and infrared spectra; (2) oxygenation products from (+)-limonene,[9] (+)-isosylvestrene,[18] (−)-caryophyllene and (−)-isocaryophyllene[19,20] by vapor phase chromatography (vpc) using Perkin-Elmer fractometers F_6-4 and F_{20} (Bodenseewerk) and 50 and 100 m Golay $4G_3$ columns for nonpolar and polar compounds, respectively. In the case of photooxygenations of 2,5-dimethylfurna and tetramethylethylene, the sensitizers were recovered after irradiation in each case in about 90% yield, except for anthracene which was dimerized to about 75% during the irradiation period, and azulene which was destroyed to an extent of about 10%.

Competitive photooxygenations of (−)-caryophyllene and (−)-isocaryophyllene were carried out in the presence of clovane as an internal standard. The disappearance of the caryophyllenes was followed by vpc by determining ratios of peak areas of the caryophyllenes and the nonoxidizable clovane.

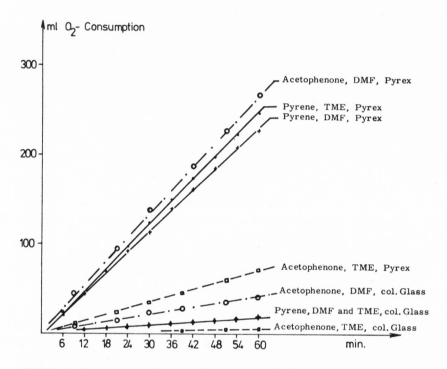

FIGURE 3. Oxygen consumption as a function of irradiation time of acetophenone- and pyrene-sensitized photooxygenation of DMF and TME in methanol at 20°C using Pyrex® and colored-glass filters.

For quantum-yield determinations, the irradiation unit shown in FIGURE 2 was used. It was connected at the gas outlet A (via an intensive reflux condenser) and the gas inlet B with a gas circulation pump and an oxygen burette in a way similar to that described earlier for an automatic oxygen-supplying device.[3] The reaction solution is poured into the irradiation unit as indicated in FIGURE 2. A vigorous oxygen-gas stream entering the solution at D provides for a rapid circulation of the solution and at the same time supplies that amount of dissolved oxygen that is consumed in the reaction vessel C during the photooxygenation. If the apparatus is filled as indicated in FIGURE 2, no oxygen bubbles appear in the reaction vessel. The irradiation source (Philips HPK, 125 w) was placed in front of the reaction vessel in a fixed position and the light was filtered through either Pyrex® glass (cut off at

283 mμ) or slightly greenish colored glass (Glashütte Wertheim, Germany, cut off at 373 mμ). Sensitizer concentrations were such that total absorption of light (optical pathlenth 2.5 cm) was achieved in the wavelength region 283–373 mμ. Oxygen consumption was measured and plotted vs. irradiation time. Actinometry was performed by using the potassium ferrioxalate actinometer since the quantum yield of Fe^{2+} formation at room temperature is nearly independent of the exciting wavelength in the region 282–373 mμ: $\Phi_{Fe2+} = 1.21$ (366 mμ); 1.23 (334 mμ); 1.24 (313, 297, 302 mμ); 1.25 (254 mμ).[12,21] Photooxygenation of 2,5-dimethylfuran (DMF) and tetramethylethylene (TME) (initial concentrations in all cases 1.67 $\times 10^{-1}$ M) and actinometry performed by using Pyrex® and colored glass filters thus allowed the determination of quantum yields of photosensitized oxygenation

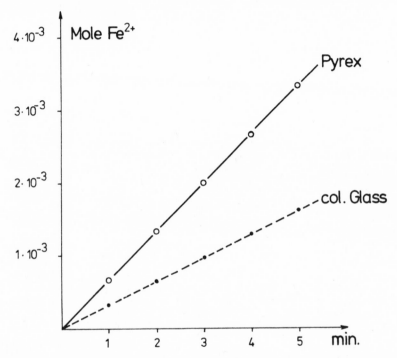

FIGURE 4. Fe^{2+} production from ferrioxalate as a function of irradiation time using Pyrex® and colored-glass filters.

reactions in the wavelength region 283–373 mμ. A few examples are given in FIGURES 3 and 4; since oxygen consumption is independent of DMF and TME concentrations in all cases, [A] $\gg k_8/k_9$ (see Eqation 11) and thus $\Phi_{AO_2} = \Phi_{1O_2}$. Quantum yields given in TABLE 1 are extreme values from at least five determinations in each case.

Results and Discussion

2.5-Dimethylfuran (DMF) and tetramethylethylene (TME, 2,3-dimethyl-2-butene) have been shown to be two of the most reactive substrates of photooxygenation[1] and hydrogen peroxide/hypochlorite[4] reactions. A 1,4-cycloaddition of oxygen to DMF occurs which gives rise to an unstable ozonide of 1,2-dimethylcylobutadiene. This reacts immediately with methanol to isolable 2,5-dimethyl-2-methoxy-5-hydroperoxy-2,5-dihydrofuran which may be hydrolyzed to the corresponding 5-hydroxy compound. An ene-reaction takes place with TME forming 2,3-dimethyl-3-hydroperoxy-1-butene.

Oxygen consumption and formation of typical oxygenation products (isolated in higher than 85% yield with respect to oxygen consumption in each case) were observed when aromatic hydrocarbons, dyes, and carbonyl compounds were used

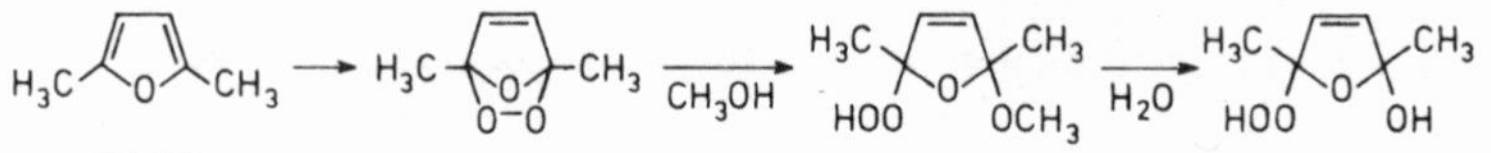

DMF

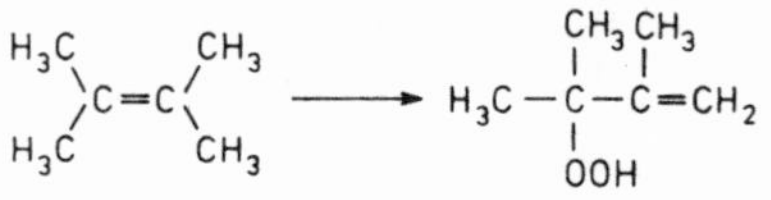

TME

TABLE 1

QUANTUM YIELDS OF SINGLET OXYGEN FORMATION IN PHOTOSENSITIZED OXYGENATION REACTIONS OF 2,5-DIMETHYLFURAN (DMF) AND TETRAMETHYLETHYLENE (TME) IN METHANOL AT 20° C.

Sensitizer	Triplet Energy* E_T, kcal/mole	$\Phi_{isc.}$*	$\Phi^1O_2[\lambda_{exc.} = 283–373m\mu]$ DMF	TME
Benzaldehyde	72.0	—	0.64–0.53	0.64–0.57
Acetophenone	73.6	0.99	0.59–0.56	0.17–0.16
Benzophenone	68.5	0.99	0.53–0.46	0.25–0.23
Fluorene	67.6	0.31	0.11–0.08	0.10–0.08
Quinoline	62.0	0.16	0.10–0.08	0.11–0.09
Napthalene	60.9	0.39	0.15–0.12	0.15–0.13
Fluorenone	53.3	0.93	0.07–0.06	0.03–0.02
Pyrene	48.7	0.18[‡]	0.62–0.57	0.65–0.60
Rose Bengal	39.5–42.2[†,¶]	0.8[§]	0.83–0.80	0.82–0.78

* Reference 12 and literature cited therein.
[†] Reference 1 and 24.
[‡] Reference 25.
[§] Reference 6.
[¶] The higher value of E_T is determined from the onset of the phosphorescence spectrum, the lower value from the first short-wavelength maximum of the phosphorescence spectrum.

as sensitizers for DMF- and TME-oxygenation reactions (TABLES 1 & 2). On electronic excitation, the aromatic hydrocarbons and the dyes give rise to[3] $(\pi, \pi*)$ triplets, whereas the carbonyl compounds (except the naphthylphenyl ketones[22] and fluorenone[23]) form $^3(n, \pi*)$ sensitizers. The triplet energies vary from about 30–85 kcal/mole.

Earlier studies have unequivocally shown that in xanthene dye-sensitized oxygenation reactions, only triplet-excited sensitizers take part in the energy transfer processes involved.[1,6] This may also be inferred for the methylene blue-[11,27] and zinc tetraphenylporphin[28]-sensitized oxygenation reactions. The triplet quantum

TABLE 2

Photosensitizers of 2,5-Dimethylfuran and Tetramethylethylene Oxygenations

Sensitizer	E_T[†]	Sensitizer	E_T[§,‖]	Sensitizer	E_T[†]
Benzene	84.5	Fluorescein	45.2–48.1	Acetone	—
Triphenylene*	66.6	Tetrachloro fluorescein	42.2–44.1	Propiophenone	74.6
Biphenyl	65.7	Eosin	43.2–46.0	Xanthone	74.2
Phenanthrene	62.2	Phloxin	39.5–42.2	Carbazole	70.1
1-Bromonaphthalene	59.0	Erythrosin	43.1–45.8	Thioxanthone	65.5
1,2-Benzanthracene	47.2	Rhodamine B	43[¶]	Flavone	62.0
Anthracene	42.0	Methylene blue	34[¶]	2-Naphtyhlphenyl ketone	59.6
Acenaphthylene	—	Zinc tetraphenyl porphin	—	1-Naphthylphenyl ketone	57.5
Azulene	38–31[‡]	Chrysazin	—	Butyraldehyde	—
Tetracene	29.4			Acrolein	—

* In methanol/benzene (1 : 1).
[†] Reference 12.
[‡] Reference 26.
[§] Reference 1 and 24.
[¶] Reference 11.
[‖] The higher value of E_T is determined from the onset of the phosphorescence spectrum; the lower value from the first short-wavelength maximum of the phosphorescence spectrum.

yields, Φ_{isc}, of nearly unity for the carbonyl sensitizers indicate very short lifetimes of the excited singlet states ($\leq 2 \times 10^{-10}$ sec for singlet-excited benzophenone[29]) which prevent these states from reacting with O_2 in bimolecular reactions even if the solutions are saturated with oxygen. In case of the aromatic hydrocarbons, however, fluorescence quenching by O_2 has been observed,[1,12,30] which may lead to energy transfer and thus to the formation of singlet oxygen. A possible mechanism for such an energy-transfer reaction has been discussed elsewhere.[1,2] According to the the postulated mechanism, interaction between the excited singlet sensitizer, $^1S_1$, and ground-state oxygen gives an excited sensitizer-oxygen pair[31] in its ^3G-state. This excited contact pair rapidly undergoes a spin-allowed transition to the energetically lower lying $^{1,3,5}F$ state, which is normally reached by interaction between excited triplet sensitizer, $^3S_1$, and $^3\Sigma_g^- O_2$. That is, reaction of ground-state oxygen with $^3S_1$ as well as with $^1S_1$ may result in the formation of the F-state contact pair, which, according to its energy, governs the production of $^1\Sigma_g^+$, $^1\Delta_g$, and $^3\Sigma_g^- O_2$ via transition into the repulsive $^1X_2$, $^1X_1$, and $^3G_0$ states, respectively. If, on the other hand, fluorescence quenching of certain aromatic hydrocarbons by ground-state oxygen should result only in enhanced internal conversion, the observed photooxygenation sensitized by such hydrocarbons would result exclusively from triplet sensitizer-oxygen interactions. Therefore, whether or not the excited singlet sensitizer takes part in the production of the oxygenating species, the triplet energy of the sensitizer should determine the probabilities with which the two singlet and the ground-state oxygen molecules are formed.

From the results obtained with the sensitizers of TABLES 1 and 2, it is inferred that only one (and the same) oxygenating intermediate is involved in the final product-producing reaction (9). Furthermore, since compounds of triplet energies below 37.7 kcal/mole act as sensitizers, and since oxygenations with the $^1\Delta_g O_2$-generating hydrogen peroxide/hypochlorite system give identical products,[4] the most likely common intermediate in all these oxygenation reactions is the singlet oxygen in its $^1\Delta_g$ state. Its formation is obviously independent of whether π, π^* or n, π^* sensitizers are used.

Quantum yields of singlet oxygen formation were determined in some cases (TABLE 1). In agreement with previous results,[6] Φ_{1O_2} was found to be about 0.8 for rose bengal. An elaborate kinetic analysis has shown that for this sensitizer, the quantum yield of photooxygenation equals that of triplet formation.[1,6] Generally, however, Φ_{1O_2} is smaller than Φ_{isc}. Φ_{1O_2} values determined for benzaldehyde, acetophenone, and benzophenone with DMF as substrate represent upper limits, since these compounds sensitize to some extent the free-radical oxidation of methanol by primary dehydrogenation of the alcohol. If Φ_{1O_2} is determined for the ketonic sensitizers by using TME as substrate, the quantum yield of photooxygenation de-decreases considerably, probably because of triplet-sensitizer quenching by TME. It is not known at present whether the singlet-excited states of fluorene, quinoline, and naphthalene take part in the energy-transfer process to ground-state oxygen or not. Direct evidence for the production of $^1\Delta_g O_2$ from triplet naphthalene[31,32] as well as from benzene,[33] however, was recently obtained. With pyrene, the quantum yield of photooxygenation is distinctly greater than that obtained for triplet production of monomeric pyrene. Further investigations are needed in order to determine whether monomeric singlet-excited pyrene or excited pyrene dimers or both are involved in the energy-transfer process to ground-state oxygen.

In order to further check on the possibility of $^1\Sigma_g^+ O_2$ and a sensitizer-oxygen contact-pair participation in the product forming step of photooxygenation reactions, (+)-limonene[9], (+)-isosylvestrene[18], (−)-caryophyllene,[19] and trimethyl ethylene[13] were employed as substrates since they are known to give characteristic

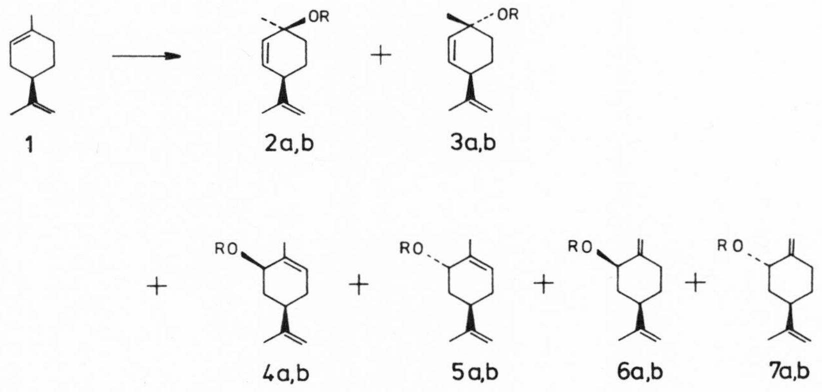

1

2a,b   +   3a,b

4a,b   +   5a,b   +   6a,b   +   7a,b

a : R = -OH
b : R = - H

TABLE 3

PHOTOSENSITIZED OXYGENATION OF (+)-LIMONENE IN METHANOL AT 20°C.

Sensitizer	Triplet Energy E_T(kcal/mole)	Alcohols from (+)-Limonene (%)*						$[\alpha]_D^{20}$ of Alcohol 5b(°)
		2b	3b	4b	5b	6b	7b	
Benzophenone	68.5	27	27	7	18	8	13	-18
Triphenylene†	66.5	38	10	3	8	20	21	-141
Quinoline	62.0	33	12	4	10	20	21	not determined
Pyrene	48.7	36	9	4	9	20	22	-176
Rose bengal	39.5–42.2	34	10	5	10	21	20	-178
Methylene blue	34.0	36	10	4	9	21	19	-178

* Product distribution after reduction of hydroperoxide mixture.
† In methanol/benzene (1 : 1).

products and product distributions in methylene blue- and rose bengal-sensitized photooxygenation reactions and, as was shown for (+)-limonene,[8] in $^1\Delta_g$ O_2-generating hydrogen peroxide/hypochlorite reactions. Sensitizers covering the triplet energy range from about 34 kcal/mole (methylene blue) to about 68 kcal/mole (triphenylene and benzophenone) were applied.

The alcohol mixtures obtained after reduction of the primarily formed hydroperoxide mixtures are practically identical when $^3(\pi, \pi^*)$ sensitizers were used which are hardly capable of effective dehydrogenation of the substrates. These

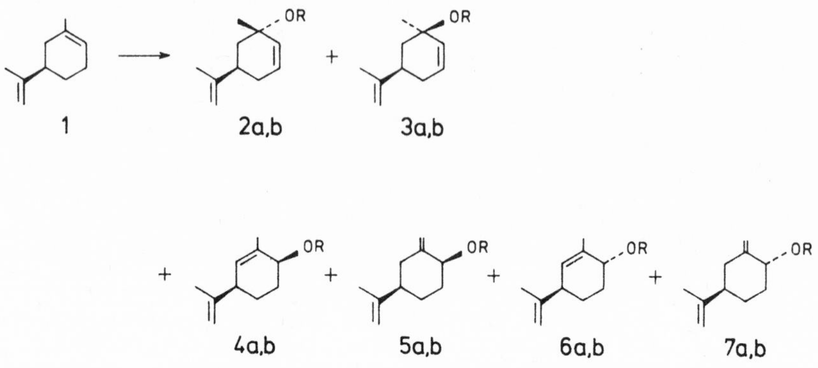

a : R = -OH
b : R = -H

TABLE 4

PHOTOSENSITIZED OXYGENATION OF (+)-ISOSYLVESTRENE IN METHANOL AT 20°C.

Sensitizer	Triplet Energy E_T(kcal/mole)	Alcohols from (+)-Isosylvestrene (%)*					
		2b	3b	4b	5b	6b	7b
Triphenylene[†]	66.5	38	6	10	17	5	24
Pyrene	48.7	38	4	11	17	5	25
Rose bengal	39.5–42.2	37	4	10	17	5	26
Methylene blue	34.0	38	4	10	17	5	26

* Product distribution after reduction of hydroperoxide mixture.
† In methanol/benzene (1 : 1).

results are again in agreement with the assumption that $^1\Delta_g$ O_2 exclusively participates in the product-forming step of photooxygenation reactions.

However, when the $^3(n, \pi^*)$-sensitizer benzophenone (of nearly the same triplet energy as the $^3(\pi, \pi^*)$-sensitizer triphenylene) was applied, the product distribution changes in a manner that indicates that free-radical oxidation reactions proceed in addition to the $^1\Delta_g$ O_2-oxygenation reaction. Whereas $^1\Delta_g$ O_2-oxygenation reactions, assumed to occur in a one-step concerted reaction via a six-membered cyclic transition state and thus to be sensitive to steric influences and the conformation of the participating allylic hydrogen atoms,[1,2,4,8-10] give rise to optically active carveols 4b and 5b from (+)-limonene, dehydrogenation in the allylic position of (+)-limonene must produce a symmetrical allylic radical from which racemic carveols

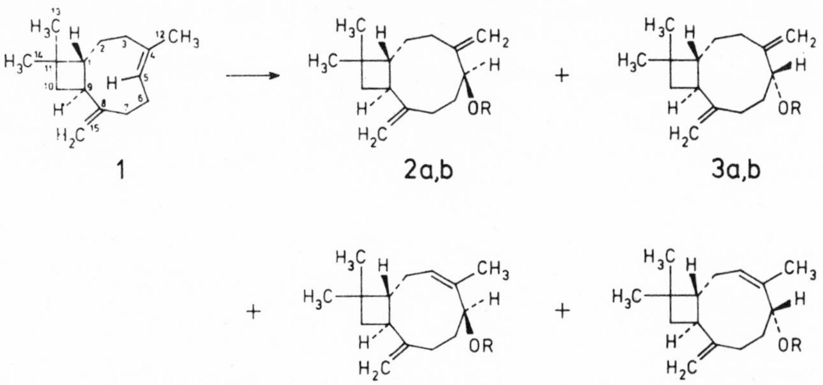

1 → 2a,b + 3a,b

4a,b + 5a,b

a: R = -OH
b: R = -H

1A ⇌ 1B

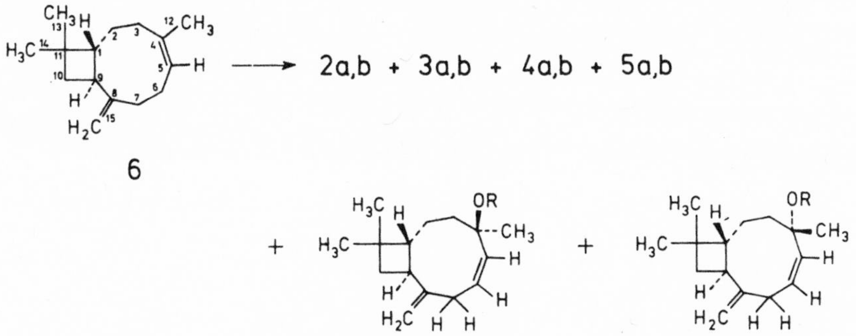

6 → 2a,b + 3a,b + 4a,b + 5a,b

+ 7a,b + 8a,b

TABLE 5
PHOTOSENSITIZED OXYGENATION OF (−)-CARYOPHYLLENE IN METHANOL AT 20°C.

Sensitizer	Triplet Energy E_T(kcal/mole)	Alcohols from Caryophyllene (%)*				$\dfrac{k_9^a}{k_9^b}$[†]
		2b	3b	4b	5b	
Benzophenone	68.5	?	7 :	13 :	2	4.0
Triphenylene[‡]	66.5	16	69	12	3	5.0
Quinoline	62.0	18	66	14	2	5.1
Naphthalene	60.9	17	72	10	1	5.1
Pyrene	48.7	17	72	9	2	5.2
Rose bengal	39.5–42.2	17	66	13	4	5.2
Methylene blue	34.0	not determined				5.8

* Product distribution after reduction of hydroperoxide mixture.
† Ratio of rates of (−)-caryophyllene and (−)-isocaryophyllene with singlet oxygen.
‡ In methanol/benzene (1 : 1).

253

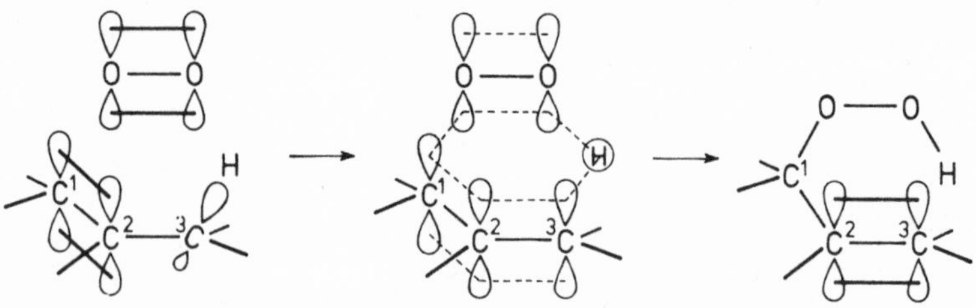

are obtained. Actually, $[\alpha]_D^{20}$ of 5b is very much reduced compared to that obtained from $^3(\pi, \pi^*)$-sensitized photooxygenations. Also, the enhanced production of 3b from (+)-limonene and the considerable decrease of 6b and 7b (secondary alcohols with exocyclic double bonds) indicate that some free-radical oxidation is occuring in the presence of benzophenone. Similarly, with (−)-caryophyllene and benzophenone as a sensitizer, production of hydroperoxides having exocyclic double bonds from C-4 to C-12 decreases drastically due to free-radical oxidation in which dehydrogenation from a methyl group is very much disfavored compared to that from a methylene group.

Competitive oxygenations were carried out with (−)-caryophyllene and (−)-isocaryophyllene. The latter gives rise to products 7a and 8a in addition to 2a through 5a as was found for rose bengal-sensitized photooxygenation reactions. Using clovane (C) as an inert internal standard, the ratios of rate constants k_9^a and k_9^b were obtained from the disappearance of (−)-caryophyllene (A) and (−)-isocaryophyllene (B):

$$^1O_2 + A \xrightarrow{k_9^a} AO_2 \tag{9a}$$

$$^1O_2 + B \xrightarrow{k_9^b} BO_2 \tag{9b}$$

$$-(d[A]/dt) = I_a \cdot {}^1O_2 \cdot \frac{k_9^a[A]}{k_9^a[A] + k_9^b[B] + k_8} \tag{15}$$

$$-(d[B]/dt) = I_a \cdot {}^1O_2 \cdot \frac{k_9^b[B]}{k_9^a[A] + k_9^b[B] + k_8} \tag{16}$$

$$-d[A]/-d[B] = (k_9^a/k_9^b) \cdot ([A]/[B]) \tag{17}$$

which leads to

$$\lg(A/C) = (k_9^a/k_9^b) \cdot \lg(B/C) + \text{const.} \tag{18}$$

with

A, B, C = peak areas (vpc) of A, B, and C.

$k_9^a/k_9^b = 5.8$, determined for the methylene blue-sensitized reaction agrees very well with that of 5.84 determined by Litt.[34] FIGURES 5 and 6 show the experimental data; k_9^a/k_9^b and the straight lines passing through the experimental points were obtained from calculations according to the method of the least squares; correlation coefficients were found to be highly significant.

The k_9^a/k_9^b values of the $^3(\pi, \pi^*)$ sensitizers are all close to 5.2–5.3, whereas that obtained with benzophenone is noticeably smaller, emphasizing again that the $^3(n, \pi^*)$-sensitized reaction is not a clean $^1\Delta_g$ O_2-oxygenation reaction.

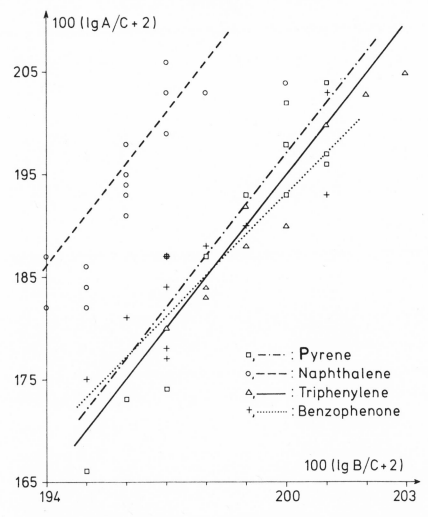

FIGURE 5. Competitive oxygenation of (−)-caryophyllene and (−)-isocaryophllene in methanol at 20°C., sensitized by benzophenone, triphenylene, quinoline, naphthalene, and pyrene.

In our opinion, the results so far obtained clearly indicate that only one oxygenating species is involved in the photosensitized oxygenation reactions in solution. This species is with all probability the $^1\Delta_g$ O_2, produced from $^3(\pi, \pi^*)$- as well as from $^3(n, \pi^*)$-sensitizers independently of the energy of these triplet-excited sensitizers. Deviations in the product distributions and ratios of rate constants when benzophenone is applied as sensitizer is very likely due to free-radical oxidations initiated by the $^3(n, \pi^*)$ sensitizer rather than by $^1\Sigma_g^+$ O_2 as was assumed in order to explain the varying product composition when cholest-4-en-3β-ol was oxygenated in the presence of sensitizers like methylene blue, rose bengal, erythrosin, eosin, fluorescein, or triphenylene.[11,35]

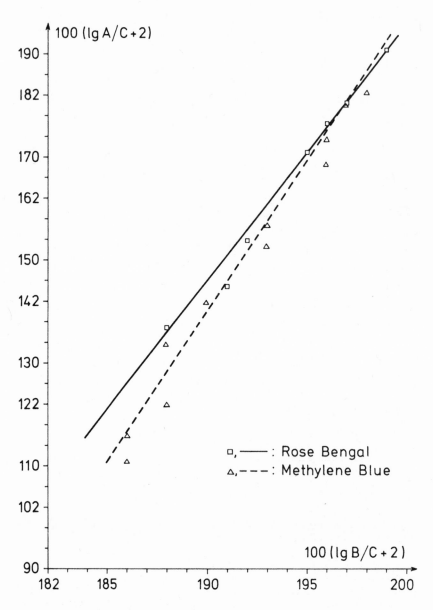

FIGURE 6. Competitive oxygenation of (−)-caryophyllene and (−)-isocaryophyllene in methanol at 20°C., sensitized by rose bengal and methylene blue.

When these $^3(\pi, \pi^*)$ sensitizers were used in the photooxygenation reaction of trimethylethylene, the ratio of the two products, a tertiary and a secondary allylic hydroperoxide, remained constant (55 : 45) in all cases (TABLE 6).

In view of our results it seems rather unlikely that reactions of $^1\Sigma_g^+$ O$_2$ produced from triplet-excited sensitizers by energy-transfer in solution have ever been observed. This conclusion is in agreement with Ogryzlo's results,[36] according to which

TABLE 6

PHOTOSENSITIZED OXYGENATION OF TRIMETHYLETHYLENE IN METHANOL AT 20°C.

Sensitizer	Triplet Energy E_T(kcal/mole)	Alcohols from Trimethylethylene (%)* tertiary	secondary
Fluorescein	45.2–48.1	55	45
Tetrachloro-fluorescein	42.2–44.1	55	45
Eosin	43.2–46.0	55	45
Tetrachloro-eosin (phloxin)	39.5–42.2	55	45
Erythrosin	43.1–45.8	55	45
Rose bengal	39.5–42.2	55	45
Rhodamin B	43.0	55	45
Zinc tetraphenylporphin	—	56	44
Methylene blue	34.0	55	45
Chrysazin	—	57	43

* Product distribution after reduction of hydroperoxide mixture.

the $^1\Sigma_g^+$ O$_2 \rightarrow {}^1\Delta_g$ O$_2$ transition is so effective that the rate constant for reactions with $^1\Sigma_g^+$ O$_2$ would have to be about 10^8 times larger than that for $^1\Delta_g$ O$_2$ reactions in order to be competitive. Furthermore, our results provide further evidence that a sensitizer-oxygen contact pair participation in the product-forming step (9) is very unlikely. These two species, excited sensitizer-oxygen contact pairs and $^1\Sigma_g^+$ O$_2$, should, however, play important roles as precursors of $^1\Delta_g$ O$_2$ in photosensitized oxygenation reactions in solution.

References

1. GOLLNICK, K. 1968. Advan. Photochem. 6: 1.
2. GOLLNICK, K. 1968. Advan. Chem. Ser. 77: 78.
3. GOLLNICK, K. & G. O. SCHENCK. 1967. Oxygen as a Dienophile. In 1,4-Cycloaddition Reactions; The Diels-Alder Reaction in Heterocyclic Syntheses. J. Hamer, Ed. : 255. Academic Press. New York, N.Y.
4. FOOTE, C. S. 1968. Acct. Chem. Res. 1: 104.
5. KHAN, A. U. & D. R. KEARNS. 1968. Advan. Chem. Ser. 77: 143.
6. GOLLNICK, K. & G. O. SCHENCK. 1964. Pure Appl. Chem. 9: 507.
7. COREY, E. J. & W. C. TAYLOR. 1964. J. Am. Chem. Soc. 86: 3881.
8. FOOTE, C. S., S. WEXLER & W. ANDO. 1965. Tetrahedron Letters 4111.
9. SCHENCK, G. O., K. GOLLNICK, G. BUCHWALD, S. SCHROETER & G. OHLOFF. 1964. Ann. Chem. 674: 93.

10. NICKON, A. & J. F. BAGLI. 1961. J. Am. Chem. Soc. 83: 1498.
11. KEARNS, D. R., R. A. HOLLINS, A. U. KHAN, R. W. CHAMBERS & P. RADLICK. 1967. J. Am. Chem. Soc. 89: 5455.
12. CALVERT, J. G. & J. N. PITTS, Jr. 1966. Photochemistry. J. Wiley & Sons. New York, N.Y.
13. FRANKEN, T. 1969. Ph.D. thesis. University of Bonn, Bonn, Germany.
14. GOLLNICK, K. & G. SCHADE. 1969. Ann. Chem. 721: 133.
15. SCHENCK, G. O. & K. H. SCHULTE-ELTE. 1958. Ann. Chem. 618: 185.
16. SCHULTE-ELTE, K. H. 1961. Ph.D. thesis. University of Göttingen, Göttingen, Germany.
17. FOOTE, C. S., M. T. WUESTHOFF, S. WEXLER, I. G. BURSTAIN, R. DENNY, G. O. SCHENCK & K. H. SCHULTE-ELTE. 1967. Tetrahedron 23: 2583.
18. GOLLNICK, K. & G. Schade. In press.
19. GOLLNICK, K. & G. SCHADE. 1968. Tetrahedron Letters 689.
20. SCHULTE-ELTE, K. H. & G. OHLOFF. 1968. Helvetica Chim. Acta 51: 494.
21. HATCHARD, C. G. & PARKER, C. A. 1956. Proc. Roy. Soc. (London) A235: 518.
22. HAMMOND, G. S. & P. A. LEERMAKERS. 1962. J. Am. Chem. Soc. 84: 207.
23. YOSHIHARA, K. & D. R. KEARNS. 1966. J. Chem. Phys. 45: 1991.
24. GOLLNICK, K. & T. FRANKEN. Unpublished.
25. PARKER, C. A. 1964. Advan. Photochem. 2: 305.
26. LAMOLA, A. A., W. G. HERKSTROETER, J. DALTON & G. S. HAMMOND. 1965. J. Chem. Phys. 42: 1715.
27. MÜLLER, W. 1956. Ph.D. thesis. University of Göttingen, Göttingen, Germany.
28. KOPP, R. 1958. Diplomarbeit. University of Göttingen, Göttingen, Germany.
29. MOORE, W. M., G. S. HAMMOND & R. P. FOSS. 1961. J. Am. Chem. Soc. 83: 2789.
30. PARMENTER, C. S. & J. D. RAU. 1969. J. Chem. Phys. 51: 2242.
31. KEARNS, D. R., A. U. KHAN, C. K. DUNCAN & A. H. MAKI. 1969. J. Am. Chem. Soc. 91: 1039.
32. WASSERMAN, E., V. J. KUCK, W. M. DELAVAN & W. A. YAGER. 1969. J. Am. Chem. Soc. 91: 1041.
33. SNELLING, D. R. 1968. Chem. Phys. Letters 2: 346.
34. LITT, F. A. 1967. Ph.D. thesis, Johns Hopkins University, Baltimore, Md.
35. KEARNS, D. R., R. A. HOLLINS, A. U. KHAN & P. RADLICK. 1967. J. Am. Chem. Soc. 89: 5456.
36. ARNOLD, S. J., M. KUBO & E. A. OGRYZLO. 1968. Advan. Chem. Ser. 77: 133.

Discussion of Paper

C. S. FOOTE: This is a very pretty piece of work and corresponds nicely with some things that we've done for which I apologize for not having published yet because we don't understand them in detail. We reinvestigated a system analogous to the one Nickon and his students originally looked at in the steroidal system. We looked at an oxidized alcohol that was closely analogous to it. We found that there was, in fact, a product distribution variation dependence on sensitizer as Nickon had reported, and as Kearns had also reported. However, this product distribution could not be explained by any simple mechanism based on two oxidizing species. To put it briefly, without going into details, there is no simple dependence on triplet energy of the sensitizer. There is a strong dependence on dye concentration. Hydrocarbons don't seem to give anomalous results, hydrocarbons as high as 55 kcal give essentially low energy behavior. It's only xanthene dyes and acridines, of the ones we looked at—we didn't look at a large number, about eight or nine, only those gave anomalous results. We, therefore, conclude that the reaction is a good deal more complex than has been suggested. The dye is certainly involved indirectly with the substrate at some stages of reaction and unfortunately, any simple series of schemes that we suggest such as hydrogen extraction by dye competition with say, $^1\Delta$ O$_2$ are still not sufficient to explain the results. But what is clear is that the simple scheme based on two oxygen singlets is not sufficient to explain the results, in fact, is inconsistent with the results that we have obtained.

D. R. KEARNS: We certainly agree that considerable doubt has been cast on the original evidence (the photooxygenation of cholesterol) for the participation of $^1\Sigma$ in singlet oxygen reactions. In particular, the quenching data of Ogryzlo make it very difficult to see how $^1\Sigma$ could have a sufficient lifetime in solution to allow it to react before it is quenched. Our prediction that with high-energy sensitizers the formation of $^1\Sigma$ is about 10 times faster than the formation of $^1\Delta$ (see Kawaoka, Khan and Kearns. 1967. J. Chem. Phys. **46**: 1842) still appears reasonable on theoretical grounds and remains to be tested experimentally.

A. NICKON: I might just add that in our studies of primary isotope effects with deuterium in place of hydrogen, using methylene blue and eosine as examples of low energy and high energy dyes, we get what seems to be the same primary isotope effect irrespective of the triplet energies of the dyes. That's negative evidence, but it fits in.

R. P. WAYNE: I think that Dr. Kearns has knowledge of something Dr. Donovan in Cambridge has been discussing about energy transfer from singlet states: The nonspin conserve reaction, that is, the fact that you would expect to get a much higher population of the $^1\Sigma$ state than of the $^1\Delta$ state in the energy transfer process.

Our investigations of quenching of both $^1\Sigma$ and of $^1\Delta$ show that $^1\Sigma$ is quenched much more effectively by nitrogen than it is by oxygen, whereas $^1\Delta$ is quenched much more effectively by oxygen than it is by nitrogen. Now, since $^1\Delta$ must go down to the ground state, which is a spin-forbidden process, you might expect that the paramagnetic species O$_2$ would be a more effective quencher than N$_2$, whereas $^1\Sigma$ can in fact be quenched to $^1\Delta$, and under those circumstances you might then look for some kind of resonance energy process. And that you can find from nitrogen but you cannot find it from oxygen. This is a nonspin conserve process.

D. S. BRESLOW: I found it rather interesting that the two isomers that you obtained from the trimethylethylene were formed in practically equal quantities. Can I infer from this that the trimethylethylene reacts at a comparable rate to tetramethylethylene?

K. GOLLNICK: Trimethylethylene is not as good a substrate as tetramethylethylene.

D. S. BRESLOW: Then how do you account for the fact that the relative reactivity of the two different carbon atoms, if this is actually an ene reaction, seem to be practically identical?

K. GOLLNICK: I think that this reaction with trimethylethylene is again a point against a perepoxide intermediate, because the perepoxide intermediate is such that you have a bridged ionic species in between. This is the formulation that was given by Sharp when he postulated this kind of mechanism.

All these cationic bridge intermediates open up in the Markovnikoff-type reaction, but we do not see, in any of these unsymmetrically substituted olefins, any Markovnikoff reactions. I think this again argues against such an intermediate. It is very nicely explained with an ene reaction. It is, of course, not very easy to know exactly how the singlet oxygen comes in on the double bond and on the hydrogen when it attacks the double bond. We get a little difference—I mean it is not a one-to-one ratio; it is 55 to 45, we know this, but we cannot explain this little difference at the moment.

D. S. BRESLOW: Well, I certainly agree that taking the reactions of the normal oxidation of a double bond, for example, that one would expect a much larger difference, depending upon whether you have a trisubstituted olefin or tetra-substituted olefin; this has been shown.

K. GOLLNICK: I haven't talked about the ratio of rates of singlet oxygen with tetramethylethylene and trimethylethylene. It was a product ratio. I think it's very hard to say anything about the tetramethylethylene because it just simply can't give you another product.

J. N. PITTS: We've been trying to unravel for well over a year the complexities of the kinetics of tetramethylethylene plus singlet oxygen in the gas phase. It's a mess. We've also measured rate constants, absolute rate constants for the two reactions, DMF and TME. Some of this work was monitored by following the quenching of the singlet oxygen.

We had an amusing experience in which we were trying to demonstrate, and in fact had demonstrated (only to follow up Snelling's work in the gas phase) that one could transfer in the gas phase from naphthalene to oxygen and then add to tetramethylethylene. We took a 40-m infrared cell, and we added tetramethyl-ethylene, benzaldehyde, and oxygen, and irradiated it. The idea is that benzaldehyde absorbs the radiation and transfers it to the oxygen, which attacks the TME and of course you get nice yields of peroxide. Dr. Coombers took out the benzaldehyde and we still found the hydroperoxide formed. We looked at this very carefully and drew the conclusion that, in fact, if we irradiate at 3,200 Å (nothing absorbs supposedly, the TME shouldn't absorb, the oxygen shouldn't absorb), we get nice production of hydroperoxide.

We feel now we have good evidence for the formation of a complex, whatever one wants to call it. If you look at the absorption curve of TME and oxygen, look back at what Mulliken did and look back at what Evans did, but do it just in simple systems relating to the atmosphere or gas phase photochemistry, you find that it might well absorb. So I think this again confirms that there is an oxygen-perturbed absorption that leads to the $^1\Delta$, which leads to very efficient formation of the hydro-peroxide. Only it's done in such small yields that one needs a trap and a 40-m path-length cell to detect the production of the hydroperoxide.

45

Reprinted from *Chem. Phys. Lett.*, **10**(2), 113–116 (1971)

FORMATION OF $O_2(^1\Sigma_g^+)$ BY 1-FLUORONAPHTHALENE SENSITIZATION

L.J. ANDREWS AND E.W. ABRAHAMSON

Department of Chemistry, Case Western Reserve Univeristy,
Cleveland, Ohio 44106, USA

Received 9 December 1970

Emission from $O_2(^1\Sigma_g^+)$ was detected following the flash photolysis of O_2 and 1-fluoronaphthalene mixtures in the gas phase. Assignment of the emitting state was confirmed by O_2, N_2 and C_2H_4 quenching studies. The rate constant for 1-fluoronaphthalene quenching of $O_2(^1\Sigma_g^+)$ was estimated to be $(7.6 \pm 0.7) \times 10^8$ M^{-1} sec^{-1}.

1. Introduction

The production of $O_2(^1\Delta_g)$ by electronic energy transfer from the lowest triplet levels (T_1) of a number of aromatic hydrocarbons has recently been demonstrated by both optical emission [1–3] and electron paramagnetic resonance [4, 5] studies. These investigations have provided direct confirmation of Kautsky's original proposal [6] that luminescence quenching of dyestuffs by oxygen is attended by energy transfer to oxygen. It has been theoretically predicted [7], however, that the second singlet level of oxygen, $O_2(^1\Sigma_g^+)$, should be formed with greater efficiency than $O_2(^1\Delta_g)$ when the sensitizer T_1 level has sufficient energy. There is an argument for the participation of $O_2(^1\Sigma_g^+)$ in certain solution photo-oxygenations [8], based on product analysis, but this has recently been challenged [9]. As the formation of $O_2(^1\Sigma_g^+)$ has eluded direct detection in systems favorable for its production because of the very rapid collisional quenching which this species undergoes [10], we accordingly sought for this state in emission using a time resolved flash photolysis technique

2. Experimental

Gaseous mixtures of O_2 and 1-fluoronaphthalene (1-FN) with and without quenchers were subjected to a 1770 Joule flash photolysis in a cylindrical quartz cell, 30 mm i.d. and 90 mm in length. The cell was surrounded with a quartz jacket, 46 mm i.d., filled with aqueous 3M $NiCl_3 \cdot 6H_2O$. This chemical filter has a window in the quartz ultraviolet and blocks radiation at 762 nm. Emission following flash photolysis was detected via a wide band interference filter centered at 760 nm (the emission $^3\Sigma_g^- \leftarrow {}^1\Sigma_g^+$ for O_2 has its 0–0 and 1–1 bands at 7619 Å and 7700 Å, respectively) and an S-20 PM tube (EMI 9558C). Using a glass filter which blocks light below 655 nm (Corning 2–64) and a light tight connection between the PM tube housing and the cell to further shield the PM tube from the flash, emission intensities were recorded as early as 75 μsec after the peak flash intensity. The concentration of 1-FN was measured by observing the change in percent transmission of the 2804 Å Hg line which occurred when the evacuated cell was saturated with 1-FN vapor. The Hg line was isolated from a 140 watt medium pressure Hg arc

(Hanovia) with a grating monochromator (Bausch and Lomb) set for a 3.3 Å band pass and detected with an S-5 PM tube (G.E. 1P28A). The extinction coefficient of 1-FN vapor at 2804 Å is $4.11 \times 10^3 \, M^{-1} \, cm^{-1}$.

3. Results and discussion

Weak emission ($S/N \approx 5$ or 6) which decayed exponentially with $\tau_{1/2} = 83 \, \mu sec$ was observed following the flash photolysis of 50 torr O_2 saturated with 1-FN vapor (0.19 torr). The emission lifetime approximately doubled when the 1-FN concentration was halved. Control experiments showed that both 1-FN and O_2 were necessary to produce the emission.

3.1. Determination of k_{N_2} and $k_{C_2H_4}$

As consistent values of the rate constants for N_2 and C_2H_4 quenching of $O_2(^1\Sigma_g^+)$, k_{N_2} and $k_{C_2H_4}$, have recently been published [11–15], it was decided that determination of these rate constants for quenching of the observed emission was a useful means of confirming the origin of the luminescence as $O_2(^1\Sigma_g^+)$.

The observed first-order luminescence decay rate constant, k_{app}, was measured as a function of quencher (Q) concentration. Plots of k_{app} versus [Q] (figs. 1 and 2) were linear for Q = N_2 and C_2H_4 and yielded $k_{N_2} = (1.5 \pm 0.2) \times 10^6 \, M^{-1} \, sec^{-1}$ and $k_{C_2H_4} = (3.4 \pm 0.9) \times 10^8 \, M^{-1} \, sec^{-1}$. The present results and those of other investigators are listed in table 1.

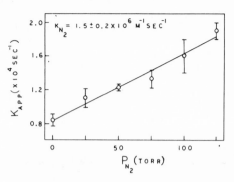

Fig. 1. Plot of k_{app} versus nitrogen pressure yielding k_{N_2}.

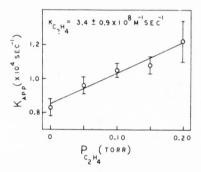

Fig. 2. Plot of k_{app} versus ethylene pressure yielding $k_{C_2H_4}$.

Table 1

k_{N_2} ($M^{-1} \, sec^{-1}$)	Ref.
1.2×10^6	Filseth et al. [11]
$(1.2 \pm 0.3) \times 10^6$	Noxon [12]
1.4×10^6	Izod and Wayne [13]
1.5×10^6	Arnold et al. [14]
$(1.5 \pm 0.2) \times 10^6$	present work
$k_{C_2H_4}$ ($M^{-1} \, sec^{-1}$)	
$(1.9 \pm 0.9) \times 10^8$	Filseth et al. [11]
$(2.6 \pm 0.6) \times 10^8$	Stuhl and Niki [15]
$(3.4 \pm 0.9) \times 10^8$	present work

It was observed qualitatively that O_2 is a poorer quencher of the emission than N_2, in agreement with quantitative studies that show $k_{O_2} \cong k_{N_2}/10$ for $O_2(^1\Sigma_g^+)$ quenching. [10, 11].

3.2. Determination of $k_{1\text{-}FN}$

Relaxation of $O_2(^1\Sigma_g^+)$ in 50 torr O_2 saturated with 1-FN vapor occurs exclusively through bimolecular quenching by O_2 and 1-FN. Wall deactivation and radiative relaxation [16] are negligible on the time scale of our measurements. Thus, we can obtain k_{app} in terms of the quenching constants for $O_2(^1\Sigma_g^+)$ by $O_2(^3\Sigma_g^-)$, k_{O_2}, and 1-FN in its ground state, $k_{1\text{-}FN}$.

$$k_{app} = k_{O_2}[O_2] + k_{1\text{-}FN}[1\text{-}FN] \,.$$

Under the above conditions, using the average of eight

determinations of k_{app}, $(0.83 \pm 0.05) \times 10^4 \text{ sec}^{-1}$, percent transmission at 2804 Å equal to 6.0% ($\therefore$ [1-FN] $= 1.03 \times 10^{-5}$ M); and the average of the values reported in ref. [10] and ref. [11] for k_{O_2} [$(1.8 \pm 0.9) \times 10^5 \text{ M}^{-1} \text{ sec}^{-1}$]; we find $k_{1\text{-FN}} = (7.6 \pm 0.7) \times 10^8$ $\text{M}^{-1} \text{ sec}^{-1}$.

3.3. Nature of the $O_2(^1\Sigma_g^+)$ precursor

There was no significant change in emission intensity when the O_2 pressure was raised from 13 to 700 torr. This implies that nearly all the sensitizing state of naphthalene is being quenched at the lower pressure, which from simple kinetic theory corresponds to a collision frequency of $\approx 3 \times 10^8 \text{ sec}^{-1}$. The possible pathways by which singlet oxygen may be produced are:

$$S_1 + O_2(^3\Sigma_g^-) \to T_1 + O_2(^1\Delta_g \text{ or } ^1\Sigma_g^+), \qquad (1)$$

$$S_1 + O_2(^3\Sigma_g^-) \to S_0 + O_2(^1\Delta_g \text{ or } ^1\Sigma_g^+), \qquad (2)$$

$$T_1 + O_2(^3\Sigma_g^-) \to S_0 + O_2(^1\Delta_g \text{ or } ^1\Sigma_g^+). \qquad (3)$$

Process (1) may be ruled out for the production of $O_2(^1\Sigma_g^+)$ because the necessary energetic requirements are not satisfied, the S_1–T_1 separation of naphthalene (and by inference 1-FN) being about 10600 cm^{-1} – 2600 cm^{-1} less than needed to populate the $^1\Sigma_g^+$ state. $O_2(^1\Delta_g)$, on the other hand, may be formed by this process since only 7900 cm^{-1} are required. Process (2) has the requisite energy to produce $O_2(^1\Sigma_g^+)$ but should be inefficient by Förester type transfer [19] because of the very small dipole moment associated with the $O_2(^3\Sigma_g^-) \to O_2(^1\Sigma_g^+)$ transition, and the relatively short lifetime of the S_1 state of 1-FN as deduced from lifetime measurements on naphthalene [17]. The most probable mechanism for the population of the $^1\Sigma_g^+$ state is clearly process (3) as the longer lived triplet state not only has the necessary energy to populate the $^1\Sigma_g^+$ state but the excitation transfer is spin allowed in a Wigner sense.

3.4. Possible secondary pathways to $O_2(^1\Sigma_g^+)$

If $O_2(^1\Delta_g)$ alone were formed in the primary energy transfer from electronically excited 1-FN, then the observed presence of $O_2(^1\Sigma_g^+)$ could be accounted for in only two ways. Firstly, $O_2(^1\Delta_g)$ could be excited to the $^1\Sigma_g^+$ level by collision with an excited 1-FN molecule in its S_1 or T_1 level. This mechanism predicts $O_2(^1\Sigma_g^+)$ luminescence to be strongly O_2 pressure dependent. Furthermore, at high O_2 pressures the probability of S_1 or T_1 surviving a large number of collisions with $O_2(^3\Sigma_g^-)$ in order to transfer energy to $O_2(^1\Delta_g)$ is vanishingly small. Secondly, two $O_2(^1\Delta_g)$ molecules could collide producing one $O_2(^1\Sigma_g^+)$ molecule and one $O_2(^3\Sigma_g^-)$ molecule in the well known energy pooling process [10]. Using the measured rate constant for this reaction ($k = 1.4 \times 10^3 \text{ M}^{-1} \text{ sec}^{-1}$) and the highest possible $O_2(^1\Delta_g)$ concentration in our system the expression $k[O_2(^1\Delta_g)]^2$ predicts an $O_2(^1\Sigma_g^+)$ formation rate which is too slow by many orders of magnitude to explain the fast buildup and observed first order luminescence decay rate constant for $O_2(^1\Sigma_g^+)$.

4. Conclusion

From the above considerations, we conclude that $O_2(^1\Sigma_g^+)$ is produced directly in energy transfer from electronically excited 1-FN, largely in its T_1 level, to ground state oxygen. It is known that $O_2(^1\Delta_g)$ is formed efficiently in this system [4, 5], but it is not known whether $O_2(^1\Sigma_g^+)$ is a precursor of $O_2(^1\Delta_g)$ or if both forms of excited oxygen are produced simultaneously in energy transfer from the 1-FN sensitizer. Experiments are now in progress to clarify this point.

Acknowledgements

We gratefully acknowledge the U.S. Atomic Energy Commission for their support to this work under contract number AT(11-1)-904. Our thanks are due also to Mr. James Davidson for his helpful criticism in the preparation of this paper.

References

[1] D.R. Snelling, Chem. Phys. Letters 2 (1968) 346.
[2] R.H. Kummler and M.H. Bortner, Environ. Sci. Tech. 3 (1969) 944.
[3] R.P. Steer, J.L. Sprung and J.N. Pitts Jr., Environ. Sci. Tech. 3 (1969) 946.

[4] D.R. Kearns, A.U. Khan, C.K. Duncan and A.H. Maki, J. Am. Chem. Soc. 91 (1969) 1039.

[5] E. Wasserman, V.J. Kuck, W.M. Delavan and W.A. Yager, · J. Am. Chem. Soc. 91 (1969) 1041.

[6] H. Kautsky and H. de Bruijn, Naturwissenschaften 19 (1931) 1043.

[7] K. Kawaoka, A.U. Khan and D.R. Kearns, J. Chem. Phys. 46 (1967) 1842.

[8] D.R. Kearns, R.A. Hollins, A.U. Khan, R.W. Chambers and P. Radlick, J. Am. Chem. Soc. 89 (1967) 5455; D.R. Kearns, R.A. Hollins, A.U. Khan and P. Radlick, J. Am. Chem. Soc. 89 (1967) 5456.

[9] C.S. Foote and S.Y. Wong, Abstr. of Papers, Am. Chem. Soc. Appl. Meeting (1969) Minneapolis, Petr. 12.

[10] R.P. Wayne, Advan. Photochem. 7 (1969) 311.

[11] S.V. Filseth, A. Zia and K.H. Welge, J. Chem. Phys. 52 (1970) 5502.

[12] J.F. Noxon, J. Chem. Phys. 52 (1970) 1852.

[13] R.P.J. Izod and R.P. Wayne, Proc. Roy. Soc. 81 (1968).

[14] S.J. Arnold, M. Kubo and E.A. Ogryzlo, Advan. Chem. Ser. 77 (1968) 133.

[15] F. Stuhl and H. Niki, Chem. Phys. Letters 5 (1970) 573.

[16] L. Wallace and D.M. Hunten, J. Geophys. Res. 73 (1968) 4813.

[17] I. Berlmann, Handbook of fluorescence spectra of aromatic molecules (Academic Press, New York, 1965).

[18] J. Ferguson, J. Chem. Soc. (1954) 304; J. Ferguson, T. Iredale and J.A. Taylor, J. Chem. Soc. (1954) 3160.

[19] T. Förster, Ann. Physik 2 (1948) 55.

46

Reprinted from *J. Chem. Phys.*, 55(12), 5822–5823 (1971)

Spectroscopic Detection of the Photosensitized Formation of $^1\Sigma$ Oxygen in the Gas Phase

CHRISTOPHER K. DUNCAN AND DAVID R. KEARNS

Department of Chemistry, University of California, Riverside, California 92502

(Received 27 September 1971)

Several years ago we carried out a theoretical examination of the oxygen quenching of excited triplet state molecules.[1] The three possible mechanisms which we considered were

$$^3S + {}^3O_2 \xrightarrow{k_1} S + {}^1\Sigma, \tag{1}$$

$$^3S + {}^3O_2 \xrightarrow{k_2} S + {}^1\Delta, \tag{2}$$

$$^3S + {}^3O_2 \xrightarrow{k_3} S + {}^3O_2. \tag{3}$$

With high energy ($E_T > 13\,500$ cm^{-1}) triplet state sensitizers theory predicts $k_1 \simeq 10 k_2 \simeq 100 k_3$. According to these results the initial quenching step should lead to the formation of both $^1\Sigma$ and $^1\Delta$ oxygen with $^1\Sigma$ formation being favored over $^1\Delta$ by a factor of 10 to 1. The quantity $\Phi_\Sigma = k_1/(k_1 + k_2)$ which measures the partitioning of the energy between $^1\Sigma$ and $^1\Delta$ is therefore expected to approach 0.9. $^1\Sigma$ oxygen is very sensitive to deactivation to delta,[2-4] and because of this it has not previously been possible to experimentally measure the partition ratio Φ_Σ. Experiments do, however, confirm the other strong theoretical prediction that $(k_1 + k_2) \gg k_3$.[5-7] In this Communication we now present spectroscopic evidence that $^1\Sigma$ is generated directly by energy transfer from triplet state sensitizers and obtain a preliminary estimate of Φ_Σ.

The experimental setup used in these measurements is depicted in Fig. 1. With this apparatus excited oxygen molecules were generated either by passing oxygen through a microwave discharge, or by photoexcitation of a sensitizer, quinoxaline, introduced downstream from the microwave discharge, or by both methods simultaneously. The concentration of $^1\Delta$ was monitored by EPR a short distance downstream from the irradiation chamber. The formation of $^1\Sigma$ in the various experiments was followed by monitoring the emission at 762 nm using a detection system consisting

of a collimator, a 7-60 Corning filter (o.d. >2 for wavelengths shorter than 580 nm), a Jarrel-Ash $\frac{1}{4}$-m monochromator, and a cooled EMI 9558 photomultiplier. When we irradiated mixtures of oxygen and gaseous quinoxaline at 313 nm with light from an AH-6 lamp, a new emission band centered at 762 nm could be observed. The same emission band was also observed when the microwave discharge was operated alone, and when both the discharge and the lamp were operated simultaneously (see Fig. 2). When no sensitizer was present in the photoexperiment, no emission at 762 nm could be detected above the scattered light background which was about the same as when quinoxaline was present. When the discharge is operated alone, the $^1\Sigma$ oxygen is produced by $^1\Delta$-$^1\Delta$ annihilation ($k_{\Delta-\Delta} = 1.2 \times 10^3 M^{-1} \cdot \sec^{-1}$).[3,8,9] In the photosensitization experiments, however, we conclude that $^1\Sigma$ is generated primarily by direct energy transfer from

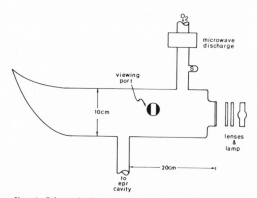

FIG. 1. Schematic diagram of the experimental setup used to observe the 762 nm emission from $^1\Sigma$ oxygen. The sensitizer, quinoxaline, was introduced downstream from the microwave discharge at the point S. The important cell dimensions are indicated.

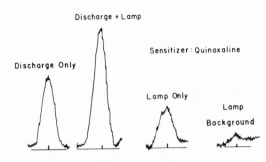

Discharge + Lamp

Discharge Only

Sensitizer : Quinoxaline

Lamp Only

Lamp
Background

FIG. 2. The 762 nm emission from $^1\Sigma$ produced by a microwave discharge, by photosensitized energy transfer from triplet state quinoxaline sensitizers, and by both sources operating together. In the experiments depicted, the O_2 pressure was 2.0 torr, and the He pressure was 3.5 torr. The concentration of $^1\Delta$ generated by the microwave discharge was $6\times10^{-6}M$, whereas in the photoexperiments the *net* $^1\Delta$ concentration exiting from the irradiation cell was $2\times10^{-6}M$. The residence time of O_2 in the irradiation cell was 4 sec and the lifetime of $^1\Delta$ was approximately 10–12 sec.

triplet state quinoxaline for the following reasons. First of all, the photoinduced emission is observed at 762 nm, which is the same position observed from $^1\Sigma$ generated by the microwave discharge. Secondly, the net concentration of $^1\Delta$ generated in these photoexperiments due to light absorption down the *entire* length of the irradiation tube is much too low to account for the observed intensity of $^1\Sigma$ emission. We estimate that if $^1\Delta$–$^1\Delta$ annihilation were responsible for the $^1\Sigma$ emission in the photosensitization experiments, it would have been 50 times smaller than that

actually observed. The experimental observation that the intensity of the $^1\Sigma$ emission varies linearly with the light intensity also rules out $^1\Delta$–$^1\Delta$ annihilation as the source of $^1\Sigma$ in the photosensitization experiments.

If we assume that deactivation of $^1\Sigma$ leads to the formation of $^1\Delta$ oxygen,[4] then it is possible to use the above experimental results in conjunction with EPR measurements of the photogenerated $^1\Delta$ concentration downstream from the irradiation cell, and a knowledge of the geometry of the detection system to obtain a rough estimate of Φ_Σ. The results of such a calculation lead to a value $\Phi_\Sigma \simeq 0.4$. Because of the various factors which enter into this calculation, it is difficult to assess the error limits on this number. It could be off by a factor of 2, but probably not much more. Accepting the above value for Φ_Σ, we conclude that the rate constant for the formation of $^1\Sigma$ by energy transfer from triplet state sensitizers is at least comparable to, and perhaps somewhat larger than, the rate constant for the sensitized formation of $^1\Delta$. It would appear, then, that most of the major features of our earlier theoretical calculations have now been verified.[1]

The support of the U.S. Public Health Service (Grant No. CA 11459) is most gratefully acknowledged.

[1] K. Kawaoka, A. U. Khan, and D. R. Kearns, J. Chem. Phys. 46, 1842 (1967).
[2] S. J. Arnold, M. Kubo, and E. A. Ogryzlo. Advan. Chem. Ser. 77, 133 (1968).
[3] R. P. Wayne, Advan. Photochem. 1, 311 (1969).
[4] D. R. Kearns, Chem. Rev. 71, 395 (1971).
[5] D. R. Kearns, A. U. Khan, C. K. Duncan, and A. H. Maki, J. Am. Chem. Soc. 91, 1039 (1969).
[6] E. Wasserman, V. J. Kuck, W. M. Delevan, and W. A. Yager, J. Am. Chem. Soc. 91, 1040 (1969).
[7] D. R. Snelling, Chem. Phys. Letters 2, 346 (1968).
[8] R. A. Young and G. J. Black, J. Chem. Phys. 44, 3741 (1966).
[9] S. J. Arnold and E. A. Ogryzlo, Can. J. Phys. 45, 2053 (1969).

Editor's Comments
on Papers 47 Through 53

CHEMICAL SOURCES OF SINGLET OXYGEN

The generation of singlet oxygen from the reaction of NaOCl and
H_2O_2 and the electric discharge both played a particularly important
role in establishing the intermediacy of singlet oxygen in photooxygenation. These sources were therefore considered separately.

Murray and Kaplan (Paper 47) found that triphenyl phosphite ozonide, formed at –78°C by the addition of ozone to the phosphite, decomposes at –35°C to yield singlet oxygen. The ozonide can be used
as a convenient source of $^1O_2$ for oxygenating various substrates. There
had been some question as to whether free $^1\Delta_g$ oxygen was generated
or whether the ozonide was acting simply as an oxygen transfer agent.[1]

This was answered by observation of the epr spectrum of $^1\Delta_g$ O_2 produced by the decomposition of the ozonide (Paper 48). Several cyclic phosphite ozonides of considerably greater stability have been reported recently.[2-4]

Moureu, Dufraisse, and Dean (Paper 5) reported the first example of a reversible addition of oxygen to an aromatic hydrocarbon in 1926. Heating rubrene peroxide regenerated the red hydrocarbon with concomitant evolution of oxygen. Wasserman and Scheffer (Paper 49) have demonstrated that the thermal decomposition of 9,10-diphenylanthracene endoperoxide produces $^1O_2$, which can be trapped by acceptors in solution.

Howard and Ingold (Paper 50) observed that the Russell termination mechanism for *sec*-peroxy radicals yields singlet oxygen. Although this is not a useful, preparative source of singlet oxygen, it does provide an interesting example of a reaction in which $^1O_2$ is generated.

Khan (Paper 51) has suggested that the dismutation reaction of superoxide anion radical yields singlet oxygen. This paper has had a considerable impact in the area of biological (enzymatic) oxidation processes.[5] Several enzyme systems produce the superoxide anion radical.

$$O_2^{\bar{\cdot}} + O_2^{\bar{\cdot}} \xrightarrow{\text{H}^+} {}^1O_2 + H_2O_2$$

Evidence has also been presented for the formation of $^1O_2$ in the base-induced decomposition of peroxyacetylnitrate (Paper 52) and in microwave discharge through CO_2 (Paper 53).

Pitts and coworkers have reported that potassium perchromate (K_3CrO_8) generates singlet oxygen upon aqueous decomposition. However, the utility of this reaction as a source of singlet oxygen is reduced by other competing oxidative processes.[6]

NOTES AND REFERENCES

1. Triphenyl phosphite ozonide also undergoes a *direct* reaction with several alkenes: P. D. Bartlett and G. D. Mendenhall, *J. Amer. Chem. Soc.*, **92**, 210 (1970), and A. P. Schaap and P. D. Bartlett, *ibid.*, **92**, 6055 (1970).
2. M. E. Brennan, *Chem. Commun.*, 956 (1970).
3. L. M. Stephenson and D. E. McClure, *J. Amer. Chem. Soc.*, **95**, 3074 (1973).
4. A. P. Schaap, K. Kees, and A. L. Thayer, *J. Org. Chem.*, **40**, 1185 (1975).
5. Kearns has recently described several experiments that seem to indicate that $^1O_2$ is not formed from the dismutation reaction of $O_2^{\bar{\cdot}}$: R. Nilsson and D. R. Kearns, *J. Phys. Chem.*, **78**, 1681 (1974). Work on this problem is continuing in several laboratories.
6. J. W. Peters, P. J. Bekowies, A. M. Winer, and J. N. Pitts, Jr., *J. Amer. Chem. Soc.*, **97**, 3299 (1975).

47

Reprinted from *J. Amer. Chem. Soc.*, **91**(19), 5358–5364 (1969)

Singlet Oxygen Sources in Ozone Chemistry. Chemical Oxygenations Using the Adducts between Phosphite Esters and Ozone[1]

R. W. Murray[2] and M. L. Kaplan

*Contribution from the Bell Telephone Laboratories, Inc.,
Murray Hill, New Jersey   07974. Received April 23, 1969*

Abstract: .The adduct between ozone and triphenyl phosphite provides a convenient method of chemical oxygenation for synthetic or mechanistic purposes.   Control experiments, esr results, and the nature of the chemistry observed lead to the conclusion that the oxygenation agent is free singlet oxygen.

The highly significant report by Foote and Wexler[3] that the reaction of sodium hypochlorite and hydrogen peroxide produces an oxidizing species which gives products identical with those of dye-sensitized oxygenations has initiated an intensive investigation of the chemistry of the species suggested[3] as being the common oxidizing agent in these apparently diverse systems, namely, $^1\Delta$-oxygen.   While many of these reports have been concerned with demonstrating further the likely involvement of $^1\Delta$-oxygen in dye-sensitized oxygenations,[4] others have described additional methods of producing $^1\Delta$-oxygen either for use in chemical oxygenation or as evidence for its intermediacy.   Other such methods reported include the use of a radiofrequency discharge in gaseous oxygen,[5,6] the reaction of bromine and hydrogen peroxide,[7] the decomposition of alkaline solutions of peracids,[7] the decomposition of photoperoxides,[8] and the self-reaction of *sec*-butylperoxy radicals.[9]

We previously reported in preliminary form that the adduct between triphenyl phosphite and ozone can be used to accomplish typical singlet oxygen oxidations both in solution[1] and in the gas phase.[1] It has also been reported that the oxygen produced in the decomposition of the triphenyl phosphite–ozone adduct contains $^1\Delta$-oxygen as demonstrated by its characteristic epr absorptions.[10] In this paper we describe further the use of phosphite ester–ozone adducts in chemical oxygenations as well as some additional evidence on the nature of the oxidizing species and the stability of the adduct. This work provides additional evidence that $^1\Delta$-oxygen is indeed the oxygenation reagent and indicates that the triphenyl phosphite–ozone adduct is a convenient and versatile source of singlet oxygen for use in chemical oxygenation. It further suggests that there may be many additional sources of singlet oxygen in ozone chemistry and that such a possibility has important con-

sequences not only to synthetic and mechanistic ozone chemistry studies but also to such areas as air pollution and biological oxidations.

Results and Discussion

A number of cases are known in which ozone reacts with a substrate in such a way that only one oxygen atom of the ozone is incorporated in the oxidized product.   The remaining two oxygen atoms of the ozone are usually evolved as molecular oxygen.   Examples include the oxidation of tertiary amines to amine oxides, phosphines to phosphine oxides, sulfides to sulfoxides, and sulfoxides to sulfones.[11]   Even in ozonolysis, where the usual reaction leads to all three oxygen atoms of ozone being incorporated in products, there are a number of cases reported where again only one of the ozone oxygen atoms is accounted for by the major reaction products.   These latter cases are usually 1-olefins with special structural features which lead to the corresponding epoxide as the ozonization product.[12–15]

A consideration of the conservation of spin principle suggested to us some time ago that the oxygen involved in some or all of these reactions probably has singlet multiplicity.   This possibility has also been considered by Corey and Taylor.[5] The recent intensive investigations of singlet oxygen beginning with the work of Foote and Wexler[3] provided us with the additional incentive to examine the general question of singlet oxygen sources in ozone chemistry.

The case examined in our initial investigation in this area was reported first in 1961 by Thompson.[16] Thompson had observed that ozone and triaryl phosphites formed 1:1 adducts which were stable at low temperatures.   When these adducts were allowed to warm the corresponding phosphates were produced and molecular oxygen was evolved.   In such cases, we would be able to purge the reaction solution of excess ozone and use reactions diagnostic of singlet oxygen to examine the evolved oxygen while avoiding the added complexity which accompanying ozone oxidations would bring to product analysis.   This latter complication still remains in other cases we have attempted or contemplated.

(1) Preliminary accounts of this work have appeared: R. W. Murray and M. L. Kaplan, *J. Am. Chem. Soc.*, **90**, 537, 4161 (1968).

(2) To whom all correspondence should be addressed at the Chemistry Department, The University of Missouri at St. Louis, St. Louis, Mo. 63121.

(3) C. S. Foote and S. Wexler, *J. Am. Chem. Soc.*, **86**, 3879 (1964).

(4) For a recent summary of this aspect of $^1\Delta$ oxygen work, see C. S. Foote, S. Wexler, W. Ando, and R. Higgins, *ibid.*, **90**, 975 (1968).

(5) E. J. Corey and W. C. Taylor, *ibid.*, **86**, 3881 (1964).

(6) A. M. Winer and K. D. Bayes, *J. Phys. Chem.*, **70**, 302 (1966).

(7) E. McKeown and W. A. Waters, *J. Chem. Soc., B*, 1040 (1966).

(8) H. H. Wasserman and J. R. Scheffer, *J. Am. Chem. Soc.*, **89**, 3073 (1967).

(9) J. A. Howard and K. U. Ingold, *ibid.*, **90**, 1956 (1968).

(10) E. Wasserman, R. W. Murray, M. L. Kaplan, and W. A. Yager, *ibid.*, **90**, 4160 (1968).

(11) P. S. Bailey, *Chem. Rev.*, **58**, 925 (1958).

(12) R. C. Fuson, M. D. Armstrong, W. E. Wallace, and J. W. Kneisley, *J. Am. Chem. Soc.*, **66**, 1274 (1944).

(13) P. D. Bartlett and M. Stiles, *ibid.*, **77**, 2806 (1955).

(14) R. Criegee, Advances in Chemistry Series, No. 21, American Chemical Society, Washington, D. C., 1959, p 133.

(15) P. S. Bailey and A. G. Lane, *J. Am. Chem. Soc.*, **89**, 4473 (1967).

(16) Q. E. Thompson, *ibid.*, **83**, 846 (1961).

In the general procedure the adduct is formed from triphenyl phosphite and ozone at −78° by passing ozone into a methylene chloride solution of the phosphite until the blue color of ozone persists. Excess ozone is then removed by nitrogen purging. A cold methylene chloride solution of the acceptor is then added. The adduct is then permitted to warm and evidence for reaction with the acceptor is then sought.

Singlet Oxygen Acceptors

Largely because of their importance in dye-photosensitized oxidations two general types of acceptor molecules have become useful for diagnosing the presence of singlet oxygen. These are (1) conjugated dienes which undergo a Diels–Alder type reaction to give 1,4-*endo*-peroxides and (2) olefins with allylic hydrogens available for undergoing an "ene" reaction to an allylic hydroperoxide.

Using the general procedure described above we have shown that when the triphenyl phosphite–ozone adduct is allowed to warm up in the presence of 1,3-cyclohexadiene a 67% yield of 5,6-dioxabicyclo[2.2.2]octene-2 ("norascaridole") is produced. This product is identical with that produced in photooxygenations[17] and with the hypochlorite–hydrogen peroxide system[4] and attributed to a singlet oxygen reaction.[4]

A number of variations on the general experimental procedure were attempted in order to optimize product yield. The yield of *endo*-peroxide was found to be consistently about 67% regardless of whether acceptor and adduct were present in equimolar amounts or whether one or the other were present in excess.

Part of the difficulty encountered in improving the yield may be the nucleophilic attack of the substrate phosphite on the phosphite–ozone adduct as suggested by Thompson.[16] Thus, while in all cases we observe a quantitative yield of phosphate, a portion of this may have been produced prior to the addition of acceptor and subsequent warm-up. This possibility is being investigated.

Using the same general procedure described above the phosphite–ozone adduct was prepared and then allowed to warm up in the presence of an equimolar amount of 2,3-dimethylbutene-2. In this case a 53% yield of the allylic hydroperoxide, 2,3-dimethyl-3-hydroperoxybutene-1, was obtained. This product had the same properties as that reported by Foote and Wexler[3] and that obtained in a photosensitized oxidation.

The phosphite–ozone adduct was capable, therefore, of giving both general types of singlet oxygen reactions.

Other systems known to undergo chemical oxygenations were then treated with the phosphite–ozone adduct. A methylene chloride solution of tetraphenylcyclopentadienone was added to a 2:1 excess of the triphenyl phosphite–ozone adduct at −78° with subsequent warm-up. This procedure gave a 38.2% yield of *cis*-dibenzoylstilbene. This product is reported to be formed in photooxygenation[18-20] and has also been ob-

(17) G. O. Schenck and W. Willmund, reported by R. Criegee in Houben-Weyl, "Methoden der organischen Chemie," Vol. VIII, E. Müller, Ed., 4th ed, Georg Thieme Verlag, Stuttgart, 1952, p 16.

(18) C. F. Wilcox, Jr., and M. P. Stevens, *J. Am. Chem. Soc.*, **84**, 1258 (1962).

(19) G. O. Schenck, *Z. Elektrochem.*, **56**, 855 (1952).

(20) N. M. Bikales and E. I. Becker, *J. Org. Chem.*, **21**, 1405 (1956).

tained using the hypochlorite–hydrogen peroxide system.[4]

While generally less reactive as a singlet oxygen acceptor than tetramethylethylene or 9,10-dimethylanthracene,[21] 9,10-diphenylanthracene has also been observed to give the *endo*-peroxide. When treated with a 2:1 excess of the triphenyl phosphite–ozone adduct in the usual manner, 9,10-diphenylanthracene gave a 77% yield of the *endo*-peroxide which had properties identical with those for the photochemically obtained material.

Finally the oxygenation of α-terpinene to ascaridole was accomplished in 60% yield using the triphenyl

(21) T. Wilson, *J. Am. Chem. Soc.*, **88**, 2898 (1966).

phosphite–ozone adduct. This reagent is thus seen to be a very useful one for accomplishing a variety of chemical oxygenations in good yield and under very mild conditions.

Use of Other Phosphites

Thompson[16] had found that only the triaryl phosphites appeared to be capable of giving stable adducts. This conclusion was based primarily on the stoichiometry of the ozone–phosphite reaction. We have attempted to detect formation of such an adduct by use of the acceptor oxygenation technique. When triethyl phosphite was ozonized at −78° and then cyclohexadiene-1,3 added with subsequent warm-up no oxygenation of the olefin was observed. However, when this same experiment was carried out at −95° a low yield (∼10%) of cyclohexadiene *endo*-peroxide could be obtained.

This result suggests that there may be a range of stabilities of ozone adducts and that under the proper conditions, *i.e.*, very low temperatures, other species which undergo one oxygen atom oxidations with ozone, may give stable adducts. Such other species might include, in principle at least, phosphines, sulfides, sulfones, tertiary amines, and perhaps even suitably substituted 1-olefins.

The relationship of such potential ozone adducts' to the moloxide concept is perhaps worth pointing out. While the recent work of Foote and Wexler[3] and others strongly suggest that the species involved in photosensitized oxygenations is free singlet oxygen as opposed to the sensitizer–oxygen complex concept originally advocated by Schönberg,[22] the ozone adducts referred to here could, formally, be regarded as moloxides which are capable of causing oxygenations. The adduct between triphenyl phosphite and ozone, for example, might also be regarded as a moloxide of triphenyl phosphate. While this adduct was observed to decompose to give molecular oxygen and triphenyl phosphate, the question remained as to whether the oxygenation reagent was the adduct itself or the molecular oxygen evolved. This point is examined further below.

Nature of the Oxygenation Reagent

In order to test for any oxygenation activity of the phosphite–ozone adduct we have treated a solution of rubrene with a 200-fold excess of the adduct. Rubrene is one of the most active singlet oxygen acceptors[21] and, in addition, its bright orange color compared with the colorless *endo*-peroxide permits visible evidence of reaction.

A −78° solution of the adduct and rubrene showed no evidence of bleaching of the rubrene color after 30-min storage at this temperature. When the solution was per-

(22) A. Schönberg, *Ann.*, **518**, 299 (1935).

mitted to slowly warm up, however, some color loss was evident at −35° and by −25° the solution was colorless. Continued warm-up to −10° gave vigorous gas evolution. Since our procedure calls for nitrogen purging of unreacted ozone prior to addition of acceptor it became necessary to verify the adequacy of the nitrogen purge. This was done by adding ozone to methylene chloride at −78°, purging with nitrogen, adding rubrene, and then warming. No bleaching of the rubrene occurred. When a cold solution of rubrene was treated with ozone the color was immediately bleached. These experiments indicate that there is little, if any, bimolecular reaction between the phosphite–ozone adduct and rubrene. On the other hand when the adduct is allowed to decompose and evolve oxygen then reaction with the rubrene is rapid.[23]

Further evidence on the nature and degree of freedom of the oxygenation species was obtained by observing gas phase reactions of the molecular oxygen evolved in the decomposition of the phosphite–ozone adduct. In an earlier experiment we attempted to transfer the gas evolved in the solution decomposition of the adduct to an adjacent flask containing a solution of rubrene. Even under conditions of vigorous gas evolution, no bleaching of the rubrene was observed. Apparently collisional deactivation of the $^1\Delta$-oxygen is far too rapid to permit transfer.

We then devised an apparatus to permit gas phase evolution and reaction of the oxygen. The phosphite–ozone adduct was prepared in dichlorodifluoromethane solvent which was subsequently removed at low temperature to give the solid adduct. It was necessary to add crushed chalk sticks to the adduct prior to solvent removal in order to moderate the subsequent decomposition of the solid adduct. The solid adduct was then attached to the apparatus shown in Figure 1. This apparatus was constructed in such a way as to prevent acceptor molecules from undergoing a heterogeneous reaction on the surface of the solid adduct.

Using this system it was possible to observe gas phase oxygenation of 1,3-cyclohexadiene to the *endo*-peroxide. The yield is small, *i.e.*, ∼0.01% based on theoretically available oxygen. Again this could be due to collisional deactivation. The pressure during the decomposition reaction phase was ∼0.2–0.6 mm. In a similar experiment α-terpinene was converted into ascaridole in ∼0.01% yield. In both experiments the products were analyzed by flame ionization glpc. In both cases the starting materials were free of products. Likewise use of ground-state molecular oxygen in control experiments gave no *endo*-peroxide products.

A similar experiment using tetramethylethylene was less conclusive because of a failure to remove completely the allylic hydroperoxide product from starting material.

The combined weight of the esr results,[10] rubrene solution results, and gas phase oxygenations leads us to conclude that the oxygenation reagent in this system is excited molecular oxygen and not the phosphite–ozone adduct itself.

(23) Professor P. D. Bartlett has observed a bimolecular reaction between the triphenylphosphite–ozone adduct and tetramethylethylene at low temperature. He has confirmed the absence of such a reaction in diene-type acceptors: P. D. Bartlett, paper presented at the 155th National Meeting of the American Chemical Society, San Francisco, Calif., March 1968, Abstract R20.

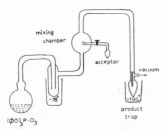

Figure 1. Diagrammatic representation of the Pyrex gas phase reactor.

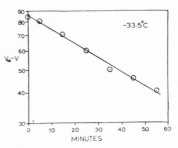

Figure 2. Typical first-order rate plot for the evolution of oxygen from the triphenyl phosphite–ozone adduct at $-33.5°$.

Nature and Stability of the Triphenyl Phosphite–Ozone Adduct

On the basis of the ^{31}P nmr spectrum Thompson[16] concluded that the triphenyl phosphite–ozone adduct must have a cyclic structure and finally settled on the four-membered ring representation used here by analogy with other pentacovalent phosphorus compounds. Thompson also pointed out that such a structure is formally analogous to the initial adduct or molozonide formed in olefin–ozone systems.

We have measured the stability of the adduct by following its rate of decomposition by means of the oxygen evolved. Oxygen evolution was found to follow a typical first-order rate law (Figure 2). This method was used to calculate the rate of decomposition of the adduct in methylene chloride at a series of temperatures as shown in Table I. These data were then used to ob-

Table I. Rate Constants at Several Temperatures for the Evolution of Oxygen from Triphenyl Phosphite–Ozone Adduct

T, °C	k, sec^{-1}
-17.0	1.58×10^{-3}
-24.0	5.97×10^{-4}
-33.5	2.27×10^{-4}
-41.0	7.90×10^{-5}
-44.0	6.24×10^{-5}

tain the activation energy for decomposition and the transition-state parameters shown in Table II. The

Table II. Activation Energy and Transition-State Parameters for Triphenyl Phosphite–Ozone Adduct Decomposition

E_a, kcal mole	14.1 ± 1.8
Log A	9.167
$\Delta F^{\ddagger}$, kcal mole $(-24°)$	17.5
$\Delta H^{\ddagger}$, kcal mole $(-24°)$	13.6
$\Delta S^{\ddagger}$, eu $(-24°)$	-3.9

value for E_a of 14.1 ± 1.8 kcal/mole would appear to be a reasonable one for the assigned structure. The four-membered ring structure containing three oxygen atoms would be expected to have considerable steric strain and resultant instability. In fact few data are available for comparison. Benson has estimated[24] the bond strength of the oxygen–oxygen bond in the 1,2,3-trioxalane produced as the initial adduct in an olefin–ozone system to

(24) S. W. Benson, Advances in Chemistry Series, No. 77, Vol. III, American Chemical Society, Washington, D. C., 1968, p 74.

be about 15 kcal/mole. Also, Criegee and Schröder[25] have found the heat of rearrangement of the initial olefin–ozone adduct in the case of *trans*-1,2-di-*t*-butylethylene to be 37 kcal/mole in pentane. These authors also reported that this value was lower in halide-containing solvents.

On the other hand it has been reported that the initial olefin–ozone adduct for hexene-1 has an E_a for decomposition in ethanol of 7 ± 2 kcal/mole.[26]

We have also attempted to determine the effect of solvent structure on the rate of decomposition of the adduct. These results are shown in Table III. As with

Table III. Effect of Solvent on Rate of Decomposition of the Triphenyl Phosphite–Ozone Adduct $(-24°)$

Solvent	Rate constant, sec^{-1}	Relative rate
CH$_2$Cl$_2$	5.97×10^{-4}	1.0
CH$_3$OH	2.07×10^{-3}	3.5
(CH$_3$)$_2$CHOH	5.73×10^{-4}	0.96
C$_6$H$_5$CH$_3$	7.40×10^{-5}	0.12
CH$_3$COCH$_3$	1.60×10^{-4}	0.27
(CH$_3$CH$_2$)$_2$O	1.55×10^{-4}	0.26

the initial adduct in the olefin–ozone system,[25] the halogen-containing solvent methylene chloride leads to a more rapid rate of decomposition of the phosphite–ozone adduct than does ether. Also the hydrogen-donating solvents methanol and isopropyl alcohol gave higher rates of decomposition of the phosphite–ozone adduct.

Nature of Singlet Oxygen

A final point regarding the nature of the oxygenation reagent in this system needs to be considered, that is, the possibility that both Σ and Δ singlet oxygen are involved. This possibility has been recently discussed for dye-photosensitized oxygenations.[27,28] An experimental discussion of this possibility requires an acceptor which is presumably able to discriminate between $^1\Sigma$- and $^1\Delta$-oxygen in product formation. Kearns and co-workers have presented such a discussion[27,28] in the case of the dye-sensitized photooxygenation of cholest-4-en-3β-ol reported earlier by Nickon and Mendelson.[29]

(25) R. Criegee and G. Schröder, *Chem. Ber.*, **93**, 689 (1960).
(26) S. D. Razumousky and L. V. Berezova, *Izv. Akad. Nauk SSSR, Ser. Khim.*, No. 1, 207 (1968).
(27) D. R. Kearns, R. A. Hollins, A. U. Khan, R. W. Chambers, and P. Radlick, *J. Am. Chem. Soc.*, **89**, 5455 (1967).
(28) D. R. Kearns, R. A. Hollins, A. U. Khan, and P. Radlick, *ibid.*, **89**, 5456 (1967).

We have attempted oxygenations of this same acceptor using the triphenyl phosphite–ozone adduct in order to examine the possibility of a distribution between $^1\Sigma$- and $^1\Delta$-oxygen in this system. While the results are as yet inconclusive it is clear that the product distribution is complex and make a definite conclusion difficult. We are presently considering other simpler acceptors to be used for this purpose. In addition we are attempting to determine the product selectivity of the singlet oxygen produced from the phosphite–ozone adduct. This method has been used effectively by Foote and coworkers.[30]

Experimental Section

All melting points and boiling points are uncorrected unless otherwise stated. Ir spectra were determined on a Perkin-Elmer Infracord spectrophotometer. Nmr spectra were obtained on a Varian Associates A-60 spectrometer, and are reported as τ values relative to external TMS.

Preparative gas chromatography utilized the Aerograph A-700 Autoprep instrument. Analytical gpc was performed with flame ionization detection cells on either the Aerograph Hi-Fi Model 600D or Aerograph Model 200. These units were used in conjunction with an Aerograph 471 digital integrator.

Ozonizations. Phosphite esters were dissolved in methylene chloride for ozonation unless otherwise stated. The solutions were placed in a two-necked, round-bottomed flask fitted with a gas inlet tube emerging near the bottom of the flask and a calcium sulfate drying tube. The flask was cooled in a Dry Ice–acetone bath at $-78°$ and then ozonized using a Welsbach Model T-23 ozonator. Ozone was supplied at the rate of approximately 100 mmoles/hr. During ozonization the temperature of the solution rose to about $-70°$. When the blue color of excess ozone was observed the ozone stream was disconnected and replaced by a stream of dry nitrogen. Nitrogen purging was carried on well beyond the time necessary for the solution to loose its blue color.

Bleaching of Rubrene. A solution of triphenyl phosphite (3.1 g, 0.01 mole) was ozonized in methylene chloride (100 ml) in a flask fitted with a gas inlet tube, a drying tube, and a low-temperature thermometer. After nitrogen purging ($\sim$20 min) a cold solution of rubrene (Eastman, 0.025 g, 4.7×10^{-5} mole) in methylene chloride (50 ml) was added. Mixing was effected with the nitrogen stream. The bright red-orange solution appeared to remain unchanged during 30 min at $-78°$. When the cooling bath was removed and the mixture permitted to warm up, a slight color loss was apparent at $-35°$ and by $-25°$ the solution was colorless. Vigorous gas evolution was evident at $-10°$. No attempt was made to identify the oxidized product.

In a separate experiment a flask containing methylene chloride (100 ml) was ozonized at $-78°$. The blue color of ozone was observed after a very short time. A nitrogen purge was continued for $\sim$25 min after which time a cold solution of rubrene (0.025 g, 4.7×10^{-5} mole) in methylene chloride (50 ml) was added. The bright color of the rubrene persisted even when the bath was removed and the solution warmed to room temperature. When the solution was recooled to $-78°$ and treated with ozone the rubrene color was immediately bleached.

Attempted Transfer of Singlet Oxygen Produced in Solution. Triphenyl phosphite (9.3 g, 0.03 mole) was ozonized in methylene chloride and purged with nitrogen. The drying tube was replaced with a gas transfer tube, the open end of which was introduced below the surface of a solution of rubrene ($\sim$10 mg) in methylene chloride. The $-78°$ bath was removed and the solution warmed to room temperature. At $-10°$ vigorous gas evolution proceeded. When the reaction was complete no color change was observed in the rubrene solution.

Oxidation of 1,3-Cyclohexadiene. A solution of triphenyl phosphite (Aldrich, 9.3 g, 0.03 mole) was ozonized in methylene chloride (100 ml). After nitrogen purging a cold solution of 1,3-cyclohexadiene (Aldrich, 1.60 g, 0.02 mole) in methylene chloride (45 ml) was added by funnel. Thorough mixing was achieved with the nitrogen stream. The Dry Ice–acetone bath was removed and

the mixture was permitted to warm to room temperature. The solution was concentrated on a rotary evaporator and distilled *in vacuo* (<0.1 torr), with a short-path still. The product, 1.51 g, was a pale yellow semisolid (yield 67.4% based on olefin). After three recrystallizations from pentane the mp was 90–91°, with some gas evolution (lit.[17] mp 88.5°). The nmr and ir spectra were identical with those obtained from photochemically produced material.[4,31]

Several other variations were attempted in order to determine optimum oxidation conditions.

Method A. Triphenyl phosphite (6.2 g, 0.02 mole) was ozonized in methylene chloride (150 ml), purged with nitrogen, and followed by the addition of cyclohexadiene-1,3 (1.60 g, 0.02 mole) in cold methylene chloride solution (25 ml). The $-78°$ bath was removed and replaced with a $-25°$ bath (methanol–ice). The reaction remained at this temperature for 1.5 hr and was then permitted to warm to room temperature. A work-up identical with that above was used and the yield of product was 1.50 g (67%).

Method B. Triphenyl phosphite (6.2 g, 0.02 mole) was ozonized and purged with nitrogen as in method A. A cold solution of cyclohexadiene-1,3 (8.0 g, 0.1 mole) in methylene chloride (25 ml) was added and the $-78°$ bath removed and the solution warmed to room temperature. A work-up identical with that used in method A gave 1.51 g (67.4%) of *endo*-peroxide.

Oxidation of 2,3-Dimethylbutene-2. A solution of triphenyl phosphite (6.2 g, 0.02 mole) in methylene chloride (100 ml) was ozonized and purged with nitrogen at $-78°$. A cold solution of 2,3-dimethylbutene-2 (1.68 g, 0.02 mole) in methylene chloride (45 ml) was added and mixing effected with the nitrogen stream. The $-78°$ bath was removed and replaced with a $-25°$ bath (ice–methanol). After about 16 hr the temperature had risen to $-10°$. At this point the reaction mixture was warmed to room temperature, concentrated on a rotary evaporator and distilled *in vacuo* on a short-path still. The product was a colorless liquid boiling at 52–54 (9 mm) (lit.[32] bp 55° (12 mm)) and its ir and nmr spectra were identical with those for the product from a photooxygenation reaction.[4,31]

Oxidation of Tetraphenylcyclopentadienone. A solution of triphenyl phosphite (3.1 g, 0.01 mole) in methylene chloride was ozonized at $-78°$ and purged with nitrogen. A cold deep red solution of tetraphenylcyclopentadienone (1.92 g, 0.005 mole) in methylene chloride (50 ml) was then added. A slow stream of nitrogen mixed the reactants and was continued after removal of the cold bath. After reaching room temperature the resulting mixture had lost most of its color. The solution was concentrated on a rotary evaporator and chromatographed on neutral alumina. A pale pink solid was eluted with benzene and was recrystallized from absolute ethyl alcohol. The yield of small pale pink needles was 0.71 g (38.2%). The mp of recrystallized material was 218–220° cor (lit.[33] mp 216–217°) and its ir spectrum was identical with that previously reported[33] for *cis*-dibenzoylstilbene.

Oxidation of 9,10-Diphenylanthracene. A solution of triphenyl phosphite (3.1 g, 0.01 mole) was ozonized in methylene chloride (100 ml) and purged with nitrogen in the usual way. A cold solution of 9,10-diphenylanthracene (1.65 g, 0.005 mole) in methylene chloride (50 ml) was then added. A nitrogen stream mixed the solution during warm-up of the reaction to room temperature. The reaction mixture was concentrated on a rotary evaporator and chromatographed on neutral alumina (Fisher). The pale yellow solid which was eluted with pentane was recrystallized twice from acetone. The nearly white needles which resulted, yield 1.33 g (77%), decomposed rapidly at $\sim$200° (lit.[34] dec pt 180°) to give a higher melting yellow solid whose ir spectrum was identical with that of 9,10-diphenylanthracene. The ir spectrum of the peroxide in a KBr pellet was identical with that for the product formed in a photooxidation.[35]

Oxidation of α-Terpinene. α-Terpinene was prepared from α-terpineol by the oxalic acid dehydration procedure of von Rudloff.[36] The products were distilled on a Nester-Faust Teflon auto-

(29) A. Nickon and W. L. Mendelson, *J. Am. Chem. Soc.*, **85**, 1894 (1963); **87**, 3921 (1965).

(30) C. S. Foote, S. Wexler, and W. Ando, *Tetrahedron Lett.*, 4111 (1965).

(31) We gratefully acknowledge the gift of photochemically produced norascaridole and 2,3-dimethyl-3-hydroperoxybutene-1 from Mrs. S. R. Fahrenholtz.

(32) G. O. Schenck and K. H. Schulte-Elte, *Ann.*, **618**, 185 (1958).

(33) D. R. Berger and R. K. Summerbell, *J. Org. Chem.*, **24**, 1881 (1959).

(34) C. Dufraisse and J. LeBras, *Bull. Soc. Chim. Fr.*, 4 [5], 349 (1937).

(35) We thank Mrs. V. Kuck for a sample of the photochemically produced product.

(36) E. von Rudloff, *Can. J. Chem.*, **39**, 1 (1961).

273

annular still and the fractions boiling at 76–77° (25 mm) were collected. Their ir spectra were identical with the published spectrum of α-terpinene.[37] Gpc analysis of the fractions on a 10 ft × ³/₈ in. 10% XE-60 column at 105° and 200 cc/min carrier gas showed them to contain from 83 to 93% α-terpinene (by comparison with retention time of analytical standard grade α-terpinene)[38] and the main impurity was shown to be *p*-cymene.

A cold solution of α-terpinene (2.72 g, 0.02 mole; material was 93% pure) in methylene chloride (50 ml) was added to the ozonized and nitrogen-purged solution of triphenyl phosphite (6.2 g, 0.02 mole) in methylene chloride (125 ml). A slow nitrogen stream was allowed to continue after the −78° bath was removed and until room temperature was reached. After concentration on a rotary evaporator a vacuum distillation on a short-path still gave 2.35 g of a pale yellow liquid (bp 40–50° (<0.4 mm)). The nmr spectrum of the crude material (aromatic impurities) showed it to be ~85% ascaridole. This corresponds to an over-all yield of ascaridol of 60%. A portion of the crude material was purified by gpc on a 10 ft × ³/₈ in. 10% XE-60 column at 135° with helium at 200 cc/min (retention time 28 min). The nmr spectrum of the pure material was identical with the published spectrum of ascaridole.[39]

Oxygenation with Triethyl Phosphite–Ozone Adduct. A solution of triethyl phosphite (Eastman Practical, 1.66 g, 0.01 mole) in dichlorodifluoromethane (Freon 12) (~100 ml) was ozonized at −95° (frozen acetone bath). After purging with nitrogen a cold solution of cyclohexadiene-1,3 (0.8 g, 0.01 mole) in Freon 12 was added. The nitrogen stream was used to mix the reactants. The cold bath was removed and the solution permitted to warm to room temperature. When all the solvent had boiled away the remaining viscous liquid was distilled *in vacuo*. The material boiling at 40–63° and 1 torr was collected. The nmr spectrum of this material showed it to be mainly oxidized triethyl phosphite with about 10% cyclohexadiene *endo*-peroxide.

When triethyl phosphite was ozonized at higher temperatures (*e.g.*, −78°) with subsequent addition of olefin no oxidation products were observed.

General Procedure for Gas Phase Reactions. The triphenyl phosphite–ozone adduct was typically prepared by ozonizing triphenyl phosphite (3.1 g, 0.01 mole) in dichlorodifluoromethane followed by nitrogen purging and then adding cold crushed chalk sticks as moderator (100 g/0.01 mole of adduct) to the ozonized solution and solvent removal by pumping on the cold (−78°) reaction vessel for ~20 hr. The flask containing the solid adduct and moderator was kept at −78° and transferred to the Pyrex gas phase reaction apparatus (Figure 1). The system was evacuated to <10⁻³ torr (while acceptor was kept frozen). A trap at −78°, between the adduct flask and mixing chamber, ensured that acceptor molecules, cyclohexadiene-1,3, α-terpinene, or 2,3-dimethylbutene-2, could not diffuse back to the adduct and possibly undergo a heterogeneous reaction. With the acceptor bleed-in valve closed the temperature of the adduct was raised gradually (by adding water to a methanol–ice bath) until it was ~0°. The pressure in the system usually rose to between 0.2 and 0.6 torr and remained there during the course of a run. When the pressure had stabilized, olefin was permitted to enter the mixing chamber by opening the needle valve leading to the acceptor flask. In the case of 1,3-cyclohexadiene and 2,3-dimethylbutene-2, the rate of introduction of the olefins was adequate when room temperature of the acceptor flask was maintained. In the case of α-terpinene some external warming was necessary. The product materials were condensed and frozen in a trap kept at liquid nitrogen temperature.

Gas Phase Oxidation of Cyclohexadiene-1,3. The cyclohexadiene-1,3 used for the gas phase oxidations was purified by passing it through Woelm neutral alumina and distilling (spinning band column) under nitrogen (bp 80°). The pure material was stored under nitrogen at −20°. The product trapping flask was tared, to allow determination of the amount of product during each run. After a run was completed the product flask was warmed to room temperature and the solution weighed and immediately analyzed by gpc[19] using a flame ionization detector. A standard solution of cyclohexadiene *endo*-peroxide in heptane was used to calibrate the detector. The glpc conditions used to analyze for *endo*-peroxide were a 5 ft × ¹/₈ in. 5% XF-1150 on Chromosorb G column, with

nitrogen carrier gas at 40 cc/min, hydrogen flow at 20 cc/min, and oven temperature of 110°. An average of ten digital integrator determinations gave 46,500 counts/μg of *endo*-peroxide. The retention time of the *endo*-peroxide was 29 min. Analysis of the product solution gave a yield figure of 116 μg (0.01% based on theoretically available oxygen). In an identical separate experiment the yield of *endo*-peroxide was 118 μg. In both cases a smaller peak was evident with a retention time of 24 min. If the product solution was permitted to stand for even a few hours this peak at 24 min grew considerably. No attempt was made to identify it. In neither of these cases was *endo*-peroxide present in the starting material. A control experiment was performed in which tank oxygen at a pressure of 0.35 mm was maintained in the oxidation system (Figure 1). Acceptor molecules were introduced into the system for 1.5 hr. Flame ionization gpc analysis of both starting material and product showed the absence of *endo*-peroxide.

Gas Phase Oxidation of α-Terpinene. Immediately before use in the gas phase oxidations α-terpinene[36] was purified by gas chromatography on a 10 ft × ³/₈ in. 10% XE-60 column at 90° and a 200 cc/min flow rate of helium. Introduction of the α-terpinene into the acceptor vessel (Figure 1) was accomplished by passing it through a neutral alumina (Woelm) column. After oxidation the product was immediately analyzed by flame ionization gpc. The column used was 5 ft × ¹/₈ in., 5% XF-1150 on Chromosorb G. The carrier gas was nitrogen at a flow of 40 cc/min, the hydrogen was at a flow of 20 cc/min, and the oven temperature was 130°. Ascaridole prepared by liquid phase oxidation of α-terpinene was purified by preparative gpc (10 ft × ³/₈ in., 10% XE-60 column at 135° and carrier gas at 200 cc/min) and made up into a standard solution with dodecane as solvent in order to calibrate the flame ionization detector. An average of several determinations gave 61,470 counts/μg of ascaridole. The retention time of ascaridole was 25 min. The analysis of the product indicated a yield of 60 μg of ascaridole (0.01% based on the α-terpinene utilized). Two minor products, one with a retention time of 8 min and one with a retention time of 39 min, were also present. The peak at 39 min is also present in a pure ascaridole solution which has stood several days at room temperature. Analysis of the starting material showed it to contain no product peaks.

Gas Phase Oxidation of 2,3-Dimethylbutene-2 (Tetramethylethylene). The olefin was purified by first passing it through neutral alumina (Woelm) and distilling on a spinning band column (bp 72°) under nitrogen. It was stored under nitrogen at −20°. Gas phase oxidation with singlet oxygen was followed by analysis by flame ionization gpc on a 10 ft × ¹/₈ in., 10% Dow 710 on a Chromosorb W column at 45°. The carrier gas (nitrogen) flow rate was 40 cc/min and the hydrogen flow was 20 cc/min. The chromatogram of the starting material was not significantly different from that of the product. It was not possible by our techniques to remove all traces of allylic hydroperoxide from the starting material.

Kinetic Experiments in the Liquid Phase. The rate of decomposition of the triphenyl phosphite–ozone adduct was measured by following the evolution of oxygen with time. The apparatus for preparation of the adduct consisted of a 300-ml, two-necked, round-bottomed Pyrex flask initially fitted with a gas bubbling tube exiting near the bottom of the flask and a calcium sulfate drying tube. The flask was charged with triphenyl phosphite (1.55 g, 0.005 mole) and 200 ml of solvent. The phosphite was ozonized to completion at −78° and then purged with dry nitrogen. The gas inlet tube was removed and replaced with a combination Stermowell gas delivery exit tube. The reaction flask was then connected to a calibrated gas buret and associated leveling tube. The gas buret was allowed to fill with mercury at the start of each experiment. When the reaction solution had reached the desired temperature[40] (the cooling baths used consisted of ice–methanol and varying amounts of Dry Ice) the flask was sealed with a Teflon-coated stopper. The system was found to be leak free. At the same instant the system was sealed a stop watch was activated. The body of the flask containing the adduct was always completely submerged in the cooling bath and shaken vigorously before each reading of oxygen volume. During the course of each run the leveling tube was used to keep the system as close to atmospheric pressure as possible and adjusted more precisely immediately before

(37) B. M. Mitzner, E. T. Theimer, and S. K. Freeman, *Appl. Spectrosc.*, **19**, 169 (1965).

(38) We gratefully acknowledge a gift of this material from the Hercules Powder Co.

(39) "High Resolution NMR Spectra Catalogue," Varian Associates, Palo Alto, Calif., 1962, Spectrum No. 276.

(40) A copper–constantan thermocouple inserted into a thermowell in the flask, in conjunction with a Leeds and Northrup Co. millivolt potentiometer, No. 8690, was used to monitor temperature throughout a run.

5364   *Journal of the American Chemical Society | 91:19 | September 10, 1969*

volume readings were made.   The volume of gas evolved at infinite time (V_∞) was determined by removing the reaction flask from the cooling bath, warming it to room temperature, and allowing it to remain there for some time (15–30 min).   The flask was then recooled to the temperature of the experiment and the gas volume read.   Plots of the log ($V_\infty - V$) *vs.* time were made and slopes were determined by a least-squares treatment of the data.   A typical first-order rate plot is shown in Figure 2.   Most runs were followed for about 2 half-lives and all were followed at least 1 half-life.   In the experiments where the amount of initial adduct was quite small, due to decomposition during the warm-up of the adduct flask to the temperature of the experiment, it was found that the slope of the line changed somewhat at the end of the run.   These points were neglected when calculating the rate constants.   Temperature control in all cases was better than $\pm 1°$ and in most cases better than $\pm 0.5°$

Determination of Transition-State Parameters.   Rate constants for the decomposition of the triphenyl phosphite–ozone adduct were determined at five different temperatures using methylene chloride as solvent.   A summary of the results of these experiments appears in Table I.   An Arrhenius plot of log k *vs.* $1/T°K$ was treated by least squares to determine the slope, the intercept, and the probable error in the slope.   From these were calculated the energy of activation (E_a) and the transition-state parameters for the decomposition of the adduct.   These results are tabulated in Table II.

Effect of Solvent.   Rate constants for the decomposition of the triphenyl phosphite–ozone adduct were determined at $-24°$ in a variety of solvents.   These results are collected in Table III.

Acknowledgment.   We wish to acknowledge helpful discussions with Dr. A. M. Trozzolo.

275

48

Reprinted from *J. Amer. Chem. Soc.*, **90**(15), 4160–4161 (1968)

Electron Paramagnetic Resonance of ¹Δ Oxygen from a Phosphite–Ozone Complex

Sir:

The identification of electronically excited oxygen as a product of organic chemical reactions is usually based on its subsequent reaction with unsaturated organic molecules.[1-3] The evidence is usually the observation of products which have also been obtained with photosensitized oxidations or with microwave discharged oxygen.[4,5] While the physical evidence is unequivocal for both ¹Δ and ¹Σ oxygen in the discharge,[6] similar support for the chemical production of these states has been confined to their electronic emission spectra arising with some inorganic reactions.[7] To determine if excited oxygen is, in fact, produced in one of the organic reactions which had been suggested as a source, we have examined the epr spectrum of the oxygen produced by the decomposition of a triphenyl phosphite–ozone complex.[3] The presence of ¹Δ oxygen has been demonstrated by its characteristic absorptions.

The epr of ¹Δ O_2 in a discharge was first observed and characterized by Falick, *et al.*[8] We may regard these absorptions as arising from the magnetic moment due to the orbital motion of the two π* electrons. In a corresponding ¹D state of an atom, the pure orbital motion produces a $g = 1$, and epr absorption would be near 6000 G at 9 Gc. In the ¹Δ O_2 the component of orbital angular momentum along the internuclear axis is $2\hbar$. This knowledge of one component of the total angular momentum J implies, by the uncertainty principle, that the direction of J is off the axis. We may view the internuclear axis as tumbling about J so that, on the average, only the component of the orbital magnetic moment parallel to J remains to interact with the external magnetic field. The reduced moment is equivalent to $g = 4/(J(J + 1))$.[8,9] For the lowest rotational state of the molecule $J = 2$, a value which arises from the orbital motion. For this state $g = {}^2/_3$ and absorption is expected at ~9000 G. The higher rotational states have a smaller g factor, and resonance will occur only at fields beyond the upper limit of the magnet (15 kG). Consequently we see only the four $\Delta m_J = 1$ transitions of $J = 2$. These transitions are split by a second-order interaction, and four absorptions at intervals of ~100 G are observed.[8]

In a typical run, 0.01 mol of the solid triphenyl phosphite–ozone complex,[3] diluted with sand to moderate the decomposition,[10] was allowed to decompose at

~−20°. The oxygen produced traversed a quartz tube which passed through a Varian V-4535 large-access cavity. Large diameters (~20 mm) were used in the system to minimize the time interval between the production of the O_2 and its entrance into the cavity. The oxygen pressure in the tube was kept in the range of 0.2–2 mm. The higher pressure was associated with collisional broadening, and most experiments were done at 0.3–0.6 mm. The intensity of the absorptions paralleled the rate of oxygen evolution by the complex. The strong signals characteristic of ³Σ O_2 were easily detected. In addition, the weaker absorptions of the ¹Δ O_2 were observed. These were identified by comparison with the spectrum observed with oxygen which had experienced a microwave discharge and then passed through the same cavity. The positions of the lines agreed within 1 G. For both the chemically and electrically produced ¹Δ O_2 the inner two lines are about 50% stronger than the outer, as expected for a ¹Δ state.[9]

The relative intensities of the ³Σ and ¹Δ were also measured for discharged and for chemically produced oxygen. Assuming that the discharge contains 10% ¹Δ,[8,11] the gas from the complex contained ~1% ¹Δ when it was in the cavity. Implicit in this determination is the assumption that both sources of ¹Δ have the same relative amount of the $J = 2$ state, the one rotational state observed in the resonance experiments. At 300°K, in thermal equilibrium, some 3% of the oxygen is in $J = 2$. Significant departures from equilibrium in one sample could invalidate the comparison, but we do not feel that large differences are likely.

Although the ¹Δ is observed, we do not wish to imply that this must be the state originally produced on decomposition of the complex. Another possibility is that ¹Σ is the first product, but would, of course, not be detectable in the epr experiment. Relaxation to the ¹Δ and also to the ³Σ state would then allow their observation. Relaxation of the original excited state could occur as it diffuses out of the reaction mixture and travels to the cavity, although direct production of ³Σ in the decomposition is also possible.

The advantage of the use of the epr spectrum is that it allows the unequivocal identification of the ¹Δ state as well as some estimate of the amount of the species present. Together with the gas-phase experiments described in the following communication,[10] the resonance measurements bring strong support to the intermediacy of excited oxygen in chemical reactions.

Acknowledgment. We wish to thank Mr. W. Delavan for his assistance in performing these measurements.

(1) H. H. Wasserman and J. R. Scheffer, *J. Am. Chem. Soc.*, **89**, 3073 (1967).
(2) J. A. Howard and K. U. Ingold, *ibid.*, **90**, 1956 (1968).
(3) R. W. Murray and M. L. Kaplan, *ibid.*, **90**, 537 (1968).
(4) C. S. Foote, S. Wexler, W. Ando, and R. Higgins, *ibid.*, **90**, 975 (1968), and references cited therein.
(5) E. J. Corey and W. C. Taylor, *ibid.*, **86**, 3882 (1964).
(6) L. W. Bader and E. A. Ogryzlo, *Discussions Faraday Soc.*, **37**, 46 (1964).
(7) J. S. Arnold, R. J. Browne, and E. A. Ogryzlo, *Photochem. Photobiol.*, **4**, 963 (1965).
(8) A. M. Falick, B. H. Mahan, and R. J. Meyers, *J. Chem. Phys.*, **42**, 1837 (1965).
(9) A. Carrington, D. H. Levy, and T. A. Miller, *Proc. Roy. Soc.*, (London), **B293**, 108 (1966).
(10) R. W. Murray and M. L. Kaplan, *J. Am. Chem. Soc.*, **90**, 4161 (1968).
(11) In our experiments the concentration of ¹Δ in the discharge was largely independent of the microwave discharge power except at the lowest power levels. A decrease of the ³Σ signal of 7–8% was observed with the initiation of the discharge.
(12) Also, Department of Chemistry, Rutgers, The State University, New Brunswick, N. J.

E. Wasserman,[12] R. W. Murray
M. L. Kaplan, W. A. Yager
Bell Telephone Laboratories, Inc.
Murray Hill, New Jersey 07974

Received May 8, 1968

49

Copyright © 1967 by the American Chemical Society

Reprinted from *J. Amer. Chem. Soc.*, **89**(12), 3073–3075 (1967)

Singlet Oxygen Reactions from Photoperoxides

Sir:

It has been known for some time that aromatic hydrocarbons such as anthracene, rubrene, and tetra-

Chart I. Oxidation of Singlet Oxygen Acceptors by 9,10-Diphenylanthracene Peroxide

cene undergo photosensitized autoxidation to form transannular peroxides,[1] and a number of recent reports have provided strong evidence indicating that singlet oxygen is involved in the formation of these "photoperoxides."[2] It is also well known that many of these peroxides undergo dissociation on heating to regenerate oxygen and the parent hydrocarbon. The ease of oxygen release from these systems depends on the polycyclic aromatic system and the nature of the substituents in the *meso* positions.[3]

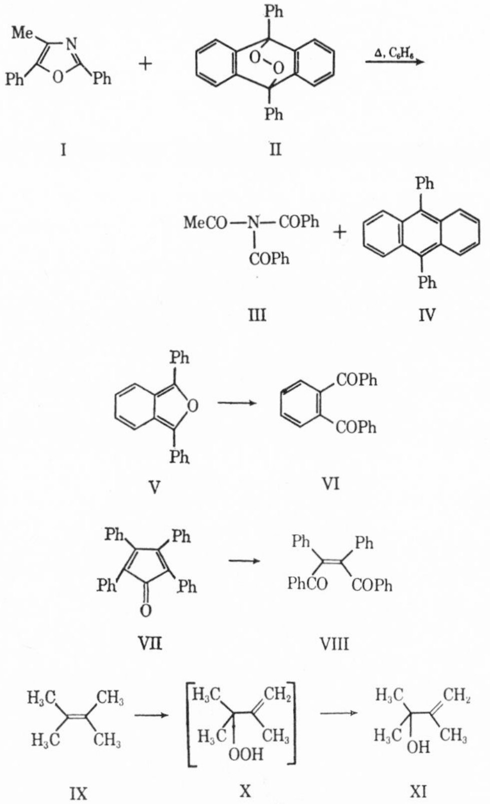

We now report that 9,10-diphenylanthracene peroxide (II) may be used to bring about typical singlet oxygen reactions when it is allowed to decompose[4,5] in the presence of a variety of known singlet oxygen acceptors.

Examples of the oxidations studied are given in Chart I. In each case the reaction was carried out by heating a mixture of peroxide and acceptor (ratio 2:1) in benzene or chloroform for 2–4 days. The products isolated were identical with those formed in parallel dye-photosensitized autoxidation reactions and were characterized by comparison (mixture melting point or vpc retention time, infrared and nmr spectra) with authentic samples. Thus, 2,5-diphenyl-4-methyloxazole (I) was converted to N-acetyldibenzamide (III)[6] (82%) along with 10% of dibenzamide. 1,3-Diphenylisobenzofuran (V) was oxidized to *o*-dibenzoylbenzene (VI)[7] in 95% yield, while treatment of tetracyclone VII with II afforded *cis*-dibenzoylstilbene (VIII) in 50% yield.[8] The intermediate hydroperoxide X from the tetramethylethylene oxidation was not isolated, but was reduced with triphenylphosphine to form the unsaturated alcohol XI. No attempt was made to optimize yields in these initial runs. Control reactions using 9,10-diphenylanthracene (IV) in place of the photoperoxide II resulted in essentially quantitative recovery of starting materials either in air or under nitrogen.

The oxidation of 2,5-diphenyl-4-methyloxazole (I)[9] by the peroxide II is outlined as a typical procedure.

(1) The first example of an aromatic transannular peroxide, rubrene peroxide, was reported by C. Moreau, C. Dufraisse, and P. M. Dean, *Compt. Rend.*, **182**, 1440, 1584 (1926). For a recent review, see Y. A. Arbuzov, *Russ. Chem. Rev.*, **34**, 558 (1965).

(2) C. S. Foote and S. Wexler, *J. Am. Chem. Soc.*, **86**, 3879, 3880 (1964); C. S. Foote, S. Wexler, and W. Ando, *Tetrahedron Letters*, **46**, 4111 (1965); E. J. Corey and W. C. Taylor, *J. Am. Chem. Soc.*, **86**, 3881 (1964); T. Wilson, *ibid.*, **88**, 2898 (1966); E. McKeown and W. A. Waters, *J. Chem. Soc., Sect. B*, 1040 (1966).

(3) For a review, see W. Bergmann and M. J. McLean, *Chem. Rev.*, **28**, 367 (1941).

(4) C. Dufraisse and L. Enderlin, *Compt. Rend.*, **191**, 1321 (1930); C. Dufraisse and J. LeBras, *Bull. Soc. Chim. France*, **4**, 349 (1937).

(5) Preliminary kinetic studies on the thermal decomposition of 9,10-diphenylanthracene peroxide in methylene chloride at 90° indicate a first-order rate constant for oxygen release of 2.5×10^{-5} sec^{-1} ($t_{1/2}$ 8 hr). This value is considerably faster than the rate of decomposition of tertiary peroxides such as *t*-butyl or trityl (which, furthermore, decomposes by peroxide bond cleavage) and suggests that 9,10-diphenylanthracene peroxide undergoes a concerted oxygen release (E. Hedaya, private communication).

(6) H. H. Wasserman and M. B. Floyd, *Tetrahedron Suppl.*, **7**, 441 (1966).

(7) A. Guyot and J. Catel, *Bull. Soc. Chim. France*, **35**, 1124 (1906); C. Dufraisse and S. Ecary, *Compt. Rend.*, **223**, 735 (1946).

(8) C. F. Wilcox, Jr., and M. P. Stevens, *J. Am. Chem. Soc.*, **84**, 1258 (1962); G. O. Schenck, *Z. Elektrochem.*, **56**, 855 (1952).

(9) Earlier studies[6] on dye-photosensitized autoxidations have shown that oxazoles are remarkably sensitive to the action of singlet oxygen. Formation of the triamide in this process appears to involve rearrangement of the intermediates XII and XIII.

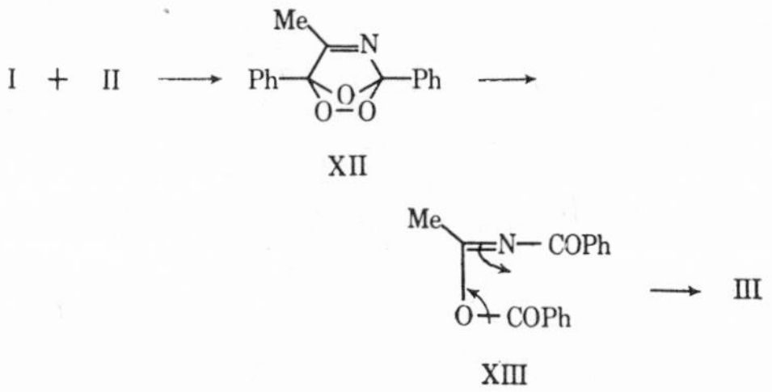

A solution of 0.724 g (0.002 mole) of 9,10-diphenylanthracene peroxide[10] and 0.235 g (0.001 mole) of 2,5-diphenyl-4-methyloxazole[11] in 50 ml of anhydrous benzene was stirred at reflux temperature in the dark for 94 hr under a positive pressure of nitrogen. Benzene was removed *in vacuo,* yieding a solid residue, 0.960 g, which, after chromatography on deactivated silica gel using ether–hexane as eluent, could be separated into a mixture of N-acetyldibenzamide[12] (0.17 g), mp 64–65°, dibenzamide (0.062 g), mp 147–149°,[13] as well as 9,10-diphenylanthracene (0.538 g) and unreacted 9,10-diphenylanthracene peroxide (0.08 g). The over-all yield of di- and triamide was 92%.

We are investigating mechanistic aspects of the process by which oxygen is transferred from photoperoxide. to acceptor, as well as the possibility that other types of cyclic peroxides may provide sources of singlet oxygen.

Acknowledgment. This work was supported in part by Grant GM-13854 from the National Institutes of Health.

(10) C. Dufraisse and A. Etienne, *Compt. Rend.,* **201,** 280 (1935).

(11) G. H. Cleland and C. Niemann, *J. Am. Chem. Soc.,* **71,** 841 (1949).

(12) Identical (mixture melting point, infrared and nmr spectra) with the product formed in the dye-photosensitized autoxidation of I.

(13) Q. E. Thompson, *J. Am. Chem. Soc.,* **73,** 5841 (1951). Dibenzamide apparently results from hydrolysis of the triamide during chromatography since this diamide could not be detected (infrared spectrum) in the crude reaction mixture before work-up.

(14) National Institutes of Health Postdoctoral Fellow, 1966–1967.

Harry H. Wasserman, John R. Scheffer[14]
Department of Chemistry, Yale University
New Haven, Connecticut 06520
Received April 6, 1967

50

Reprinted from *J. Amer. Chem. Soc.*, **90**(4), 1056–1058 (1968)

The Self-Reaction of *sec*-Butylperoxy Radicals. Confirmation of the Russell Mechanism[1,2]

Sir:

Rate constants for chain termination in the oxidation of hydrocarbons giving primary or secondary peroxy radicals are generally considerably faster than for hydrocarbons giving tertiary peroxy radicals.[3] About 10 years ago Russell proposed[4] that the self-reaction of secondary peroxy radicals involved a cyclic transition state. The recent discovery that di-*t*-butyl tetroxide

is a stable species below −85°[5] supports Russell's suggestion[4] that "I may actually be an intermediate that is formed rapidly and reversibly and which decomposes slowly in an irreversible manner." The decomposition of I will not violate the Wigner spin-conservation rule if the oxygen is eliminated in the singlet state,[6] either $^1\Sigma_g^+$ or $^1\Delta_g$.[7] Alternatively, the oxygen could be eliminated in its triplet ground state, $^3\Sigma_g^-$, if the ketone is also formed in its excited triplet state.[8] We have now identified singlet oxygen in the self-reaction of *sec*-butylperoxy radicals and have thus obtained strong experimental support for the Russell mechanism of termination. The following gives some additional experimental support for this mechanism.

(1) Absolute Rate Constants for Hydrocarbon Oxidation. IX· For part VIII see J. A. Howard, K. U. Ingold, and M. Symonds' *Can. J. Chem.*, in press.

(2) Issued as N.R.C. No. 9992.

(3) J. A. Howard and K. U. Ingold, *Can. J. Chem.*, **45**, 793 (1967).

(4) G. A. Russell, *J. Am. Chem. Soc.*, **79**, 3871 (1957).

(5) P. D. Bartlett and G. Guaraldi, *ibid.*, **89**, 4799 (1967).

(6) The present experiments were started after this point was brought to the authors' attention by Professor Russell.

(7) The over-all reaction is exothermic by over 100 kcal/mole (Russell[4] estimates 157 kcal/mole) and there is therefore plenty of energy available to form either electronically excited singlet state of oxygen. The $^1\Sigma_g^+$ state is 37.5 kcal and the $^1\Delta_g$ state is 22.5 kcal above the triplet ground state.

(8) Organic materials undergoing autoxidation luminesce very weakly with a maximum emission generally in the range 420–450 mμ.[9] The emitter of this bluish luminescence appears to be an electronically excited carbonyl compound (n,π*).[9] Transitions from both the triplet and singlet states are observed in the oxidation of methyl ethyl ketone but, with hydrocarbons, emission appears to come mainly from the triplet state.[9] The excited carbonyl compound is generally believed to be formed in the chain-termination process.[9,10] However, emission from the triplet state does not necessarily imply that triplet ketone and triplet oxygen are produced together in termination since singlet ketone is very rapidly converted to the triplet. An emission in the red region of the spectrum has also been observed and has been assigned to the $^1\Sigma_g^+ \longrightarrow {}^3\Sigma_g^-$ transition, the excited oxygen being produced in termination[11] (*cf.*, however, ref 10). If this luminescence is indeed homogeneous rather than heterogeneous (as appears likely in hydrocarbon oxidation[10]) it is more probably derived from a dimer of the $^1\Delta_g$ state of oxygen.[12] In any case, the over-all chemiluminescence quantum yield (*i.e.*, the light emitted per chain terminated) is so extremely small (10^{-7}–10^{-10}) that the nature of the emitted light provides little help in determining the principle mechanism of termination.

Our general technique was similar to that employed by McKeown and Waters.[13] The formation of singlet oxygen was inferred by the isolation of 9,10-diphenylanthracene 9,10-*endo*-peroxide from a reaction carried out in the presence of 9,10-diphenylanthracene (DPA), a compound known to be a fairly efficient trap for singlet oxygen.[13–15] The peroxy radicals were produced from *sec*-butyl hydroperoxide[16] by oxidation with ceric ion.

$$C_2H_5(CH_3)C(H)OOH + Ce^{4+} \longrightarrow$$
$$C_2H_5(CH_3)C(H)OO\cdot + H^+ + Ce^{3+}$$

DPA (1 g) in 50 ml of benzene containing 20 g of *sec*-butyl hydroperoxide (0.22 mole) was floated on 100 ml of water. The mixture was stirred gently so that the boundary between the two layers was not too badly broken and, in the dark, excess ceric ammonium nitrate (183 g (0.33 mole) in 200 ml of water) was added to the aqueous layer through a capillary tube over ~2 hr. The organic layer was separated, washed with water, evaporated under reduced pressure, and chromatographed through alumina (150 g), initially with benzene–*n*-hexane (1:6) and finally with benzene. All the DPA had been consumed in the reaction. The *endo*-peroxide (0.040 g)[17] was the second crystalline compound eluted.[18] This compound was identical (melting point[19] and infrared and X-ray spectra) with authentic *endo*-peroxide prepared both by oxidizing DPA in sunlight[20] and by the procedure of McKeown and Waters of oxidizing alkaline hydrogen peroxide with bromine in a similar two-phase system.[13] In this last reaction, McKeown and Waters obtained 0.5 g of *endo*-peroxide from 1 g of DPA in 50 ml of chlorobenzene and 0.32 mole of oxygen,[21] but in repeating this experiment we obtained only 0.35 g. The ceric

(9) R. F. Vasil'ev, *Progr. Reaction Kinetics*, **4**, 305 (1967).

(10) For a different view see: R. A. Lloyd, *Trans. Faraday Soc.*, **61**, 2173, 2182 (1965).

(11) R. F. Vasil'ev and I. F. Rusina, *Izv. Akad. Nauk SSSR, Ser. Khim.* 1728 (1964).

(12) S. J. Arnold, E. A. Ogryzlo, and H. Witzke, *J. Chem. Phys.*, **40**, 1769 (1964); R. J. Browne and E. A. Ogryzlo, *Proc. Chem. Soc.*, 117 (1964); A. M. Viner and K. D. Bayes, *J. Phys. Chem.*, **70**, 302 (1966); A. U. Khan and M. Kasha, *J. Am. Chem. Soc.*, **88**, 1574 (1966).

(13) E. McKeown and W. A. Waters, *J. Chem. Soc., Sect. B*, 1040 (1966).

(14) (a) T. Wilson, *J. Am. Chem. Soc.*, **88**, 2898 (1966); (b) E. J. Corey and W. C. Taylor, *ibid.*, **86**, 3881 (1964).

(15) It is worth noting that the spin conservation rule also applies to the reverse of this reaction, that is, decomposition of this *endo*-peroxide gives DPA and singlet oxygen (H. H. Wasserman and J. R. Scheffer, *ibid.*, **89**, 3073 (1967)).

(16) H. R. Williams and H. S. Mosher, *ibid.*, **76**, 2987 (1954).

(17) As crystals 0.030 g was isolated and another 0.010 g was estimated to be present in mother liquors containing several minor components by thin layer chromatography. A great many minor components, which were mostly noncrystalline, were observed by tlc.

(18) The first, 0.70 g of white needles from benzene–hexane, mp 120–145° (decomposes with gas evolution); nmr and infrared spectra consistent with *cis*-9,10-di-*sec*-butylperoxy-9,10-dihydro-9,10-diphenylanthracene. *Anal.* Calcd for $C_{34}H_{36}O_4$: C, 80.28; H, 7.13; mol wt, 508.7. Found: C, 80.10; H, 7.23; mol wt, 502.

(19) Mp 130–180° with decomposition to DPA.[20]

(20) P. F. Southern and W. A. Waters, *J. Chem. Soc.*, 4340 (1960).

(21) From 35 ml of 30% H_2O_2.

oxidation of 0.22 mole of *sec*-butyl hydroperoxide produces 0.11 mole of oxygen. The singlet oxygen from the *sec*-butylperoxy radicals was therefore trapped at about one-third to one-fourth of the efficiency with which it was trapped from the Br_2–H_2O_2 reaction, assuming, of course, that only singlet oxygen is produced in both reactions. The *endo*-peroxide is decomposed rather slowly by ceric ion under the experimental conditions. Its lower yield from the peroxy radical reaction may possibly be due to a solvent effect[11b] or to the quenching of singlet oxygen to its triplet ground state by the cerium ions. However, the low yield is more probably due to the destruction of DPA in other reactions which presumably involve free radicals[22] since free radicals are not involved in the H_2O_2–Br_2 reaction. In the H_2O_2–Br_2 reaction the DPA is accounted for almost quantitatively as *endo*-peroxide and unreacted DPA.[13]

No *endo*-peroxide was detected in the ceric oxidation of *t*-butyl hydroperoxide or hydrogen peroxide under conditions similar to those employed with *sec*-butyl hydroperoxide.[23] The limit of detection (tlc after chromatography) is estimated to be about 1 mg in both cases.[25] The self-reactions of tertiary peroxy radicals

$$2(CH_3)_3COO\cdot \rightleftarrows (CH_3)_3COOOOC(CH_3)_3 \longrightarrow$$

$$(CH_3)_3CO\cdot + {}^3O_2 + \cdot OC(CH_3)_3$$

and hydroperoxy radicals

$$HOO\cdot + HOO\cdot \longrightarrow HOOH + {}^3O_2$$

therefore do not yield singlet oxygen in significant quantities.

(22) 9,10-Diphenylanthracene is quite reactive toward the free radicals present in oxidizing hydrocarbons: A. M. Turner and W. A. Waters, *J. Chem. Soc.*, 879 (1956). The *meso* positions are probably most reactive: A. L. J. Beckwith, R. O. C. Norman, and W. A. Waters, *ibid.*, 171 (1958).

(23) The H_2O_2 reaction yielded essentially a single product, 0.94 g of white needles from benzene, mp 198–199° (decomposition at 250°); nmr and infrared spectra consistent with *cis*-9,10-dihydroxy-9,10-dihydro-9,10-diphenylanthracene; lit.[24] mp 195–196° for one crystal form of the *cis* isomer of this compound. *Anal.* Calcd for $C_{26}H_{20}O_2$: C, 85.69; H, 5.53; mol wt, 364.4. Found: C, 85.47; H, 5.51; mol wt, 370.

A large number of products were formed in the *t*-butyl hydroperoxide reaction, probably because of the presence of the highly reactive *t*-butoxy radical in this system. There were two main crystalline products: first, 0.12 g of white needles from hexane, mp 191–208° (decomposition with gas evolution); nmr and infrared spectra consistent with that of *cis*-9,10-di-*t*-butylperoxy-9,10-dihydro-9,10-diphenylanthracene (*Anal.* Calcd for $C_{34}H_{36}O_4$: C, 80.28; H, 7.13; mol wt, 508.7. Found: C, 80.30; H, 7.38; mol wt, 515); second, 0.24 g of white needles from benzene, mp 181–190° (decomposition with gas evolution); nmr and infrared spectra consistent with that of 9-*t*-butylperoxy-10-hydroxy-9,10-dihydro-9,10-diphenylanthracene (*Anal.* Calcd for $C_{30}H_{28}O_3$: C, 82.54; H, 6.46; mol wt, 436.6. Found: C, 82.58; H, 6.55; mol wt, 439).

(24) C. Dufraisse and J. Le Bras, *Bull. Soc. Chim. France*, **4**, 1037 (1937); C. Dufraisse and J. Houpillart. *Compt. Rend.*, **205**, 740 (1937).

(25) No *endo*-peroxide was detected from tetralin hydroperoxide either, but in this case the detection limit was not less than 100 mg because the large quantity of nonvolatile products from the hydroperoxide (ketone and alcohol) prevented purification and concentration of the *endo*-peroxide by chromatography.

J. A. Howard, K. U. Ingold
Division of Applied Chemistry
National Research Council, Ottawa, Canada
Received December 8, 1967

51

Reprinted from *Science*, **168**, 476–477 (Apr. 1970)

Singlet Molecular Oxygen from Superoxide Anion and Sensitized Fluorescence of Organic Molecules

Abstract. *The superoxide anion, $O_2{}^-$, evolves singlet molecular oxygen in dimethylsulfoxide solution. Pronounced water quenching of superoxide-sensitized luminescence is indicative of the preferential generation of the $^1\Sigma_g{}^+$ state. Recent identification of $O_2{}^-$ in the xanthine oxidase system suggests that the generation of singlet oxygen may also occur in enzymatic systems.*

Within the few years since its first identification in a chemical system (*1, 2*), singlet molecular oxygen has been implicated either by direct or by circumstantial evidence as the active agent in a large number of physical, chemical, and biological systems. I report here the generation of singlet molecular oxygen from potassium superoxide in dimethylsulfoxide (DMSO) solution. The presence of singlet oxygen was detected by the chemical scavenger (*3*) and by the fluorescence sensitization (*2*) techniques. In the luminescence sensitization experiment a pronounced quenching effect of water was observed, as would be expected were $^1\Sigma_g{}^+$ molecular oxygen to be involved. The work reported here indicates that the superoxide ion may be a direct source for $^1\Sigma_g{}^+$ oxygen; peroxides, in contrast, yield $^1\Delta_g$ oxygen.

For the fluorescence sensitization experiment a saturated solution of potassium superoxide (Alfa Inorganics, Inc.) in DMSO (Baker Chemical Company, reagent grade) and a solution of an appropriate fluorescing molecule in DMSO were mixed in a flow apparatus in the dark. The characteristic fluorescence of the dye was then observed. In this manner fluorescence from anthracene, esculin, eosine, and violanthrone was detected.

For the detection of singlet oxygen by chemical means, 2,5-dimethylfuran (City Chemical Corporation) purified by gas chromatography was used as a scavenger. Five milliliters of 2,5-dimethylfuran were added to 100 ml of DMSO saturated with potassium

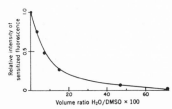

Fig. 1. Water quenching of potassium superoxide–sensitized fluorescence of anthracene in DMSO.

superoxide. By following the procedure outlined by Foote *et al.* (*4*), I obtained a residue that gave the characteristic peroxide test with potassium iodide.

In order to study the effect of water on the sensitized fluorescence intensity, the total light output was recorded with a 1P 28 photomultiplier recorder setup as a function of the percentage of water in the system. A set of solutions was prepared, each containing 10 ml of DMSO saturated with potassium superoxide. Another set of solutions was prepared, each containing 50 ml of $10^{-4}M$ anthracene dissolved in a mixture of DMSO and water; the relative proportions of the components of the solvent mixture were allowed to vary in these solutions. The potassium superoxide solution was added to the anthracene solution, and Fig. 1 shows the result of a study of the quenching of sensitized anthracene fluorescence as a function of the volume of water added.

Potassium superoxide crystal is an ionic lattice of K^+ and $O_2{}^-$ ions, and from these experiments it is concluded that in the one-electron transfer reaction

$$O_2{}^- \rightarrow O_2{}^* + e^-$$

singlet molecular oxygen is generated; however, the detailed nature of this chemical reaction is not yet clearly understood. The energetic requirements of anthracene fluorescence (see Fig. 2) necessitate the generation of the $^1\Sigma_g{}^+$ state. Bader and Ogryzlo (*5*) have shown that this species is sensitive to water quenching, a result which is consistent with the findings reported here. The detailed mechanism of the quenching of the $^1\Sigma_g{}^+$ state by water has not yet been elucidated. A kinetic study of the quenching of sensitized violanthrone fluorescence by water was undertaken to resolve this question but, although violanthrone fluorescence shows a quenching effect similar to that of anthracene fluorescence, the solvent sensitivity of the electronic spectra has complicated the results until now.

Complementary fluorescence sensi-

tization results were found by Legg and Hercules (*6*) when they electrochemically generated $O_2{}^-$ in DMSO solution, and the fluorescing compound *N*-methylacridone, derived from lucigenin, was luminescent in DMSO but failed to show any luminescence in aqueous solution. McCapra and Hann (*7*) suggested the necessity for *N*-methylacridone chemiluminescence of an "oxetane" intermediate derived from 10,10'-dimethyl-9,9'-biacridylidene in a hydrogen peroxide and sodium hypochlorite reaction mixture. The "oxetane" intermediate was postulated because this system did not exhibit water quenching, although the energetic requirements suggested by our chemiluminescence mechanism (*2*) would require the participation of the water-sensitive $^1\Sigma_g{}^+$ state. In the study reported here the 10,10'-dimethyl-9,9'-biacridylidene system has not been investigated. However, in view of the results of Legg and Hercules, McCapra and Hann's use of pyridine as a solvent may explain their failure to observe a water quenching effect; their result can be attributed to the strong pyridine-water association.

The potassium superoxide–DMSO system provides a direct and controlled source of singlet oxygen which is expected to be very useful experimentally. This system, in contrast to previously explored systems that were found to generate mostly $^1\Delta_g$ oxygen, also provides a possible source for the $^1\Sigma_g{}^+$ state of molecular oxygen. Fridovich and Handler (*8*) suggested that the xanthine oxidase–catalyzed oxidation of xanthine by molecular oxygen involves the superoxide anion, $O_2{}^-$, and recent

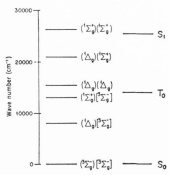

Fig. 2. Energetic disposition of the simultaneous transitions in double molecule states of metastable molecular oxygen and the electronically excited lowest singlet and the triplet states of the anthracene molecule.

experimental evidence, including a direct demonstration with electron paramagnetic resonance spectroscopy (9), has shown the presence of a substantial quantity of the O_2^- ion in this system. By analogy, it is expected that singlet molecular oxygen would be generated in this enzymatic system, and this expectation is strongly supported by the observed chemiluminescence of a number of organic molecules present in the xanthine oxidase system (10). Since a large number of biological oxidative processes are thought to proceed by way of an "electron transfer" step, the possibility of the involvement of singlet oxygen in such systems needs careful evaluation.

The reaction of potassium superoxide with water is the single most important *chemical* source of oxygen for breathing purposes in hospitals (11), mines, submarines (12), and space capsules (13). Even if only a small fraction of singlet oxygen survives quenching, it could prove to be a serious health hazard.

AHSAN U. KHAN
*Institute of Molecular Biophysics,
Florida State University,
Tallahassee 32306*

References and Notes

1. A. U. Khan and M. Kasha, *J. Chem. Phys.* **39**, 2105 (1963); *Nature* **204**, 241 (1964); *J. Amer. Chem. Soc.*, in press; M. Kasha and A. U. Khan, *Ann. N.Y. Acad Sci.*, in press; S. J. Arnold, E. A. Ogryzlo, H. Witzke, *J. Chem. Phys.* **40**, 1769 (1964); R. J. Browne and E. A. Ogryzlo, *Proc. Chem. Soc. London* **1964**, 117 (1964); J. S. Arnold, R. J. Browne, E. A. Ogryzlo, *Photochem. Photobiol.* **4**, 963 (1965).
2. A. U. Khan and M. Kasha, *J. Amer. Chem. Soc.* **88**, 1574 (1966).
3. C. S. Foote, *Accounts Chem. Res.* **1**, 104 (1968).
4. C. S. Foote, M. T. Wuesthoff, S. Wexler, I. G. Burstain, R. Denney, G. O. Schenck, K. H. Schulte-Elte, *Tetrahedron* **23**, 2583 (1967).
5. L. W. Bader and E. A. Ogryzlo, *Discuss. Faraday Soc.* **37**, 46 (1964).
6. K. D. Legg and D. M. Hercules, *J. Amer. Chem. Soc.* **91**, 1902 (1969).
7. F. McCapra and R. A. Hann, *Chem. Commun.* **1969**, 443 (1969).
8. I. Fridovich and P. Handler, *J. Biol. Chem.* **233**, 1578 (1958); *ibid.*, p. 1581; *ibid.* **236**, 1836 (1961); *ibid.* **237**, 916 (1962).
9. V. Massey, S. Strickland, S. G. Mayhew, L. G. Howell, P. C. Engel, R. G. Matthews, M. Schuman, P. A. Sullivan, *Biochem. Biophys. Res. Commun.* **36**, 891 (1969); D. Ballou, G. Palmer, V. Massey, *ibid.*, p. 898; W. H. Orme-Johnson and H. Beinert, *ibid.*, p. 905.
10. J. R. Totter, E. C. de Dugros, C. Riveiro, *J. Biol. Chem.* **235**, 1839 (1960).
11. W. A. Noyes, Jr., Ed., *Science in World War II: U.S. Office of Scientific Research and Development, Chemistry* (Little, Brown, Boston, 1948), vol. 6, p. 363.
12. J. Clarke, *J. Amer. Chem. Soc. Nav. Eng.* **68**, 105 (1956).
13. R. M. Bovard, *Aerospace Med.* **31**, 407 (1960).
14. Supported by a contract between the Division of Biology and Medicine, U.S. Atomic Energy Commission, and Florida State University. I thank Professor M. Kasha for his suggestions and encouragement, Dr. M. G. Nair for his help with the chemical scavenger experiment, and Dr. R. Light for bringing to my attention the biological references.

24 November 1969

52

Copyright © 1969 by Pergamon Press

Reprinted from *Tetrahedron Lett.*, No. 43, 3765-3767 (1969)

THE BASE-INDUCED DECOMPOSITION OF PEROXYACETYLNITRATE

R. P. Steer, K. R. Darnall* and J. N. Pitts, Jr.

Department of Chemistry, University of California

Riverside, California 92502

(Received in USA 21 July 1969; received in UK for publication 15 August 1969)

Peroxyacetylnitrate (PAN), a highly reactive, biologically important photo-chemical air pollutant (1), decomposes in the presence of base (2,3) according to reaction 1. Whereas this stoichiometrically correct reaction in no way sug-

$$CH_3-C\overset{O}{\underset{OONO_2}{\diagdown}} + 2\ OH^- \rightarrow CH_3C\overset{O^-}{\underset{O}{\diagdown}} + H_2O + O_2 + NO_2^- \qquad \underline{1}$$

gests a mechanism for the decomposition, on the basis of spin conservation it is possible that molecular oxygen in either the $^1\Sigma_g^+$ or $^1\Delta_g$ excited state might be formed. We wish to report that, in fact, singlet oxygen is a product of reaction 1, and to suggest that singlet oxygen may, therefore, be important in oxidations involving PAN and its homologues, the peroxyacylnitrates.

Spectroscopic investigations of the emissions from $O_2(^1\Sigma_g^+$ and $^1\Delta_g)$ have been undertaken by Bowen and Lloyd (4), Khan and Kasha (5) and Ogryzlo et al. (6). In our study, the presence of $O_2(^1\Delta_g)$ was determined by monitoring the emission at 1.27 μ due to process 2. The near infrared detection system consisting of a

$$O_2(^1\Delta_g) \rightarrow O_2(^3\Sigma_g^-) + h\nu \qquad \underline{2}$$

chopper, monochromator and liquid-nitrogen-cooled germanium photodiode, has been previously described (7). Reactions were carried out in one cm square cuvettes placed in front of the entrance slit of the monochromator. Dilute solutions of PAN (0.2 - 0.3M in benzene) were treated with aliquots of 0.4M potassium hydroxide in benzene-methanol (9:1). An immediate, vigorous and short-lived evolution of

---

*Public Health Service Air Pollution Special Postdoctoral Fellow.

oxygen occurred and simultaneously emission at 1.27 μ was observed. Although the decomposition of PAN and the subsequent quenching of the singlet oxygen were too rapid to permit the entire emission spectrum to be obtained, point-by point scanning between 1.1 and 1.4 μ showed that emission was observed only at 1.27 μ. The observation of $O_2(^1\Delta_g)$ in this system does not exclude the possibility that $O_2(^1\Sigma_g^+)$ may be formed initially in reaction 1 since deactivation from the $^1\Sigma_g^+$ state to the $^1\Delta_g$ state is possible (8).

The use of methanolic potassium hydroxide to effect the decomposition cau a significant attenuation of the emission intensity. This result is not unexpected if the correlation (9) between the efficiency of quenching of $O_2(^1\Sigma$ and the magnitude of the intermolecular potential between $O_2(^1\Sigma_g^+)$ and the quen ing species may be extended to $O_2(^1\Delta_g)$ systems. Attempts to quench the emissi with tetramethylethylene (TME), and efficient scavenger (10) of $O_2(^1\Delta_g)$, gave inconclusive results because of a competing reaction involving only PAN and TM Similar results were found for ethanethiol; further work on all of these processes is in progress.

The extreme reactivity of PAN in chemical and in biological _in vitro_ and _in vivo_ systems is well documented, but little understood. PAN is known to be and important cause of eye irritation and plant damage in polluted urban atmos pheres (1b,11) and studies with simpler systems have shown that it can act as both an acetylating and oxidizing agent (12). The results presented here sugg that singlet molecular oxygen may have to be considered not only as a possible environmental oxidant (13) but also in the elucidation of the mechanism(s) of oxidation by PAN.

Acknowledgement:

We are grateful to Mr. R. E. Antower, Mr. E. A. Cardiff and Dr. O. C. Tay of the University of California Statewide Air Pollution Research Center for providing solutions of PAN. This work was supported by Grant AP 00771, Research Grants Branch, National Air Pollution Control Administration, Consumer Protecti and Environmental Health Service, U. S. Public Health Service.

References:

1. (a) E. F. Darley, K. A. Kettner and E. R. Stephens, _Anal. Chem._, _35_, 589 (1963); (b) E. R. Stephens, E. F. Darley, O. C. Taylor and W. E. Scott, _Intern. J. Air Pollution_, _4_, 79 (1961).

2. S. W. Nicksic, J. Harkins and P. K. Mueller, _Atmos. Environ._, _1_, 11 (1967); E. R. Stephens, _ibid._, _1_, 19 (1967).

3. E. R. Stephens, "The Formation, Reactions, and Properties of Peroxyacyl Nitrates (PANs) in Photochemical Air Pollution", in Advances in Environmental Sciences, Vol. I, J. N. Pitts, Jr. and R. A. Metcalf, Eds., Wiley-Interscience, New York, 1969, in press.

4. E. J. Bowen and R. A. Lloyd, _Proc. Roy. Soc._, A, _275_, 465 (1963); _Proc. Chem. Soc._, 305 (1963); E. J. Bowen, _Nature_, _201_, 180 (1964).

5. A. U. Khan and M. Kasha, _J. Chem. Phys._, _39_, 2105 (1963); _40_, 605 (1964); _Nature_, _204_, 241 (1964).

6. R. J. Brown and E. A. Ogryzlo, _Proc. Chem. Soc._, 117 (1964); S. J. Arnold, R. J. Brown and E. A. Ogryzlo, _Photochem. Photobiol._, _4_, 963 (1965).

7. R. P. Wayne and J. N. Pitts, Jr., _J. Chem. Phys._, _50_, 3644 (1969).

8. R. P. Wayne, _Advan. Photochem._, _7_, 400 (1969).

9. S. J. Arnold, Ph.D. Thesis, University of British Columbia, 1966, p. 118 ff.

10. C. S. Foote, _Accounts Chem. Res._, _1_, 104 (1968); C. S. Foote, _Science_, _162_, 963 (1968); A. D. Broadbent, W. S. Gleason, J. N. Pitts, Jr., and E. Whittle, _Chem. Commun._, 1315 (1968).

11. W. M. Dugger, Jr. and I. P. Ting, _Phytopath._, _58_, 1102 (1968) and references cited therein.

12. J. B. Mudd, _J. Biol. Chem._, _241_, 4077 (1966); J. B. Mudd and W. M. Dugger Jr., _Arch. Biochem. Biophys._, _102_, 52 (1963).

13. J. N. Pitts, Jr., A. U. Khan, E. B. Smith and R. P. Wayne, _Environ. Sci. Technol._, _3_, 241 (1969); J. N. Pitts, Jr., "Photochemical Air Pollution: Singlet Molecular Oxygen as an Environmental Oxidant", in Advances in Environmental Sciences, Vol. I, J. N. Pitts, Jr. and R. A. Metcalf, Eds., Wiley-Interscience, New York, 1969, in press.

53

Reprinted from *Tetrahedron Lett.*, No. 11, 857-860 (1973)

MIKROWELLENENTLADUNG VON CO_2 : EINE NEUE, ERGIEBIGE QUELLE FÜR
SINGULETT-SAUERSTOFF, $O_2(^1\Delta_g)$

Klaus Gollnick und Gerhard Schade

Organisch-Chemisches Institut der Universität München, D-8 München 2, Karlstr

und Max-Planck-Institut für Kohlenforschung, Abteilung Strahlenchemie,

D-433 Mülheim-Ruhr 1, Stiftstrasse 34-36

(Received in Germany 30 January 1973; received in UK for publication 3 February 1973)

Setzt man einen Sauerstoffstrom $[O_2, (^3\Sigma_g^-); 3\text{-}4 \text{ Torr}]$ einer Mikrowellenentladung [1] aus, so beobachtet man in der Entladungszone eine rote Emission [6340 Å-Bande, $O_4^*(^1\Delta_g {}^1\Delta_g) \rightarrow 2\,O_2(^3\Sigma_g^-) + h\nu$] und bei Durchleiten des Gases durch Lösungen, welche geeignete Akzeptoren für Singulett-Sauerstoff enthalten, die entsprechenden Oxygenierungsprodukte [2]; vgl. auch Tabelle 2 und Figur 1.

Setzten wir unter gleichen Bedingungen einen O_2-freien CO_2-Strom [3] der Mikrowellenentladung aus, so beobachteten wir eine fahl-blaue Emission in der Entladungszone, wie sie auch bei der zwischen Elektroden stattfindenden Gasentladung durch CO_2 [4] entsteht. Hinter der Mikrowellen-Entladungszone liess sich nun im CO_2-Strom sowohl CO als auch O_2 gaschromatographisch [5] nachweisen. Leiteten wir diesen Strom durch Lösungen, welche die in Tabelle 1 aufgeführten 1,3-Diene bzw. Olefine enthielten, so konnten wir ausschließlich die für Singulett-Sauerstoff-Reaktionen typischen Oxygenierungsprodukte [6] in guten Ausbeut isolieren (vgl. auch Tabelle 2 und Figur 1).

Die fahl-blaue Emission weist auf das Vorhandensein von elektronisch-ange regten CO_2-Molekülen hin [4]. Für die Bildung dieser Moleküle bei der zwischen Elektroden stattfindenden Gasentladung durch CO_2 sowie für die Entstehung der sogenannten CO-Flammenbande [4,8] wird die Reaktionsfolge (1) diskutiert [4,9]:

(1) $O(^3P) + CO(^1\Sigma^+) \rightarrow CO_2^*(^1B_2) \rightarrow CO_2(^1\Sigma^+) + h\nu.$

Singulett-D-Sauerstoffatome, $O(^1D)$, sollen durch CO_2 nur katalytisch in Triplett-P-Sauerstoffatome, $O(^3P)$, umgewandelt werden [10, 11], während "heisse $O(^1D)$-Atome (mit Überschuss-Translationsenergie) mit CO_2 zu Singulett-angeregt CO_3^*-Molekülen reagieren sollen, die dann entweder in CO und $O_2(^1\Delta_g)$ zerfallen oder durch Stoss mit CO_2 zu $2\,CO_2 + O(^3P)$ reagieren [12,13]. Allerdings konnte die Reaktion (2),

Tab. 1: Mikrowellenentladung (2450 MHz) von O_2-freiem CO_2: Produktbildung mit
typischen Singulett-Sauerstoff-Akzeptoren in Lösung

1,3-Dien bzw. Olefin	g	Solvens ml	Temp. ^{o}C	Reakt.- Dauer(Std)	Produkte	Ausbeute[a] %
	2,5	Benzyl-acetat 5	-50	8	Askaridol vgl.Fig.1	35
Ph = C_6H_5	0,2	Benzyl-acetat 5	-30	16		95
	2,5	Methanol 20	-50	16	[b]	48
	2,5	Methanol 20	-60	16		45
(+)-Limonen	3,0	Methanol 10	-50	8	Alkohole [b] vgl. Tab. 2	3

a) bez. auf eingesetzte Diene bzw. Olefine, die jedoch unter den Reaktionsbe-
dingungen (3-4 Torr, Ölpumpe) in unterschiedlichem Ausmaß abgezogen werden;
b) nach Reduktion mit wässriger Natriumsulfit-Lösung

Tab. 2: Alkohole aus (+)-Limonen nach Reduktion des primär gebildeten Hydro-
peroxidgemischs (Produktverteilung in % des Alkoholgemischs)

A	41	9	18	7	3	22
B	30	12	8	13	9	28
C	34	10	20	10	5	21

A: 3,0 g (+)-Limonen in 10 ml Methanol bei -50°; 8 Stdn. Mikrowellenentladung
von O_2 (Umsatz: 7%). B: wie A, aber Mikrowellenentladung von CO_2 (Umsatz: 3%);
C: Durch Rose Bengale photosensibilisierte Oxygenierung von (+)-Limonen in
Methanol bei 20°; vgl. Lit. 6,7).

287

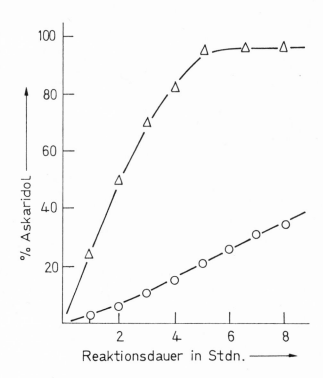

Fig. 1: Askaridolausbeute als Funktion der Reaktionsdauer;
Reaktion von α-Terpinen in Benzylacetat bei -50° mit
Singulett-Sauerstoff $[O_2(^1\Delta_g)]$, dargestellt durch
Mikrowellenentladung (2450 MHz) von O_2 (Δ) bzw. von CO_2

(2) $O(^1D) + CO_2 \rightarrow CO + O_2(^1\Delta_g)$,

die mit 15 kcal/Mol exotherm ist, bisher weder bei der direkten oder Edelgas-
sensibilisierten Photolyse von CO_2 noch bei der UV-Photolyse von O_3 in Gegenwart von CO_2 beobachtet werden [10,12].

Wir vermuten, daß auch bei der Mikrowellenentladung von CO_2 Singulett-
Sauerstoff-Atome entstehen

(3) $CO_2 \xrightarrow{\text{MWE}} CO + O(^1D)$,

die dann sowohl nach (2) (eventuell als "heisse" Atome und über CO_3^*) mit CO_2 zu
CO und $O_2(^1\Delta_g)$ reagieren als auch teilweise durch CO_2 zu $O(^3P)$-Atomen umgewandelt werden. Letztere geben in der Entladungszone Anlaß zur fahl-blauen
Emission (vgl. (1)) oder werden durch das im System vorhandene Quecksilberoxyd
abgefangen.

Mit dem Studium der Mikrowellenentladung von CO_2 (und anderen O-enthaltenden
asen wie N_2O, NO_2, SO_2 usw.) und der Bildung von molekularem Singulett-Sauer-
toff, $O_2(^1\Delta_g)$, sind wir zur Zeit beschäftigt.

iteraturverzeichnis

) Apparative Anordnung: J.R.Scheffer und M.D.Ouchi, Tetrahedron Letters 1970,
   223. Zur Entfernung von O-Atomen diente Quecksilberoxyd; vgl. auch Lit. 2).

) K.Furukawa, E.W.Gray und E.A.Ogryzlo, Ann. N.Y. Acad. Sci. 171, 175 (1970).

) Dargestellt aus Marmor und HCl im Kippschen Apparat oder durch Verdampfen
   von festem CO_2.

) a) R.N.Dixon, Proc. Roy. Soc. (London) 275 A, 431 (1963).

   b) R.N.Dixon, Disc. Faraday Soc. 35, 105 (1963).

   c) G. Herzberg, Molecular Spectra and Molecular Structure. III. Electronic
      Spectra and Electronic Structure of Polyatomic Molecules; D. van Nostrand
      Comp., Inc., New York, 1966, S. 219 und 500.

) Perkin-Elmer 900; 2 m Molekularsieb; 50°C, 0.35 atü He.

) z.B. K.Gollnick, Adv. Photochem. 6, 1 (1968).

) G.O. Schenck, K.Gollnick, G.Buchwald, S.Schroeter und G.Ohloff, Ann. Chem.
   674, 93 (1964).

) A.G.Gaydon, Proc. Roy. Soc. (London) 176 A, 505 (1940).

) B.H.Mahan und R.B.Solo, J. Chem. Phys. 37, 2669 (1962).

) D.Katakis und H.Taube, J. Chem. Phys. 36, 416 (1962).

) W.B.DeMore und C.Dede, J. Phys. Chem. 74, 2621 (1970).

) L.J.Stief, V.J.DeCarlo und W.A.Payne, J. Chem. Phys. 51, 3336 (1969).

) Die Bildung von molekularem Sauerstoff durch Reaktion von elektronisch-
   angeregtem CO_2^* mit CO_2 (im Grundzustand) konnte ausgeschlossen werden[12].

Editor's Comments
on Papers 54 Through 59

1,2 - DIOXETANES FROM THE CYCLOADDITION OF SINGLET OXYGEN TO ALKENES

Singlet oxygen undergoes 1,4-cycloaddition with conjugated dienes to yield 1,4-epidioxides (Papers 4–6, 10, and 11) and the "ene" reaction with alkenes to give allylic hydroperoxides (Papers 16 and 17). A third mode of reaction of singlet oxygen with electron-rich alkenes has recently been reported by several investigators: 1,2-cycloaddition to form 1,2-dioxetanes (2). These four-membered ring peroxides had been of interest as possible intermediates in chemiluminescent reactions,[1] as well

$$\underset{(1)}{\overset{}{\diagup}C=C\overset{}{\diagdown} + {}^1O_2} \longrightarrow \underset{(2)}{-\overset{\overset{O-O}{|\quad|}}{C}-\overset{}{C}-} \xrightarrow{\text{heat}} \underset{(3)}{2\ \overset{\overset{O}{\|}}{C} + h\nu}$$

290

as in the photoxidative cleavage of alkenes.[2] In 1969, Kopecky and Mumford described the first isolation of a stable 1,2-dioxetane prepared by the base-catalyzed cyclization of an a-bromohydroperoxide.[3] The thermal decomposition of 1,2-dioxetanes produces carbonyl-containing compounds in an electronically excited state. The chemiluminescent decomposition of 1,2-dioxetanes has become the subject of considerable investigation. [4a-d]

In 1968, Foote and Lin (Paper 54) and Huber (Paper 55) independently observed that dye-sensitized photooxygenation of enamines gave carbonyl-cleavage products in high yield. 1,2-Dioxetanes were considered as possible intermediates. Schaap and Barlett subsequently described the first isolation of a 1,2-dioxetane from the photooxygenation of an alkene.[5] Singlet oxygen adds stereospecifically to *cis*- and *trans*-1,2-diethoxyethylene to give the 1,2-dioxetanes (Paper 56). Mazur and Foote (Paper 57) found that an isolable 1,2-dioxetane is formed by the addition of singlet oxygen to tetramethoxyethylene. In 1971, Rio and Berthelot (Paper 58) described the stereospecific addition of singlet oxygen to *cis*- and *trans*-a, β-dimethoxystilbene. A 1,2-dioxetane of unusual stability has been prepared by the addition of $^1O_2$ to adamantylideneadamantane (Paper 59).

A theoretical treatment of several reactions of singlet oxygen with alkenes and dienes including the 1,2-cycloaddition reaction has been described by Kearns.[6] A number of selection rules are derived from consideration of molecular orbital diagrams.

NOTES AND REFERENCES

1. F. McCapra, *Quart. Rev. (London)*, **20**, 485 (1966), and *Chem. Commun.*, 155 (1968).
2. K. Gollnick, *Advan. Photochem.*, **6**, 1 (1968).
3. K. R. Kopecky and C. Mumford, *Can. J. Chem.*, **47**, 709 (1969).
4. (a) E. H. White, J. Wiecko, and D. R. Roswell, *J. Amer. Chem. Soc.*, **91**, 5194 (1969); (b) T. Wilson and A. P. Schaap, *ibid.*, **93**, 4126 (1971); (c) N. J. Turro, P. Lechtken, N. E. Schore, G. Schuster, H. C. Steinmetzer, and A. Yekta, *Accounts Chem. Res.*, **7**, 87 (1974); (d) G. Barnett, *Can. J. Chem.*, **52**, 3837 (1974).
5. Preliminary results reported at the International Symposium on Singlet Oxygen, New York Academy of Sciences, New York, Oct. 23, 1969; *Ann. N.Y. Acad. Sci.*, **171**, 79 (1970).
6. D. R. Kearns, *J. Amer. Chem. Soc.*, **91**, 6554 (1969). See also the following papers: S. Inagaki, S. Yamabe, H. Fujimoto, and K. Fukui, *Bull. Chem. Soc. Jpn.*, **45**, 3510 (1972); S. Inagaki and K. Fukui, *ibid.*, **46**, 2240 (1973); M. J. S. Dewar and W. Thiel, *J. Amer. Chem. Soc.*, **97**, 3978 (1975); M. J. S. Dewar, A. C. Griffin, W. Thiel, and I. J. Turchi, *ibid.*, **97**, 4439 (1975); S. Inagaki and K. Fukui, *ibid.*, **97**, 7480 (1975).

54

Reprinted from *Tetrahedron Lett.*, No. 29, 3267-3270 (1968)

CHEMISTRY OF SINGLET OXYGEN. VI.

PHOTOOXYGENATION OF ENAMINES: EVIDENCE FOR AN INTERMEDIATE (1).

Christopher S. Foote (2) and John Wei-Ping Lin (3)

Contribution No. 2237 from the Department of Chemistry,
University of California, Los Angeles, California 90024

(Received in USA 11 March 1968; received in UK for publication 24 April 1968)

Photooxidation of heterocyclic bases is of interest because of the relationship to the

mechanism of "Photodynamic Action", the toxic effect on organisms of certain photosensitizers,

light, and oxygen (4). Since the only isolable photooxidation products of many heterocyclic

compounds are extensively degraded (4,5), we are studying enamines as models for more complex

systems, in the hope of elucidating the chemistry of the primary photooxidation steps.

Enamines (I) prove to be excellent substrates for dye-sensitized photooxygenation; the C-C

double bond is cleaved to give a ketone or aldehyde (II) and an amide (III) (6,7a).

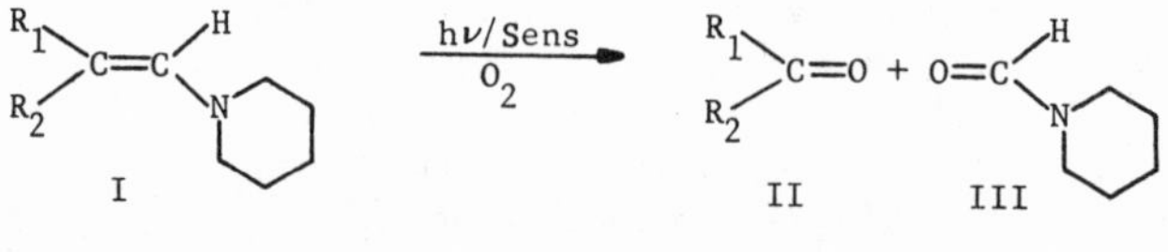

(a) R_1 = R_2 = CH_3

(b) R_1 = R_2 = C_6H_5

(c) R_1 = C_6H_5, R_2 = H

Thus Ia produced acetone and N-formylpiperidine (III) in 96 and 100% yield respectively,

Ib gave benzophenone and III (90 and 90%), and Ic gave benzaldehyde and III (48 and 49%):

unidentified non-volatile compounds are also produced in the latter case. Photooxygenations

were carried out in benzene using zinc tetraphenylporphine as sensitizer; the apparatus has

been described (8). Oxygen uptake was rapid and ceased abruptly after one mole O_2 per mole
of enamine had been taken up. Absolute yields quoted are based on enamine and were determined by g.l.c. using an internal standard; products were characterized by spectroscopic
and gas-chromatographic comparison with authentic materials. Addition of 0.1 M 2,6-di-$\underline{t}$-
butylphenol, a good free-radical inhibitor, produced no effect either on O_2 uptake rate or on
product yield in photooxidations of Ia or Ic.

Under the same conditions, enamine (IV)(7b), and β-methoxystyrene (V) (7c), both of which
have double bonds which are less electron-rich than those of enamines Ia-c, failed to produce
cleavage. However, N,N-diethyl-1-propynylamine (VI) (7d) smoothly took up one mole of oxygen
to give N,N-diethylpyruvamide (VII) in a reaction analogous to the cleavage of Ia-c.

$$CH_3O_2C \diagdown \atop H \diagup C=C \diagup ^{CO_2CH_3} \diagdown _{N} \qquad\qquad C_6H_5CH=CHOCH_3$$

IV   V

$$CH_3-C\equiv C-N(C_2H_5)_2 \quad \xrightarrow[\text{Sens}]{h\nu/O_2} \quad CH_3-\overset{O}{\overset{\|}{C}}-\overset{O}{\overset{\|}{C}}-N(C_2H_5)_2$$

VI   VII

Photooxygenation of β-piperidinostyrene (Ic) was carried out at -60° in $CH_3OH-CH_2Cl_2$;
NaBH$_4$ was added to the cold solution, and the resulting suspension allowed to warm to room
temperature. From this solution, a 72% yield (after sublimation and recrystallization) of
2-N-piperidino-1-phenylethanol (VIII), m.p. 69.5-70° (reported (9) 67-69°) was isolated. If
the photooxygenation was carried out at room temperature and the resulting solution reduced
with sodium borohydride, no appreciable amount of VIII could be detected.

Photooxygenation of N,N-dimethyl isobutenylamine (IX) was carried out at about -60° in
CDCl$_3$ until reaction was complete. The nmr spectra of the resulting solution showed no gross
change between -60° and -21°, except that broad resonances split into sharp multiplets. At
-21°, the nmr spectrum showed, along with resonances (2.22 δ, 2.98 δ, 8.12 δ) caused by a
calculated 6.5% conversion to the final products acetone and dimethylformamide, new resonances
at 4.5-5.0 δ (1H) 2.4-2.6 δ (6H), and 1.1-1.7 δ (6H). On warming the solution to room temperature, this spectrum was replaced within twenty minutes by a spectrum which showed only
resonances attributable to acetone and dimethylformamide (~90% yield by peak areas), along
with minor resonances attributable to 10% of a byproduct. Similar behavior was observed with
photooxidized solutions of enamines Ia and Ic, but the spectra were more complex.

293

These experiments demonstrate that an intermediate is formed which retains the C-C bond and is oxygenated at least on the position β to the nitrogen. This intermediate is stable at low temperatures and breaks down to the observed products near room temperature. The nmr spectrum of the intermediate from IX is not consistent with either of the two most likely structures X and XI. The resonances near 5 and 2.5δ are at too high field for the aldimmonium C-H and $\overset{+}{N}CH_3$ in X, while the complexity of each of the resonances is inconsistent with the sole presence of compound XI, in which all resonances should be sharp singlets near the observed positions, except that the two C-CH$_3$ groups would be magnetically nonequivalent. Possibly dimeric or polymeric species are present, perhaps formed from X or XI; a possible structure for a dimer would be XII, which has several configurational and conformational isomers, and could account for the complexity of the spectra and their temperature dependance (10). Further experiments will obviously be required.

A zwitterionic intermediate analogous to X has been suggested by Matsuura to account for unusual cleavage products in the photooxygenation of purine derivatives (11); either XI or XII should cleave in the same way; such cleavage has ample analogy (12). Formal 1,2 cycloaddition of this sort contrasts with the normal mode of photooxidative addition of oxygen to olefins which produces allylic hydroperoxides, apparently by a cyclic, non-polar transition state similar to that of the "ene" reaction (13). However, enamines are known to react with many dienophiles to give net 1,2 cycloaddition (14).

REFERENCES

1. Paper V: R. Higgins, C.S. Foote, and H. Cheng, American Chemical Society Advances in

    Chemistry Series, in press. Work was supported by National Science Foundation Grant

    GP-5835, and University of California Cancer Research Grant. Some preliminary experi-

    ments were carried out by M. Brenner.

2. A.P. Sloan fellow,1965-67; J.S. Guggenheim fellow, 1967-68.

3. A non-resident tuition grant from funds provided by the U.S. Rubber Co. is gratefully

    acknowledged.

4. J.D. Spikes and R. Straight, <u>Ann. Rev. Phys. Chem</u>. <u>18</u>, 409 (1967); H.F. Blum,

    <u>Photodynamic Action and Diseases Caused by Light</u>, Reinhold, New York (1941).

5. S. Gurnani, M. Arifuddin, and K.T. Augusti, <u>Photochem. Photobiol</u>. <u>5</u>, 495 (1966);

    C.A. Beñassi, E. Scoffone, G. Galiazzo, and G. Iori, <u>ibid</u>. <u>6</u>, 857 (1967); S. Gurnani

    and M. Arifuddin, <u>ibid</u>. <u>5</u>, 341 (1966); J.S. Sussenbach and W. Berends, <u>Biochim</u>.

    <u>Biophys. Acta</u> <u>95</u>, 184 (1965); T. Matsuura and I. Saito, <u>Chem. Comm</u>. 693 (1967);

    H.H. Wasserman and M.B. Floyd, <u>Tetrahedron Suppl. 7</u>, 441 (1966).

6. J. Huber, <u>Tetrahedron Letters</u> No.<u>29</u>, 3271 (1968) reports a similar reaction. We are

    grateful to Dr. Huber for communication of unpublished results.

7. (a) Enamines Ia-c were prepared from the corresponding carbonyl compound by standard

    methods (C. Mannich and H. Davidsen, <u>Ber. Dtsch. Chem. Ges</u>. <u>69</u>, 2106 (1936)); (b) IV:

    R. Huisgen and K. Herbig, <u>Ann</u>. <u>688</u>, 98 (1965); (c) V: M.Ch. Moureu, <u>Bull Soc. Chim</u>.

    <u>(France)</u> [<u>1</u>] <u>31</u>, 526 (1904); (d) VI was purchased from Fluka A.G.

8. C.S. Foote, S. Wexler, W. Ando, and R. Higgins, <u>J. Am. Chem. Soc</u>. <u>90</u>, 975 (1968).

9. M. Häring, <u>Helv. Chim. Acta</u> <u>42</u>, 1916 (1959).

10. We are grateful to Professor Frank Anet for discussion of this point.

11. T. Matsuura and I. Saito, <u>Tetrahedron Letters</u> No.<u>29</u>, 3273 (1968); we are grateful to

    Professor Matsuura for communication of unpublished results.

12. W.H. Richardson, J.W. Peters, and W.P. Knopka, <u>ibid</u>. <u>No. 45</u>, 5531 (1966); L. Dulog and

    W. Vogt, <u>ibid</u>. <u>No. 42</u>, 5169 (1966); M. Schultz, A. Rieche, and K. Kirschke, <u>Chem. Ber</u>.

    <u>100</u>, 370 (1967); F. McCapra and Y.C. Chang, <u>Chem. Comm</u>. 522 (1966).

13. C.S. Foote, <u>Accts. of Chem. Res</u>., in press.

14. J. Ciabattoni and E.C. Nathan, III, <u>J. Am. Chem. Soc</u>. <u>89</u>, 3081 (1967); M.E. Kuehne and

    T. Kitagawa, <u>J. Org. Chem</u>. <u>29</u>, 1270 (1964).

55

Reprinted from *Tetrahedron Lett.*, No. 29, 3271-3272 (1968)

PHOTOOXYGENATION OF ENAMINES - A

PARTIAL SYNTHESIS OF PROGESTERONE

Joel E. Huber

The Upjohn Company

Kalamazoo, Michigan   49001

(Received in USA 15 March 1968; received in UK for publication 24 April 1968)

The growing interest in singlet oxygen has resulted in a number of examples demonstrating
the usefulness of this reagent as a synthetic tool (1). We wish now to report a further exampl
of this utility. We have found that certain enamines which were stable to ground state oxygen
were cleanly cleaved by photooxygenation to form two carbonyl fragments.

3-Ketobisnor-4-cholen-22-al was selectively converted to the 22-morpholine enamine by a
standard procedure (2). Photooxygenation (3) of a slurry of 5.0 g. of the pure enamine in 25 m
of DMF at 15° using rose bengal as the sensitizer afforded, after 17.5 hrs., a quantitative
yield of progesterone, m.p. 126-128°, $[\alpha]_D$ +176° (dioxane). This product was identified by
comparison of its n.m.r. spectrum and mobility in both t.l.c. and g.l.c. with that of an authen
tic sample. The other fragment from this cleavage was shown by direct g.l.c. analysis of the

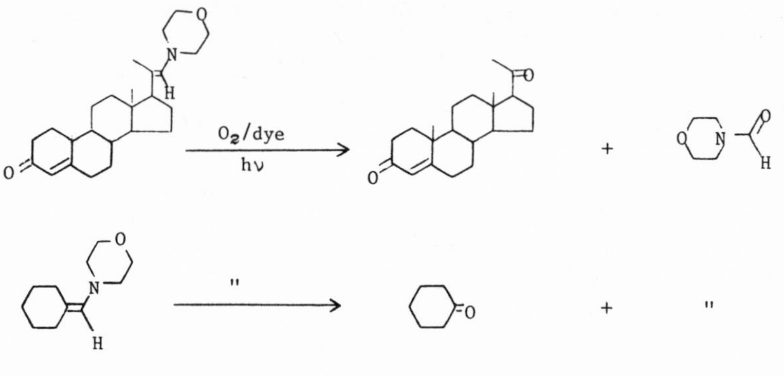

296

reaction mixture to be 4-morpholinecarboxaldehyde. Similarly, the morpholine enamine of cyclo-

hexanecarboxaldehyde (b.p. 100-106°/8 mm., n_D^{25} 1.5020) upon photooxygenation (methanol solution,

rose bengal dye) produced cyclohexanone and 4-morpholinecarboxaldehyde as the only products as

determined by direct g.l.c. analysis. The cyclohexanone in the mixture was isolated as the

2,4-dinitrophenylhydrazone, m.p. 159-160°. In both of these examples, no experimental evidence

(t.l.c. or g.l.c.) for any intermediates could be found.

This cleavage of the carbon-carbon double bond is envisioned simply as the 1,2-cyclo-

addition of the singlet ($'\Delta$ g) oxygen molecule to the enamine and subsequent decomposition of

this postulated intermediate to the two carbonyl fragments.

Proposed Mechanism:

REFERENCES

1. See for example: T. J. Katz, V. Balogh and J. Schulman, J. Am. Chem. Soc., 90, 734

    (1968) and J. A. Marshall and A. R. Hochstetler, J. Org. Chem., 31, 1020 (1966).

2. M. E. Herr and F. W. Heyl, J. Am. Chem. Soc., 74, 3627 (1952).

3. The photo-reactor used in these examples was similar to that described by Nickon and Bagli

    [J. Am. Chem. Soc., 83, 1498 (1961)] except that it was fitted with a cold finger. This

    was illuminated with four 15-watt fluorescent tubes. No reaction would take place in

    these examples if either the light or dye was omitted.

Reprinted from *J. Amer. Chem. Soc.*, **92**(10), 3223–3225 (1970)

Stereospecific Formation of 1,2-Dioxetanes from cis- and trans-Diethoxyethylenes by Singlet Oxygen

Sir:

To the two modes of reaction of singlet oxygen, conjugate addition to dienes and allylic hydroperoxide formation,[1] a third mode, formation of 1,2-dioxetanes, has recently been added.[2] This reaction requires alkenes specially activated, as by amino[2a] or alkoxy[2f-i] groups, and the absence of very active allylic hydrogen in the molecule. Dioxetanes, whether prepared from singlet oxygen[2g-i] or by cyclization of halohydroperoxides,[3] have been observed to decompose cleanly to carbonyl compounds with chemiluminescence.[4] We report here evidence supporting stereospecific *cis* addition of singlet oxygen to vinylene diethers to give isolable dioxetanes as products.

cis-Diethoxyethylene (**1**), freed from its *trans* isomer by preparative vapor phase chromatography, 0.211 g, was dissolved in 7 ml of fluorotrichloromethane (Freon 11) which had been dried over molecular sieves. The solution, containing 10^{-4} *M* tetraphenylporphin, was held in an acetone–Dry Ice bath at $-78°$ in a dewar with Pyrex windows and irradiated through a Corning uv filter with a 500-W lamp for 25 min in a stream of oxygen. Under these conditions white crystals began to form in 15–20 min, and soon filled the vessel. The crystalline dioxetane **2** was isolated by removing the Freon under vacuum through a sintered glass disk. A small sample of the product, on warming to room temperature, melted and exploded.

When the photooxidation of **1** was monitored by nmr, the ethylenic singlet of **1** at δ 5.12 gave way to a new singlet at δ 5.91. At the same time the methylene quartet of the ethyl group initially at δ 3.73 in **1** was further split because of the proximity of the new asymmetric carbon atom in **2**, the signal in **2** being centered at δ 3.88. The 100-MHz nmr spectrum of **2** is shown in Figure 1. The position of the methyl triplet was shifted only from δ 1.25 to 1.29. Under the conditions described the ethyl formate previously reported[2g] to accompany the dioxetane is not produced. Ethyl formate (**5**) appears either on prolonged irradiation in the presence of sensitizer or on warming to temperatures

(1) C. S. Foote, *Accounts Chem. Res.*, **1**, 104 (1968), and references therein.
(2) (a) C. S. Foote and J. W.-P. Lin, *Tetrahedron Lett.*, **29**, 3267 (1968); (b) J. Huber, *ibid.*, **29**, 3271, (1968); (c) W. Fenical, D. R. Kearns, and P. Radlick, *J. Amer. Chem. Soc.*, **91**, 3396 (1969); (d) W. Fenical, D. R. Kearns, and P. Radlick, *ibid.*, **91**, 7771 (1969); (e) G. Rio and J. Berthelot, *Bull. Soc. Chim. Fr.*, **10**, 3609 (1969); (f) R. S. Atkinson, *Chem. Commun.*, 177 (1970); (g) P. D. Bartlett, paper presented at the International Symposium on Singlet Oxygen, New York Academy of Sciences, New York, N. Y., Oct 23, 1969; (h) P. D. Bartlett, G. D. Mendenhall, and A. P. Schaap, *Ann. N. Y. Acad. Sci.*, in press; (i) S. Mazur and C. S. Foote, *J. Amer. Chem. Soc.*, **92**, 3225 (1970).
(3) K. R. Kopecky, J. H. van de Sande, and C. Mumford, *Can. J. Chem.*, **46**, 25 (1968).
(4) (a) K. R. Kopecky and C. Mumford, *ibid.*, **47**, 709 (1969); (b) E. H. White, J. Wiecko, and D. R. Roswell, *J. Amer. Chem. Soc.*, **91**, 5194 (1969).

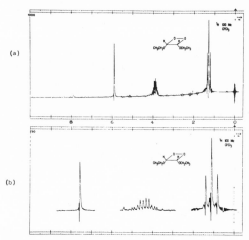

Figure 1. (a) 100-MHz proton nmr spectrum of recrystallized *cis*-diethoxydioxetane in CFCl$_3$; (b) same, expanded scale.

above 50° ($k_1 = 3.17 \times 10^{-3}$ sec^{-1} at 54°) which converts the nmr spectrum of the solution completely to

that of **5** (singlet at 7.92, quartet at 4.18, triplet at δ 1.29). During the decomposition in benzene the broad ir absorption at 847 cm^{-1} decreased as the typical carbonyl stretch of ethyl formate appeared.[5]

A molecular weight determination by isothermal distillation in Freon 11 at 2.6° against a dimethyl phthalate standard showed the dioxetane **2** to be mono-

(5) A similar band at 850 cm^{-1} was ascribed to O–O stretch in a cannabinoid dioxetane isolated by R. K. Razdan and V. V. Kane, *J. Amer. Chem. Soc.*, **91**, 5190 (1969), footnote 5.

meric (calcd volume of solution for monomer, 10.7 ml; for dimer, 7.3 ml; initial volume, 8.1 ml; volume after 30 min, 9.3 ml). The nmr spectrum showed no decomposition immediately after the molecular weight determination, but quantitative conversion to **5** after 30 min at 60°.

A quantitative study of the luminescence quantum yield of the thermal decomposition of *cis*-diethoxydioxetane **2** has been carried out and will be reported shortly.[6]

Identical products resulted from photooxidation of **1** with tetraphenylporphin in Freon 11 and with Rose Bengal in deuterioacetone, as shown by the single nmr spectrum of a mixture of the two product solutions.

When *trans*-diethoxyethylene (**3**) was photooxidized in deuterioacetone at −78°, a solution of **4** was obtained whose nmr spectrum had its singlet at 5.89 instead of the 6.15 observed for **2**. Spectra of mixtures showed both peaks, but the photooxidation in deuterioacetone of each pure isomer yielded its own dioxetane unaccompanied, within the sensitivity of the nmr method, by the other.[7] Each dioxetane decomposed quantitatively on warming to ethyl formate, $k_{cis}^{50°}/k_{trans}^{50°} = 1.40$.

First-order orbital symmetry theory leads to the prediction that singlet oxygen should not add suprafacially 1,2 to an alkene in a concerted manner.[8] One anticipates therefore that dioxetane formation will be either stepwise or antarafacial ($2_s + 2_a$).[9] A stepwise *biradical* addition of singlet oxygen would be expected to lead to loss of configuration, but stepwise cycloadditions by way of *dipolar ions* show much more retention of configuration.[10-12] The solvent susceptibility which we have reported[2g,h] for dioxetane formation (k_{addn}/k_{ene} greater by a factor of 59 for acetonitrile than for benzene) is inconclusive, being of a magnitude possible for a radical reaction with substantial polar

character,[13] or a concerted reaction with some charge transfer at the transition state.[14,15,15a]

A well-studied case of stereospecific 2 + 2 cycloaddition is that of ketenes.[16-18] Singlet oxygen shares with ketene its unhindered linearity, which allows it to undergo antarafacial addition to an alkene. Although oxygen lacks the strongly electrophilic center which can be discerned in ketenes and vinyl cations,[19] it has one property that must predispose it to this mode of reaction: any peroxide structure has such a strong preference for a nonplanar conformation that the contorted transition state (**6**) for the $2_s + 2_a$ addition may bring hardly any more strain than does the dioxetane ring itself.

6

Acknowledgment. This research was supported by grants from the National Science Foundation and the National Institutes of Health.

(13) J. C. Martin, D. L. Tuleen, and W. G. Bentrude, *Tetrahedron Lett.*, 229 (1962).

(14) R. Huisgen, L. A. Feiler, and P. Otto, *ibid.*, 4485 (1968); *Chem. Ber.*, **102**, 3444 (1969).

(15) The proposal (Kearns, ref 8, footnote 51) of a peroxirane as a common intermediate in the formation of dioxetanes and allylic hydroperoxides might also accommodate the polar solvent dependence of dioxetane formation, although it does not predict the manner[2g,h] in which partition between these two products responds to the medium.

(15a) NOTE ADDED IN PROOF. The substantial solvent effect on k_{addn}/k_{ene} for dihydropyran is not observed in the rate competition between two symmetrical molecules, tetramethylethylene, which gives only ene reaction, and *cis*-diethoxyethylene, which gives only dioxetane (competition ratio in benzene, 0.82; in acetonitrile, 1.10). The most significant difference between these cases may be the symmetry of the substrate double bond.

(16) R. Montaigne and L. Ghosez, *Angew. Chem.*, **7**, 221 (1968).

(17) G. Binsch, L. A. Feiler, and R. Huisgen, *Tetrahedron Lett.*, 4497 (1968).

(18) R. Huisgen, L. A. Feiler, and G. Binsch, *Angew. Chem. Int. Ed. Engl.*, **3**, 753 (1964); *Chem. Ber.*, **102**, 3460 (1969).

(19) Reference 9, p 847.

Paul D. Bartlett, A. Paul Schaap
Converse Memorial Laboratory, Harvard University
Cambridge, Massachusetts 02138
Received March 11, 1970

(6) A. P. Schaap and T. Wilson, forthcoming publication.

(7) Photooxidation of **3** with TPP in Freon 11 at −78° gave irreproducible ratios of **2** and **4**. The reason for this is not known, but it may be the result of photosensitized isomerization of **2**.

(8) D. Kearns, *J. Amer. Chem. Soc.*, **91**, 6554 (1969).

(9) R. B. Woodward and R. Hoffmann, *Angew. Chem. Int. Ed. Engl.*, **8**, 808 (1969).

(10) B. D. Kramer, Ph.D. Thesis, Harvard University, Cambridge, Mass., 1968.

(11) C. J. Dempster, Ph.D. Thesis, Harvard University, Cambridge, Mass., 1967.

(12) P. D. Bartlett, *Quart. Rev., Chem. Soc.*, in press.

Reprinted form *J. Amer. Chem. Soc.*, **92**(10), 3225–3226 (1970)

Chemistry of Singlet Oxygen. IX. A Stable Dioxetane from Photooxygenation of Tetramethoxyethylene[1]

Sir:

There has been much recent interest in 1,2-dioxetanes as possible intermediates in chemiluminescent reactions[2] and in olefin photooxygenations.[3] Electron-rich olefins are well known to undergo 1,2 cycloadditions with dienophiles,[4] and enamines were found to undergo photosensitized oxygenation[3a,d] to produce unstable intermediates which decomposed cleanly to carbonyl fragments. However, these intermediates were shown to be not solely monomeric 1,2 cycloadducts.[3a] The report by Hoffman and Häuser that the reactive olefin 1,1,2,2-tetramethoxyethylene (**1**) undergoes 1,2 cyclo-addition[5] suggested that photooxygenation of this compound might produce the dioxetane 1,2-dioxa-3,3,-4,4-tetramethoxycyclobutane (**2**).

Photooxygenation of **1** in ether, at −70°, sensitized by either zinc tetraphenylporphine or dinaphthalenethiophene with visible light[6] proceeded rapidly, and ceased abruptly after uptake of 1 equiv of oxygen. Both sensitizer and light were found to be essential. Evaporation of solvent (−78°) and evaporative distillation of the residue (25°) yielded a clear, pale-yellow liquid in 94% yield. This material, which contained (nmr) approximately 10% of dimethyl carbonate, was further purified by low-temperature crystallization from pentane–ether. The pure product (mp −8 to −9°) was found to be remarkably stable ($t_{1/2}$ = 102 min at 56°), although its decomposition appears to be catalyzed by zinc tetraphenylporphine.[7] The nmr spectrum (in Freon 12, 100 MHz) consisted of a single sharp resonance at 3.45 ppm at temperatures as low as −118°. The ir spectrum (neat) had principal bands at 2980 (m), 2870 (w), 1440 (m), 1345 (m), 1210 (s), 1130 (s), 1065 (s), 1020 (m), 975 (w), 910 (m), 875 (w), and 830 cm^{-1}. Elemental analysis was consistent with the formula

$C_6H_{12}O_6$ (*Anal.* Calcd: C, 40.01; H, 6.71. Found: C, 40.02; H, 6.74). The molecular weight was determined to be 185 ± 6 (cryoscopic, benzene) and 198 (vapor pressure osmometry); the calculated value is 180. Iodometric determination showed 87% of one peroxidic oxygen. When the product was heated at 56° in benzene in an nmr tube, a smooth decomposition occurred, giving dimethyl carbonate (identified by ir, nmr, and vpc comparison with authentic material) as the sole product in quantitative yield.

These data are entirely consistent with dioxetane structure **2**. Although alternate structures **3** and **4** can be envisioned, they may be discounted on the following grounds. Structure **3** would be expected to show two nonequivalent methoxyl resonances in the nmr (assuming oxygen inversion to be slow on the nmr time scale at −118°, the lowest temperature so far investigated).[8] The equilibrium **4a** ⇌ **4b**, even if it were rapid enough to result in a single methoxyl resonance, should produce an averaged chemical shift considerably downfield from the observed position.[9]

One of the most intriguing aspects of structure **2** is its close relationship to the CO_2 dimer (**5**), which has been implicated as the key intermediate in the chemiluminescent reaction of oxalyl halides and esters with H_2O_2,[10] and to the suggested dioxetane interme-

(1) Contribution No. 2544; supported by HEW-NAPCA Grant No. AP-00681 and NSF Grant No. GP-8293. Paper VIII: C. S. Foote and M. Brenner, *Tetrahedron Lett.*, 6941 (1968).

(2) (a) F. McCapra, *Quart. Rev., Chem. Soc.*, **20**, 485 (1966); (b) F. McCapra, *Chem. Commun.*, 155 (1968); (c) K. R. Kopecky and C. Mumford, *Can. J. Chem.*, **47**, 709 (1969); (d) E. H. White, J. Wiecko, and D. R. Roswell, *J. Amer. Chem. Soc.*, **91**, 5194 (1969).

(3) (a) C. S. Foote and J. W.-P. Lin, *Tetrahedron Lett.*, 3267 (1968); (b) D. R. Kearns, *J. Amer. Chem. Soc.*, **91**, 6554 (1969); (c) W. Fenical, D. R. Kearns, and P. Radlick, *ibid.*, **91**, 3396, 7771 (1969); (d) J. Huber, *Tetrahedron Lett.*, 3271 (1968).

(4) Reviews: R. Gompper, *Angew. Chem. Int. Ed. Engl.*, **8**, 312 (1969); R. W. Hoffman, *ibid.*, **7**, 754 (1968).

(5) R. W. Hoffman and H. Häuser, *ibid.*, **3**, 380 (1964).

(6) See C. S. Foote, S. Wexler, W. Ando, and R. Higgins (*J. Amer. Chem. Soc.*, **90**, 975 (1968)) for details of the technique.

(7) The dioxetane reported by Kopecky[2c] appears to be comparably stable, but no evidence for its structure or molecular weight has so far been published.

(7a) NOTE ADDED IN PROOF. The dioxetane has been characterized (E. H. White, J. Wiecko, and C. C. Wei, *J. Amer. Chem. Soc.*, **92**, 2167 (1970)).

(8) An approximate analogy may be found in the inversion rate of O-isopropylethyleneoxonium fluoroborate, for which the coalescence temperature is −50°: J. B. Lambert, *ibid.*, **90**, 1349 (1968).

(9) The average of the methoxyl shifts of trimethyl orthoformate (3.23 ppm) and of the dimethoxycarbonium ion [HC(OCH$_3$)$_2$]$^+$ (4.64 ppm: see R. I. Borch, *ibid.*, **90**, 5303 (1968)) would be 3.93 ppm whereas the observed value is 3.45 ppm.

(10) M. M. Rauhut, *Accounts Chem. Res.*, **2**, 80 (1969); H. F. Cordes, H. P. Richter, and C. A. Heller, *J. Amer. Chem. Soc.*, **91**, 7209 (1969).

3

4a   4b

diate (**6**) in the chemiluminescent oxidation of tetrakis(dimethylamino)ethylene.[11]

5   6

When **2** was heated to 80° in dilute benzene solution with several different fluorescent hydrocarbons, distinct

luminescence was observed; the color corresponded to the hydrocarbon fluorescence.

This represents the first preparation of a 1,2-dioxetane by photooxygenation[12] but by no means requires that 1,2 cycloaddition of singlet oxygen be concerted;[3b] in particular, dipolar ions such as **4** may well be intermediate.[4] Furthermore, these experiments do not provide evidence either for or against the assertion[3c] that dioxetanes may be intermediate in the formation of allylic hydroperoxides in the ene-type photooxygenation.

Acknowledgment. Miss Heather King carried out microanalyses; Dr. Gwendolyn Chmurny recorded the 100-MHz nmr spectra.

(11) J. P. Paris, *Photochem. Photobiol.*, **4**, 1059 (1965).
(12) See P. D. Bartlett and A. P. Schaap, *J. Amer. Chem. Soc.*, **92**, 3223 (1970), for similar results. We thank Professor Bartlett for communicating these results, reported in part at the International Conference on Singlet Oxygen, New York, N. Y., Oct 1969 (*Ann. N. Y. Acad. Sci.*, in press).

Stephen Mazur, Christopher S. Foote
Department of Chemistry, University of California
Los Angeles, California   90024
Received February 23, 1970

58

Reprinted from *Bull. Soc. Chim. France*, No. 10, 3555–3557 (1971)

N° 599. — **Photoxydation sensibilisée d'éthyléniques :
étude des deux diméthoxy-3,4 diphényldioxétanes,**

par Guy Rio et Jacques Berthelot.

(Université de Paris VI, et Laboratoire de chimie organique des hormones, Collège de France, Paris, 5ᵉ.)
(*Manuscrit reçu le 30.4.71.*)

Photoxydation des αβ-diméthoxystilbènes-(Z) et -(E); isolement des deux dioxétanes-1,2. Réduction en benzile. Décomposition en benzoate de méthyle et, pour l'isomère *trans*, en αβ-diméthoxystilbène-(E) (en très faibles proportions).

On sait maintenant que beaucoup de liaisons éthyléniques sont susceptibles d'être rompues par oxydation sous l'influence de l'oxygène excité à l'état singulet, $^1\Delta_g$, à condition qu'aucun groupe électroaccepteur ne soit fixé sur l'un des carbones éthyléniques (1). Au contraire, les groupes donneurs d'électrons, du type alcoxyle, par exemple, facilitent la réaction (1), et permettent parfois l'isolement des peroxydes intermédiaires, ou dioxétanes-1,2 (2, 3, 4). C'est le cas, notamment, des diéthoxy-3,4 dioxétanes, lesquels, toutefois, ne sont stables qu'à basse température (3).

Les deux αβ-diméthoxystilbènes, **1Z** et **1E**, forment des peroxydes, **2c** et **2t**, dont la stabilité à température ambiante est suffisante pour permettre d'étudier quelques-unes de leurs propriétés.

1) L'αβ-diméthoxystilbène décrit (5) a la configuration Z, **1Z**, comme on le montre ci-dessous. La photoisomérisation non sensibilisée donne le composé **1E** avec un rendement élevé.

Les configurations des deux isomères sont déduites, comme dans un cas semblable (6), de la comparaison des spectres. D'une part, le coefficient d'extinction de la bande d'absorption du spectre ultraviolet est plus grand pour l'isomère **1E**. D'autre part, dans le spectre de RMN de celui-ci, l'ensemble des raies correspondant aux hydrogènes aromatiques est étalé, alors que le spectre de **1Z** comporte un singulet.

2) La photoxydation sensibilisée des deux isomères **1Z** et **1E** est très rapide (quelques minutes) si le solvant est le chloroforme et le sensibilisateur le bleu de méthylène (avec un filtre optique), ce qui montre l'influence très favorable des deux méthoxyles [durée de la photoxydation 24-26 h pour les deux stilbènes non substitués (1)]. Le produit est le benzoate de méthyle.

Les peroxydes **2c** et **2t** sont obtenus par photoxydation à température ambiante de **1Z** et de **1E**, respectivement,

en solution dans un mélange éther-méthanol, en présence de bleu de méthylène; d'autres sensibilisateurs sont moins favorables.

La durée de la photoxydation est plus grande pour l'isomère *trans*, **2t** (45 mn environ) que pour le *cis*, **2c** (14 mn).

3) Les deux peroxydes, à l'état solide, se décomposent à température ordinaire en benzoate de méthyle, plus rapidement pour l'isomère **2t** (réaction totale en 5 h environ) que pour **2c** (8 h environ). A chaud, la décomposition est explosive; elle est chimiluminescente au-dessus de 180° environ.

En solution, la vitesse de décomposition varie notablement avec la nature du solvant. Pour l'isomère *cis*, **2c**, à température ambiante, elle est assez faible avec l'éther, le méthanol et l'éthanol (décomposition complète en 24 h environ), plus grande avec les hydrocarbures et la propanone, et surtout le dichlorométhane et le chloroforme (décomposition en 5 mn).

On s'explique ainsi que l'on n'avait pu mettre en évidence antérieurement ceux des dioxétanes dont la formation, trop lente, nécessitait l'emploi du chloroforme, qui est l'un des meilleurs solvants pour la photoxydation des éthyléniques (1).

Pour un solvant donné, la décomposition est un peu plus lente pour l'isomère *cis*, **2c**, que pour le *trans*, **2t**. C'était l'inverse qui avait été observé pour les deux diéthoxy-3,4 dioxétanes (3).

Avec le second isomère, **2t**, on peut classer les solvants comme ci-dessus : à température ambiante, dans l'éther, la décomposition est à peu près totale en 7 h, alors qu'elle ne demande que 3 mn dans le chloroforme.

De plus, à température ambiante, à l'état solide aussi bien qu'en solution, dans l'éther ou le méthanol, mais non dans le chloroforme, un autre mode de décomposition est observé : la dissociation en l'éthylénique correspondant, **1E**, et oxygène (lequel n'est pas mis en évidence). Certes, le rendement en **1E** est très faible (moins de 1 %), mais la réaction est reproductible; une décomposition analogue ne se produit pas avec l'autre peroxyde, **2c**.

Il s'agit là du premier exemple, à notre connaissance, de dissociation d'un dioxétane-1,2 en éthylénique et oxygène. Il est vraisemblable qu'elle est facilitée dans ce cas, car l'éthylénique **1E** doit être plus stable que **1Z**, si l'on se réfère au stilbène-(E) non substitué, dont l'énergie de résonance dépasse de 4 à 5 kcal.mole⁻¹ celle de son isomère (7).

Les acides, les bases, l'azoture de sodium, la lumière, avec ou sans sensibilisateur, accélèrent également la décomposition en benzoate de méthyle.

Il en est de même des agents réducteurs. Cependant, certains de ceux-ci (sulfite, thiourée, triphénylphosphine) provoquent en outre la réduction des deux peroxydes en benzile :

La proportion de benzile est nettement plus faible avec le dioxétane *trans*, **2t**; celui-ci, dans ces conditions également, se dissocie en l'éthylénique **1E**, en très faibles proportions.

Avec les hydrures mixtes, la réduction va jusqu'au diol (*méso*-hydrobenzoïne).

PARTIE EXPÉRIMENTALE.

Techniques expérimentales : cf. (1).

Dans presque tous les cas, l'éluant utilisé pour les chromatographies en couche mince (CCM) ou épaisse (CCE) est un mélange acétate d'éthyle 15 %, cyclohexane 85 %. Les réactions sont suivies par CCM.

Certains spectres UV sont dus à M. VAIRON et à Mme CONTASSOT (Université de Paris VI). Les spectres de RMN sont enregistrés sur un appareil Perkin-Elmer R 12 ou Varian HA 100; l'un est dû à Mme L. LACOMBE (Collège de France). Les analyses sont dues à M. DORME (Université de Paris VI).

αβ-Diméthoxystilbène-(Z), 1Z, $C_{16}H_{16}O_2$ (5b).

F_{inst} 126-127° [F 127° (5a); 125,5-126,5° (5b)].
Spectre IR (KBr) : ν(C — H) 2 810 (OCH₃); ν(C — O) 1 010 cm⁻¹.
Spectre UV (éther) : λ_{max} 299 nm (ε 12 100) [295 (log ε 4,08) (5b)].
Spectre de RMN (CDCl₃) : singulets à δ 3,6 (6 H des deux groupes OCH₃) et à 7,2 ppm (10 H aromatiques).

Photoxydation de l'αβ-diméthoxystilbène-(Z), 1Z.

Chaque essai est effectué à température ambiante sur 150 mg de **1Z**, en solution à 1 °/₀₀ contenant 0,01 °/₀₀ de sensibilisateur. Le dispositif a été décrit (1); le filtre est une solution à 2 % de K₂CrO₄, sauf pour le dinaphtylènethiophène (NaNO₂ saturé).

1) Sensibilisateur : bleu de méthylène; solvant : éther 85 %, méthanol 15 %; durée 14 mn. Après évaporation sous vide, à température ambiante, extraction à l'éther, lavages à l'eau et évaporation sous vide, la cristallisation de **2c** est complétée par addition d'un peu d'éthanol; Rdt 83 %.

2) Autres conditions.

Dans les essais donnant C₆H₅COOCH₃, celui-ci est séparé par CCE.

Sensibilisateur	Solvant	Temps (mn)	Rdt (%)
bleu de méthylène	éther 50 %, CH₃OH 50 %	25	**2c** : 75
— —	éther 85 %, CH₃OH 15 %; — 50°	45	**2c** : 80
rose bengale	éther 50 %, CH₃OH 50 %	25	**2c** : 75
dichlorhydrate d'hématoporphyrine	éther 50 %, CH₃OH 50 %	40	**2c** : 50
bleu de méthylène	CHCl₃ (1)	3	ΦCOOCH₃ : 93
— —	(CH₃)₂CO	10	ΦCOOCH₃ : 75
— —	CS₂ 85 %, éther 5 %, CH₃OH 10 %	25	ΦCOOCH₃ : 35 (et résines)
dinaphtylène-thiophène. bleu de méthylène, sans filtre	CHCl₃	15	ΦCOOCH₃ : 70
	éther 85 %, CH₃OH 15 %	25	ΦCOOCH₃ et résines
néant, sans filtre	—	16 h	ΦCOOCH₃
— —	— (conc. en **1Z** : 0,05 °/₀₀)	36 h	—
dinaphtylène-thiophène.	éther	90	**1Z**

Diméthoxy-3,4 diphényldioxétane-cis, **2c**, $C_{16}H_{16}O_4$.

F_{inst} 67-68°.

Analyse : Calc. % : C 70,57 H 5,92
        Tr. :   70,39   6,09
Poids moléculaire (cryoscopie, benzène) : calc. 272; tr. 242, 225.
Spectre IR (KBr) : ν(C — O) 1 010; absence de bande correspondant à C = O ou à O — H.
Spectre de RMN (C₆D₆) : singulet à 3,4 (6 H des deux OCH₃); multiplets entre 6,7 et 7,0 (6 H aromatiques en *méta* et *para*) et entre 7,1 et 7,45 (4 H aromatiques en *ortho*).

En CCM, quel que soit l'éluant, **2c** ne peut pas être distingué du benzoate de méthyle.

Décomposition à l'état solide.

A température ordinaire, variation du PF avec le temps : 0,5 h, 64-65° (odeur de C₆H₅COOCH₃); 1 h, 60° [spectre IR : ν(C = O) 1 730 cm⁻¹ faible]; 2 h, 53-57°; 4 h, pâteux; 8 h, décomposition totale (spectre IR voisin de celui de C₆H₅COOCH₃).
A — 20° : 1 j, odeur de C₆H₅COOCH₃; 3 j, F 65-66°; 5 j 58-60°; 8 j, pâteux.
A la fusion (67-68°) décomposition en quelques minutes; au-dessus de 180° environ, explosion et émission d'une lueur bleu pâle visible dans l'obscurité.

Décomposition en solution.

On ne peut pas la suivre en CCM : on évapore les prises d'essai et prend le PF.

1) A température ambiante.
— Éther, méthanol ou éthanol : 2 h, F 67-68°; 4 h, 55-60°; 8 h, peu de cristaux; 15 h, beaucoup de C₆H₅COOCH₃; 24 h, décomposition totale.
— Autres solvants. — Durées approximatives de décomposition : CS₂, 9 h; C₆H₆, 6 h; cyclohexane, 3 h; propanone, 2 h; pentane, 1,5 h.
— Dichlorométhane, chloroforme : 1 mn, F 55-59°; 3 mn, peu de cristaux; 5 mn, décomposition complète.

2) A chaud (au reflux).
— Éther : 3 mn, F 60°; 5 mn, peu de cristaux; 10 mn, décomposition totale.
— Méthanol, benzène : 2 mn, peu de cristaux.

Décomposition par la lumière.

Rdt 92 % en C₆H₅COOCH₃ (CCE) après irradiation de solutions de 30 mg de **2c** en tubes scellés sous vide, avec des lampes à incandescence (3 kW, cf. (1)] : dans 3 cm³ d'éther, sans sensibilisateur, durée 1 h; ou dans 5 cm³ de mélange éther 80 %, CH₃OH 20 % contenant 0,5 mg de bleu de méthylène, durée 1,5 h.

Décomposition par des réactifs.

Rdt 60-80 % en C₆H₅COOCH₃ (CCE) par action, à température ambiante, des réactifs suivants sur 35 mg de **2c** (durée 3 à 15 mn) : CH₃CO₂H pur, ou contenant 50 % d'eau, ou des traces de HCl; MgI₂ en solution dans l'éther; KOH en solution dans CH₃OH; NaN₃ en solution dans un mélange CH₃OH 50 %, éther 50 %.

Réduction.

— Na₂SO₃ : 1 dg de **2c** et 1,8 dg de Na₂SO₃, 6 cm³ de mélange éther 33 %, CH₃OH 67 %; durée 45 mn. Par CCE (deux élutions) : C₆H₅COOCH₃ 28 mg (Rdt 28 %), benzile 40,5 mg (52 %); un autre essai donne 29 et 50 %, respectivement.
La photoxydation de **1Z** (éther 85 %, CH₃OH 15 %; bleu de méthylène) en présence de Na₂SO₃ solide dure 50 mn; Rdt 75 % en C₆H₅COOCH₃, 12 % en benzile.
— Thiourée : 135 mg de **2c**, 150 mg de thiourée, 6 cm³ d'éthanol absolu; durée 15 mn. La CCE livre 28 mg (20 %) de C₆H₅COOCH₃ et 69,5 mg (65 %) de benzile. Autre essai : Rdt 18 et 66 %, respectivement.
— Triphénylphosphine : 135 mg de **2c**, 520 mg de P(C₆H₅)₃, 6 cm³ d'éther anhydre; durée 30 mn. Par CCE : 54 mg (40 %) de C₆H₅COOCH₃ et 43 mg (41 %) de benzile. Autre essai : Rdt 39 et 43 %, respectivement; CH₃OH est mis en évidence (CPV).
La photoxydation de **1Z** (bleu de méthylène; éther 80 %, CH₃OH 20 %) en présence de P(C₆H₅)₃ dure 75 mn; Rdt 62 % en C₆H₅COOCH₃; 10 % en benzile.
— KBH₄ : 135 mg de **2c**, 110 mg de KBH₄, 8 cm³ de mélange dioxane 50 %, CH₃OH 50 %; durée 45 mn. Après hydrolyse et extraction, la CCE donne 19 mg (14 %) de C₆H₅COOCH₃ (éluant acétate d'éthyle 15 %, cyclohexane 85 %) et 80 mg (73 %) de *méso*-hydrobenzoïne (éluant acétate d'éthyle 15 %, cyclohexane 50 %). Autre essai : Rdt 15 % et 71 %, respectivement.
— LiAlH₄ : 135 mg de **2c**, 80 mg de LiAlH₄, 6 cm³ d'éther anhydre; durée 45 mn. Après hydrolyse, on sépare par CCE 28 mg (20 %) de C₆H₅COOCH₃ et 76 mg (68 %) de *méso*-hydrobenzoïne. Autre essai : Rdt 19 % et 68 %, respectivement.

αβ-*Diméthoxystilbène-(E)*, **1E**, $C_{16}H_{16}O_2$.

On irradie pendant 26 h, à l'aide de lampes à incandescence [3 kW, cf. (1)], sans filtre, une solution de 150 mg de **1Z** dans 30 cm³ de benzène exempt de thiophène. Après concentration de la solution, on sépare par CCE (deux élutions) : 34 mg (Rdt 22 %) de **1Z** et 108 mg (72 %) de **1E** (un peu moins polaire que **1Z**).
Un mélange semblable est obtenu par irradiation (25 h) d'une solution de 150 mg de **1E** : 30 mg (20 %) de **1Z** et 111 mg (74 %) de **1E**.
F$_{inst}$ 97-98° (méthanol).
    Analyse: Calc. % : C 79,97   H 6,71
         Tr.   :   79,94   6,67.
Spectre IR (KBr) : ν(C — H) 2 810 ; ν(C — O) 1 030.
Spectre UV (éther) : λ$_{max}$ 290,5 (ε 19 600).
Spectre de RMN (CDCl₃) : singulet à 3,4 (6 H des deux OCH₃) ; multiplets entre 7,25 et 7,6 (6 H aromatiques en *méta* et *para*) et entre 7,65 et 7,95 (4 H aromatiques en *ortho*).

Photoxydation de l'αβ-*diméthoxystilbène-(E)-*, **1E**.

Mêmes conditions que celles de **1Z** (bleu de méthylène ; éther 85 %, CH₃OH 15 %) ; pour 1 dg, durée 45 mn à température ambiante, 75 mn à — 10° ; Rdt 75 % en **2t**. Même durée si la concentration est de 0,3 °/₀₀ au lieu de 1 °/₀₀.
Avec le bleu de méthylène et le méthanol, ou le rose bengale, et le mélange éther 50 %, méthanol 50 %, la durée est de 90 mn ; obtention d'un mélange de **2t** et de C₆H₅COOCH₃, difficiles à séparer.
Avec le bleu de méthylène et CHCl₃, durée 5 mn à température ambiante ; Rdt 90 % en C₆H₅COOCH₃.

Diméthoxy-3,4 diphényldioxétane-trans, **2t**, $C_{16}H_{16}O_4$.

F$_{inst}$ 59-61° (décomposition rapide).
    Analyse: Calc. % : C 70,57   H 5,92
         Tr.   :   70,79   6,04
Spectre IR (KBr) : ν(C — O) 1 015.
Spectre de RMN (C₆D₆) : singulet à 2,9 (6 H des deux OCH₃) ; multiplets entre 7,0 et 7,3 (6 H aromatiques en *méta* et *para*) et entre 7,55 et 7,75 (4 H aromatiques en *ortho*).
Révélé en CCM ; plus polaire que **2c** (et C₆H₅COOCH₃).

Décomposition à l'état solide.

La décomposition à température ambiante est suivie par CCM et PF : 3 mn, odeur de C₆H₅COOCH₃ ; 15 mn, F 55-60° ; 30 mn, liquide, forte proportion de C₆H₅COOCH₃ ; 2 h, décomposition presque complète, apparition de **1E** ; 5 h, décomposition totale.
40 mg de **2t** pur [ne contenant pas de **1E** (CCM)], après 5 h, donnent, par CCE, 34 mg (Rdt 85 %) de C₆H₅COOCH₃ et 0,22 mg (dosage par spectrophotométrie UV) (Rdt 0,55 %) de **1E**.
A — 20° : 1 j, F 50-55° ; 2 j, pâteux.
A chaud, au-dessus de 180°, explosion avec émission d'une lueur bleu pâle, d'intensité plus faible que celle de **2c**.

Décomposition en solution.

A température ambiante, dans CHCl₃ : décomposition totale en 3 mn, en C₆H₅COOCH₃ ; absence de **1E** (CCM et spectre UV).
Dans l'éther, le méthanol et le pentane : vitesses de décomposition voisines ; 5 mn, F 58-60°, apparition de C₆H₅COOCH₃ ; 15 mn, apparition de **1E** ; 20 mn, F 55° ; 45 mn, pâteux ; 1 h, peu de cristaux ; 1,5 h, liquide, proportion importante de C₆H₅COOCH₃ ; 7 h, liquide, décomposition totale.
40 mg de **2t** pur, en solution dans 5 cm³ de méthanol, après 7 h à température ambiante, donnent par CCE : 35 mg (87 %) de C₆H₅COOCH₃ et 0,15 mg (0,4 %) de **1E**.
A — 10° : 30 mn, F 58-60°, apparition de **1E** ; 75 mn, F 55°.
A l'ébullition, dans l'éther, le méthanol, l'éthanol, le pentane : décomposition instantanée, absence de **1E**.

Décomposition par des réactifs.

Décomposition instantanée en solution dans l'éther contenant une trace de HCl : C₆H₅COOCH₃, absence de **1E**.
NaN₃ : la solution provenant de la photoxydation à — 10° de 50 mg de **1E** (éther 85 %, CH₃OH 15 %), et ne contenant pas de **1E** (CCM), est additionnée de 40 mg de NaN₃ ; après 3 h à 0°, la CCE livre 34 mg de C₆H₅COOCH₃ (Rdt 85 % par rapport à **2t**, formé avec un Rdt de 75 %), et 0,2 mg (0,5 %) de **1E**.

Réduction.

— Thiourée : le peroxyde **2t** brut (exempt de **1E**), provenant de la photoxydation à — 10° de 50 mg de **1E**, est dissous dans 6 cm³ d'éthanol absolu contenant 60 mg de thiourée ; on maintient à 0° pendant 3 h. La CCE donne : 32 mg (80 % par rapport à **2t**) de C₆H₅COOCH₃ ; 1 mg (3 %) de benzile, et 0,16 mg (0,4 %) de **1E**.
A température ambiante, réaction complète en 30 mn, résultats comparables.
— Triphénylphosphine : la solution provenant de la photoxydation à — 10° de 50 mg de **1E** est additionnée de 2 dg de P(C₆H₅)₃ et maintenue à 0° pendant 3 h. Par CCE, on sépare 28 mg (70 %) de C₆H₅COOCH₃ et 2,5 mg (8,5 %) de benzile ; **1E** et P(C₆H₅)₃ ne sont pas séparables.

BIBLIOGRAPHIE.

(1) G. Rio et J. Berthelot, *Bull. Soc. chim.*, 1969, 3609 et références 4 à 9.
(2) J.-J. Basselier et J.-P. Le Roux, *Comptes rendus*, 1970, **270** C, 1366.
(3) P. D. Bartlett et A. P. Schaap, *J. amer. chem. Soc.*, 1970, **92**, 3223, 6055.
(4) S. Mazur et C. S. Foote, *Ibid.*, 3225.
(5) a) H. S. Staudinger et A. Binkert, *Helv. chim. Acta*, 1922, **5**, 703 ; — b) H. Kunimoto, *J. Soc. chim. japon.*,1962, **83**, 1279 ; *Chem. Abstr.*, 1963, **59**, 11309d.
(6) G. Rio et D. Masure, *Bull. Soc. chim.*, 1971, 3232.
(7) G. W. Wheland, *Resonance in organic chemistry*, Wiley, New York, 1955, 80.

59

Reprinted from *Tetrahedron Lett.*, No. 2, 169–172 (1972)

ADAMANTYLIDENEADAMANTANE PEROXIDE,

A STABLE 1,2-DIOXETANE.

J.H. Wieringa, J. Strating and Hans Wynberg,[*]

Department of Organic Chemistry, The University,

Zernikelaan, Groningen, The Netherlands.

and

Waldemar Adam,

Department of Chemistry, University of Puerto Rico,

Rio Piedras, Puerto Rico 00931.

(Received in UK 23 November 1971; accepted for publication 9 December 1971)

Proposed intermediates in bromine and chlorine addition reactions to olefins were stabilized when adamantylideneadamantane (III) was the substrate.[1,2] This unique property of the adamantane moiety warrants further study.

Singlet oxygen has the property of adding to electron rich olefins to form 1,2-dioxetanes as the principal products.[3] Reasonably stable tetramethoxy- and 3,4-dimethoxy-1,2-dioxetanes have been obtained in the photooxygenation of tetramethoxyethylene[4] and 1,2-diethoxyethylene[5] respectively. Attempts to cycloadd singlet oxygen to olefins bearing allylic hydrogens, in order to prepare alkylated 1,2-dioxetanes (I), resulted in the formation of allylic hydroperoxides of rearranged structure (II).[6,15]

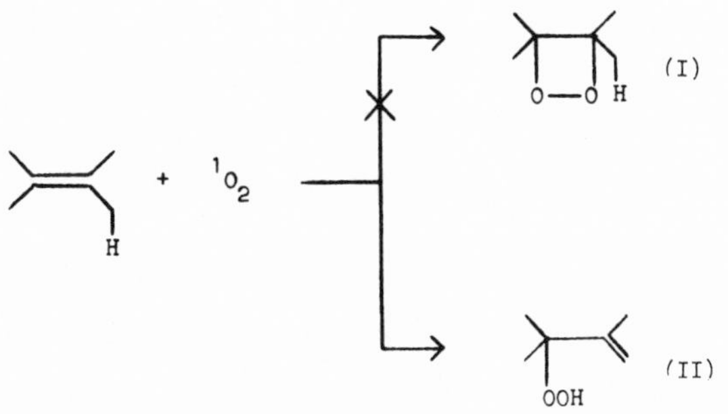

305

To achieve this goal these conditions must be met:

  i. The olefin must be sufficiently electron-rich,

 ii. No allylic hydrogens must be available (the word allylic implies real allylic activity),

iii. Steric hindrance should not inhibit the reaction.

We have found that adamantylideneadamantane (III) meets these three conditions in contrast to all other alkylated olefins studied thusfar.

        When singlet oxygen was generated by Methylene Blue sensitization of molecular oxygen, the cyclic peroxide (IV) could be isolated by evaporation of the solvent after absorbing the dye on charcoal. Using the triphenylphosphite-ozone complex[7] as source of singlet oxygen, the cycloadduct IV was readily isolated by column chromatography on florisil with carbontetrachloride as solvent. The yield in both cases must be nearly quantitative since 85% of pure material was obtained after recrystallization from methanol.

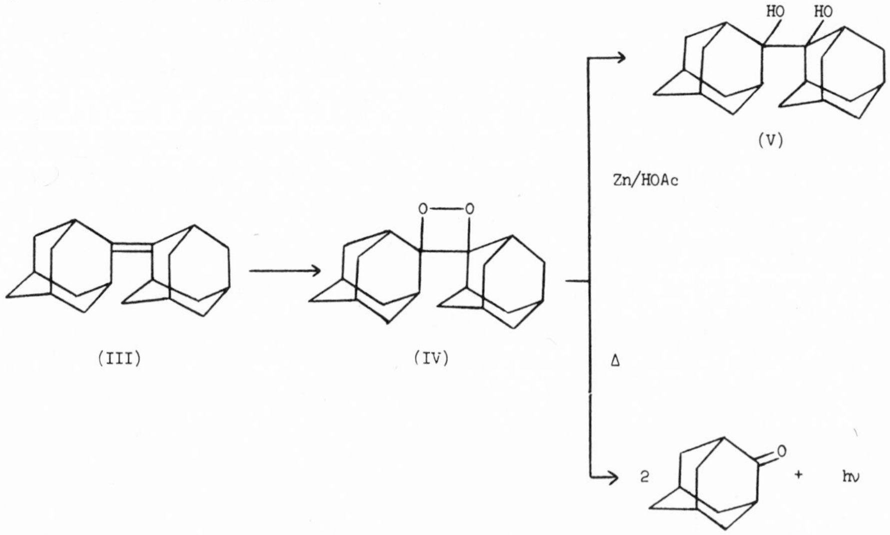

The cyclic peroxide (IV) exhibited the following spectral properties:

IR-spectrum: no carbonyl absorption, multiple strong absorptions at 1100, 1065, 1010, 1000, 975 and
        923 cm^{-1}.

Mass-spectrum: no molecular ion at m/e 300 ($\Delta v = 70$ V, source: $170°$); base peak at m/e 150 $(C_{10}H_{14}O)^+$
        peaks at m/e 284 $(C_{20}H_{28}O)^+$ and at m/e 268 $(C_{20}H_{28})^+$.

^{1}H-NMR-spectrum (CCl$_4$): complex multiplet at τ = 7.75-8,70 (24H) and complex multiplet at

τ = 7,25-7,60 (4H).

^{13}C-NMR-spectrum[8] (CCl$_4$): in accordance with the proposed peroxide structure.[9]

The elemental analysis and molecular weight (osmometrical) are in accordance with the peroxide structure.

The thermal properties of the compound are also typical for a peroxide structure.

Heated above its melting point (163-164°), IV explodes vigorously at about 240°, the product being adamantanone. Gentle heating in ethyleneglycol till reflux causes a beautiful, very bright chemiluminescence.[10,11] Work up gave an 80% yield of pure adamantanone.

Conclusive chemical support for the 1,2-dioxetane structure of IV was found in the zinc/acetic reduction to V. This glycol, formed in 80% yield, was identical in all respects to that previously prepared in this laboratory.[12]

Several other reduction procedures failed to provide definitive structure proof:

i. Treatment of IV with sodiumborohydride for 62 hrs. at room temperature gave only starting material.

ii. Reduction of IV with LAH for 15 hrs. at room temperature led to quantitative formation of 2-adamantanol.

iii. Catalytic hydrogenation with Pd/C gave a mixture of 2-adamantanol and adamantanone. Omission of hydrogen led only to adamantanone, indicating that the metal catalyzed the breakdown.

In sharp contrast to the behaviour of adamantylideneadamantane, olefins VI[13], VIIa[14],b[9],c[9] were inert under identical oxygenation conditions.

(VI)

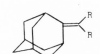

(VII)

a) R = H

b) R = C$_6$H$_5$

c) R = tert.-C$_4$H$_9$

REFERENCES

1. J. Strating, J.H. Wieringa and H. Wynberg, Chem. Comm., 1969, 907.

2. J.H. Wieringa, J. Strating and H. Wynberg, Tetrahedron Letters, 1970, 4579.

3. D.R. Kearns, Chem. Revs., 71, 395 (1971).

4. S. Mazur and C.S. Foote, J. Amer. Chem. Soc., 92, 3225 (1970).

5. P.D. Bartlett and A.P. Schaap, ibid., 92, 3223 (1970).

   A.P. Schaap and P.D. Bartlett, ibid., 92, 6055 (1970).

6. C.S. Foote and R.W. Denny, ibid., 93, 5162, 5168 (1971).

7. Q.E. Thompson, ibid., 83, 845 (1961).

8. We thank Dr. J. Lugtenburg and the staff of JEOL-USA for recording and interpreting this spectrum.

9. Details may be found in the forthcoming dissertation of J.H. Wieringa.

10. 1,2-Dioxetanes are known to luminesce on thermal decomposition. See refs. 4 and 5.

11. The elucidation of the thermochemistry and kinetics of the thermal decomposition of IV by DSC (differential scanning calorimetry) is in progress. See: W. Adam and J.C. Chang, Int. J. Chem. Kin., 1, 487 (1969).

12. H. Wynberg, E. Boelema, J.H. Wieringa and J. Strating, Tetrahedron Letters, 1970, 3613.

13. P. v. R. Schleyer, E. Funke and S.H. Liggero, J. Amer. Chem. Soc., 91, 3965 (1969).

14. P. v. R. Schleyer and R.D. Nicholas, ibid., 83, 182 (1961).

15. For recent analogous oxygenations in biological systems see H.W.-S. Chan, ibid., 93, 4632 (1971).

Editor's Comments
on Papers 60 Through 69

QUENCHING OF SINGLET OXYGEN

Several types of compounds are able to deactivate physically (quench) singlet oxygen without undergoing oxidation. Ouannès and Wilson (Paper 60) observed that singlet oxygen is quenched by tertiary amines such as 1,4-diazobicyclo [2.2.2]octane (DABCO). Matheson and Lee (Paper 61) subsequently demonstrated that the quenching rates of tertiary amines in solution correlate with the ionization potential of the amine. These results suggest that the quenching mechanism involves a charge-transfer interaction.

Foote and Denny (Paper 62) reported that singlet oxygen is efficiently quenched by β-carotene. These authors indicate that β-carotene and similar carotenoids may serve to protect photosynthetic plant systems from oxidative damage by singlet oxygen. It is interesting that β-carotene has been used clinically to treat erythropoetic protoporphyria, a disease characterized by a high level of porphyrins that render a person sensitive to strong light (see Papers 67–69).

Many investigators have also found β-carotene and DABCO useful in mechanistic studies. Inhibition of an oxygenation reaction by these quenchers has been taken as evidence for the intermediacy of singlet oxygen.

Metal chelates have also been found to quench singlet oxygen (Papers 64 and 65).[1] Farmilo and Wilkinson have likewise shown that the quenching of singlet oxygen by β-carotene occurs by energy transfer to give the β-carotene triplet.

In 1972, Grams and Eskins (Paper 66) reported that a-tocopherol reacts with singlet oxygen. This reactivity was compared with the vitamin E activity of a-tocopherol. These workers suggested that one function of vitamin E may be to protect biological systems from the damaging effects of singlet oxygen. It was subsequently shown (Papers 67–69) that a-tocopherol quenches singlet oxygen at a much faster rate than it undergoes chemical reaction with $^1O_2$. These three papers point out the necessity of considering physical quenching as well as chemical trapping in a kinetic treatment of $^1O_2$ reactions.

NOTES AND REFERENCES

1. See also D. J. Carlsson, T. Suprunchuk, and D. M. Wiles, *Can. J. Chem.*, **52**, 3728 (1974).

60

Copyright © 1968 by the American Chemical Society

Reprinted from *J. Amer. Chem. Soc.*, **90**(22), 6527–6528 (1968)

Quenching of Singlet Oxygen by Tertiary Aliphatic Amines. Effect of DABCO

Sir:

While studying the reactions of amines with singlet oxygen ($^1O_2$) generated by an electrodeless radiofrequency discharge,[1] we observed that tertiary aliphatic amines not only yield no reaction products, but act as inhibitors of the oxidation of known reactive acceptors of singlet oxygen. Thus, no reaction product could be detected[2] after treating 2 mmol of 1,4-diazabicyclo[2.2.2]octane (DABCO, 1) in 25 ml of bromobenzene with $^1O_2$ during 2 hr at 0°. Yet when a small amount of DABCO was added to a solution of 1,3-

1

diphenylisobenzofuran (2), the complete oxidation[3] of 2 took much longer than in the absence of DABCO. The oxidation of dibenzyl sulfide and rubrene was also inhibited (see Table Ia). This inhibitory property is not limited to that tertiary aliphatic amine. All those tried, ethyldiisopropylamine, N,N'-tetramethylethylenediamine, and N-allylpiperidine, were also found to be apparently unreactive toward $^1O_2$, yet to retard the oxidation of 2.

It therefore seemed interesting to find out if an actual quenching of singlet oxygen took place. Indeed, reactions involving $^1O_2$ generated by entirely different techniques were found to be retarded or suppressed by DABCO. In its presence, not only was the rate of rubrene photooxidation[4] reduced (Table Ib), but a correspondingly smaller volume of oxygen was absorbed by the solution. If the quenching of the ru-

brene photooxidation had been due to the competitive faster oxidation of the quencher, the rate of oxygen intake would have been unchanged or increased by DABCO. Thus the DABCO is not oxidized. Chemically generated singlet oxygen in aqueous solution is also quenched by DABCO. 2,5-Dimethylfuran, one of Foote's most efficient acceptors,[5] was chosen for this test (Table Ic).

Quenching of $^1O_2$ by DABCO must be very efficient since it affects the rate of oxidation of 2, the most reactive substrate known,[6] when 2 and DABCO are present at concentrations of the same order. A gas-phase experiment supports this conclusion. A small amount of DABCO was sublimed into a stream of oxygen rich in $^1O_2$ (at a total pressure of 6 torr), while the intensity of the emission band at 635 mμ[7] which is proportional to the square of the concentration of $O_2(^1\Delta_g)$[8] was being monitored downstream. At an estimated flow rate of ~4.5 × 10^{-3} mol/min, the gradual introduction of <3 × 10^{-4} mol of DABCO into the gas stream resulted in >100-fold decrease in the intensity of the 635-mμ band, lasting 15 min, suggesting that each molecule of DABCO is able to deactivate several molecules of $O_2(^1\Delta_g)$.[9] This nonreactive quenching of $O_2(^1\Delta_g)$ is remarkable in view of its known stability;[10a] for comparison, naphthalene, water, pyridine, and bromobenzene, introduced in the oxygen stream in the same manner as DABCO, caused no or negligible quenching of $O_2(^1\Delta_g)$ (the 635-mμ band).

(1) E. J. Corey and W. C. Taylor, *J. Am. Chem. Soc.*, **86**, 3881 (1964). In the present experiments, a mixture of 10% oxygen in helium, at a total pressure of 50 torr, passed through the discharge tube, then bubbled through the reacting solution in a vessel with a reflux condenser at Dry Ice temperature.

(2) By nmr, ir, and tlc.

(3) Indicated by the discoloration of 2.

(4) At 540 mμ DABCO does not interfere with light absorption by the rubrene. Besides, DABCO had no measurable quenching effect on the fluorescence of rubrene, even at a concentration of 4 × 10^{-2} *M*, greater than in the oxidation experiments. (No effort was made to remove oxygen in the fluorescence measurements, because of the obvious presence of oxygen in the oxidation experiments.)

(5) C. S. Foote, M. T. Wuesthoff, S. Wexler, I. G. Burstain, R. Denny, G. O. Schenck, and K. H. Schulte-Elte, *Tetrahedron*, **23**, 2583 (1967).

(6) T. Wilson, *J. Am. Chem. Soc.*, **88**, 2898 (1966).

(7) Through a Baird-Atomic interference filter, Standard Visible Type B-3, with peak wavelength at 6350 Å, by means of an EMI 9558B photomultiplier tube.

(8) L. W. Bader and E. A. Ogryzlo, *Discussions Faraday Soc.*, **37**, 46 (1964); S. H. Whitlow and F. D. Findlay, *Can. J. Chem.*, **45**, 2087 (1967).

(9) Assuming that about 10% of the oxygen is in the $^1\Delta_g$ stage. See L. Elias, E. A. Ogryzlo, and H. I. Schiff, *ibid.*, **37**, 1680 (1959).

Table I. Effect of DABCO on Oxidation Reactions *via* $^1O_2$

Source of $^1O_2$	Substrate	Initial concn $\times$ 10^3 M		Results
		Substrate	DABCO	
a. Electric discharge	2[a]	8	36	100% oxidn after 45 min
		8	0	100% oxidn after 6 min
	Dibenzyl sulfide[a]	40	160	No detectable product after 3.5 hr
		40	0	75% oxidn after 3.5 hr
	Rubrene[b]	0.72	43	14% rubrene oxidized in 1 min[f]
		0.72	16	20% rubrene oxidized in 1 min
		0.72	0	84% rubrene oxidized in 1 min
b. Photochemical	Rubrene[c]	3.9	13	10% rubrene oxidized in 16 min
		3.9	3.4	40% rubrene oxidized in 16 min
		3.9	0	100% rubrene oxidized in 16 min
	Rubrene[d]	3.6	13	12% rubrene oxidized in 6 min
		3.6	0	100% rubrene oxidized in 6 min
c. Chemical (H_2O_2 + NaOCl)	2,5-Dimethylfuran[e]	130	400	25 mg of hydroperoxide[g]
		130	0	400 mg of hydroperoxide[h]

[a] In bromobenzene. [b] In *o*-dichlorobenzene. [c] In pyridine. [d] In toluene. [e] In methanol and water (initial amount of substrate 1 g). [f] Calculated from initial and final concentrations of rubrene measured photometrically. [g] 2-Methoxy-5-hydroperoxy-2,5-dimethyldihydrofuran. [h] In excellent agreement with the yield reported by Foote.[13]

DABCO also completely suppresses the red luminescence from a stream of $^1O_2$ bubbled into a solution of violanthrone in bromobenzene. This emission corresponds to the violanthrone fluorescence and depends on the square of the concentration of $O_2(^1\Delta_g)$.[10]

This quenching by DABCO may provide a useful test of involvement of singlet oxygen. Exploratory experiments show, for example, that the rate of autoxidation of tetralin,[11] a well-established free-radical chain reaction, is only negligibly reduced by DABCO, even at a concentration 0.1 M, far greater than that at which it markedly inhibits singlet oxygen reactions.[12]

The observations reported here throw no light on the mechanism of the deactivation of singlet oxygen by DABCO. No permanent chemical reaction takes place, and the existence of a low-lying triplet state of DABCO (below 22.5 kcal) to which $O_2(^1\Delta_g)$ could transfer its energy seems unlikely. The availability of the lone electron pairs of DABCO suggests that the quenching of $^1O_2$ may occur through a charge-transfer process.[14]

Acknowledgments. We are grateful to Professors P. D. Bartlett and E. J. Corey for helpful discussions. This work was supported by the Petroleum Research Fund, administered by the American Chemical Society, and by a grant from the Milton Fund of Harvard University.

(10) (a) E. A. Ogryzlo, International Oxidation Symposium, San Francisco, Calif., Aug 1967; Advances in Chemistry Series, American Chemical Society, Washington, D. C., in press. (b) E. A. Ogryzlo and A. E. Pearson, *J. Phys. Chem.*, **72**, 2913 (1968). A preprint is gratefully acknowledged. Although it was noted that the quenching of the fluorescence of violanthrone, *i.e.*, its singlet state, was too small (only a factor of ~3) to account for the effect of DABCO on the luminescence caused by $^1O_2$ (a factor of ~10³), one cannot rule out a more effective quenching by DABCO of the violanthrone triplet state, which appears to be involved in the luminescence.
(11) Initiated by azobisisobutyronitrile at 60° in *o*-dichlorobenzene.
(12) 2,6-Di-*t*-butylphenol, 0.04 M, stops the autoxidation. Conversely, this free-radical inhibitor does not suppress singlet oxygen oxygenation, according to Foote,[13] although it does seem to have a small inhibitory effect (at high concentration: 0.3 M) on the photooxidation of rubrene.
(13) C. S. Foote, S. Wexler, and W. Ando, *Tetrahedron Letters*, 4111 (1965).
(14) *Cf.* quenching of the fluorescence of perylene by amines: H. Leonhardt and A. Weller in "Luminescence of Organic and Inorganic Materials," H. P. Kallmann and G. M Spruch, Ed., John Wiley & Sons, New York, N. Y., 1962, p 74; W. R. Ware and H. P. Richter, *J. Chem. Phys.*, **48**, 1595 (1968).
(15) Chargée de Recherches au C.N.R.S., Paris.
(16) Senior Fellow of the Radcliffe Institute, 1966–1967.

Catherine Ouannès,[15] **Thérèse Wilson**[16]
Converse Memorial Laboratory, Harvard University
Cambridge, Massachusetts 02138
Received July 29, 1968

61

Reprinted from *J. Amer. Chem. Soc.*, **94**(10), 3310–3313 (1972)

Quenching of Photophysically Formed Singlet ($^1\Delta_g$) Oxygen in Solution by Amines[1]

I. B. C. Matheson and John Lee*

Contribution from the Department of Biochemistry, University of Georgia, Athens, Georgia 30601. Received September 25, 1971

Abstract: The quenching by amines of singlet ($^1\Delta_g$) oxygen formed in Freon-113 solution by laser excitation within the $^1\Delta_g + 1v \leftarrow {}^3\Sigma_g$ oxygen absorption band envelope has been studied by inhibition of the oxidation of tetraphenylcyclopentadienone (tetracyclone). The quenching rate constants range from 3.1×10^4 l. mol^{-1} sec^{-1} for ethylamine to 2.1×10^6 l. mol^{-1} sec^{-1} for triethylamine and correlate with the ionization potential of the quencher.

If a solution of oxygen in Freon-113 is irradiated in the $^1\Delta_g + 1v \leftarrow {}^3\Sigma_g^-$ absorption band envelope with the 1.065-μm output from a Nd-YAG laser, singlet oxygen ($^1\Delta_g$) is produced unambiguously and in quantity.[2] Other techniques for producing singlet oxygen

are low-pressure microwave discharge in the gas phase and in solution by chemical means or dye photosensitization.[3] It is possible that in these other techniques extraneous chemical species may be produced and cause reactions which could be mistaken for those of singlet oxygen.

(1) Presented at the 6th International Conference on Photochemistry, 1971.

(2) I. B. C. Matheson and John Lee, *Chem. Phys. Lett.*, **7**, 475 (1970).

(3) For recent years' reviews of the generation and properties of singlet oxygen, see *Ann. N. Y. Acad. Sci.*, **171**, 1 (1970).

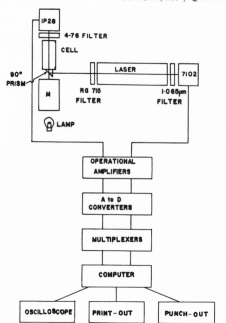

Figure 1. Schematic of apparatus for generating $O_2(^1\Delta_g)$ in a high pressure cell by absorption of 1.065-μm radiation from a Nd-YAG laser. M is a Beckmann DU monochromator for monitoring the concentration of acceptor.

Direct photophysical generation of $^1\Delta_g$ oxygen in solution has the advantage over chemical and dye sensitization methods in that the only species present are $^1\Delta_g$ and ground state oxygen. Reactions of the latter may be eliminated by blank experiments.

Ouannès and Wilson[4] have reported that aliphatic tertiary amines inhibit the reaction of $^1\Delta_g$ oxygen with known acceptors. This occurs without chemical change to the amine. Ogryzlo and Tang[5] confirmed these results and went on to show that in the gas phase the absolute quenching rate constants of a number of aliphatic amines correlated with their ionization potential.

In this present report we have studied the quenching of laser-produced $^1\Delta_g$ oxygen in Freon-113 solution under high oxygen pressure by both aliphatic and aromatic amines. In a like manner to the gas-phase results the quenching rate constants strongly correlate with the ionization potentials but their absolute values are about an order of magnitude lower than the estimates of Ogryzlo and Tang.

Experimental Section

A diagram of the apparatus is shown in Figure 1.

The reaction vessel was a high-pressure absorption cell of 50-mm path length which could withstand up to 140 atm of pressure. The cell was mounted on a modified Beckman DU single beam spectrophotometer. The 1.065-μm beam from the Nd-YAG laser (Holobeam 250 IRT) run, usually multimode C.W. at 4–6 W, was deflected into the absorption cell by means of a 7-mm right-angle prism mounted at the cell entrance. A Schott RG-715 color filter,

(4) C. Ouannès and T. Wilson, *J. Amer. Chem. Soc.*, **90**, 6527 (1968).
(5) E. A. Ogryzlo and C. W. Tang, *ibid.*, **92**, 5034 (1970).

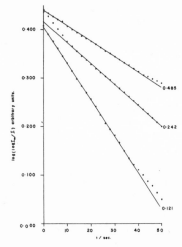

Figure 2. Decay of tetracyclone absorbance (580 nm) as a function of time caused by reaction with $O_2(^1\Delta_g)$. The concentration of ethylamine which inhibits the reaction is shown alongside each line.

which transmitted only above 700 nm, was placed at the laser beam entrance to prevent laser pump radiation and room light from entering the spectrophotometer cell area. The spectrophotometer RCA 1P21 photomultiplier was protected from direct laser irradiation by a Corning 4-76 color filter, which transmits only in the range 350–650 nm. The tungsten lamp of the spectrophotometer was run off a regulated dc supply to minimize lamp energy fluctuations during the course of an experiment. The laser power was monitored by a red sensitive photomultiplier (RCA 7102, S-1 response) which viewed the laser back beam through a 1.065-μm transmitting interference filter. The output of the 7102 photomultiplier was calibrated frequently relative to the laser main beam power by direct calorimetric measurement of the main beam power output. A 4% power loss at each optical surface was assumed in calculating the number of photons traversing the high-pressure cell.

The outputs of the 1P21 and 7102 photomultipliers were amplified by operational amplifiers, and the signal was then fed to analog to digital converters and then via a multiplexer to the Data General Nova computer. Timing was carried out relative to the internal computer clock. The outputs of both photomultipliers could be observed as a function of time on a display oscilloscope. The completion of a reaction could be observed when the photomultiplier output ceased rising with time. At this point the value for the 1P21 current was recorded, I_0, and the computer instructed to calculate the logarithm of the tetracyclone absorbance, log [log (I_0/I)], as a function of time. The computer then displayed the results in the form of time, log [log (I_0/I)] and relative laser power (as measured by the 7102 photomultiplier) were either printed out or punched out on paper tape. The individual results were then corrected for laser power fluctuations and an average first-order rate constant was calculated.

Oxygen was added to the solution by applying full cylinder pressure of about 2000 psi. The oxygen concentration was determined at the end of the experiment by measuring the absorbance of the solution at 1.26 μm. Data connecting this absorbance to concentration have been previously determined.[6]

Results

The singlet Δ oxygen acceptor used for all experiments was tetracyclone,[7] K & K Laboratories, Inc., maximum initial concentration 10^{-4} M. Its rate of decay was measured spectrophotometrically at 580 nm. Figure 2 shows the effect of ethylamine on the absorption change

(6) I. B. C. Matheson and John Lee, *Chem. Phys. Lett.*, **8**, 173 (1971).
(7) T. Wilson, *J. Amer. Chem. Soc.*, **88**, 2898 (1966).

Table I. Quenching Rate Constants for $O_2(^1\Delta_g)$ by Amines[a]

Quencher	Slope/intercept	$k_Q(\text{soln})$, l. mol^{-1} sec^{-1}	$k_Q(\text{gas})$, l. mol^{-1} sec^{-1} [b]	Ionization potential, eV[c]
Ethylamine	$2.4 \pm 0.5 \times 10^1$	$3.1 \pm 0.6 \times 10^4$	$3.3 \pm 0.3 \times 10^5$	9.19
Diethylamine	$4.4 \pm 0.7 \times 10^2$	$5.7 \pm 0.9 \times 10^5$	$2.0 \pm 0.2 \times 10^6$	8.44
Triethylamine	$1.6 \pm 0.3 \times 10^3$	$2.1 \pm 0.3 \times 10^6$	$1.9 \pm 0.2 \times 10^7$	7.85
N-Methylaniline	$2.9 \pm 0.2 \times 10^1$	$3.8 \pm 0.3 \times 10^4$		7.84
N,N-Dimethyl-aniline	$1.6 \pm 0.3 \times 10^2$	$2.0 \pm 0.4 \times 10^5$		7.14

[a] All results were at 23°. [b] From ref 5. [c] From ref 8.

which under the conditions of the experiment has a pseudo-first-order decay constant.

Laser irradiation of oxygen generates $O_2(^1\Delta_g)$ in a bimolecular process[6]

$$O_2 + O_2 \xrightarrow[\sigma E]{1.065\ \mu m} O_2(^1\Delta_g) + O_2$$

where E is the laser intensity ($h\nu$, sec^{-1}) and σ the absorption cross section at 1.065 μm.

There are three loss processes given by the reactions

$$O_2(^1\Delta_g) + A \xrightarrow{k_A} AO_2$$

$$O_2(^1\Delta_g) + O_2 \xrightarrow{k_{O_2}} O_2 + O_2$$

$$O_2(^1\Delta_g) + Q \xrightarrow{k_Q} O_2 + Q$$

The first is chemical reaction with the acceptor A (tetracyclone), the second quenching by ground state oxygen, and the last quenching by the added quencher, Q.

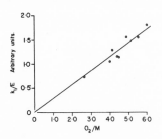

Figure 3. Variation of the tetracyclone–$O_2(^1\Delta_g)$ reaction rate with oxygen concentration.

Applying the stationary state assumption for the concentration of singlet Δ oxygen, $[O_2(^1\Delta_g)]$

$$\frac{d}{dt}[O_2(^1\Delta_g)] = \sigma E[O_2]^2 - k_A[A][O_2(^1\Delta_g)] -$$

$$k_{O_2}[O_2][O_2(^1\Delta_g)] - k_Q[Q][O_2(^1\Delta_g)] = 0$$

$$[O_2(^1\Delta_g)] = \frac{\sigma E[O_2]^2}{k_A[A] + k_{O_2}[O_2] + k_Q[Q]} =$$

$$\frac{\sigma E[O_2]^2}{k_{O_2}[O_2] + k_Q[Q]}$$

since $k_A[A] \ll k_{O_2}[O_2] + k_Q Q$ (see ref 2).

The rate of change of acceptor concentration is

$$d[A]/dt = k_A[A][O_2(^1\Delta_g)]$$

so that the pseudo-first-order decay constant for the disappearance of 580-nm absorption is

$$k_1 = k_A[O_2(^1\Delta_g)] = \frac{\sigma E[O_2]^2 k_A}{k_{O_2}[O_2] + k_Q[Q]}$$

This may be rearranged to give

$$\frac{E}{k_1}[O_2] = \frac{k_{O_2}}{\sigma k_A} + \frac{k_Q}{\sigma k_A} \frac{[Q]}{[O_2]} \quad (1)$$

Equation 1, on elimination of the term in [Q], may be rearranged to give

$$k_1/E = (\sigma k_A/k_{O_2})[O_2] \quad (2)$$

This suggests that in the absence of added quenchers k_1/E should be a linear function of $[O_2]$. Such a plot is shown in Figure 3. Plots corresponding to eq 1 are shown in Figure 4.

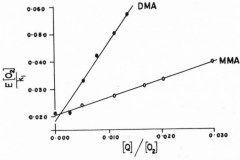

Figure 4. Quenching of the tetracyclone–$O_2(^1\Delta_g)$ reaction by N,N-dimethylaniline and N-methylaniline.

It will be noted that the ratio of slope to intercept for these plots gives k_Q/k_{O_2}. These ratios for all the amines studied are given in the first column of Table I. The second column lists the absolute value of the quenching constant obtained by multiplying the ratio by k_{O_2}. The gas-phase k_Q obtained by Ogryzlo and Tang are also listed for comparison.

Discussion

The physical quenching rates for $O_2(^1\Delta_g)$ in solution for quenchers such as these amines span about the same range as the chemical quenchers.[2] The highest rate belongs to a physical quencher, β-carotene,[9,10] but of

(8) V. I. Vedeneyev, L. V. Gurvich, V. N. Kondrat'yev, V. A. Medvedev, and Ye. L. Frankevich, "Bond Energies, Ionization Potentials and Electron Affinities," Arnold, London, 1966.
(9) C. S. Foote and R. W. Denny, *J. Amer. Chem. Soc.*, **90**, 6223 (1968).
(10) I. B. C. Matheson and John Lee, *Bull. Amer. Phys. Soc.*, **17**, 327 (1972).

these amines the fastest is about 10^4 times less than the diffusion-controlled rate, roughly paralleling the gas-phase case where the fastest is 10^4 times less than the bimolecular collision frequency.

The estimated values of the solution phase quenching rate constants k_Q in Table I are approximately an order of magnitude lower than the gas phase ones. These k_Q-(soln) values were calculated using the low-pressure gas-phase value for k_{O_2} of 1.3×10^3 l. mol^{-1} sec^{-1}.[11,12] Although there seems no reason to doubt the extrapolation of this value to high pressure, this rate has not so far been amenable to direct measurement in our apparatus. Attempts to measure oxygen emission from either single molecule or dimol states for both solution or gas-phase oxygen at high pressure have been unsuccessful. The experiments were done using multimode short cavity operation of the laser at up to 30 W and with a photomultiplier detection system capable of measuring a dimol (633 nm) emission rate of 10^6 photons sec^{-1}, about one-hundred times less than predicted rate of emission. This suggests the values for the radiative and quenching rate constants obtained from low-pressure gas may not be applicable at high pressure where third body effects may predominate.

Figure 5 shows the strong correlation between log k_Q(soln) and the quencher ionization potential. This relationship supports the suggestion[4,5] of a charge-transfer quenching mechanism. The aliphatic and aromatic amines, however, lie on different lines. This would be expected if the quenching mechanism involves a charge-transfer interaction since the aliphatic and aromatic amines belong to chemically different classes of electron donors.

In order to gain additional support for this mechanism, we have studied the formation of charge-transfers complexes between some of these quenchers and ground state oxygen in Freon-113.[13] These were visible as an increase in the long wavelength absorption of the donors. They were, however, of dominant stoichiometry QO_4 and weak with termolecular association constants of 0.5–3 l.^{2} mol^{-2}. As such they seem to be largely collisional. This situation does not exclude the possibility of complexes of stoichiometry $QO_2(^1\Delta_g)$,

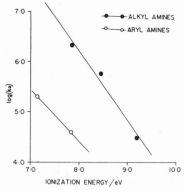

Figure 5. Relation between the quenching constant k_Q(soln) and the amine ionization potential.

since termolecular collisions appear to be excluded by the high values of the gas-phase quenching constants. The temperature dependence of k_Q(gas) indicates that the quenching complexes must be weak and largely collisional.[5]

In conclusion, the dependence of the quenching rate of amines for $O_2(^1\Delta_g)$ on the ionization potential of the amine and its electron donor class adds support to the charge-transfer quenching mechanism. The difference in absolute magnitude of the quenching rates between the low-pressure gas and high-pressure liquid condition may be the result of an unexpected enhancement of the rate of quenching by ground state oxygen at high pressure or may represent a true enhancement of the amine quenching rates on going into solution. The details of these mechanisms, however, remain obscure.[14]

Acknowledgments. We thank John E. Wampler for assistance with the computer programming. This research was supported in part by grants from the Research Corporation and the National Science Foundation, GP-24008.

(11) R. P. Wayne, *Advan. Photochem.*, **7**, 311 (1969).
(12) K. H. Becker, W. Groth, and U. Schurath, *Chem. Phys. Lett.*, **8**, 259 (1971).
(13) I. B. C. Matheson and J. Lee, unpublished results.
(14) NOTE ADDED IN PROOF. Traces of atomic oxygen apparently led to the higher values of the singlet oxygen amine quenching constants in the gas phase reported in ref 5. A revised value of k_Q(gas) is 2.0×10^6 l. mol^{-1} sec^{-1} for triethylamine [K. Furukawa and E. A. Ogryzlo, *Chem. Phys. Lett.*, **12**, 370 (1971)] in good agreement with the k_Q(soln) in Table I above.

62

Reprinted from *J. Amer. Chem. Soc.*, **90**(22), 6233–6235 (1968)

Chemistry of Singlet Oxygen. VII. Quenching by β-Carotene[1]

Sir:

The detrimental effects of photosensitizing dyes, light, and oxygen upon various organisms, including the human, are well known. These effects are referred to as "photodynamic action," and are the result of photosensitized oxidation of certain cell constituents.[2] Photosensitized oxygenation of olefins, dienes, and heterocyclic compounds is believed to proceed *via* the intermediacy of singlet oxygen, formed by energy transfer from triplet sensitizer.[3,4] Chlorophyll is among the most effective sensitizers for dye-sensitized photooxygenations of organic substrates.[4] Photosynthetic organisms, however, are protected from the lethal effects of their own chlorophyll by carotenoids; mutants lacking certain carotenoids are readily killed by light and oxygen; carotenoids also protect organisms against the effects of exogenous photosensitizers such as methylene blue.[5] The mechanism of this protective action has not been established, although it is known that carotenes quench triplet sensitizers efficiently.[6] However, since oxygen also quenches triplet sensitizers at a diffusion-controlled rate (to give singlet oxygen),[4]

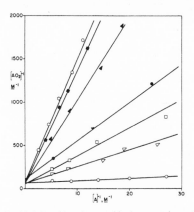

Figure 1. Methylene blue photosensitized oxygenation. β-Carotene concentrations: ○, 0.0 M; △, 3.10 × 10⁻⁵ M; ●, 3.54 × 10⁻⁵ M; □, 6.11 × 10-5 M; ▲, 8.61 × 10⁻⁵ M; ■, 8.95 × 10⁻⁵ M; ○, 9.90 × 10⁻⁵ M.

(1) Supported by National Science Foundation Grants GP-5835 and GP-8293.

(2) J. D. Spikes and R. Straight, *Ann. Rev. Phys. Chem.*, **18**, 409 (1967).

(3) C. S. Foote, *Accounts Chem. Res.*, **1**, 104 (1968).

(4) K. Gollnick, *Advan. Photochem.*, in press.

(5) R. Y. Stanier, *Brookhaven Symp. Biol.*, **11**, 143 (1959); W. A, Maxwell, J. D. Macmillan, and C. O. Chichester, *Photochem. Photobiol.*. **5**, 567 (1966); M. M. Mathews, *Nature*, **203**, 1092 (1964).

(6) M. Chessin, R. Livingston, and T. G. Truscott, *Trans. Faraday Soc.*, **62**, 1519 (1966); E. Fujimori and R. Livingston, *Nature*, **180**, 1036 (1957).

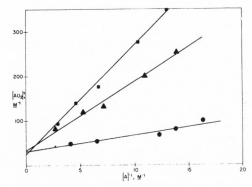

Figure 2. OCl⁻–H₂O₂ oxygenation. β-Carotene concentrations: ●, 0.0 M; ▲, 4.50 × 10⁻⁵ M; ■, 9.31 × 10⁻⁵ M.

quenching of triplet chlorophyll by carotenes cannot be responsible for the protective effects unless the local concentration of carotenes greatly exceeds that of oxygen.

We wish to report that singlet oxygen (generated by methylene blue photosensitization or by reaction of NaOCl and H₂O₂)[3] is quenched very effectively by low concentrations of β-carotene; 95% of the photooxygenation of 0.1 M 2-methyl-2-pentene is inhibited by 10⁻⁴ M β-carotene.

Solutions in benzene–methanol (4:1, v:v) containing *ca.* 4 × 10⁻⁵ M methylene blue (MB), various amounts of β-carotene (C), and various amounts of 2-methyl-2-pentene (A) were irradiated in an immersion irradiation apparatus at 25° for a known time. The resulting mixture of hydroperoxides (AO₂) was reduced, and the total amount of product alcohols was determined gas chromatographically using an internal standard.[7] Results are shown in Figure 1.

Chemical oxygenations were carried out on 100-ml solutions in 1:1:1 methanol–benzene–diglyme (the latter added for solubility) 0.26 M in H₂O₂, containing various amounts of A. To these solutions were added 7.2 × 10⁻³ equiv of aqueous 22% NaOCl. The solutions were reduced and analyzed as above; the results are shown in Figure 2.

Both methods give results which fit the following scheme

$$^3O_2 \xleftarrow{k_d} \ ^1O_2 \xrightarrow{k_A}{} AO_2$$
$$C \downarrow k_C$$
$$^3O_2$$

in which

$$[AO_2]^{-1} = \text{constant}\left(1 + \left[\frac{k_d}{k_A} + \frac{k_C}{k_A}[C]\right]\frac{1}{[A]}\right)$$

(7) R. Higgins, C. S. Foote, and H. Cheng, International Oxidation Symposium, San Francisco, Calif., 1967; Advances in Chemistry Series, No. 77, American Chemical Society, Washington, D. C., in press.

The best fit of the data to this equation gives $k_C/k_A = 1.6 \times 10^4$ and 0.7×10^4 for the photochemical and nonphotochemical oxygenations, respectively. Because these values depend critically on the values of the intercepts of each of the lines in Figures 1 and 2, and since these intercepts are difficult to determine accurately, the values of k_C/k_A are subject to an uncertainty of at least a factor of two. However, the following conclusions can be drawn. (1) The inhibitory effect of β-carotene in the photochemical reaction is *not* caused by quenching of triplet sensitizer, since the intercepts of the lines in Figure 1 are constant within the experimental uncertainty. (Light absorption by β-carotene resulting in a "shadowing effect" is also ruled out by this observation, as well as by the fact that the absorption spectra of β-carotene and methylene blue do not overlap.) (2) The values of k_C/k_A for the two methods are in reasonable agreement, allowing for the difference in reaction conditions and the uncertainty. As a further check, a second series of experiments was carried out in which solutions containing a constant amount of A were oxygenated at different carotene concentrations; plots of $[AO_2]^{-1}$ *vs.* [C] gave straight lines from which the values of $k_C/k_A = 2.2 \times 10^4$ and 2.9×10^4 for the two methods were determined. These methods would be subject to error if quenching of triplet sensitizer were occurring in the photochemical case or if production of singlet oxygen were being inhibited by β-carotene in the chemical case (since "constant" in the above expression would vary); however, the fact that constant intercepts were obtained in Figures 1 and 2 makes this unlikely. The best value of k_C/k_A is therefore probably around 2×10^4. In none of these experiments was there appreciable consumption of β-carotene; in fact, one molecule of β-carotene must quench as many as 250 molecules of singlet oxygen in some runs.

An attractive mechanism for the quenching of singlet oxygen by β-carotene would involve energy transfer in a process which is the reverse of the reaction which produces singlet oxygen by energy transfer from triplet sensitizer to oxygen.[3] This mechanism is tenable

$$^1O_2 + C \longrightarrow {}^3C + {}^3O_2$$

only if the triplet energy of β-carotene is below or near that of singlet oxygen (the $^1\Delta_g$ state is at 22.5 kcal). Unfortunately, the triplet energy does not seem to be known, but such a low energy would not be unreasonable. However, other mechanisms for the quenching which do not involve consumption of β-carotene cannot be excluded.

From the large value of k_C/k_A, it is seen that the quenching is very efficient. If k_C is taken to be the diffusion-controlled rate, roughly 5×10^9 M^{-1} sec^{-1},[8] then the decay rate of singlet oxygen in solution, k_d, can be calculated to be about 10^4 sec^{-1}, from the fact that k_d/k_A is 0.07 M, as determined from Figures 1 and 2. This value of k_d is an upper limit and is considerably lower than that estimated by Schenck and Koch by an indirect method for the reactive intermediate in the photooxygenation.[9]

From the results reported here, it is evident that quenching of singlet oxygen by β-carotene may be an important protective mechanism in plant biochemistry. How relevant these results are to *in vivo* systems will require further study. It will also be of interest to see what other quenchers for singlet oxygen can be found.[10]

(8) P. J. Wagner and I. Kochevar, *J. Am. Chem. Soc.*, **90**, 2232 (1968).

(9) G. O. Schenck and E. Koch, *Z. Electrochem.*, **64**, 170 (1960); see also C. S. Foote, S. Wexler, W. Ando, and R. Higgins, *J. Am. Chem. Soc.*, **90**, 975 (1968).

(10) G. O. Schenck and K. Gollnick, *Angew. Chem.*, **70**, 509 (1958), report nicotine quenches the reactive intermediate (presumably singlet oxygen) in photosensitized oxygenations. See also ref 4 for some other possible cases.

(11) John Simon Guggenheim Fellow, 1967–1968.

Christopher S. Foote,[11] **Robert W. Denny**
Contribution No. 2264 from the Department of Chemistry
University of California, Los Angeles, California 90024
Received August 19, 1968

63

Copyright © 1970 by the American Chemical Society

Reprinted from *J. Amer. Chem. Soc.*, **92**(17), 5216-5218 (1970)

Chemistry of Singlet Oxygen. X. Carotenoid Quenching Parallels Biological Protection[1]

Sir:

β-Carotene efficiently quenches singlet oxygen, generated either by dye sensitization or by the NaOCl–H₂O₂ reaction.[2] This quenching bears on the mechanism of the protective action of carotenoids against photodynamic damage in living organisms; this protective action may be the "universal function of carotenoid pigments."[3]

We now report that *the rate of quenching is a sensitive function of the length of the conjugated polyene chain and parallels the protective action of natural compounds.* The techniques used were similar to those previously reported.[2,4] The compounds used as quenchers were all-*trans*-retinol (**1**, 5 conjugated C=C) and two synthetic carotene analogs, a C₃₀ hydrocarbon (**2**, 7 conjugated C=C) and a C₃₅ hydrocarbon (**3**, 9 conjugated C=C).[5]

In the previous paper, it was shown that when [AO₂]⁻¹ is plotted against [A]⁻¹ at constant [Q], straight lines result; the intercept on the [AO₂]⁻¹ axis is a measure of the amount of singlet oxygen formed, and the increase in ratio of slope to intercept compared to plots with no Q is a measure of quenching of singlet oxygen (a kinetic scheme for this system is given in the accompanying communication).[6] Thus quenching of triplet sensitizer but not singlet oxygen gives an increase in intercept, but no change in slope/intercept; quenching of singlet oxygen but not of sensitizer gives an increase only in slope, and the intercept remains constant. β-Carotene showed the latter behavior.[2]

The behavior of the present three compounds shows a dramatic change with chain length: whereas β-carotene is an effective inhibitor of photooxidation at 10⁻⁴ *M*, **3** is an order of magnitude less effective; however,

both of these compounds quench singlet oxygen at concentrations well below those at which triplet sensitizer could be quenched, and no increase in intercept is observed at the highest concentrations used (10⁻⁴ *M* for β-carotene and 8 × 10⁻⁴ *M* for **3**). Compound **2** is less efficient as a quencher, and both singlet oxygen and triplet sensitizer are quenched (at 9 × 10⁻³ *M*, there is a fivefold increase in intercept and a fourfold increase in slope/intercept ratio). With retinol at 6.7 × 10⁻³ *M*, the intercept increased by a factor of 3, but there was no detectable increase in slope/intercept, so that only triplet quenching is detectable.[7] Under the conditions of the experiment, none of the quenchers was destroyed appreciably, although a small amount of retinol (∼10%) was lost by spectrophotometric assay.[9]

Table I. Quenching Parameters for Carotenoid Inhibition of 2-Methyl-2-pentene Photooxygenation, Sensitized by Methylene Blue

Quencher	$k_Q/k_A{}^a$ (singlet oxygen)	$k_Q{}^T$, M^{-1} sec^{-1} ^{a} (³sensitizer)
1	≤9ᶜ	4 × 10⁹ ᵈ
2	57	7 × 10⁹ ᵈ
3	1900	ᵉ
β-Caroteneᵇ	10⁴	ᵉ

ᵃ See ref 6 for kinetic scheme. ᵇ Reference 2. ᶜ Upper limit, twice the probable error in the determination. ᵈ Assumes [O₂] = 10⁻² *M* and O₂ quenching rate 1.2 × 10⁹ M^{-1} sec⁻¹; see ref 10. ᵉ Not measurable at concentrations used.

The kinetic parameters[6] determined for the three new quenchers are listed in Table I, with the value

(1) Paper IX: S. Mazur and C. S. Foote, *J. Amer. Chem. Soc.*, **92**, 3225 (1970). Contribution No. 2557; supported by a grant from the USPHS-NAPCA (No. AP-00681).

(2) C. S. Foote and R. W. Denny, *J. Amer. Chem. Soc.*, **90**, 6233 (1968); C. S. Foote, R. W. Denny, L. Weaver, Y. Chang, and J. Peters, *Ann. N. Y. Acad. Sci.*, in press.

(3) N. I. Krinsky, *Photophysiology*, **3**, 123 (1968).

(4) Solutions in benzene–methanol (80:20) containing known amounts of the photooxygenation acceptor 2-methyl-2-pentene (A) and the quencher (Q) were irradiated for a constant time, sufficient to produce a readily measurable amount of product (AO₂) without oxidizing a significant fraction of A (<7%). The sensitizer was methylene blue (MB); solutions under pure O₂ were irradiated with a tungsten-halogen lamp through a K₂Cr₂O₇ filter (3 g/100 ml of H₂O, 2.5-cm path length) with shortwave cutoff (1% transmission) of 510 nm, which ensured that light was absorbed only by the sensitizer, not by the quencher. Although this filter was not used in previous experiments,[2] controls and subsequent experiments with filtered light established that light absorbed by β-carotene had no effect on the observed quenching or on the production of AO₂. Photooxygenated solutions were reduced with NaBH₄, internal standard was added, and the product alcohols were determined gas chromatographically.

(5) The C₃₀ and C₃₅ hydrocarbons were kindly supplied by Dr. H. Pommer, BASF, Ludwigshafen, Germany. For the purposes of this discussion, the endocyclic double bonds in all carotenoids are counted as conjugated.

(6) C. S. Foote, Y. C. Chang, and R. W. Denny, *J. Amer. Chem. Soc.*, **92**, 5218 (1970).

(7) Both retinol and β-carotene are reported to quench triplet sensitizers at a high rate;[8] it is presumed that all four compounds used here quench triplet methylene blue efficiently; however, this quenching, even if diffusion controlled, cannot compete with oxygen quenching of triplet methylene blue below about 10⁻³ *M*.

(8) R. Livingston and A. C. Pugh, *Discuss. Faraday Soc.*, **27**, 144 (1959); E. Fujimori and R. Livingston, *Nature (London)*, **180**, 1036 (1957); A. Sykes and T. G. Truscott, *Chem. Commun.*, 929 (1969).

(9) Quenching values were also determined for lycopene (11 C=C) and β-*apo*-8′-carotenol (9 C=C) and found to be similar to those for β-carotene and **3**, respectively, but the rates could not be determined accurately because substantial destruction of these carotenoids occurred under the conditions.

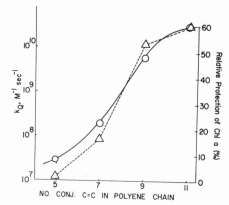

Figure 1. $^1O_2$ quenching rates (k_Q, O—O) and protective action against photobleaching of chlorophyll a[11] (Δ---Δ) as a function of length of conjugated system.

for β-carotene[2] for comparison. The ratio k_Q/k_A is calculated from the increase in slope/intercept, and is the ratio of the rate at which singlet oxygen is quenched (k_Q) to its reaction rate with A (k_A, see ref 2); the value k_Q^T is the rate of quenching of triplet methylene blue, estimated from the increase in intercept. The parameters are estimated to have a probable error of a factor of 1.5 because the intercept values are difficult to determine accurately.

If the singlet oxygen quenching by β-carotene is assumed to be diffusion controlled, and a rate of 3×10^{10} M^{-1} sec^{-1} is assigned,[10] the quenching rates for all quenchers can be calculated. These rates are plotted

against the number of conjugated double bonds in Figure 1. It is immediately obvious that the rates fall off sharply with decreasing chain length, the sharpest drop occurring between 7 and 9 conjugated double bonds.[10a]

This behavior is remarkably similar to the protective action of natural carotenoids against chlorophyll photobleaching[11] (also shown in Figure 1) which also falls off most sharply between 7 and 9 conjugated double bonds. This protection against chlorophyll photobleaching is one of the mechanisms by which carotenoids protect against photodynamic damage in living systems.[3,11] The similarity of the effects suggests that (1) carotenoids protect living organisms from photodynamic damage at least partly by quenching singlet oxygen and (2) singlet oxygen is at least one cause of photodynamic damage. Although enzymes, nucleic acids, and cell membranes are also damaged in photodynamic action, carotenoids also protect against these effects,[3] presumably by the same mechanism.

The behavior of these carotenoids in triplet MB quenching also serves to explain the fact that shorter carotenoid chains are effective in protecting against *anaerobic* chlorophyll photoreduction, since triplet chlorophyll a is probably intermediate in this reaction, and quenching of this triplet is effective even with carotenoids of only 5 conjugated double bonds.[8]

The implications of these observations for the mechanism of singlet oxygen quenching by carotenoids are discussed in the accompanying communication.[6]

(10) The rate of O_2 quenching of naphthalene fluorescence in benzene: W. R. Ware, *J. Phys. Chem.*, **66**, 455 (1962). The rate used in ref 2 was an O_2 *triplet* quenching rate, and is too low because of a spin-statistical factor of 9. The absolute rate of $^1O_2$ decay in solution is calculated, based on this rate, to be 1.0×10^{-5} sec^{-1} and is a lower limit. The rate of triplet quenching by O_2 used in the calculation, 1.2×10^9 M^{-1} sec^{-1}, was taken from K. Gollnick, *Advan. Photochem.*, **6**, 1 (1968).

(10a) NOTE ADDED IN PROOF. Quenching rates for a C_{50} and a C_{60} carotenoid (15 and 19 conjugated C=C) are essentially the same as that of β-carotene, confirming the suggestion that the latter rate is diffusion controlled.

(11) (a) H. Claes, *Biochem. Biophys. Res. Commun.*, **3**, 585 (1960); (b) H. Claes and T. O. M. Nakayama, *Z. Naturforsch. B*, **14**, 746 (1959).

* To whom correspondence should be addressed.

Christopher S. Foote,* Yew C. Chang, Robert W. Denny
Department of Chemistry, University of California
Los Angeles, California 90024
Received March 19, 1970

Reprinted from *J. Amer. Chem. Soc.,* **94**(25), 8960–8962 (1972)

Singlet Oxygen (¹Δg) Quenching in the Liquid Phase by Metal(II) Chelates

Sir:

Relatively few classes of compounds have been shown to quench the reactions of singlet oxygen (¹Δg), ¹O₂. Known quenchers include various dialkyl sulfides[1] and amines,[2,3] some of which are also oxidized by ¹O₂,[1,2b] and carotenoids.[1] Since Pfeil[4] has shown that the ²E state of several Cr(III) complexes can be quenched by oxygen (³Σg-) to give ¹O₂, we have investigated the occurrence of the reverse process, ¹O₂ quenching by metal chelates, and have found that some Ni(II) compounds are efficient ¹O₂ quenchers.

Quenching experiments were performed with rubrene (9,10,11,12-tetraphenylnaphthacene) as the ¹O₂ receptor; bleaching of the rubrene absorption at 520 nm was used to follow the rubrene oxidation to the *endo*-peroxide. ¹O₂ was generated by three separate methods. In method a, air-saturated rubrene solutions· were photolyzed at 520 ± 20 nm,[5] and initial rates of rubrene consumption calculated. In method b, a solution of triphenyl phosphite ozonide[6] at −70° was added to a rubrene solution (at 25°) to give an initial ozonide concentration in the range 2.4×10^{-3}–0.94×10^{-3} M. The rubrene solution contained small percentages of methanol and pyridine so as to catalyze a virtually instantaneous ¹O₂ evolution.[7] In method c, oxygen (6 Torr) was fed through a microwave discharge (2450 MHz, 70 W) in a quartz tube and atomic oxygen, ozone, and ¹Σg⁺ oxygen were removed.[8] The ¹O₂ stream was split into two identical gas streams each of which passed over a stirred solution (2.0 ml), one with quencher and one quencher free. This procedure allowed parallel experiments to be performed so as to compensate for fluctuations in the ¹O₂ concentration. Effective quenchers and the solvents used for each method are listed in Table I.

For methods b and c, integrated forms of the rate equations were used in the derivation of the pseudo-unimolecular rate constants (k_d) for the decay of ¹O₂ and the ¹O₂ quenching rate constants (k_q), whereas the direct rate equation was employed for method a.[9] The rate

constant for the ¹O₂ oxidation of rubrene (k_{ox}) was assumed to be 7×10^7 M^{-1} sec⁻¹.[10] In method b, a quantitative yield of ¹O₂ was assumed from the ozonide[7] and this then allowed the k_d values in each solvent to be calculated directly from the rubrene concentration changes in the absence of quenchers. For methods a and c, k_d had to be derived indirectly from 1,4-diazabicyclo[2.2.2]octane (Dabco) quenching data, using k_q for Dabco calculated from the results of method b (Table I) which was assumed solvent independent.[14] The k_d values for methods a, b, and c are shown in Table I together with k_q values for quenching by the effective compounds, calculated from the appropriate k_d values and the above assumptions.

The variations in ¹O₂ decay constants with solvent type (Table I) are in good agreement with published ¹O₂ lifetimes.[16] Similarly, the k_q values for Dabco (Table I) are also in good agreement with the literature value in benzene–methanol (1.6 × 10⁷ M^{-1} sec⁻¹).[1] ¹O₂ quenching did not appear to involve chelate reaction since exposure (30 min) of the effective chelates in rubrene-free solutions to ¹O₂ from the microwave discharge resulted in negligible change in the uv spectra of the chelates. Although all three methods of ¹O₂ generation are open to question,[17] the fair agreement in observed k_q values for the various methods (Table I) indicates that nickel(II) chelates are true quenchers of ¹O₂. The k_q values derived for several of the metal chelates (Table I) are approaching diffusion-controlled values, similar to that found for *N*-phenyl-*N′*-cyclohexyl-*p*-phenylenediamine[3] ($k_q \sim 5 \times 10^9$ M^{-1} sec⁻¹ if $k_d \sim 4 \times 10^5$ sec⁻¹, as found for 2-butoxyethanol, Table I) but still somewhat slower than β-carotene ($k_q \sim 3 \times 10^{10}$ M^{-1} sec⁻¹).[1]

From the data in Table I, it is possible to speculate on mechanisms for ¹O₂ quenching by the chelates. The quenching effect appears to be a property of the Ni(II) chelate as a whole, or of the Ni(II) nucleus, since the

---

(1) C. S. Foote, R. W. Denny, L. Weaver, Y. Chang, and J. Peters, *Ann. N. Y. Acad. Sci.,* **171**, 139 (1970).

(2) (a) C. Ouannès and T. Wilson, *J. Amer. Chem. Soc.,* **90**, 6527 (1968); (b) W. F. Smith, *ibid.,* **94**, 186 (1972).

(3) J. P. Dalle, R. Magous, and M. Mousseron-Canet, *Photochem. Photobiol.,* **15**, 411 (1972).

(4) A. Pfeil, *J. Amer. Chem. Soc.,* **93**, 5395 (1971).

(5) T. Wilson, *ibid.,* **88**, 2898 (1966).

(6) R. W. Murray and M. L. Kaplan, *ibid.,* **91**, 5358 (1969).

(7) G. D. Mendenhall, Ph.D. Thesis, Harvard University, 1970.

(8) M. L. Kaplan and P. G. Kelleher, *Science,* **169**, 1207 (1970).

(9) Method a is

$$k_q = \frac{k_{ox}[R\bar{u}]_0 + k_d}{[Q]}\left[\frac{(d[R\bar{u}]/dt)^0}{(d[R\bar{u}]/dt)_0{}^Q} - 1\right]$$

Method b is

$$k_q = \frac{k_{ox}}{[Q]}\left(\frac{[(PhO)_3PO_3]_0 - [R\bar{u}]_0 + [R\bar{u}]_t}{\ln [R\bar{u}]_0/[R\bar{u}]_t}\right) - \frac{k_d}{k_{ox}}$$

Method c is

$$k_q = \frac{k_{ox}([R\bar{u}]_t{}^Q - [R\bar{u}]_t{}^0) + k_d \ln [R\bar{u}]_t{}^Q/[R\bar{u}]_t{}^0}{[Q]\ln [R\bar{u}]_0/[R\bar{u}]_t{}^Q}$$

where [Q] = quencher concentration, $[(PhO)_3O_3]_0$ = initial ozonide concentration, $[R\bar{u}]_0$ = initial rubrene concentration (unchanged by the presence of quencher), $[R\bar{u}]_t$ = final rubrene concentration, and $-(d[R\bar{u}]/dt)_0$ = initial rate of rubrene loss. Superscripts Q and 0 indicate the presence or absence of quencher.

(10) Calculated from the following literature values: k_{ox}(rubrene) = 3.0k_{ox}(2,3-dimethyl-2-butene),[11] k_{ox}(2,3-dimethyl-2-butene) = 50k_{ox}(2-methyl-2-pentene),[12] and k_{ox}(2-methyl-2-pentene) = 5 × 10⁵ M^{-1} sec⁻¹, since k_q(β-carotene)/k_{ox}(2-methyl-2-pentene) = 2 × 10⁴[13] in benzene–methanol solutions and quenching by β-carotene is diffusion controlled ($k_q = 1 \times 10^{10}$ M^{-1} sec⁻¹ from the modified Debye equation).

(11) B. E. Algar and B. Stevens, *J. Phys. Chem.,* **74**, 3029 (1970).

(12) C. S. Foote, *Accounts Chem. Res.,* **1**, 104 (1968).

(13) C. S. Foote and R. W. Denney, *J. Amer. Chem. Soc.,* **90**, 6233 (1968).

(14) Combination of the k_d/k_q values of Foote, *et al.*,[15] for Dabco quenching in methanol and in benzene with the ¹O₂ lifetimes in these solvents derived experimentally by Merkel and Kearns[16] indicates that k_q for Dabco is solvent independent.

(15) C. S. Foote, E. R. Peterson, and K.-W. Lee, *J. Amer. Chem. Soc.,* **94**, 1032 (1972).

(16) P. B. Merkel and D. R. Kearns, *ibid.,* **94**, 1029 (1972).

(17) Method a because of possible quenching of the rubrene triplet (which is probably the chief precursor of ¹O₂ formation[11]), method b because of some direct reaction between the ozonide and the chelates, and method c because of the possibility of residual oxygen atoms being present in the discharge.[18]

(18) K. Furukawa and E. A. Ogryzlo, *Chem. Phys. Lett.,* **12**, 370 (1971).

Table I. Singlet Oxygen ($^1\Delta_g$) Decay (k_d) and Quenching Rate Constants (k_q) in Various Solvents[a]

Quencher	Rubrene $\xrightarrow{h\nu}$ [b]		(PhO)$_3$P–O$_3$ decompn[c]		Microwave discharge[d]	
	Isooctane	CH$_2$Cl$_2$	Isooctane[e]	CH$_2$Cl$_2$[f]	Hexadecane	2-Butoxy-ethanol[g]
	$k_d \times 10^{-4}$ sec^{-1}					
None	4.0[h]	N.D.[i]	5.0	0.73	9.0[h]	38[h]
	$k_q \times 10^{-8}$ M^{-1} sec^{-1}					
Dabco			0.35	0.33		
Nickel(II) bis(2-hydroxy-5-methoxy-phenyl-*N*-*n*-butylaldimine)		>10	35		2.0	
Nickel(II) di-*n*-butyldithiocarbamate	70			>10[j]	9.0	
Nickel(II),*n*-butylamine[2,2'-thiobis-(4-*tert*-octyl)phenolate]	1.8		2.7		0.80	2.8
Nickel(II) bis[2,2'-thiobis(4-*tert*-octyl)phenolate]	1.3		2.0		1.3	
Nickel(II) acetylacetonate						0.75
Nickel(II) bis(butyl-3,5-di-*tert*-butyl-4-hydroxybenzyl phosphonate)		0.14		0.10		0.34
NiCl$_2$·6H$_2$O						3.1
CoCl$_2$·6H$_2$O						0.48
MnCl$_2$·4H$_2$O						<0.01

[a] At 25° unless specified. [b] Initial rubrene concentration 1.0×10^{-4} M; 50% rubrene oxidation in ~3.5 min in absence of quencher. Quencher concentrations 4.0×10^{-3} M. [c] Quencher concentration 2.4×10^{-3} M. [d] Initial rubrene concentration 1.9×10^{-4} M. Exposures 5 or 10 min. Without quencher, ~50% rubrene consumption in ~3 min. Quencher concentrations 4.0×10^{-3} M. [e] Methanol–pyridine–isooctane mixture (3:3:94% by volume); rubrene concentration 4.0×10^{-4} M. [f] Methanol–pyridine–methylene chloride mixture (3:3:94% by volume); rubrene concentration 6.0×10^{-4} M. [g] At 0°. [h] Calculated from k_q(Dabco) = 3.4×10^7 M^{-1} sec^{-1}, from method b. [i] Assumed 0.73×10^4 sec^{-1}, from method b. [j] Interference by chelate–ozonide reaction to give colored products.

ligand 2,2'-thiobis(4-*tert*-octylphenol) failed to quench $^1O_2$ whereas Ni(II) chelates of this ligand are effective (Table I). The chelate coordination and magnetism also appear to be unimportant, since an octahedral (paramagnetic) complex, NiCl$_2$·6H$_2$O,[19] quenches $^1O_2$ with a similar k_q to that of a square planar (diamagnetic) complex (the nickel aldimine[19] or nickel di-*n*-butyldithiocarbamate,[20] for example), yet nickel acetylacetonate (an octahedral trimer)[19] is less efficient. Heavy atom effects have previously been precluded in $^1O_2$ quenching.[1] The marked increase in k_q on going from Mn(II) to Co(II) to Ni(II) for the hydrated chlorides (Table I) is probably significant since both [Ni(H$_2$O)$_6$]$^{2+}$ and [Co(H$_2$O)$_6$]$^{2+}$ have electronic absorptions corresponding to energy levels of 8000–9000 cm^{-1},[19] whereas all transitions of the unpaired d^5 electrons of [Mn(H$_2$O)$_6$]$^{2+}$ are forbidden.[19] Since the fluorescence emission for the $^1\Delta_g \rightarrow\ ^3\Sigma_g{}^-$ transition occurs at ~8000 cm^{-1} (22.5 kcal/mol),[1] transfer from $^1O_2$ to the Ni(II)

and Co(II) chelates is possible either *via* a collisional process as suggested for β-carotene,[1] or by a long-range energy-transfer mechanism. However, other possible $^1O_2$ quenching mechanisms might include charge or electron transfer processes (*cf.* amines[21]), which occur between some chelates and oxy radicals,[22] and chelate–$^1O_2$ complex formation with the Ni(II) d^8 chelates, analogous to those reported to be formed with ground-state oxygen by Ni(0) complexes[23] or by d^8 complexes of iridium.[24]

(19) F. A. Cotton and G. Wilkinson, "Advanced Inorganic Chemistry," 2nd ed, Interscience, New York, N. Y., 1966, Chapter 28.
(20) C. K. Jorgensen, "Inorganic Complexes," Academic Press, London, 1963.
(21) I. B. C. Matheson and J. Lee, *J. Amer. Chem. Soc.*, **94**, 3310 (1972).
(22) T. V. Liston, H. G. Ingersoll, and J. Q. Adams, *Amer. Chem. Soc., Div. Petrol. Chem., Prepr.*, **14** (4), A83 (1969).
(23) S. Otsuka, A. Nakamura, Y. Tatsuno, and M. Miki, *J. Amer. Chem. Soc.*, **94**, 3761 (1972).
(24) L. Vaska, *Accounts Chem. Res.*, **1**, 335 (1968).
(25) Issued as NRCC No. 12965.

D. J. Carlsson,* G. D. Mendenhall
T. Suprunchuk, D. M. Wiles
Division of Chemistry, National Research Council of Canada
Ottawa, Canada K1A 0R9[25]

Received July 13, 1972

322

65

Copyright © 1973 by Pergamon Press

Reprinted from *Photochem. Photobiol.*, **18**, 447–450 (1973)

ON THE MECHANISM OF QUENCHING OF SINGLET OXYGEN IN SOLUTION

A. Farmilo and F. Wilkinson

University of East Anglia, School of Chemical Sciences, University Plain,
Norwich NOR 88C, England

(*Received 5 February 1973; accepted 29 May 1973*)

Abstract—Bimolecular rate constants for the quenching of singlet oxygen $O_2^*(^1\Delta_g)$, have been obtained for several transition-metal complexes and for β-carotene. Laser photolysis experiments of aerated solutions, in which triplet anthracene is produced and quenched by oxygen, yielding singlet oxygen which then sensitizes absorption due to triplet carotene, firmly establishes diffusion-controlled energy transfer from singlet oxygen as the quenching mechanism in the case of β-carotene. The efficient quenching of singlet oxygen by two trans-planar Schiff-base Ni(II) complexes, which have low-lying triplet ligand-field states, most probably also occurs as a result of electronic energy transfer, since an analogous Pd(II) complex and ferrocene, which both have lowest-lying triplet states at higher energies than the $O_2^*(^1\Delta_g)$ state, quench much less effectively.

In a recent laser photolysis study (Adams and Wilkinson, 1972), we describe how the lifetime of singlet molecular oxygen in solution can be estimated from a kinetic analysis of the disappearance of 1,3-diphenylisobenzofuran, DPBF, which reacts with $O_2^*(^1\Delta_g)$ produced by energy transfer from triplet-state energy donors. Such measurements made in the presence of various concentrations of molecules which quench $O_2^*(^1\Delta_g)$ can be used to determine quenching rate constants (e.g. see Merkel and Kearns, 1972).

It has been pointed out by Young *et al.* (1973) that, to a good approximation, the measured decay constant for singlet oxygen in the absence of quencher is composite, and is given by

$$k_D = k_1 + k_R[DPBF]_{AV} \qquad (1)$$

where k_1 is the decay constant for $O_2^*(^1\Delta_g)$ and k_R is the bimolecular rate constant for the reaction between DPBF and $O_2^*(^1\Delta_g)$. $[DPBF]_{AV}$ represents the average concentration of DPBF used in the experiment, i.e. $\frac{1}{2}([DPBF]_0 + [DPBF]_\infty)$. Although in our earlier work no correlation of k_D with the initial concentration of DPBF was found, more careful analysis, using a numerical integration procedure on a computer, has confirmed a concentration dependence similar to that found by Young. Our present values for k_1 and k_R in benzene are $3.9 \pm 0.4 \times 10^4\,s^{-1}$ and $1.5 \pm 0.5 \times 10^9\,l.mol^{-1}\,s^{-1}$ respectively. These values are in good agreement with the results of other workers, e.g. the results of Merkel and Kearns (1972) yield $k_1 = 4.2 \times 10^4\,s^{-1}$.

In the presence of a molecule, Q, that quenches singlet oxygen, equation (1) becomes

$$k_D = k_1 + k_R[DPBF]_{av} + k_Q[Q] \qquad (2)$$

Thus, provided $[DPBF]_{av}$ is constant, a plot of k_D vs. [Q] will give a straight line of slope k_Q (see Fig. 1). The values of k_Q given in Table 1 were obtained from plots of the type shown in Fig. 1, using anthracene as the sensitizer of singlet oxygen. (N.B. At least four traces were analysed and averaged for each point shown.)

In fact in this work the initial concentration of DPBF was held constant in each quenching experiment and consequently $[DPBF]_{av}$ increases slightly

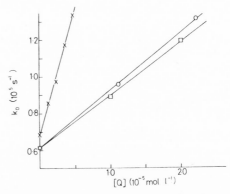

Figure 1. Variation in singlet oxygen $O_2^*(^1\Delta_g)$ decay constant k_D, obtained from photo-oxidation of DPBF, as a function of the concentration of the quenchers Negopex A (○), Negopex B (□), and β-carotene (×).

323

Table 1. Quenching of $O_2^*(^1\Delta_g)$ in benzene at 25°C

Quencher	$k_Q(\text{l.mol}^{-1}\text{s}^{-1})$
β-carotene	$(1\cdot3\pm0\cdot2)\times10^{10}$
Negopex A(I)	$(3\cdot0\pm0\cdot3)\times10^9$
Negopex B(II)	$(2\cdot7\pm0\cdot3)\times10^9$
Pd complex(III)	$(6\cdot0\pm0\cdot5)\times10^7$
Ferrocene	$\leqslant 5\times10^6$

	R	R'	Ring position of R'	M
(I) Negopex A	CH_3	CH_3	para to oxime group	Ni(II)
(II) Negopex B	$C_{11}H_{23}$	CH_3	para to oxime group	Ni(II)
(III) Pd complex	$C_{17}H_{35}$	tC_4H_9	para to hydroxy group	Pd(II)

at higher concentrations of Q. However corrections to allow for this were always less than 5 per cent and were therefore small compared with the total errors involved, which are estimated as 10 per cent. Furthermore, analysis by the numerical method gives quenching constants which agree with these values to within ±5 per cent. There was no change in the absorption spectrum of any of the quenchers used even after they were subjected to laser photolysis fifty times in the presence of anthracene as sensitizer and DPBF as the oxidisable acceptor, demonstrating that quenching by chemical reaction is negligible for these molecules.

The quenching constant obtained for β-carotene, which Foote *et al.* (1970) showed to be a very efficient quencher, can be compared with the value obtained by Merkel and Kearns (1972) of $2\times10^{10}\text{l.mol}^{-1}\text{s}^{-1}$. The rate constants for Ni(II) complexes are similar to the values obtained for the more efficient Ni(II) singlet oxygen quenchers determined by Carlsson *et al.* (1972). For β-carotene, Negopex A(I), and Negopex B(II), the quenching constants are close to those expected for diffusion-controlled reactions, which is consistent with quenching due to the energy-transfer process:

$$O_2^*(^1\Delta_g)+{}^1Q \rightarrow O_2(^3\Sigma_g^-)+{}^3Q^*, \qquad (3)$$

since literature estimates of the energies of the lowest triplet states indicate that this process would be exothermic for these quenchers. Furthermore, the lowest triplet levels of the palladium complex (III)

(Allsopp and Wilkinson, 1973) and of ferroce (Scott and Becker, 1965) lie at higher energies th $O_2^*(^1\Delta_g)$, and these compounds quench much le efficiently. The mechanism operating for these tw quenchers may well involve heavy atom quenchir in which case k_Q would be expected (McClur 1959) to depend on ξ^2 where ξ is the atomic spi orbit coupling parameter of the heaviest nucleus the compound, and $k_Q(\text{Pd complex})/k_Q(\text{ferrocen}$ should be ~ 12, which is in fair agreement with th experimental value of ~ 7.

Since there is considerable biological interest the protective role played by carotenoids in phot dynamic oxidation, it is important to establi beyond doubt whether or not β-carotene quench $O_2^*(^1\Delta_g)$ as a result of energy transfer to form i lowest triplet state.

Oxygen-saturated, aerated and deaerated be zene solutions of anthracene ($\sim 10^{-3} M$) and ρ carotene ($4-36\times10^{-6} M$) were subjected to pulse laser excitation at 347 nm, and the decay of tripl anthracene, ${}^3A^*$, absorbing at 425 nm (Bensasso and Land, 1971) and the build-up and decay of tri let carotene, ${}^3C^*$, which absorbs at 515 nm (Chess *et al.*, 1966) and has a lifetime of $\sim 9\,\mu s$ in deae ated solutions (Land *et al.*, 1970), was studied. F the higher concentrations of β-carotene, the deca of triplet anthracene was monitored in a thin ce (1 mm), otherwise 1 cm^2 cells were use throughout.

With our apparatus, no triplet carotene is ob served when β-carotene is subjected to direct lase otolysis, indicating a low quantum yield of triplet oduction. In deaerated solutions containing anracene, the sensitized production of triplet rotene is observed (Mathis, 1972) due to the process:

$$^3A^* + {}^1C \xrightarrow{k_{TC}} {}^1A + {}^3C^* \qquad (4)$$

ith $k_{TC} = (1 \cdot 3 \pm 0 \cdot 2) \times 10^{10}$ l.mol^{-1}s^{-1} and a triplet rotene lifetime of $7 \cdot 7 \, \mu$s.

In the presence of oxygen, anthracene solutions ntaining β-carotene also show sensitized triplet rotene production, which may arise as a result of e following sequence of reactions in addition to ocess (4):

$$^3A^* + O_2(^3\Sigma_g^-) \xrightarrow{k_{TO}} O_2^*(^1\Delta_g) + {}^1A \qquad (5)$$

$$O_2^*(^1\Delta_g) + {}^1C \xrightarrow{k_{\Delta C}} {}^3C^* + O_2(^3\Sigma_g^-) \qquad (6)$$

$$O_2^*(^1\Delta_g) \xrightarrow{k_1} O_2(^3\Sigma_g^-) \qquad (7)$$

$$^3C^* + O_2(^3\Sigma_g^-) \xrightarrow{k_{CO}} {}^1C + O_2(^3\Sigma_g^-) \cdot \qquad (8)$$

$$^3C^* \xrightarrow{k_{DC}} {}^1C. \qquad (9)$$

In fact there are a number of reasons which demstrate unequivocally that in this work very little iplet carotene is formed by direct energy transfer om triplet anthracene in aerated solutions, i.e. via ocess (4). Firstly, the decay of triplet anthracene aerated solutions is unaffected (within experiental error) by the addition of even the highest ncentration $(3 \cdot 6 \times 10^{-5} \, M)$ of β-carotene used. his is because the concentration of oxygen $(1 \cdot 5 \times$ $^{-3} M)$ in aerated benzene solutions (Pringsheim, 49) is such that $k_{TO}[O_2] \gg k_{TC}[C]$, for according to ir measurements $k_{TO}[O_2] = 5 \cdot 05 \times 10^6$ s^{-1}, and thus, en in the presence of $3 \cdot 6 \times 10^{-5} \, M$ β-carotene, ily ~ 8 per cent of the triplet anthracene olecules formed initially will transfer energy directly to β-carotene in aerated solutions. Also, the aximum absorbance of sensitized triplet carotene r aerated solutions containing $3 \cdot 6 \times 10^{-5} \, M$ β-rotene was $0 \cdot 16$, which is much greater than 8 per nt of $0 \cdot 47$, the maximum absorbance in the same eaerated solution. Furthermore Truscott et al. 973) have shown that triplet carotene is quenched oxygen with a rate constant of $\sim 3 \times$ ^{9} l.mol^{-1}s^{-1}, and if most of the sensitized triplet rotene was produced via process (4) this would inconsistent with the observed long decay time r triplet carotene in aerated solution, which was ways $> 2 \, \mu$s. On the other hand the general fea-

tures observed are all explainable if energy transfer from a relatively long-lived intermediate to produce triplet carotene occurs by a rate-determining step, and it can be shown that this intermediate is $O_2^*(^1\Delta_g)$, i.e., the results are consistent with sensitization via steps (5) and (6), where (6) is a slow step.

Quantitative kinetic analysis of steps (4) to (9) gives

$$\frac{d[^3C^*]}{dt} = k_{\Delta C}[C][^1O_2^*] + k_{TC}[^3A^*][C]$$

$$- k_{CO}[O_2][^3C^*] - k_{DC}[^3C^*] \qquad (10)$$

$$-\frac{d[^1O_2^*]}{dt} = (k_1 + k_{\Delta C}[C])[^1O_2^*]$$

$$- k_{TO}[^3A^*][O_2] \qquad (11)$$

$$-\frac{d[^3A^*]}{dt} = (k_{TO}[O_2] + k_{TC}[C])[^3A^*]. \qquad (12)$$

The exact solution of these three equations gives

$$[^3C^*] = P(e^{-k_A t} - e^{-k_C t}) + Q(e^{-k_B t} - e^{-k_C t}) \qquad (13)$$

where

$$k_A = k_{TO}[O_2] + k_{TC}[C],$$
$$k_B = k_1 + k_{\Delta C}[C],$$

and

$$k_C = k_{CO}[O_2] + k_{DC} \approx k_{CO}[O_2],$$

since k_{DC} is small compared with $k_{CO}[O_2]$,

$$P = \frac{[^3A^*]_0(k_{TC}[C](k_B - k_A) + k_{\Delta C}[C]k_{TO}[O_2])}{(k_B - k_A)(k_C - k_A)}$$

and

$$Q = \frac{[^3A^*]_0 k_{\Delta C}[C]k_{TO}[O_2]}{(k_A - k_B)(k_C - k_B)}.$$

At times greater than 1 μs, $e^{-k_A t}$ and $e^{-k_C t}$, are so small that they become negligible and equation (13) becomes

$$[^3C^*] = Qe^{-k_B t}$$

i.e. the theory predicts a first-order decay for triplet carotene, as found experimentally (see Fig. 2), with a decay constant

$$k_B = k_1 + k_{\Delta C}[C].$$

This linear dependence of k_B on $[C]$ is confirmed

325

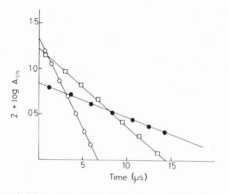

Figure 2. First order plots of the decay of absorbance at 515 nm, A_{515}, due to sensitized triplet β-carotene in aerated benzene solutions containing anthracene ($\sim 10^{-3} M$) and β-carotene: (○) 36·0; (□) 12·1; (●) $4·3 \times 10^{-6} M$.

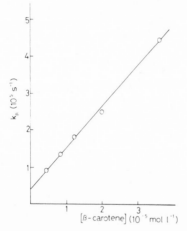

Figure 3. Variation in triplet β-carotene decay constant k_B, for aerated benzene solutions containing anthracene as sensitizer as a function of β-carotene concentration.

by experiment, as shown in Fig. 3, and the values k_1 and $k_{\Delta C}$ obtained from this plot are $(4·1 \pm 0·3) \times 10^4 \, s^{-1}$ and $(1·1 \pm 0·1) \times 10^{10} \, l.mol^{-1} \, s^{-1}$ respectively. These values are in excellent agreement with the values obtained using DPBF as an oxidisable singlet oxygen acceptor, as reported earlier in the paper.

These measurements establish beyond doubt that quenching of singlet oxygen by β-carotene is due to electronic energy transfer, which results in production of the triplet state of β-carotene, as suggested by Foote et al. (1970). The large quenching constant of $1·3 \times 10^{10} \, l.mol^{-1} \, s^{-1}$ suggests that triplet carotene has an energy lower than that of $O_2^*(^1\Delta_g)$ i.e. $< 8000 \, cm^{-1}$.

REFERENCES

Adams, D. R. and F. Wilkinson (1972) *J. C. S. Farad. II* **68**, 586–593.

Allsopp, S. R. and F. Wilkinson (1973) *Chem. Phys. Lett.* **19**, 535–537.

Bensasson, R. and E. J. Land (1971) *Trans Farad. Soc.* **67**, 1904–1915.

Carlsson, D. J., G. D. Mendenhall, T. Suprunchuk and M. Wiles (1972) *J. Am. Chem. Soc.* **94**, 8960–8962.

Chessin, M., R. Livingston and T. G. Truscott (1966) *Trans Faraday Soc.* **62**, 1519–1524.

Foote, C. S., R. W. Denny, L. Weaver, Y. Chang and Peters (1970) *Ann. N.Y. Acad. Sci.* **171**, 139–145.

Land, E. J., A. Sykes and T. G. Truscott (1970) *Chem. Commun.* 332.

McClure, D. S. (1959) *Solid St. Phys.* **9**, 399–525.

Mathis, P. (1969) *Photochem. Photobiol.* **9**, 55–63.

Merkel, P. B. and D. R. Kearns (1972) *J. Am. Chem. Soc.* **94**, 7244–7253.

Pringsheim, P. (1949) *Fluorescence and Phosphorescence*, p. 332. Interscience, New York.

Scott, D. R. and R. S. Becker (1965) *J. Organomet. Chem.* **4**, 409–411.

Truscott, T. G., E. J. Land and A. Sykes (1973) *Photochem. Photobiol.* **17**, 43–51.

66

Reprinted from *Biochemistry*, 11(4), 606–608 (1972)

Dye-Sensitized Photooxidation of Tocopherols. Correlation between Singlet Oxygen Reactivity and Vitamin E Activity[*]

G. W. Grams† and K. Eskins

ABSTRACT: The singlet oxygen reactivity of α-, β-, γ-, and δ-tocopherol was determined in methanol with methylene blue as the photosensitizer. The disappearance of tocopherol was followed colorimetrically according to the Emmerie–Engel method. Of the four tocopherols, α was the most reactive ($\beta = 1.4 \times 10^{-4}$ M) and δ was the least ($\beta = 13.5 \times 10^{-4}$ M). α-Tocopherol is one of the most reactive compounds toward singlet oxygen reported in the literature. The reactivity of each tocopherol (α, β, γ, and $\delta = 1, 0.50, 0.26$, and 0.10) correlates well with its vitamin E activity.

$\mathbf{P}$eroxidation *via* a singlet oxygen mechanism has been proposed as the initial step in autoxidation of unsaturated lipids (Rawls and van Santen, 1970; Khan, 1971). Foote *et al.* (1970a,b) recently showed that β-carotene is an extremely good quencher of singlet oxygen and in this role may protect unsaturated lipids from the effects of photodynamic action. Previously, we found that α-tocopherol photooxidized in the presence of a dye sensitizer yielded as major products 4a,5-epoxy-8a-methoxy-α-tocopherol and α-tocoquinone 2,3-oxide (Grams *et al.*, 1972). Further investigation in our laboratory with chemically generated singlet oxygen has shown that these compounds characterize the reaction of α-tocopherol with singlet oxygen (G. W. Grams, 1972). We wanted to determine whether reactivity of members of the tocopherol family toward singlet oxygen could be correlated with their vitamin E activity.

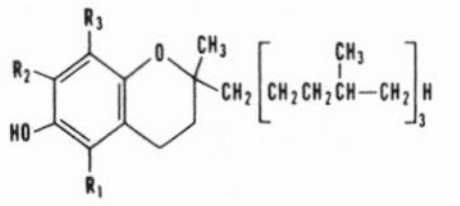

α-Tocopherol: $R_1 = R_2 = R_3 = CH_3$
β-Tocopherol: $R_1 = R_3 = CH_3$; $R_2 = H$
γ-Tocopherol: $R_2 = R_3 = CH_3$; $R_1 = H$
δ-Tocopherol: $R_1 = R_2 = H$; $R_3 = CH_3$

Experimental Section

Materials

d-α-, *d*-β-, *d*-γ-, and *d*-δ-Tocopherol were purchased from Eastman Organic Chemicals[1] and were used as received after a purity check by thin-layer chromatography. Methanol from Matheson Coleman, & Bell was spectroquality. Methylene blue (basic blue 9, Matheson, Coleman, & Bell) was dissolved in Spectroquality methanol, diluted to a concentration of 4×10^{-4} M, and kept in the dark until needed. Ethanol was prepared according to the Analytical Methods Committee, Society for Analytical Chemistry, London (1959).

Methods

A photolysis apparatus was constructed of plexiglass with five 10-ml Pyrex tubes sealed into a water reservoir. The temperature was maintained at 25° with a Hooke water-circulating pump. The light source was a GE 150-W reflector-flood lamp attached to a variable transformer (Matheson Scientific Co.) for controlling the light intensity. Constant line voltage was maintained with a Sola Electric Co. constant-voltage transformer.

To each of four tubes in the photoreactor was added an identical portion of a standardized tocopherol solution in methanol, 400 μl of the dye solution, and the amount of methanol required to bring the reaction volume to 2.00 ml. Oxygen gas was saturated with methanol in the central tube of the photolysis apparatus and passed in series through the four sample tubes. In this way each methanol solution was saturated with oxygen. After photolysis for a convenient time (α and β, 3 min; γ and δ, 5 min), the reaction mixture from the last tube in the series was removed and diluted with ethanol to a convenient volume for tocopherol analysis by the Emmerie–Engel method (Analytical Methods Committee, 1959). Photolysis was continued and the procedure was repeated after each successive photolysis time interval. A blank containing no tocopherol and a sample not photolyzed were also analyzed with each series of photolyzed samples.

Results and Discussion

Reactivity (β) of the tocopherols was determined in methanol by a modification of the method of Higgins *et al.* (1968). A reaction scheme for dye-sensitized photooxidation (Scheme I) was outlined by Young *et al.* (1970), where MB = methylene blue and T = tocopherol. If $\beta = k_d/k_T$, then $v = -dc/dt = k_{1O_2}[[T]/([T] + \beta)]$ where β is equivalent to the concentration of tocopherol at which half of the singlet oxygen is converted to product. When $[T] \gg \beta$, zero-order kinetics should be observed. For tocopherols in the concentration range of 5×10^{-4} to 20×10^{-4} M, zero-order kinetics was approximated

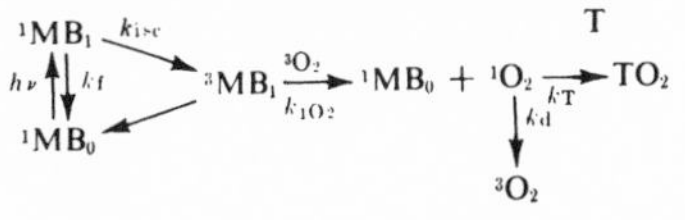

* From the Northern Regional Research Laboratory, Agricultural Research Service, U. S. Department of Agriculture, Peoria, Illinois 61604. *Received August 18, 1971.*
† To whom to address correspondence.
1 The mention of firm names or trade products does not imply that they are endorsed or recommended by the Department of Agriculture over other firms or similar products not mentioned.

SCHEME I

TABLE I: Photooxidation of Tocopherols in Methanol at 25°.

Tocopherol Compound	$c_0 \times 10^4$ (M)	$v_0 \times 10^4$ (M min^{-1})	$1/c_0 \times 10^{-4}$ (M^{-1})	$1/v_0 \times 10^{-4}$ (M^{-1} min)	Equation	$\beta \times 10^4$ (M)
α	3.42	0.183	0.292	5.5	$1/v_0 = 5.27/c_0 \times (3.86 \times 10^4)$	1.4 (± 0.4)
	6.86	0.234	0.146	4.3		
	10.33	0.224	0.097	4.5		
	20.64	0.236	0.048	4.2		
				rel std dev = 8%		
β	5.11	0.101	0.196	9.9	$1/v_0 = 18.06/c_0 + (6.34 \times 10^4)$	2.8 (± 0.2)
	7.57	0.116	0.132	8.6		
	10.05	0.121	0.100	8.3		
	20.08	0.139	0.050	7.2		
				rel std dev = 13%		
γ	4.94	0.092	0.202	10.9	$1/v_0 = 28.19/c_0 + (5.20 \times 10^4)$	5.4 (± 0.3)
	7.36	0.110	0.136	9.1		
	9.95	0.127	0.101	7.9		
	19.66	0.149	0.051	6.7		
				rel std dev = 4%		
δ	5.42	0.038	0.185	26.3	$1/v_0 = 101.11/c_0 + (7.50 \times 10^4)$	13.5 (± 0.4)
	7.80	0.049	0.128	20.4		
	10.70	0.060	0.093	16.7		
	22.31	0.082	0.045	12.2		
				rel std dev = 8%		

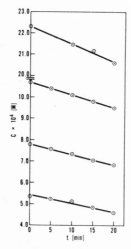

FIGURE 1: Photooxidation of δ-tocopherol.

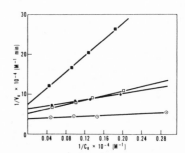

FIGURE 2: Photooxidation of tocopherols. α, ○; β, ●; γ, □; δ, ■.

when conversions were low (<25%). The initial reaction velocity, v_0, at different initial tocopherol concentrations, c_0, was determined by following the disappearance of tocopherol colorimetrically according to the Emmerie–Engel method (Analytical Methods Committee, 1959). The results for δ-tocopherol are shown in Figure 1. Data for all four are summarized in Table I.

At $t = 0$, $v_0 = -dc_0/dt = k_{1O_2} \times c_0/(c_0 + \beta)$ and $1/v_0 = \beta/c_0 k_{1O_2} + 1/k_{1O_2}$. Plotting $1/v_0$ vs. $1/c_0$ should give a straight line whose intercept is $1/k_{1O_2}$ and slope is β/k_{1O_2}. From Figure 2 the value of β can be calculated as the ratio of the slope to

the intercept of that line. The value of β for each tocopherol is given in Table I. α-Tocopherol was the most reactive ($\beta = 1.4 \times 10^{-4}$ M)[2] while δ was the least ($\beta = 13.5 \times 10^{-4}$ M). The reactivity of each tocopherol relative to the α isomer can be calculated from $k_T/k_\alpha = (k_d/k_\alpha) \times (k_T/k_d) = \beta_\alpha/\beta_T$. The value of k_d was estimated to be 1×10^5 sec^{-1} (Higgins et al., 1968). With k_d established, then the rate constant k_α can be calculated: $k_\alpha = k_d/\beta = 1 \times 10^5$ sec$^{-1}/1.4 \times 10^{-4}$ M $= 7 \times 10^8$ M^{-1} sec^{-1}. This rate constant approaches that for a diffusion controlled reaction, 10^{10} M^{-1} sec^{-1}.

Reactivities of tocopherols toward singlet oxygen correlate well with their biological activity (see Table II). When singlet

---

[2] At the suggestion of a reviewer we repeated the β value determination for α-tocopherol, replacing pure oxygen with air. Since the same β value was obtained ($\beta \times 10^4 = 1.5 \pm 0.2$ M), the reaction of α-tocopherol with dye triplet does not lead to product formation. On this time scale the reaction of α-tocopherol with singlet oxygen appears to be the only observed reaction.

TABLE II: Reactivities of Tocopherols: Correlation between Singlet Oxygen Reactivity and Biological Activity.

Toco-pherol Compound	$\beta \times 10^4$	$k_T/k_\alpha{}^b$ $\times 100$ (%)	Biological Activity[a]			
			Respiratory Decline in Rat Livers		Erythrocyte Hemolysis	
			In Vitro (%)	In Vivo (%)	In Vitro (%)	In Vivo (%)
α	1.4	100	100	100	100	100
β	2.8	50	46	55	40	23
γ	5.4	26	26	5	30	3–17
δ	13.5	10	18	4	20	2

[a] Century and Horwitt (1965). [b] Rel std dev = 30%.

oxygen reactivity of each tocopherol is correlated with respiratory decline in rat livers, the reactivity of each tocopherol falls between *in vivo* and *in vitro* biopotency. The correlation between singlet oxygen reactivity and erythrocyte hemolysis is similar except for β-tocopherol, for which the reactivity falls slightly above the range of reported biopotency.

Conclusion

We have demonstrated that oxidation of tocopherols with singlet oxygen in methanol is a good model reaction for certain functions of vitamin E. This correlation suggests that one function of vitamin E may be to protect membranes and lipids from the damaging effects of "active" oxygen.

References

Analytical Methods Committee (1959), *Analyst 84*, 356.
Century, B., and Horwitt, M. K. (1965), *Fed. Proc., Fed. Amer. Soc. Exp. Biol. 24*, 906, and references therein.
Foote, C. S., Chang, Y. C., and Denny, R. W. (1970a), *J. Amer. Chem. Soc. 92*, 5216.
Foote, C. S., Denny, R. W., Weaver, L., Chang, Y., and Peters, J. (1970b), *Ann. N. Y. Acad. Sci. 171*, 130.
Grams, G. W. (1972), *Tetrahedron Lett.* (in press).
Grams, G. W., Eskins, K., and Inglett, G. E. (1972), *J. Amer. Chem. Soc.* (in press).
Higgins, R., Foote, C. S., and Cheng, H. (1968), *Advan. Chem. Ser., No. 77*, 102.
Khan, N. A. (1971), *Fette, Seifen, Anstrichm. 73*, 109.
Rawls, H. R., and van Santen, P. J. (1970), *J. Amer. Oil Chem. Soc. 47*, 121.
Young, R. H., Chink, N., and Mallon, C. (1970), *Ann. N. Y. Acad. Sci. 171*, 130.

329

67

Reprinted from *Photochem. Photobiol.*, **20**(6), 505–509 (1974)

ON THE QUENCHING OF SINGLET OXYGEN BY α-TOCOPHEROL

S. R. Fahrenholtz, F. H. Doleiden, A. M. Trozzolo and A. A. Lamola

Bell Laboratories, Murray Hill, New Jersey 07974, U.S.A.

(*Received* 27 *March* 1974; *accepted* 14 *June* 1974)

Abstract—D-α-tocopherol was found to be an effective quencher of $^1O_2$ molecules ($k = 2·5 \times 10^8$ $\mathscr{L}$mol^{-1} s^{-1} in pyridine) by measuring its effect on the autosensitized photooxidation of rubrene. The quenching process was shown to be almost entirely 'physical', that is, α-tocopherol deactivated about 120 $^1O_2$ molecules before being destroyed. The results suggest that this process may be a mechanism for the protective effect of α- tocopherol in photodynamic action.

INTRODUCTION

The red blood cells of patients with erythropoietic protoporphyria (EPP) are hemolyzed upon irradiation with visible light due to photooxidation of membrane components sensitized by the large amount of free protoporphyrin in the cells (Harber *et al.*, 1964; Tschurdy *et al.*, 1971; Hsu *et al.*, 1971; Schothurst *et al.*, 1972; Goldstein and Harber, 1972). Ludwig *et al.*, (1967) have reported that preincubation of EPP red cells with α-tocopherol (**I**) inhibits the oxidative damage and hemolysis which result from irradiation of the cells. Goldstein and Harber (1972) reported that incubation of EPP red cells with α-tocopheryl acetate affords similar protection. As part of our studies of the molecular basis of the photohemolysis of EPP red blood cells (Lamola *et al.*, 1973), we decided to investigate the mechanism of α-tocopherol protection. The inhibition of photohemolysis of EPP red blood cells by α-tocopherol can, in principle, result from at least two modes of action: (1) inhibition of photochemical processes, and (2) inhibition of dark chemical

I
(Vitamin E)

processes which are a consequence of the ph◦ chemical events.

That α-tocopherol can act by the latter mod◦ clear, since α-tocopherol is an excellent anti◦ dant (inhibitor of radical chain oxidation) biomembranes (Green, 1972; Lucy, 1972; Tap◦ 1972). The vitamin effectively inhibits hemolysi◦ red blood cells due to oxidative membrane dam◦ (e.g. that induced by added hydrogen perox◦ Freedman *et al.*, 1958). We have observed ◦ α-tocopherol strongly inhibits the hemolysis ◦ duced by incorporation into the erythrocyte m◦ brane of 3β-hydroxy-5α-hydroperoxy-Δ6-cho◦ tene, the product of singlet oxygen attack u◦ cholesterol. The cholesterol hydroperoxide cau◦ hemolysis by initiating radical chain oxida◦ (Yamane and Lamola, in preparation).

In the present study, we wished to determ◦ whether or not α-tocopherol can inhibit ph◦ chemical processes associated with the ph◦ hemolysis. In particular we wanted to kno◦ α-tocopherol is an effective scavenger of sin◦ oxygen $^1O_2$ (the $^1\Delta_g$ state) in the red-blood ◦ membrane. We were interested in this quest◦ because the photohemolysis of EPP erythroc◦ might involve porphyrin-sensitized generation ◦ $^1O_2$ (Lamola *et al.*, 1973; Hsu *et al.*, 1971). ◦ reports that α-tocopherol reacts with $^1O_2$ (Gram◦ *al.*, 1972) encouraged our investigation.

Through kinetic experiments, we found that ◦ pyridine solution, the rate constant for α-tocophe◦ quenching of $^1O_2$ is $2·5 \times 10^8$ $\mathscr{L}$ mol^{-1} s^{-1}, and◦ rate constant for chemical reaction between ◦ tocopherol and $^1O_2$ is 2×10^6 $\mathscr{L}$ mol^{-1} s^{-1}. That is, α-tocopherol can deactivate about 120 ◦ molecules before being destroyed.*

---

*C. S. Foote and B. Stevens have also determined the quenching rate constant, k_q, for α-tocopherol; they find $2·5 \times 10^8$ $\mathscr{L}$mol^{-1} s^{-1} (in benzene) and $1·6 \times 10^8$ $\mathscr{L}$mol^{-1} s^{-1} (in benzene, respectively. (We thank Professors Foote and Stevens for relating these data).

68

Reprinted from *Photochem. Photobiol.*, **20**(6), 511–513 (1974)

CHEMISTRY OF SINGLET OXYGEN—XVIII. RATES OF REACTION AND QUENCHING OF α-TOCOPHEROL AND SINGLET OXYGEN*

CHRISTOPHER S. FOOTE†, TA-YEN CHING and GEORGE G. GELLER

Department of Chemistry, University of California, Los Angeles, California, 90024, U.S.A.

(*Received* 5 *March* 1974; *accepted* 14 *June* 1974)

Abstract—α-Tocopherol scavenges singlet oxygen (produced by methylene blue photosensitization in methanol) by a combination of chemical reaction ($4 \cdot 6 \times 10^7 \ M^{-1}s^{-1}$) and quenching ($6 \cdot 2 \times 10^8 \ M^{-1}s^{-1}$). The total rate of scavenging ($6 \cdot 7 \times 10^8 \ M^{-1}s^{-1}$) makes it an effective protective agent against photooxidation mediated by singlet oxygen.

The role of α-tocopherol as a biological protective agent has received widespread attention (Tappel, 1972; Pryor, 1973). Most interest has centered on its importance as a free-radical inhibitor, but its reaction with singlet oxygen, generated both chemically and photochemically, has been demonstrated and suggested to be a mechanism by which α-tocopherol inhibits lipid peroxidation (Grams, 1971; Grams et al., 1972; Grams and Eskins, 1972.

The importance of singlet oxygen as a damaging agent in photodynamic diseases (Foote, 1968; Wilson and Hastings, 1970; Foote et al., 1970), particularly porphyria, and the demonstration that lipid peroxidation is involved in the photohemolysis of porphyric blood and is partly inhibited by tocopherol (Lamola et al., 1973; Goldstein et al., 1972) makes the rate of reaction of α-tocopherol with singlet oxygen a matter of considerable importance.

Several studies have shown that photooxidation of phenols can proceed by way of singlet oxygen, though other mechanisms can also operate (Grams, 1971; Saito et al., 1970; Matsuura et al., 1972; Thomas, 1973). The reaction with singlet oxygen may involve both oxidation of the phenol and quenching of singlet oxygen by the phenol without reaction (Thomas, 1973; Foote and

Thomas, unpublished). The rate of reaction of singlet oxygen with several tocopherols has been measured by a technique which measures the sum of reaction and quenching rates; the two rates were not separated, nor was the possibility of quenching without product formation recognized (Grams and Eskins, 1972).

MATERIALS AND METHODS

d-α-Tocopherol (used as received) was supplied by Eastman Kodak; diphenylfuran (DPF) was recrystallized before use; diphenylisobenzofuran (DPBF) was purchased from Aldrich. Methanolic solutions of tocopherol, DPBF ($2–4 \times 10^{-6} \ M$), and methylene blue ($10^{-5} \ M$) were photolyzed in a kinetic apparatus similar to that described by Young et al. (1971). The photolyzing light, absorbed by methylene blue, was a Sylvania 500 Q/Cl Tungsten–Halogen lamp operated at 35 V with a Corning 3–68 filter (cut-off 540 nm). The fluorescence was excited by a 200 W Xe–Hg lamp (Hanovia, powered by a Schoeffel LPS 251 power supply) through a Jarrell–Ash model 82-410 monochromator at 406 nm (0·25 mm slit) with fluorescence emission at 475 nm selected by a second monochromator (slit 1·0 mm) and detected by a 1P-28 photomultiplier in a Heath EU-703-31 Photometric Readout system.

Competition experiments were carried out using diphenylfuran ($3–50 \times 10^{-6} \ M$) methylene blue ($10^{-5} \ M$) and tocopherol ($2 \cdot 5–50 \times 10^{-5} \ M$) in methanol. The apparatus described above was used to photolyze the solutions to the point where diphenylfuran was exactly 50 per cent photooxidized, as shown by the decrease in its fluorescence (excitation λ 330 nm, slit 0·25 mm; emission λ 373 nm, slit 0·25 mm). Tocopherol concentrations in these solutions were measured by a modification of the Emmerie–Engel method (Grams and Eskins, 1972; Analytical Methods Committee, 1959).

*Contribution No. 3279. Paper XVII: C. S. Foote, S. Mazur, P. A. Burns and D. Lerdal (1973) *J. Am. Chem. Soc.* **95**, 586.

†To whom correspondence should be addressed.

335

RESULTS AND DISCUSSION

The kinetic scheme for reaction of α-tocopherol (T), which can either react with (k_R) or quench (k_Q) singlet oxygen, in the presence of a second fluorescing acceptor, F (reaction rate k_F) is shown below.

$$F + {}^1O_2 \xrightarrow[k_F]{} FO_2 \text{ (non-fluorescent product)}$$

$$T + {}^1O_2 \xrightarrow[k_R]{} TO_2 \text{ (reaction)}$$

$$T + {}^1O_2 \xrightarrow[k_Q]{} T + {}^3O_2 \text{ (quenching)}$$

$${}^1O_2 \xrightarrow[k_D]{} {}^3O_2$$

In the technique of Young *et al.* (1971), the fluorescent acceptor is present in extremely low concentration, so that its disappearance is cleanly first-order. Its fluorescence is used to monitor its rate of loss on photooxidation, and the inhibition of this photooxidation by tocopherol can be easily studied. The steady-state treatment then gives, where K is the rate of formation of singlet oxygen:

$$\frac{-dF}{dt} = K\left(\frac{k_F[F]}{k_F[F] + k_R[T] + k_Q[T] + k_D}\right)$$

$$\approx K\left(\frac{k_F[F]}{k_F[T] + k_Q[T] + k_D}\right)$$

Under conditions where T is not appreciably photooxidized, this technique gives linear plots of log fluorescence vs time whose slopes (S) are given

by

$$S = K\left(\frac{k_F}{k_R[T] + k_Q[T] + k_D}\right)$$

and a plot of S_0/S_T (slopes in absence and presen of T) vs T has a slope $= (k_R + k_Q)/k_D$. Th technique, like the treatment used by Grams a Eskins (1972b), measures the sum of $(k_R + k_Q)$ onl

Concentrations of α-tocopherol ranging from to $50 \times 10^{-5}\,M$ were used in experiments w diphenylisobenzofuran (DPBF) as fluorescer. sample plot is shown in Fig. 1. The value $(k_Q + k_R)/k_D$ obtained in this way was 4800 ± 4 M^{-1}. This value agrees reasonably well with that 7100 M^{-1} obtained by Grams and Eskins (1972). average k_D (Adams and Wilkinson, 1972; Merkel al., 1972) for singlet oxygen in methanol (1·4 $10^5\,s^{-1}$) gives $(k_Q + k_R)$ for α-tocopherol the rath high value of $6·7 \pm 0·6 \times 10^8\,M^{-1}s^{-1}$.

The validity of this treatment requires that t tocopherol concentration remain constant duri photooxidation of DPBF. That this requireme was at least approximately met was demonstrat by the fact that excellent first-order plots we observed for decay of DPBF. In order to establi the validity of this assumption with certainty, a to obtain an independent value for k_R, a separate s of experiments was carried cut in which loss both fluorescer and α-tocopherol could be indepe dent monitored. The equation of Higgins *et* (1968) for the reaction of two competing su strates (F and T) is shown below. This equatic gives the ratio of k_R for each substrate, indepe dent of whether one or both substrates als quenches singlet oxygen; the subscripts f and

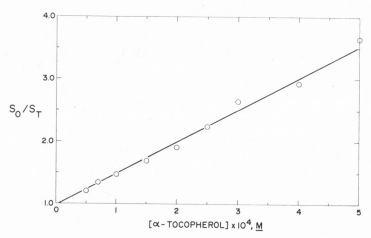

Figure 1. Slope of first-order plot of disappearance of diphenylisobenzofuran (DPBF) in absence of α-tocopherol (S_0) divided by slope in its presence (S_T), as a function of α-tocopherol concentration.

·efer to final and initial concentrations, respec-
·vely.

$$\frac{k_R{}^F}{k_R{}^T} = \frac{\log(F_f/F_0)}{\log(T_f/T_0)}$$

Diphenylfuran (DPF) was used as the fluorescent
ubstrate, and its disappearance monitored as
efore. The amount of tocopherol lost at the time
·hen 50 per cent of the DPF had disappeared was
etermined by treatment with Fe^{3+}/dipyridyl reag-
nt (Analytical Methods Committee, 1959; Grams
nd Eskins, 1972). Using this technique and the
alue of k_R for DPF (Young et al., 1972,
ecalculated with $k_D = 1\cdot4 \times 10^5\,s^{-1}$: $1\cdot56 \times$
$0^8\,M^{-1}s^{-1}$), the value of k_R for α-tocopherol was
ound to be $4\cdot6 \pm 1 \times 10^7\,M^{-1}s^{-1}$. This value is 15
mes smaller than $(k_R + k_Q)$; thus almost all of the
inglet oxygen is scavenged by a quenching
rocess. The situation is thus similar to that of
,4,6-triphenylphenol, which quenches singlet oxgen rapidly but does not react with it appreciably
Thomas, 1973; Foote and Thomas, in preparation).
he small magnitude of k_R also confirms the
ssumption of the treatment in the first part of this
aper, since, when 50 per cent of DPBF is
onverted, less than 2 per cent of the tocopherol
as reacted (rate data from Young et al., 1972).
hese results should be compared with those of
·ahrenholtz et al. (1974): $k_Q = 2\cdot5 \times 10^8\,M^{-1}s^{-1}$,

$k_R = 2 \times 10^6\,M^{-1}s^{-1}$ in pyridine, and of Stevens et al.
(1974): $k_Q = 1\cdot7 \times 10^8\,M^{-1}s^{-1}$, $k_R = 1\cdot7 \times 10^6\,M^{-1}s^{-1}$
in benzene. Previous work (Thomas, 1973) has
supported an electron-transfer process for both
reaction and quenching by phenols, and these
solvent effects are consistent with this mechanism.

The results reported here show that α-tocopherol
is one of the bast scavengers for singlet oxygen yet
found; its total rate for removal of singlet oxygen is
only 50 times slower than that of the diffusion-
controlled quencher, β-carotene. (Foote et al.,
1970).

While it is certain that much of the ability of
α-tocopherol to inhibit oxidative damage in biological systems derives from its ability to inhibit
free-radical chain autoxidation, its ability to
scavenge singlet oxygen should not be neglected.
This ability to inhibit both radical and singlet
oxygen oxidations is shared by several metal
complexes, and probably by a variety of other
compounds (Farmilo and Wilkinson, 1973; Carlson
et al., 1972, 1973; Flood et al., 1973).

Acknowledgements—This work was supported by Public
Health Service Grant No. GM 20080-01. G. G. G.
acknowledges an undergraduate research participation
fellowship from the Association of Western
Universities—UCLA Nuclear Science Summer Research
Program. We thank Professor Stevens and Drs. Lamola
and Trozzolo for prepublication copies of their manuscripts.

REFERENCES

Adams, D. R. and F. Wilkinson (1972) J. Chem. Soc. Faraday Trans. 68, 586–593.
Analytical Methods Committee (1959) Analyst 84, 356–372.
Carlson, D. J., G. D. Mendenhall, T. Suprunchuk and D. M. Wiles (1972) J. Am. Chem. Soc. 94, 8960–8962.
Carlson, D. J., T. Suprunchuk and D. M. Wiles (1973) Polymer Letts. Edition 11, 61–65.
Fahrenholtz, S. R., F. H. Doleiden, A. M. Trozzolo and A. A. Lamola (1974) Photochem. Photobiol. 20, 505–509.
Farmilo, A. and F. Wilkinson (1973) Photochem. Photobiol. 18, 447–450.
Flood, J., K. E. Russell, and J. K. S. Wan (1973) Macromolecules 6, 669–671.
Foote, C. S. (1968) Science 162, 963–970.
Foote, C. S., Y. C. Chang and R. W. Denny (1970) J. Am. Chem. Soc. 92, 5216–5219.
Goldstein, B. C. and L. C. Harber (1972) J. Clin. Invest. 51, 891–902.
Grams, G. W. (1971) Tetrahedron Letters 4823–4826.
Grams, G. W., K. Eskins, and G. E. Inglett (1972) J. Am. Chem. Soc. 94, 866–868.
Grams, G. W., and K. Eskins (1972) Biochemistry 11, 606–608.
Higgins, R., C. S. Foote, and H. Cheng (1968) Advan. Chem. Ser. 77, part III, 102–117.
Lamola, A. A., T. Yamane, and A. M. Trozzolo (1973) Science 179, 1131–1133.
Matsuura, T., N. Yoshimura, A. Nishinaga, and I. Saito (1972) Tetrahedron 28, 4933–4938.
Merkel, P. G., and D. R. Kearns (1972) J. Am. Chem. Soc. 94, 1029–1030.
Pryor, W. (1973) Fed. Procs. 32, 1862–1869.
Saito, I., S. Kato and T. Matsuura (1970) ·Tetrahedron Letters 239–242.
Stevens, B., R. D. Small, Jr. and S. R. Perez (1974) Photochem. Photobiol. 20, 515–517.
Tappel, A. (1972) Ann. N. Y. Acad. Sci. 203, 12–28.
Thomas, M. (1973) Dissertation, U.C.L.A.
Wilson, T. and J. W. Hastings (1970) Photophysiology 5, 49–95.
Young, R. H., K. Wehrly, and R. L. Martin (1971) J. Am. Chem. Soc. 93, 5774–5779.
Young, R. H., and D. T. Feriozi (1972) Chem. Commun. 841–842.

69

Reprinted from *Photochem. Photobiol.,* **20**(6), 515–517 (1974)

THE PHOTOPEROXIDATION OF UNSATURATED ORGANIC MOLECULES—XIII. O₂¹Δg QUENCHING BY α-TOCOPHEROL*

B. Stevens, R. D. Small, Jr. and S. R. Perez
Department of Chemistry, University of South Florida, Tampa, Florida 33620, U.S.A.

(Received 20 March 1974; accepted 14 June 1974)

Although the need for vitamin E in human nutrition is established, there is some disagreement as to whether it serves generally as an antioxidant or as a specific cofactor (Green, 1972). The evidence indicates however that its presence in the cell is necessary to control lipid peroxidation, which may proceed via a free radical mechanism (Gunstone and Hilditch, 1945, 1950) or involve $O_2^1\Delta g$ as a reactive intermediate which is known to produce hydroperoxides with allylic groups presumably present in the unsaturated lipid fraction (Rawls and Van Santen, 1970). An ideal $O_2^1\Delta g$ scavenger from the biological viewpoint is one which physically quenches the excited oxygen molecule, an example being β-carotene, which has been suggested (Foote *et al.*, 1970), as a protector of plants from destructive photodynamic effects during photosynthesis. It is therefore of interest to examine the possibility that vitamin E can play a similar protective rôle in animals, where, however, a source of $O_2^1\Delta g$ has yet to be established.

The $O_2^1\Delta g$ quenching efficiency of D-α-tocopherol(T) (Eastman) was estimated from its inhibition of the self-sensitised photoperoxidation of rubrene (R) at 546 nm where α-tocopherol does not absorb. The relevant processes are summarised in the following scheme, which accommodates inhibition by α-tocopherol quenching of the rubrene triplet state (process 3) which is the $O_2^1\Delta g$ precursor:

$$R + h\nu \rightarrow {}^1R*$$

$${}^1R* + O_2^3\Sigma \rightarrow {}^3R + O_2^3\Sigma \tag{}$$

$${}^3R + O_2^3\Sigma \rightarrow R + O_2^1\Delta \tag{}$$

$${}^3R + T \rightarrow R + T \tag{}$$

$$R + O_2^1\Delta \rightarrow RO_2 \tag{}$$

$$T + O_2^1\Delta \rightarrow \text{quenching} \tag{}$$

$$O_2^1\Delta \rightarrow O_2^3\Sigma \tag{}$$

Here RO_2 denotes rubrene peroxide, the so product of the uninhibited reaction, and process includes both physical quenching and chemic reaction. Under photostationary conditions, pre cesses (1)–(6) provide the expression

$$\gamma_0/\gamma = \{1 + k_3[T]/k_2[O_2]\}\{1 + k_5[T]/(k_4[R] + k_6)\} \tag{}$$

for the reduction in quantum yield of RC formation from γ_0 to γ in the presence of inhibite at concentration $[T]$; this leads to three distinguis able alternatives:

(i) γ_0/γ is a linear function of $[T]$ but independent of rubrene concentration if triple state quenching (process 3) is solely responsible fo inhibition:

(ii) γ_0/γ is a non-linear function of $[T]$ if th additive quenches both $O_2^1\Delta g$ and the rubrer triplet state;

(iii) If the linear dependence of γ_0/γ on $[T]$ increases with a reduction in rubrene concentra tion, inhibition may be attributed to additiv quenching of $O_2^1\Delta g$ only (process 5).

Relative quantum yields of rubrene autoperox

---

*Part XII: B. Stevens In *Excited States of Biological Molecules*, Edited by J. B. Birks, Wiley, London, 1974. In press.

ation were estimated as relative initial rates of brene consumption in air-saturated benzene or cyclohexane at 25°C, obtained from continuous ecordings of the optical density of rubrene at the ctinic wavelength (546 nm) as a function of exposure time in the presence of different concentrations of α-tocopherol. This procedure is based on the relationship

$$\gamma_0/\gamma = \frac{(d[R]_0/dt)}{(d[R]_0/dt)_T} \frac{(I_a)_T}{(I_a)} = \frac{(dOD_0/dt)}{(dOD_0/dt)_T}$$

where subscript T denotes the presence of inhibitor and for a constant initial rubrene concentration the absorbed light intensity $(I_a)_T = (I_a)$.

As shown in Fig. 1 γ_0/γ varies directly with

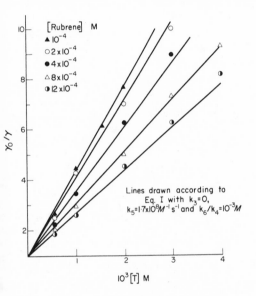

Figure 1. α-tocopherol inhibition of rubrene peroxidation at 546 nm in air-saturated benzene at 25°C. Rubrene concentration $10^{-4} M$ ▲; $2 \times 10^{-4} M$ ○; $4 \times 10^{-4} M$ ●; $8 \times 10^{-4} M$ △ and $12 \times 10^{-4} M$ ◑. Lines drawn according to Eq. I with $k_3 = 0$, $k_5 = 1.7 \times 10^8 M^{-1} s^{-1}$ and $k_6/k_4 = 10^{-3} M$.

concentration of α-tocopherol, and the extent of inhibition $d(\gamma_0/\gamma)/d[T]$ increases with a reduction in rubrene concentration consistent with additive quenching of $O_2^1\Delta g$ (process 5) as the predominant inhibiting process (alternative iii above); accordingly, Eq. I may be rearranged to

$$1/\beta_T = \{d(\gamma_0/\gamma)/d[T]\}\{1 + [R]/\beta_R\} \qquad (II)$$

where $\beta_T (= k_6/k_5)$ and $\beta_R (= k_6/k_4)$ are the respective $O_2^1\Delta g$ reactivity indices for α-tocopherol and rubrene. Use of Eq. II and reported values of β_R and k_6 in the solvents used provides the tabulated data for k_5.

The values of k_5 are of the order of magnitude reported for α-tocopherol quenching of $O_2^1\Delta g$ in pyridine and in methanol (see Table 1) and are higher than those obtained for $O_2^1\Delta g$ addition to rubrene, one of the most reactive $O_2^1\Delta g$ acceptors, but are some two orders of magnitude below the rate constant for $O_2^1\Delta g$ quenching by energy transfer to β-carotene, which is virtually diffusion-limited.

The fraction α of $O_2^1\Delta g$ quenching encounters (process 5) that lead to chemical change may be estimated from spectrophotometric measurements of the rate of α-tocopherol consumption relative to that of an established $O_2^1\Delta g$ acceptor of similar reactivity which exhibits weak absorption in the region of 300 nm and strong absorption at longer wavelengths; 9,10-diphenylanthracene(D) proved to be suitable for this purpose. Integration of the relative rate equations (Wilson, 1966).

$$\frac{-d[T]/dt}{-d[D]/dt} = \frac{d[T]}{d[D]} = \frac{\alpha k_5[T][O_2^1\Delta]}{k_4'[D][O_2^1\Delta]}$$

yields

$$\frac{\log([T_0]/[T])}{\log([D_0]/[D])} = \frac{\alpha\beta_D}{\beta_T}$$

Table 1. Reactivity indices β_T and rate constants k_5 for α-tocopherol quenching of $O_2^1\Delta g$ at 25°C

Solvent	$10^3\beta_R (M)$*	$10^4\beta_T (M)$	$10^{-4}k_6(s^{-1})$†	$10^{-8}k_5(M^{-1} s^{-1})$	$10^{-6}\alpha k_5(M^{-1} s^{-1})$
Benzene‡	1·0	2·7 ± 0·4	4·2	1·7 ± 0·3	—
Cyclohexane‡	1·4	6·8 ± 0·8	5·9	0·9 ± 0·2	1·0
Pyridine§	—	—	—	2·5	2·0
Methanol‖	—	—	—	6·7	46

*Stevens and Perez, 1974.
†Merkel and Kearns, 1972.
‡This work.
§Fahrenholtz et al., 1974.
‖Foote et al., 1974.

where $\beta_D = k_6/k'_4$, and k'_4 denotes the rate constant for $O_2^1\Delta g$ addition to 9,10-diphenylanthracene. In practice the absorption spectrum of a solution of α-tocopherol and diphenylanthracene in cyclohexane was recorded before and after several exposures to the Hg line at 365 nm, and the concentration changes computed from changes in optical density at 300 nm (α-tocopherol) and 373 nm diphenylanthracene) with appropriate corrections for absorption at 300 nm by diphenylanthracene and its peroxide. The mean of six determinations is expressed as

$$\alpha\beta_D/\beta_T = 0.9 \pm 0.1$$

which with $\beta_D = 0.056\,M$ (in cyclohexane*) leads to

$$\beta_T/\alpha = 0.06\,M$$

or

$$\alpha k_5 = 1.0 \times 10^6 M^{-1} s^{-1}$$

Accordingly $\alpha \simeq 0.01$, indicating that the α-tocopherol quenching of $O_2^1\Delta g$ (process 5) may be regarded as essentially (99 per cent) a physical

---

*Estimated from β_D and values of k_6 in benzene and cyclohexane (Merkel and Kearns, 1972) on the assumption that k'_4 is independent of solvent [Stevens and Perez, 1974].

process. In contrast to quenching by β-carotene however, this is unlikely to involve the energy transfer process 7

$$T + O_2^1\Delta \rightarrow {}^3T + O_2^3\Sigma \qquad (7$$

insofar as the α-tocopherol triplet state ${}^3T$ is not expected to lie below $O_2^1\Delta g$ at 8000 cm^{-1}. On the other hand, the change in spin angular momentum attending the overall process 8

$$T + O_2^1\Delta \rightarrow T + O_2^3\Sigma \qquad (8$$

requires the intervention of either a charge-transfer or biradical intermediate; in the former case the quenching rate constant k_5 should be sensitive to solvent polarity, whereas a biradical intermediate may be susceptible to solvent scavenging, leading to an increase in the extent of chemical quenching in H-donating solvents. In this respect it is of interest to to note that Foote et al. (1974) find that $\alpha = 0.07$ in methanol.

Acknowledgements—The continued support of the National Science Foundation under Research Grand No. GP28331X is gratefully acknowledged, as is the the provision of preprints of papers by Professor Foote and Dr. Lamola.

REFERENCES

Farenholtz, S. R., F. H. Doleidein, A. M. Trozzolo and A. A. Lamola (1974) *Photochem. Photobiol.* 20, 505–509.
Foote, C. S., R. W. Denny, L. Weaver, T. Chang and J. Peters (1970) *Ann. New York Acad. Sci.* 171, 139.
Foote, C. S., Ta-Yen Ching and G. G. Geller (1974) *Photochem. Photobiol.* 20, 511–513.
Green, J. (1972) *Ann. New York Acad. Sci.* 203, 29.
Gunstone, F. D. and T. P. Hilditch (1945) *J. Chem. Soc. (London)* 836; (1950) *Nature* 116, 558.
Merkel, P. B. and D. R. Kearns (1972) *J. Am. Chem. Soc.* 94, 7244.
Rawls, H. R. and P. J. van Santen (1970) *Ann. New York Acad. Sci.* 171, 135.
Stevens, B. and S. R. Perez (1974) *Mol. Photochem.* 6, 1.
Wilson, T. (1966) *J. Am. Chem. Soc.* 88, 2898.

Editor's Comments
on Papers 70 Through 77

SOLVENT EFFECTS ON SINGLET OXYGEN REACTIONS — LIFETIME OF $^1\Delta_g\ O_2$ IN SOLUTION

Kinetic investigations of photooxygenation have demonstrated that the reactive intermediate $^1O_2$ or a sensitizer–oxygen complex (these mechanisms are kinetically equivalent) undergoes reaction with an acceptor A or is deactivated to ground-state oxygen.[1]

$$^1O_2 \xrightarrow{k_1} {}^3O_2$$

$$A + {}^1O_2 \xrightarrow{k_2} AO_2$$

Steady-state treatment of this scheme gives

$$\Phi_{AO_2} = \Phi_{{}^1O_2} \frac{k_2[A]}{k_2[A] + k_1}$$

where Φ_{AO_2} is the quantum yield for product formation and $\Phi_{{}^1O_2}$ is the quantum yield for $^1O_2$ production. The fraction of the reactive intermediate that is trapped is given by $[A]/\{(k_1/k_2)+[A]\}$. Several investigators (Papers 14, 15, 26, and 70) have expressed the reactivity of an acceptor as the ratio of rate constants k_1/k_2, that is, the ratio of the rate of decay of $^1O_2$ to its rate of reaction with A. This ratio β, which can be determined experimentally, represents the concentration of the acceptor A at which half the $^1O_2$ gives product AO_2. The relative reactivities of a series of acceptors toward $^1O_2$ in the same solvent could be readily obtained by comparing the β values for the acceptors. However, the absolute values for k_2 and the inherent solvent effect on k_2 could not be determined without knowing the lifetime of $^1O_2$ in the solvent.

Foote (Paper 62) had observed that β-carotene is an efficient quencher for singlet oxygen. In Paper 71, Young and coworkers described a series of investigations on the solvent effects in dye-sensitized photooxygenation. By assuming that the rate of quenching of singlet oxygen by β-carotene was diffusion controlled, they were able to estimate values for k_2 in various solvents.

Merkel and Kearns (Paper 72) reported the first direct measurement of the lifetime of $^1\Delta_g$ O_2 in solution. Using a laser pulse technique, they were able to determine a value for the lifetime in methanol of 7 μs. With this information the absolute rate constants for the reaction of $^1O_2$ with 1,3-diphenylisobenzofuran and tetramethylethylene were calculated.[2] Adams and Wilkinson (Paper 73) were able to show that k_1 is independent of the sensitizer used.

Wilkinson and Farmilo[3] find the following values for the lifetime of $^1\Delta_g$ O_2 in solution: benzene 25, 26 μs; benzene-d$_6$, 36 μs; ethanol, 10 μs; 95 percent ethanol, 5 μs; methanol, 9 μs.

It has now been established that the lifetime of $^1O_2$ in solution is strongly dependent on the solvent (Papers 72–77). The tenfold increase in τ in D_2O compared to H_2O is particularly interesting and may find application in investigations of biological oxidations involving singlet oxygen.

Lifetime of singlet oxygen ($^1\Delta_g$) in various
solvents (from Paper 76 and reference 4)

Solvent	μs
H_2O	2
D_2O	20
CH_3OH	7
CH_3CH_2OH	12
C_6H_{12}	17
C_6H_6	24
CH_3COCH_3	26
CH_3CN	30
$CHCl_3$	60
CS_2	200
$CDCl_3$	300
C_6F_6	600
CCl_4	700
CF_3Cl (Freon II)	1000

NOTES AND REFERENCES

1. This kinetic scheme applies to those substrates which react with $^1O_2$ but do not give appreciable physical quenching of $^1O_2$. See reference 2 and Papers 67–69.
2. P. B. Merkel and D. R. Kearns, *J. Amer. Chem. Soc.*, **97**, 462 (1975).
3. Private communication from F. Wilkinson; Ph.D. thesis of A. Farmilo, University of East Anglia.
4. C. A. Long and D. R. Kearns, *J. Amer. Chem. Soc.*, **97**, 2018 (1975).

70

Reprinted from *J. Amer. Chem. Soc.*, 93(20), 5168–5171 (1971)

Chemistry of Singlet Oxygen.
XIII. Solvent Effects on the Reaction with Olefins[1]

Christopher S. Foote* and Robert W. Denny

*Contribution No. 2681 from the Department of Chemistry,
University of California, Los Angeles, California 90024.
Received August 31, 1970*

Abstract: Solvent effects on the product-forming step of the dye-sensitized photooxygenation of 2-methyl-2-pentene have been measured. The variation in β (ratio of the decay rate of singlet oxygen to its rate of reaction) is very small, amounting to less than a factor of four in the solvents studied. No effect of solvent polarity, internal pressure, viscosity, or heavy atoms can be discerned. The results of these studies and those previously reported are consistent with a concerted reaction mechanism involving a six-center transition state, but do not completely exclude a perepoxide intermediate.

As part of a continuing study of the chemistry of singlet oxygen, the solvent effect on the reaction with olefins was needed to provide further information about the transition state. Because dye-sensitized photooxygenations involve several steps and competing reactions, separation of the oxygenation step requires careful kinetic analysis.[2,3] Detailed kinetic studies of the reaction have been carried out by many groups, and the steps which are important under usual conditions (dye sensitizer with reasonably high intersystem crossing efficiency, absence of powerful hydrogen donors, solution in equilibrium with air or oxygen at 1 atm) are as follows[2]

$$\text{sens} \xrightarrow{h\nu} {}^1\text{sens} \xrightarrow{\varphi_{isc}} {}^3\text{sens} \xrightarrow[O_2]{} {}^1O_2 \xrightarrow{k_2} AO_2$$
$$\downarrow k_1$$
$${}^3O_2$$

where sens is sensitizer, excited singlet and triplet states are indicated by appropriate superscripts, ${}^1O_2$ is singlet molecular oxygen (probably ${}^1\Delta_g$), and A is the acceptor. The term φ_{isc} is the intersystem crossing quantum yield for the sensitizer, and k_1 and k_2 are the rates of decay and reaction with acceptor of singlet oxygen, respectively. Under the above conditions, all ^{3}sens gives ${}^1O_2$,[2] and the expression for the overall quantum yield of product (φ_{AO_2}) is given by

$$\varphi_{AO_2} = \varphi_{isc} \frac{k_2[A]}{k_2[A] + k_1}$$

This expression is equivalent to the following, in which I_{abs} is the number of mole quanta per liter absorbed during the irradiation time, and $[AO_2]$ is the final concentration of the product peroxide produced. Plots

$$[AO_2]^{-1} = (I_{abs}\varphi_{isc})^{-1}(1 + (k_1/k_2)[A]^{-1})$$

of $[AO_2]^{-1}$ *vs.* $[A]^{-1}$ give straight lines,[3] for which the ratio of the slope to the intercept on the $[AO_2]^{-1}$ axis is k_1/k_2, the ratio of the decay rate of ${}^1O_2$ to the rate of its reaction with the acceptor, a parameter

called β.[2,3] The intercept of such plots is the amount of singlet oxygen formed during the irradiation, and depends on sensitizer and many other factors. The parameter β, however, depends only on acceptor (it is the concentration of acceptor at which half the singlet oxygen is trapped). Thus determination of β for the same acceptor in different solvents is not affected by changes in absorption efficiency or triplet quantum yield of the dye, which in general will change with solvent.[2–4] Of course, it would be desirable to find the solvent effect on k_2 rather than on the ratio k_1/k_2 but no technique for determining absolute values of k_2 has been reported.

Several measurements of solvent effects on β have been reported. Schenck and Gollnick found only a small change in the value of β for citronellol in methanol (0.16 M), *n*-butyl alcohol (0.12 M), and 70% aqueous methanol ($\beta = 0.06$ M).[2,7] Similar small effects on the photooxygenation of 2-methyl-2-pentene were reported in *tert*-butyl alcohol (0.051 M) and 50% methanol-*tert*-butyl alcohol (0.13 M).[3] In contrast, large effects on β were reported in self-sensitized photooxygenations of anthracene by Bowen,[8] and in chlorophyll-sensitized oxidation of allylthiourea by Livingston and Owens.[9] Consideration of these reports will be deferred until the Discussion.

Results

2-Methyl-2-pentene was chosen as the acceptor for the present studies because previous work had shown that gas-chromatographic analysis of products from this olefin is convenient, and its solubility properties

(4) That β is not a function of sensitizer has been documented in many cases (see ref 2 for details). In particular, the β value for 2-methyl-2-pentene is the same with chemically generated singlet oxygen and with Rose Bengal sensitization.[3] The same value is obtained also with methylene blue, zinc tetraphenylporphine, and chlorophylls *a* and *b*, as will be apparent from the results of this study. Relative rates of reaction of several different substrate pairs are also independent of sensitizer.[2,5,6]

(5) K. R. Kopecky and H. J. Reich, *Can. J. Chem.*, **43**, 2265 (1965).

(6) T. Wilson, *J. Amer. Chem. Soc.*, **88**, 2898 (1966).

(7) G. O. Schenck and K. Gollnick, "Forschungsbericht des Landes Nordrhein-Westfalen," Nr. 1256, Westdeutscher Verlag, Köln and Opladen, 1963.

(8) E. J. Bowen, *Discuss. Faraday Soc.*, **14**, 143 (1953); E. J. Bowen, *Advan. Photochem.*, **1**, 23 (1963); E. J. Bowen and D. W. Tanner, *Trans. Faraday Soc.*, **51**, 475 (1955); R. Livingston in "Autoxidation and Antioxidants," W. O. Lundberg, Ed., Vol. I, Wiley-Interscience, New York, N. Y., 1961, p 249.

(9) R. Livingston and K. E. Owens, *J. Amer. Chem. Soc.*, **78**, 3301 (1956).

(1) Paper XII: C. S. Foote and R. W. Denny, *J. Amer. Chem. Soc.*, **93**, 5162 (1971); supported by National Science Foundation Grant No. GP-5835 and GP-8293; taken in part from R. W. Denny, Ph.D. Thesis, University of California, Los Angeles, 1969.

(2) K. Gollnick, *Advan. Photochem.*, **6**, 1 (1968).

(3) R. Higgins, C. S. Foote, and H. Cheng, *Advan. Chem. Ser.*, No. 77, 102 (1968).

in various solvents are desirable. Product analysis is more reliable than measurement of acceptor disappearance or oxygen uptake, because side reactions do not interfere.[3] However, because of difficulties in carrying out chemical generation of singlet oxygen in solvents other than alcohol,[10] only the photosensitized oxygenation was studied. The techniques of ref 3 were used, in which solutions containing weighed amounts of acceptor in various solvents and with various sensitizers were photooxygenated for a constant time, then reduced; the product alcohols were determined gas chromatographically, using an internal standard. The ratio of the two products, **1** and **2**, did not change measurably in different solvents from the previously reported[2,3,11] value of roughly 1:1.

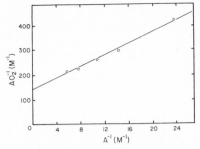

Figure 1. Determination of β for 2-methyl-2-pentene in cyclohexanol.

A typical plot is shown in Figure 1 for cyclohexanol. The values of β obtained in various solvents (see the Experimental Section for data treatment) are listed in Table I. The standard deviations listed are those

Table I. Value of β (k_1/k_2) for 2-Methyl-2-pentene in Various Solvents at 25

Solvent	β, M	Std dev	Intercept, M^{-1} i	Dielectric constant, D, 25°, D	Viscosity, η, 25°, cP
$CH_3CO_2CH_3{}^a$	0.04	0.02	293	6.02	0.44
$C_5H_5N^a$	0.05	0.01	175	12.3	0.97^g
$C_6H_5Br^h$	0.05	0.03	177	5.40	0.99^h
Cyclohexanol^a	0.07	0.01	222	15.5	57
$(CH_3)_2SO^a$	0.07	0.01	204	48.9	2.0
$(CH_3)_2CO^a$	0.08	0.01	156	20.7	0.32
$CS_2{}^{b,i}$	0.08	0.01	330	2.64^g	0.36^g
$C_6H_6{}^b$	0.10	0.01	230	2.27	0.65^g
$C_2H_5I^b$	0.11	0.03	192	7.56^h	0.59^g
$C_6H_5OCH_3{}^b$	0.13	0.02	213	4.47^i	1.32^g
m-$C_6H_4(OCH_3)_2{}^b$	0.15	0.04	2300	5.36	
CH_3OH^a	0.16	0.03	140	32.6	0.59^g
CH_3OH^c	0.13	0.01	k		
CH_3OH^d	0.20	0.08	k		
$tert$-$BuOH^{a,j}$	0.051	0.003	k	10.9^h	5.9
CH_3OH-C_6H_6; 20:80^c	0.05	0.005	k		
CH_3OH-$tert$-$BuOH$; 1:1^{a,j}	0.13	0.02	k		

^{a} Rose Bengal sensitizer. ^{b} Zinc tetraphenylporphine sensitizer. ^{c} Chlorophyll b sensitizer. ^{d} Chlorophyll a sensitizer. ^{e} Methylene blue sensitizer. Temperature other than 25°. ^{f} 0°. ^{g} 20°. ^{h} 30°. ^{i} 15°. ^{j} Corrected to 4-min irradiation if carried out for longer time. ^{k} Not comparable because run under different conditions of light intensity. ^{l} Reference 3.

for each data set; multiple determinations in several cases showed the values to be generally reproducible to ±20%.

A value for dimethylformamide of 0.03 M was also obtained, but was not considered reliable enough for inclusion in Table I because the analysis was complicated by the presence of products of unknown origin.

(10) C. S. Foote, S. Wexler, W. Ando, and R. Higgins, *J. Amer. Chem. Soc.*, **90**, 975 (1969).
(11) C. S. Foote, *Accounts Chem. Res.*, **1**, 104 (1968).

The values of the intercept of plots of $[AO_2]^{-1}$ vs. $[A^{-1}]$ (a measure of the quantum yield of singlet oxygen formation) are also listed in Table I. Because of differences in sensitizer and possible variation in sensitizer concentration, absorption spectrum, and triplet quantum yield with solvent, these values are not expected to remain completely constant. Nevertheless, the variation is small, indicating that the quantum yield of singlet oxygen formation varies only slightly in the different solvents, with the exception of m-dimethoxybenzene, for which the intercept is tenfold larger than for the other solvents, which probably indicates substantial quenching of one of the excited states of the dye in this solvent.

Discussion

The results show that the value of β for 2-methyl-2-pentene varies by at most a factor of four in the various solvents, since it would be an exceptional coincidence if both k_1 and k_2 were to vary by a large but exactly parallel amount in various solvents, and excited state lifetimes do not appear to vary much with solvent in any case (except when the solvent contains heavy atoms or when a charge-transfer interaction is possible). For example, Porter and Windsor observed only a small solvent effect on the lifetimes of triplet anthracene or naphthalene.[12] For these reasons, the variation in k_1 is not likely to be large, and thus the variation in k_2 is also probably not large. Because experimental error makes up a large proportion of the observed variation in β (although the differences between the extreme values are real), any rationalization of the variations is hazardous. However, the *lack* of correlation between the results and certain solvent properties is significant.

The most obvious property of solvent which might influence the rate would be the dielectric constant (or one of the other empirical measures of solvent polarity).[13] That no such correlation exists is immediately evident from Table I; for instance, the values of β in DMSO, acetone, CS_2, and benzene (D = 48.9, 20.7, 2.64, and 2.27) are all the same, within experimental error.

The rates of several "electroneutral" reactions (in which all participating species are neither ionic nor

(12) G. Porter and M. W. Windsor, *Proc. Roy. Soc., Ser. A*, **245**, 238 (1958); *Discuss. Faraday Soc.*, **17**, 178 (1954).
(13) See E. M. Kosower, "An Introduction to Physical Organic Chemistry," Wiley, New York, N. Y., 1968, for a discussion of various solvent polarity parameters.

highly polar) have been correlated with the cohesive energy density of the solvent (its "internal pressure").[14] However, pyridine, DMF, and methanol, which fall at the high end of the internal pressure scale, have β values which bracket those for benzene and acetone, which fall at the low end, so that no correlation with this parameter is discernible.

Livingston and Owens reported that chlorophyll-sensitized photooxygenation of allylthiourea (ATU) showed a marked dependence on solvent viscosity, with the value of β being 0.1 M in methanol ($\eta = 0.59$ cP), but ≤ 0.001 M in cyclohexanol ($\eta = 57$ cP).[9] If the reaction of singlet oxygen with an acceptor is diffusion controlled, an increase in solvent viscosity should cause a marked *decrease* in reaction rate (k_2), which should be inversely proportional to the viscosity.[12] Although some decrease of the decay rate (k_1) might also be expected (Porter and Windsor found an effect proportional to the $^1/_2$ and $^1/_3$th power of viscosity on the lifetimes of triplet anthracene and naphthalene, respectively[12]), the net effect should be an *increase* in β (k_1/k_2) with viscosity, the opposite of the reported effect. However, since the rate of quenching of singlet oxygen by β-carotene is 2×10^4 times larger than the rate of its reaction with 2-methyl-2-pentene,[13] it is obvious that the rate of reaction is far less than diffusion controlled with this olefin, and thus also with ATU which has a comparable β value in methanol.[9] Thus the only effect of viscosity to be expected would be a possible small decrease in the decay rate and thus in the β value; comparison of the β values for 2-methyl-2-pentene in CH_3OH (0.16 M) and cyclohexanol (0.07 M) shows that the effect in this case is indeed small, and may not exceed the experimental error. The reported effects on the chlorophyll-sensitized oxidation of ATU are in apparent contradiction to these results. However, several features of the Livingston–Owens paper are puzzling; they also report a kinetic term which would require a diffusion-controlled quenching of singlet oxygen by chlorophyll, but neither chlorophyl *a* nor *b* appears to quench singlet oxygen appreciably.[16] Furthermore, there was evidence that at least two oxidation mechanisms were competing in the chlorophyll-sensitized oxidation of ATU, since there was a temperature-dependent and a temperature-independent component.[9] More recent studies of sensitized oxidations of ATU have shown that several mechanisms are operative, and it is not certain that a singlet oxygen mechanism is one of them, although it seems likely.[17]

Several reports of dramatic solvent effects on self-sensitized oxygenations of various anthracene derivatives have appeared; several were carried out with kinetic analysis which would separate sensitizer and β-value effects.[8] The most remarkable effect is a decrease in β by a factor of 12–60 in going from aromatic hydrocarbon solvents to CS_2.[8] No such effect was observed in this study, values of β in benzene and CS_2 being virtually identical (0.10 and 0.08 M, respectively.[17a] The reason for the reported effects in the anthracene cases is not certain; CS_2 strongly quenches the fluorescence of anthracene,[2,8] and these reactions will bear reinvestigation.[18]

Heavy atom solvents often markedly increase the rate of spin interconversion by spin–orbit coupling; for example, the rate of intersystem crossing in naphthalene increases by a factor of 50 on going from EPA to propyl iodide (glass, 77°K).[19] However, Eisenthal and El-Sayed found that charge-transfer complex formation with heavy atom containing acceptors did not measurably affect the *non*radiative transition probability, although a large effect on the radiative transition probability of triplet naphthalene and phenanthrene was found.[20] Any heavy-atom effect on the lifetime of singlet oxygen would be expected to increase β by increasing k_1. However, neither bromobenzene ($\beta = 0.05$ M) nor ethyl iodide ($\beta = 0.11$ M) shows such an effect when compared with benzene ($\beta = 0.10$ M).

The effect of dimethoxybenzene is interesting. No significant quenching of singlet oxygen is observed ($\beta = 0.15$ M); however, the tenfold increase in intercept in this solvent is an indication of strong quenching of *sensitizer* excited states, probably *via* a charge-transfer mechanism.[21] Evidently singlet oxygen is not quenched efficiently by this mechanism, although this solvent was chosen to test for this effect. Quenching of sensitizer but not singlet oxygen has also been observed with dimethylaniline in benzene and with *p*-*N*,*N*-dimethylaminotrimethylstyrene.[16a,22]

The observed range of solvent effects of less than a factor of four is comparable to that reported by Huisgen and Pohl for the "ene reaction" between 1,3-diphenylpropene and diethyl azodicarboxylate, which was interpreted by them as evidence against a dipolar mechanism.[23] The observed effects certainly argue against any large charge separation or highly polar species in the transition state for photooxygenation of 2-methyl-2-pentene. The lack of solvent effects coupled with lack of directing effects of substituents[1] (which argues against *localized* charge or radical character) are most simply explained by a concerted mechanism. The six-center transition state **3** similar to that advanced for the ene reaction[23,24] and previously suggested for the photooxygenation on the basis of steric and conforma-

(14) A. P. Stefani, *J. Amer. Chem. Soc.*, **90**, 1694 (1968).

(15) C. S. Foote and R. W. Denny, *ibid.*, **90**, 6233 (1968).

(16) (a) C. S. Foote and R. W. Denny, unpublished results. (b) The upper limit for any quenching was at least 40 times lower than the rate of β-carotene quenching. However, some self-quenching of chlorophyll *a* triplet was found, and both chlorophyll *a* and *b* react slowly with singlet oxygen.

(17) C. S. Foote, *Science*, **162**, 963 (1968), M. Nemoto, Y. Usui, and M. Koizumi, *Bull. Chem. Soc. Jap.*, **40**, 1035 (1967); H. E. A. Kramer and A. Maute, *Ber. Bunsenges. Phys. Chem.*, **72**, 1092 (1968).

(17a) NOTE ADDED IN PROOF. Recent redeterminations have given a value of β in CS_2 of 0.02 M (average of seven independent determinations). A study of the diffusion-controlled quenching by β-carotene suggests that the lifetime of $^1O_2$ is prolonged by a factor of 10 in CS_2.

(18) The dinaphthalenethiophene-sensitized photooxygenation of anthracene shows an effect of CS_2 on β identical with that reported for the self-sensitized reaction: E. Peterson and C. S. Foote, unpublished results.

(19) S. P. McGlynn, M. J. Reynolds, G. W. Daigre, and N. D. Crystodoyleas, *J. Phys. Chem.*, **66**, 2499 (1962).

(20) K. B. Eisenthal and M. A. El-Sayed, *J. Chem. Phys.*, **66**, 2499 (1962).

(21) H. Leonhardt and A. Weller, *Z. Phys. Chem.*, (*Frankfurt am Main*), **29**, 277 (1961); H. Leonhardt and A. Weller, *Ber. Bunsenges. Phys. Chem.*, **67**, 791 (1963).

(22) However, DABCO and certain tertiary amines do quench singlet oxygen with moderate efficiency, in all probability by this mechanism: C. Oannés and T. Wilson, *J. Amer. Chem. Soc.*, **90**, 6528 (1968); C. S. Foote, R. W. Denny, L. Weaver, Y. Chang, and J. Peters, *Ann. N. Y. Acad. Sci.*, **171**, 139 (1970).

(23) R. Huisgen and H. Pohl, *Chem. Ber.*, **93**, 527 (1960).

(24) J. A. Berson, R. G. Wall, and H. O. Perlmutter, *J. Amer. Chem. Soc.*, **88**, 187 (1966); R. T. Arnold and J. F. Dowdall, *ibid.*, **70**, 2590 (1948); R. K. Hill and M. Rabinowitz, *ibid.*, **86**, 965 (1964).

tional considerations,[2,11,25] is entirely consistent with the results reported here and in the accompanying paper. However, a concerted 1,2 cycloaddition (four-center transition state **4**) recently advanced by Fenical, Radlick, and Kearns as an alternative to the ene reaction[26] is also consistent with the solvent and electronic effects reported here. However, a substituent effect on the direction of opening of dioxetane **5** might have been expected; this would have produced a variation in the product ratio for the trimethylstyrenes which was not observed.[1] Furthermore, several dioxetanes have now been prepared, and have been found to cleave to give carbonyl products.[27] In particular, that from 2-methyl-2-butene cleaves, and does not give the ene products,[27a] so that the dioxetane mechanism appears to be ruled out.

3

4   5

A variation on the dioxetane mechanism, originally put forward by Sharp[28] and, more recently, by Kopecky and Reich,[5] has recently received experimental support from the Kearns group.[29] Photooxygenations carried out in the presence of sodium azide produced azidohydroperoxides (**6**) instead of the normal ene product. It was suggested that **6** was produced by opening an intermediate perepoxide (**7**), which, if not trapped, rearranged to the normal ene product.

7

6

(25) A. Nickon and W. L. Mendelson, *J. Amer. Chem. Soc.*, **87**, 3921 (1965), and earlier papers; G. O. Schenck, K. Gollnick, G. Buchwald, S. Schroeter, and G. Ohloff, *Justus Liebigs Ann. Chem.*, **674**, 93 (1964).

(26) W. Fenical, D. R. Kearns, and P. Radlick, *J. Amer. Chem. Soc.*, **91**, 3396 (1969); D. R. Kearns, *ibid.*, **91**, 6554 (1969).

(27) (a) K. R. Kopecky and C. Mumford, *Can. J. Chem.*, **47**, 709 (1969); (b) P. D. Bartlett and A. P. Schaap, *J. Amer. Chem. Soc.*, **92**, 3223 (1970); (c) S. Mazur and C. S. Foote, *ibid.*, **92**, 3225 (1970).

(28) D. B. Sharp, Abstracts, 139th National Meeting of the American Chemical Society, New York, N. Y., Sept 1960, p 79p.

(29) W. Fenical, D. R. Kearns, and P. Radlick, *J. Amer. Chem. Soc.*, **91**, 7771 (1969).

While this mechanism cannot be ruled out, the following facts argue against it. (1) Polar solvents would have been expected to increase the rate of formation of the zwitterion **7** at least modestly, although this cannot be stated with certainty. (2) Substituents should have had a marked effect on the direction of opening of **7** to ene product, with electron donors facilitating cleavage of the C–O bond to which they were attached. (3) The azide trapping experiments appear to be more complex than originally believed. While our investigation of the azide reaction is not complete, there is a very large decrease in the overall reaction rate, apparently caused by $^1O_2$ quenching.[30] At a minimum, this result requires that the role of azide is more than simply a trap for perepoxide **7**; it is not ruled out that azidohydroperoxides are formed by a different mechanism entirely.

In conclusion, the results reported here and in the accompanying paper are most easily accommodated by the cyclic transition state **3**; this transition state is also consistent with the known stereochemical requirements[25] of the reaction, but other mechanisms cannot be rigorously excluded. The reaction behaves in all respects like the ene reaction.[23,24]

Experimental Section

General conditions are as in ref 1. Peak areas were measured by planimeter. The data were fit by a least-squares program method for estimation of a rectangular hyperbola[31] translated into PL-1 language by Mr. Phil Bernard.

β-**Value of 2-Methyl-2-pentene in Cyclohexanol.** A solution containing 1.4413 g (0.017 mol) of 2-methyl-2-pentene, 7 mg (7 × 10^{-6} mol) of Rose Bengal, and 99 ml of cyclohexanol was poured into the photolysis apparatus previously described and regulated at a temperature of 25°.[1] After flushing with oxygen, the solution was photolyzed for exactly 6 min with a 625-W Sylvania Sungun incandescent lamp set at 50 V and regulated by a Sola constant voltage transformer.

After irradiation, 1.2 ml of trimethyl phosphite was added quickly and the solution was stirred for 5 min in the dark, then left overnight in a dark refrigerator; 0.0512 g (5.8 × 10^{-4} mol) of isoamyl alcohol was added as internal standard, and the thawed solution was again stirred for 20 min in the dark. The solution was then injected at 105° onto a 6 ft × $^1/_8$ in. column containing 15% Ucon-2000 polar on 80–100 Chromosorb W. From an average of three chromatograms, a mole ratio of 1:2.19 was determined for 2-methyl-3-penten-2-ol and isoamyl alcohol. After multiplication by a factor of 1.96 to include the 2-methyl-1-penten-3-ol isomer, a value of 45.8 × 10^{-5} mol of product was obtained from 1.717 × 10^{-2} mol of starting material, indicating a total conversion of 2.67%.

The above procedure was repeated for 1.1002, 0.7787, 0.5850, and 0.3553 g of 2-methyl-2-pentene and 43.6 × 10^{-5} (3.33%), 37.4 × 10^{-5} (4.03%), 33.3 × 10^{-5} (4.78%), and 23.6 × 10^{-5} mol (5.58%) of product were obtained. A least-squares fit of the data[31] indicated a slope of 11.1 (±0.5), a correlation coefficient of 0.995, and an intercept of 148 M^{-1} (±7.1), which gave a β value of 0.07 M with a standard deviation of ±0.01 M. Runs in other solvents were carried out in a similar manner.

(30) C. S. Foote and Y. C. Chang, unpublished results.

(31) K. R. Hanson, E. A. Havir, and R. Ling, *Biochem. Biophys. Res. Commun.*, **29**, 194 (1967).

71

Copyright © 1971 by the American Chemical Society

Reprinted from *J. Amer. Chem. Soc.*, 93(22), 5774–5779 (1971)

Solvent Effects in Dye-Sensitized Photooxidation Reactions

Robert H. Young,* Kathy Wehrly, and Robert L. Martin[1]

Contribution from the Department of Chemistry, Georgetown University, Washington, D. C. 20007. Received November 9, 1970

Abstract: The problem of interpretation of solvent effects in photooxidation reactions has been due to the inability of researchers to distinguish between solvent effects on the rates of decay of singlet oxygen and the rate constants for the reactions of singlet oxygen with organic compounds. Up to this time only a ratio of these values (β value) was experimentally possible to obtain. This causes confusion as to whether differing solvents change the rate of decay, the rate constant for the reaction, or both. We have recently shown a relation between solvent polarity and Hammett ρ values for a series of solvents. This evidence suggests that there is a relationship between the rate constant for the reaction of organic compounds with singlet oxygen and the polarity of the solvent used. In an effort to clarify the situation we have developed a new method to determine β values (k_d/k_{rx}) and have adapted this to calculate the β value of a very reactive species, β-carotene (β in the order of 3–6×10^{-6} depending upon the solvent). Assuming that the rate of reaction (k_{rx}), or the rate of quenching of singlet oxygen by the β-carotene, is diffusion controlled, we have been able to obtain absolute rates of decay (k_d) of singlet oxygen in different solvents. If the β value (k_d/k_{rx}) is known for the reaction of a specific compound with singlet oxygen in a solvent where the rate of decay of the excited singlet oxygen species has been determined, then the absolute rate constant for the reaction can be calculated.

Solvent effects in photooxidation reactions have puzzled us and other researchers for some time. Foote has written that for the photooxidation of 2-methyl-2-pentane in a series of solvents that "no trend in the results (β values) with polarity is observed..." and "further no effect of viscosity is observed...."[2] Gollnick in his excellent review of photooxidation reactions more than once refers to the confusing effect of solvents on photooxidation reactions and the lifetimes of singlet oxygen in various solvents.[3] The reason in both cases is that the only measure of reactivity of organic compounds with singlet oxygen is the measurement of β values (rate of decay/rate of reaction or k_d/k_{rx}). The solvent may affect either or both the rate of reaction (k_{rx}) and the rate of decay of the singlet oxygen (k_d).

We have previously indicated that there does seem to be a regular solvent effect on the rate of photooxidation reactions as determined by changing Hammett ρ values in a series of solvents.[4] A good correlation was obtained for a plot of $1/\epsilon$ (ϵ is the dielectric constant for each solvent) *vs.* these Hammett ρ values.

To our knowledge this was the first case in which some solvent parameter was correlated with photooxidation reactions. However, the relative β values were not consistent with the polarity of the solvents. It appeared that the value of the $1/\beta$ (rel) was anomalous in glycol solvent. The reason may be that the rate of decay of singlet oxygen in the more viscous glycol solution was changed substantially from that of other solvents.

For these reasons we have expanded our studies on the effect of solvents on the decay rate of singlet oxygen and the rate constants for dye-sensitized photooxidation reactions.

Results and Discussion

The most currently acceptable scheme for dye-sensitized photooxidation reactions is outlined here

where D represents the dye-sensitizer (in our case rose bengal) and F represents any organic compound which reacts with singlet oxygen ($^1\Delta$, $^1O_2$).

The quantum yield for the disappearance of the organic species is given by

$$\Phi_{-F} = K\frac{k_7[F]}{k_7[F] + k_6}$$

or

$$\Phi_{-F} = K\frac{[F]}{[F] + \beta}$$

where

$$\beta = \frac{k_6}{k_7}$$

Usually this β value of the photooxidation reaction is determined by plotting $1/\Phi_{-F}$ *vs.* $1/[F]$ when the values of [F] and β are comparable.[2] The errors in this method have been estimated to be $\pm 30\%$.[2] Other methods also have been used.[5]

These β values in essence are an index of reactivity of particular organic compounds with singlet oxygen and as such are important in such fields as air pollution, oxidations of biologically related compounds, etc. For these reasons we have had an interest in developing other methods of measuring the reactivity of such organic compounds with singlet oxygen, especially for compounds where the standard methods are not applicable or accurate..

Method for Determination of β Values. Our method involves a quenching of the oxidation reaction of a standard compound (1,3-diphenylisobenzofuran (F)) by the organic compound for which a β value is to be

(1) Holder of an NDEA Fellowship, 1969–1971.
(2) C. S. Foote, *Accounts Chem. Res.*, **1**, 104 (1968).
(3) K. Gollnick, *Advan. Photochem.*, **6**, 1 (1968).
(4) R. H. Young, N. Chinh, and C. Mallon, *Ann. N. Y. Acad. Sci.*, **171**, 130 (1970).
(5) R. Higgins, C. S. Foote, and H. Cheng, *Advan. Chem. Ser.*, No. 77, 102 (1968).

Table I. β Values Obtained in the Photooxidation of Standard Compounds

Compound (X)	Exptl[a] β value	Solvent	Lit.[b] β value
Tetramethylethylene	0.0023	MeOH–*tert*-BuOH (50:50)	0.0027[c]
2-Methyl-2-pentene	0.15	MeOH	0.17[c]
			0.13[d]
	0.12	MeOH–*tert*-BuOH (50:50)	0.14[c]
	0.08	*tert*-BuOH	0.051[c]
cis-4-Methyl-2-pentene	8.0	MeOH–*tert*-BuOH (50:50)	10[c]
2,5-Dimethylfuran	0.00016	MeOH–*tert*-BuOH (50:50)	0.001[c]
	0.00028	MeOH	~0.0002[d]
2-Methylfuran	0.0011	MeOH	~0.0038[d]
2,5-Diphenylfuran	0.00095	MeOH	

[a] Slopes were calculated by a linear regression analysis with a least-squares fit. Errors in the β values from the Stern–Volmer slopes were calculated to be 5–10%. Reproducibility of β values is within the indicated error limits. [b] Errors ±30% or larger. [c] See ref 2. [d] See E. Koch, *Tetrahedron*, **29**, 6295 (1968).

measured. A slight variation of the previous kinetic scheme gives

$$^1D_0 \underset{k_2}{\overset{h\nu}{\rightleftharpoons}} {}^1D_1 \xrightarrow{k_3} {}^3D_1 \xrightarrow{[{}^3O_2]} {}^1D_0 + {}^1O_2 \begin{cases} \xrightarrow[k_7]{[F]} FO_2 \\ \xrightarrow[k_8]{[X]} XO_2 \\ \xrightarrow{k_6} {}^3O_2 \end{cases}$$

where X is any reactive organic compound. The kinetics of this scheme would be

$$-\frac{d[F]}{dt} = K_{[{}^1O_2]}\left(\frac{k_7[F]}{k_7[F] + k_8[X] + k_6}\right)$$

where $K_{[{}^1O_2]}$ would be a constant provided the light intensity, dye concentration, etc., are constant for all reactions. $K_{[{}^1O_2]}$ is the rate of formation of singlet oxygen.

We use a very low concentration of [F] (about 10^{-6} M) in order to get simplified first-order kinetics. Thus

$$-\frac{d[F]}{dt} = K_{[{}^1O_2]}\left(\frac{k_7[F]}{k_8[X] + k_6}\right)$$

where the slope of the first-order plots will be

$$S = K_{[{}^1O_2]}\left(\frac{k_7}{k_8[X] + k_6}\right)$$

A typical Stern–Volmer relationship results in

$$S_0/S_x = 1 + (k_8/k_6)[X]$$

and a Stern–Volmer plot of S_0/S_x *vs.* [X] results in a slope of $k_8/k_6 = 1/\beta$.

The standard compound, F, used was 1,3-diphenylisobenzofuran

F

and the organic compounds used to check the method were tetramethylethylene, 2,5-dimethylfuran, 2-methyl-2-pentene, and *cis*-4-methyl-2-pentene. It should be noted that in order for this method to work a substantial fraction of the 1,3-diphenylisobenzofuran, F, must react with little change in the concentration of compound X. This can be easily confirmed by obtaining linear first-order plots. The standard compound, F, is one of the most reactive with singlet oxygen, with a reported β value of about 10^{-4}. Thus it is ideal for such measurements. The method involves following the disappearance of F by measuring changes in the fluorescence intensity at λ_{max} (excitation 405 nm, emission 458 nm).

Good first-order plots were obtained and the relative slopes used directly in the Stern–Volmer relationship to obtain β values. The β values for a series of compounds are reported in Table I.

The value for dimethylfuran (DMF) (2×10^{-4}) is lower than that reported by Foote[2] (1×10^{-3}). However, Koch[6] reported a value of 2×10^{-4} and the ratio of rate constants of reaction of singlet oxygen with tetramethylethylene (TME)–dimethylfuran (DMF) obtained by Ogryzlo[7] was 1:3.4, by Herron[8] was 1:17, and by Timmons[9] was 1:6 all for gas-phase results, compared to our solution ratio of 1:12–15. Certainly the ratio of these rate constants does vary considerably. It should be noted that DMF is not pure when obtained commercially and it does decompose and should be purified immediately prior to use in order to get good results.

It is conceivable that as we are really measuring the ability of these compounds to quench singlet oxygen and we are assuming that this is *via* reaction, we may be observing some quenching other than reaction quenching. This appears to be unlikely as Timmons reports that the chemical reactions of TME and DMF are much greater than any collisional deactivation process.[9]

The other β values of the photooxidation reactions of singlet oxygen with the various alkenes show excellent correlation with those reported in the literature; thus we feel the method is reasonably reliable and efficient. The method is somewhat similar to that used by Wilson.[10] However, in her case only relative β values could be obtained from the quenching reactions, much like the results of Foote and others.[2,5] In our case, due to the unique conditions of low concentration of reactant (1,3-diphenylisobenzofuran, $\approx 10^{-6}$ M) and method of analysis (fluorescence), we were able to obtain absolute β values directly from the slope of the

(6) See Table I, footnote *d*.
(7) K. Furukawa, E. W. Gray, and E. A. Ogryzlo, *Ann. N. Y. Acad. Sci.*, **171**, 175 (1970).
(8) J. T. Herron and R. E. Huie, *ibid.*, **171**, 229 (1970); *J. Chem. Phys.*, **51**, 4164 (1969).
(9) G. A. Hollinder and R. B. Timmons, *J. Amer. Chem. Soc.*, **92**, 4181 (1970).
(10) T. Wilson *ibid.*, **88**, 2898 (1966).

Stern–Volmer quenching plots. It would appear that if there is no interference in the absorption or fluorescence regions of the standard compound that it can be used to give accurate β values in a minimum of time. It probably is not applicable to some of the polynuclear aromatic compounds. However, these can be related rather easily and accurately to known compounds.[4]

Solvent Effect on Decay Rates of Singlet Oxygen ($^1\Delta_g$). Our initial studies to explore the effects of solvents on the lifetime of singlet oxygen involved the use of a solvent pair (methanol–glycol). Because of the similarity of their dielectrics (methanol 32.6 and glycol 37.7) this solvent system should show the effects of viscosity only. We used solutions of 1,3-diphenyl-isobenzofuran at low concentrations and obtained relative β values by our normal technique. An absolute β value was determined by the method described in the previous section for methanol solvent only.

Viscosities of solvents were calculated by the normal method at 28°.[11] The viscosity values and relative β values obtained are given in Table II.

Table II. Solvent Effects on the β Value of 2,5-Diphenylfuran

Solvent Methanol–glycol, %[a]	Viscosity[b]	$1/\beta$(rel)[c–f]
100	0.538	1.00
75	1.15	1.73
50	2.42	2.82
25	6.17	3.60
0	15.6	2.73

[a] Volume per cent methanol. [b] Viscosity measurements at 28° using an adaptation of an Ostwald type viscometer. [c] No cut-off or interference filter used, rose bengal used as the dye sensitizer. [d] Values for $1/\beta$(rel) were calculated from the absolute β value of 2,5-diphenylfuran in methanol (see ref 4 for method). [e] Obtained from relative first-order plots using a low concentration of the diphenylfuran, followed by changes in fluorescence intensity (excitation 311 nm, emission 374 nm). [f] Includes small correction for changes in absorption of light by the rose bengal.

It can be seen from Table II that there is a regular decrease in β values, as expected for a decrease in the rate of decay of singlet oxygen with increasing viscosity of solvents. The slight anomalous result for 100% glycol could be due to either an effect on the β value through a significant change in the rate of reaction (k_{rx}) caused by polarity effects or to a specific solvent effect on either k_d or k_{rx} by the glycol.

In an effort to increase the understanding of these results further studies were carried out. A variation of the new method described above was used to obtain β values for a very "reactive" species, β-carotene. It has been suggested that β-carotene quenches singlet oxygen without chemical reaction; i.e., the reactivity with singlet oxygen is due to quenching only.[12,13]

Previously we said that using this method for determination of β values the standard must react faster (have a lower β value) than the compound being mea-

sured; i.e., 1,3-diphenylisobenzofuran is the best choice. However, as β-carotene quenches without reaction, this requirement is no longer necessary. Our choice of 2,5-diphenylfuran as the new standard reagent was based on the necessity of having a compound whose excitation and emission wavelengths are not obscured by the absorption of β-carotene. 2,5-Diphenylfuran is a good choice for this reason.[14]

The β value obtained for β-carotene from the slope of the Stern–Volmer quenching plot is equal to the rate of decay (k_8 or k_d) divided by the rate of quenching (k_8 or k_q) of the singlet oxygen by the carotene.

$$1/\text{slope} = \beta = k_d/k_8$$

Foote has previously used β-carotene to obtain an approximate decay rate for singlet oxygen in benzene–methanol (4:1).[12] He also suggested that the rate of quenching by β-carotene could be diffusion controlled.[13] His elaborate method, however, resulted in rather large errors in the results. We feel our method is experimentally easier, quicker, and more versatile.

If the rate of quenching of singlet oxygen by β-carotene is diffusion controlled then the rate constant for diffusion in the solvents we wish to use is necessary. As the Stokes–Einstein equation has been shown to be in error when one of the diffusing species is small, we used the data reported by Ware to obtain approximate rates of diffusion.[15] Ware obtained the rates of diffusion of oxygen in various solvents (including a number of alcohols of varying viscosity) via quenching of fluorescence of some aromatic molecules. These rates were later shown to be quite good by other workers.[16] All other factors being equal (solvent interactions with the diffusing molecules as well as size) it is probable that the rate of diffusion of singlet oxygen may well be the same as ground-state oxygen. If this is true then Ware's data can be used to obtain rates of diffusion of singlet oxygen and hence rates of decay from the β value of the quenching process of singlet oxygen by β-carotene.

Although methanol was the only solvent we used in which the rate of diffusion of oxygen had been measured, Ware did find a qualitative correlation between the reciprocal viscosity values for a series of solvents (11) and the rate of oxygen diffusion. Assuming a straight line for the viscosity range used ($k_d = 1.81 \times 10^{10} + 0.638 \times 10^{10} \times \eta^{-1}$; $r = 0.93$) we obtained estimated rates of diffusion of oxygen in our solvents and these are recorded in Table III.

Although there are problems with the determination of rate constants for diffusion-controlled reactions[17] these diffusion rates used for singlet oxygen are probably the best possible at this time.[18]

Also given in Table III are the β values (k_d/k_q) for β-carotene and the calculated rate of decay assuming diffusion rate control for the rates of quenching of singlet oxygen.

(14) We have recently been using rubene (filters must be used) with methylene blue to obtain similar results: R. H. Young and D. Feriozi, unpublished research, Georgetown University.

(15) W. R. Ware, *J. Phys. Chem.*, **66**, 455 (1962).

(16) (a) B. Stevens and B. E. Algar, *ibid.*, **72**, 2582 (1968); (b) L. K. Patterson, G. Porter, and M. R. Topp, *Chem. Phys., Lett.*, **7**, 612 (1970).

(17) P. J. Wagner, Symposium on the Frontiers in Photochemistry, New York, N. Y., Oct 1970.

(18) We wish to thank one of the referees for pointing out Ware's results to us.

(11) A. C. Merrington, "Viscometry," Edward Arnold & Co., London, 1949, p 20.

(12) C. S. Foote and R. W. Denny, *J. Amer. Chem. Soc.*, **90**, 6233 (1968).

(13) C. S. Foote, Y. C. Chang, and R. W. Denny, *ibid.*, **92**, 5216, 5218 (1970).

350

Table III. Lifetimes of Singlet Oxygen in Various Solvents

Solvent	k_q (rate of diffusion)[a] $\times 10^{-10}$	$1/\eta$ (viscosity) (cP, 25°)	$\beta(k_d/k_q)^b \times 10^6$	$k_d \times 10^{-4}$	τ, sec, $^1O_2$
Methanol[c]	2.98	1.83[d]	6.1	18	0.55×10^{-5}
n-Butyl alcohol	2.06	0.385[d]	5.5	11	0.91×10^{-5}
tert-Butyl alcohol	1.95	0.226[f]	3.8	7.4	1.35×10^{-5}
Benzene–methanol (4:1)	2.94	1.78[e]	3.9	12	0.83×10^{-5}

[a] From our calculations based on Ware's results, ref 15. [b] No autoxidation reactions were observed. [c] Corrected for slight sensitization of the reaction by the β-carotene. Corrections were made for each reaction for each concentration of the carotene used. This was a very small correction. [d] "Handbook of Chemistry and Physics," 46th ed, Chemical Rubber Publishing Co., Cleveland, Ohio, 1965, p F33. Sometimes interpolated for the temperature given. [e] Determined in our laboratories. [f] J. Timmermanns, "Physico-Chemical Constants of Binary Systems in Concentrated Solutions," Vol. 4, Interscience, New York, N. Y., 1960, p 236.

There are a number of interesting facts which are evident from the results given in Table III. It can be seen that the rate of decay (or the lifetime of singlet oxygen, $^1\Delta_g$) is related to the viscosity of the solvent from methanol (0.547 cP) to tert-butyl alcohol (4.43 cP). The increased stabilization of the excited singlet state of oxygen in the more viscous solvent is quite reasonable and perhaps expected. These may be compared to changes in triplet state lifetimes of organic compounds. This also correlates well with the direction of the effect we obtained from the use of methanol–glycol solvent pairs.

It can also be seen from Table III that the lifetime of singlet oxygen in benzene–methanol of 0.8×10^{-5} compares favorably with that reported by Foote of approximately 1×10^{-5} for the same solvent.[19] Our value differs substantially from that obtained by Koch in methanol for K_d (5×10^6 sec^{-1}).[6]

One of the intriguing factors in the results is the increased lifetime of singlet oxygen in the benzene–methanol solution over that of pure methanol, although they both have almost the same viscosity. Explanations include differences in rate of diffusion from that calculated due to a specific interaction of singlet oxygen with the solvent which does not occur with ground-state oxygen, a deactivating effect of a hydroxyl group, or perhaps the phenyl ring stabilizes singlet oxygen by formation of a weak excimer. Recent preliminary results indicate that it may also be due to a larger decay rate in the more polar solvent.[20]

We have shown that there is a change in the lifetime of singlet oxygen, $^1\Delta_g$, in different solvents and that this change is related to the viscosity of the solvent. The difference in the rate constant for decay in various solvents is sufficiently large to explain the problems in obtaining meaningful results from solvent effects on β values (k_d/k_{rx}). Knowing the absolute rate of decay of singlet oxygen in any specific solvent it is now possible to obtain absolute rate constants for reactions of singlet oxygen with organic compounds in photooxidation reactions. This was done for a series of compounds.

Solvent Effects on the Rate Constant for Dye-Sensitized Photooxidation Reactions. Absolute β values for some compounds were obtained by the method outlined above. Relative β values were found experimentally by the method outlined in a previous publication.[4] Two methods were employed. Limited

information was gained by the use of an appropriate solvent pair, methanol–water. This solvent pair was originally chosen because both solvents have about the same viscosity. If this were the only factor affecting the rate of decay of singlet oxygen then the relative β values should reflect only the change in the rate constant for the reaction due to changes in the polarity of the solvent. The results are given in Table IV.

Table IV. Effects of Solvent Polarity on the Photooxidation of 2,5-Diphenylfuran

Methanol in water[a]	$1/\beta^c$ (rel)	Dielectric (ϵ)[b]
100	1.00	32.6
81.37	1.24	36.8
65.79	1.61	41.6
39.79	2.24	53.1
11.5	2.28	69.9
0	2.08	78.5

[a] Mole per cent methanol. [b] Values from J. Timmerman, "Physico-Chemical Constants of Binary Systems in Concentrated Solutions," Vol. 4, Interscience, New York, N. Y., 1960, p 236. [c] β value for the 2,5-diphenylfuran in 100% methane was found to be 9.53×10^{-4}. The others were obtained from relative pseudo-first-order rate constants compared to this value. See ref 4 for method.

A linear correlation between log $1/\beta$(rel) values and $1/\epsilon$ for the photooxidation of 2,5-diphenylfuran in methanol–water was obtained. A specific solvent effect on either k_1 or k_d may be causing a slight problem at high water concentrations.

In our effort to further explore the solvent effects on the rate constant for photooxidation reactions, β values for a series of furans in a series of solvents were obtained. Knowing the rates of decay of singlet oxygen for these solvents it was possible to obtain absolute rate constants. The results are recorded in Table V.

A number of interesting observations can be made from the results recorded in Table II. There is an excellent relationship between increasing polarity of solvent and increasing rates of reaction for all compounds. Also, as expected, the faster the rate of reaction (the larger the rate constant) of the organic species with singlet oxygen, the smaller is the solvent effect. The solvent effect changes from a relatively small ratio of 1:1.7 for the 1,3-diphenylisobenzofuran to 1:32 for the p-chlorophenylfuran (tert-butyl alcohol–methanol).

A good correlation between log k_{rx}(rel) in tert-butyl alcohol, n-butyl alcohol, and methanol and the relative rates in methanol–water mixtures vs. $1/\epsilon$ for all these solvents was obtained for 2,5-diphenylfuran. This gives a polarity range of the solvents from a dielectric

(19) C. S. Foote, R. W. Denny, L. Weaver, Y. Chang, and J. Peters, Ann. N. Y. Acad. Sci., 171, 139 (1970).

(20) R. H. Young, R. L. Martin, K. Wehrly, and D. Feriozi, Amer. Chem. Soc., Div. Petrol. Chem., Prepr., 16, A89 (1971); presented at the 162nd National Meeting of the American Chemical Society, Washington, D. C., Sept 1971.

Table V. β Values, Rate Constants, and Relative Rate Constants for a Series of Compounds in a Series of Solvents

Compound	β values $\times 10^4 (k_d/k_{rx}, 25°)$			$k_{rx} \times 10^{-8}$ [a]			k_{rx} (rel)[b,c]		
	MeOH	n-BuOH	tert-BuOH	MeOH	n-BuOH	tert-BuOH	MeOH	n-BuOH	tert-BuOH
2-(p-Chloro-phenyl)furan	27	34	350	0.67	0.32	0.021	32	15	1.00
2-Phenylfuran	17	20	34	1.1	0.55	0.22	5.0	2.5	1.00
2,5-Diphenylfuran	(9.5)[d,e]	9.8	8.0	1.9	1.1	0.93	2.0	1.2	1.00
1,3-Diphenyliso-benzofuran	0.73	0.71	0.49	25	15	15	1.7	1.0	1.00
2-Methyl-2-pentene[e]	1500		810	0.012		0.091	1.3		1.00
Tetramethyl-ethylene[e,f]	46	44	29	0.39	0.25	0.25	1.6	1.0	1.00
2-Methylfuran[e,f]	11	12	6.9	1.6	0.92	1.1	1.5	0.84	1.00

[a] $k_{rx} = k_d/\beta$; k_d (25°) in methanol = 18×10^4; n-butyl alcohol = 11×10^4; tert-butyl alcohol = 7.4×10^4 sec^{-1}. [b] Each set relative to its own rate constant in tert-butyl alcohol. [c] Dielectric of methanol = 32.6; n-butyl alcohol = 17.8; tert-butyl alcohol = 10.9. [d] Determined as an absolute β value. Others for furans are relative to this value as a standard. [e] β value determined by our standard method. [f] Carried out at 23° using a different apparatus, but same method as for e.

of 10 to 80. These results agree with that expected from our previous results where we obtained a correlation between Hammett ρ values and 1/ε for a series of solvents.[4]

For the arylfurans there does appear to be a substantial solvent effect which is related to the rate constants for their reactions with singlet oxygen. However, for the "ene" reactions there is very little if any solvent effect regardless of the rates of their reaction. The same is true for one of the simple furans, 2-methylfuran. It is conceivable that the aryl group as a substituent is affecting the symmetry of the approach of the singlet oxygen moiety either by steric interaction or more probably by electronic effects. There continues to be published evidence for concerted Diels-Alder reactions.[21] However, there is some recent work which indicates that in some cases there may be an asymmetric approach of the dienophile on the diene.[22] This would be consistent with our findings on the reactions of furans with singlet oxygen.

Conclusion

A new method for the determination of β values for dye-sensitized photooxidation reactions was developed and confirmed experimentally. This method was applied to determine β values for β-carotene which has been proposed to react (by quenching) with singlet oxygen at the diffusion rate. Making this assumption we obtained absolute rate constants for the decay of singlet oxygen in a variety of solvents. Absolute rate constants for reaction of organic compounds with singlet oxygen can then be calculated with a knowledge of their decay rates and experimental β values.

Experimental Section

Materials. All chemicals were checked for purity by either melting point or glpc prior to use. In most cases the solvents were used directly and where possible spectrograde solvents were used. All solvents were checked for interference of absorption and/or fluorescence and found to be adequate.

Diphenylfuran was obtained from Dr. A. Trozolo of Bell Laboratories which we acknowledge with thanks. The 2-(p-chlorophenyl)furan and the 2-phenylfuran were prepared by the method used by Ayres.[23]

(21) R. A. Grieger and C. A. Eckert, *J. Amer. Chem. Soc.*, **92**, 4149 (1970).
(22) (a) Y. Kobuke, T. Fueno, and J. Furukawa, *ibid.*, **92**, 6548 (1970); (b) M. J. S. Dewar and R. S. Pyron, *ibid.*, **92**, 3099 (1970).
(23) D. C. Ayres and J. R. Smith, *Chem. Commun.*, 886 (1967); D. C. Ayres, private communications.

Kinetic Studies. Kinetic studies were carried out with a SPF-125 spectrophotofluorometer and a Sargent log linear recorder for direct recording of the first-order plots. The experimental conditions for any reaction series were kept constant, that is, the same light intensity, the same concentration of the dye sensitizer (rose bengal) in the same reaction apparatus. A GE 100-W tungsten lamp was used and the light reflected directly into the constant temperature cuvette chamber of the spectrophotofluorometer at right angles to the excitation and emission light paths. The light was absorbed through 1 cm of solution and was less than 30% absorbed in order to minimize mixing problems. In all cases standard reactions were carried out and these indicated no dark, reversible, or autoxidation occurring.

Determination of β Values. One-milliliter solutions of 10^{-6} M 1,3-diphenylisobenzofuran (DPBF) were placed in a 1-cm spectrofluorometric cell. Five microliters of a 0.72 mg/ml methanol solution of rose bengal was added. The rate of disappearance of the DPBF was followed by monitoring the decrease in fluorescence intensity of the DPBF throughout the course of the photooxidation reaction (excitation, 405 nm; emission, 458 nm). Irradiation for the photooxidation was with a Tungsten lamp and a cut-off filter was used which passed only light above 550 nm. In order to determine the β values, various concentration of each compound were added to the DPBF-rose bengal solutions. It was found that the best results were obtained when these compounds were added in a concentration such that the photooxidation of the DPBF was quenched by no more than 75%. Good first-order plots were obtained for all reactions. It is important that the concentration of the compound whose β value is being measured does not change much throughout the course of the photooxidation of the DPBF. If such a change in concentration occurred, the kinetic results would be invalid. Curvature in first-order plots would be suggestive of a change in the quencher concentration. However, straight lines were obtained in all cases due to the extreme reactivity of the standard, DPBF.

Stern-Volmer quenching plots were obtained for each compound in each solvent. An inverse of the slope of these plots gave the β value directly. The results are recorded in Table I. The β values for the standard compounds, tetramethylethylene and 2,5-dimethylfuran, were reproducible within the error limits reported as determined by experiments carried out independently by two different workers. The Stern-Volmer slopes were calculated by linear regression analysis with a least-squares fit. Errors in the β values were calculated to be 5-10% as determined from the error in the slope of the line.

Use of Solvent Pairs to Study Solvent Effects in Photooxidation Reactions. A. Methanol-Glycol (Viscosity Effects). Methanol-ethylene glycol solution mixtures with 100, 75, 50, 25, and 0% methanol by volume were used. Viscosity measurements at 28° were determined by using an adaption of an Ostwald type viscometer. They were determined by using the formula

$$\eta' = \frac{\eta \rho' t^1}{\rho t}$$

where η' is the viscosity of the unknown liquid; η the viscosity of the standard (methanol); ρ', ρ and t', t are the density and flow time for the unknown and methanol. The results are recorded in Table II.

352

Diphenylfuran (DPF) was used as a standard. A $10^{-4} M$ (10 μl) solution of DPF in methanol and 5 μl of a 0.72 mg/ml solution of rose bengal in methanol were added to 1.0 ml of each of the solvent mixtures. No light filter was used in the apparatus and the disappearance of the DPF was followed by a decrease in the fluorescence intensity (excitation, 311 nm; emission, 374 nm).

Corrections for changes in the rate of singlet oxygen produced due to solvent changes in the absorption of light by the rose bengal sensitizer were determined by measurement of the amount of light transmitted through the reaction solution in the region of rose bengal absorption. This was integrated over the absorption wavelength for the dye. Changes in fluorescence of the rose bengal due to solvent changes were experimentally determined to have little effect on the rate of reaction.

Relative rates of the dye-sensitized photooxidation of the diphenylfuran in different solvent mixtures, corrected for changes in the light absorption by the sensitizer, are recorded in Table II.

B. Methanol–Water (Polarity Effects). A series of methanol–water solvent mixtures was prepared as indicated in Table IV. The same procedure as above was used and corrections for changes in the absorption of light by the rose bengal were made although they were small. The corrected relative rates and dielectric constants for each solvent mixture are recorded as well.

Solvent Effects on the Rate of Singlet Oxygen Decay. Quenching of the photooxidation reaction of diphenylfuran (DPF) by β-carotene was carried out to determine β values of the quencher in different solvents. Fresh (benzene) solutions of carotene at 10^{-3} and $10^{-4} M$ were prepared daily. From 0 to 20 μl of these solutions was added to $10^{-6} M$ diphenylfuran and 2 μl of the previously mentioned rose bengal solution in 1 ml of some organic solvent. The final carotene concentrations were irradiated through an orange cut-off filter (550 nm) or a Wrattan 8 filter (470 nm) depending upon the reaction series. Since carotene absorbs slightly in the region of the filter cut-offs it was necessary to check each solvent at each concentration of carotene for possible sensitization. Slight corrections were necessary only in methanol. Decrease in the fluorescence intensity of the DPF as before was followed at 311-nm excitation and 374-nm emission. In the pure aromatic solvents the rose bengal solubility was questionable. For this reason 20% by volume of methanol in benzene was used. Various concentrations of β-carotene were used to quench the photooxidation reaction of singlet oxygen with the DPF. The relative slopes of good first-order rate plots were used to obtain Stern–Volmer

quenching plots. The slope of these Stern–Volmer plots gave $1/\beta$ values. In all cases at least five points were obtained. The values of β obtained in this manner are recorded in Table III. Errors are in the order of 10% or less as determined by a least-squares fit of the Stern–Volmer plot.

In the solvents used aggregation of the carotene probably did not occur as shown by a linear Beer's plot over the concentrations of β-carotene used.

Solvent Effects on the Rate Constant for Reaction of Singlet Oxygen with an Organic Compound. The rate of photooxidation reactions at low concentrations is given by the following equation

$$\frac{d[x]}{dt} = K_{{}^1O_2} \frac{k_{rx}[X]}{k_d}$$

where $K_{{}^1O_2}$ is the rate of formation of singlet oxygen, and k_{rx} is the rate constant for the decay of singlet oxygen. As $k_d/k_{rx} = \beta$-(rel), β values can be obtained for different compounds from the slope of first-order plots if $K_{{}^1O_2}$ remains constant. This technique was used to obtain relative β values for a series of compounds at very low concentrations in different solvents. First-order plots were obtained for all compounds. The reactions were followed by monitoring the change in the fluorescence intensity for all compounds. β values for these compounds are obtained by relating the relative values with an absolute value obtained previously for diphenylfuran. Absolute rate constants are then obtained by dividing these β values into the rate constant for the decay of singlet oxygen for each solvent. The data are recorded in Table V.

β values for 2-methyl-2-pentene, tetramethylethylene, and 2-methylfuran were carried out using the new method, outlined above. In the latter two cases methylene blue was used as the dye sensitizer and a filter fluorimeter with a Heathkit recorder was used to monitor the rate of disappearance of the diphenyliso-benzofuran in the quenching reactions.

Acknowledgments. We wish to thank the American Instrument Company (Silver Spring, Maryland) for their generosity in the loan of a SPF-125 spectrophotofluorometer without which this research could not have been carried out. We would also like to thank the Research Corporation for partial financial support.

72

Reprinted from *Chem. Phys. Lett.*, **12**(1), 120-122 (1971)

DIRECT MEASUREMENT OF THE LIFETIME OF ¹Δ OXYGEN IN SOLUTION

P.B. MERKEL and D.R. KEARNS

Department of Chemistry, University of California, Riverside, California 92502, USA

Received 17 September 1971

Direct measurement by a laser pulse technique provides a value for the lifetime of ¹Δ molecular oxygen in methanol of $\tau = 7 \pm 1$ μsec. Absolute reaction rate constants of $k_A = 7 \pm 1 \times 10^8$ M^{-1} sec^{-1} and $4 \pm 1 \times 10^7$ M^{-1} sec^{-1} are obtained for the singlet oxygen acceptors, 1,3-diphenylisobenzofuran and tetramethylethylene respectively. Additional acceptor rate constants may be calculated by reference to published β values.

Much of the current interest in the properties of singlet oxygen is due to the discovery that electronically excited singlet oxygen molecules, produced by energy transfer from triplet state sensitizers to oxygen, are the reactive intermediates in dye sensitized photo-oxygenation of unsaturated hydrocarbons [1–4]. Although *both* ¹Σ and ¹Δ oxygen are generated by the quenching of triplet state sensitizers [5, 6], only ¹Δ is important for reactions in solution, due to the rapid decay of ¹Σ (estimated lifetime $\lesssim 10^{-10}$ sec) [7].

Under usual photo-oxygenation conditions the two important competitive decay modes for ¹Δ are

$$^1\Delta \xrightarrow{1/\tau} {}^3\Sigma, \tag{1}$$

$$^1\Delta + A \xrightarrow{k_A} AO_2, \tag{2}$$

where A denotes a chemical acceptor and AO_2 a peroxide product. Since the lifetime of ¹Δ in solution was unknown, reactivities have usually been reported in terms of $\beta = (1/\tau k_A)$. Recently, Foote and co-workers [8] observed that β-carotene was a highly efficient inhibitor of singlet oxygen reactions and concluded that this was due to quenching of ¹Δ. By assuming that the quenching was diffusion controlled they estimated that the lifetime of ¹Δ is about ·10⁻⁵ sec or larger in a benzene–methanol solution. We wish to report here a direct measurement of the lifetime of singlet oxygen in solution using kinetic spectroscopy.

A 1J, 20 nsec pulse of 694 nm photons from a ruby laser was used to excite the sensitizer methylene blue in an oxygenated methanol solution. The methylene blue triplets thus formed are quenched by ground state oxygen to produce approximately 5×10^{-5} M ¹Δ in less than 10^{-7} sec *. As a probe of the singlet oxygen concentration the colored ($\lambda_{max} = 410$ nm) acceptor 1,3-diphenylisobenzofuran ($\beta \approx 10^{-4}$) [9] was added to the solution, and the rate of decrease in absorption accompanying peroxide formation was monitored synchronously with the laser pulse. An acceptor with a very low β value is required since the maximum fractional bleaching can be shown to equal $[^1\Delta]_0/\beta$, $[^1\Delta]_0$ being the concentration of singlet oxygen produced by the pulse.

Under conditions where [A] does not change substantially during the experiment, solution of the kinetic equations describing processes (1) and (2) yields

$$[AO_2]_\infty - [AO_2] =$$

$$\frac{k_A [A] [^1O_2]_0}{(1/\tau + k_A [A])} \exp \{-(1/\tau + k_A [A])t\},$$

where $[AO_2]_\infty$ is the concentration of peroxide after all singlet oxygen has decayed. From this it follows

---

* The lifetime of triplet methylene blue in deoxygenated methanol is 7 μsec and the second order rate constant for quenching of the triplet by oxygen is 2.2×10^9 M^{-1} sec^{-1}.

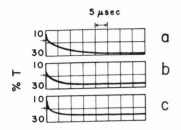

Fig. 1. Bleaching in an oxygenated solution of methylene blue (10^{-4} M) in methanol of (a) 2×10^{-5} M, 1,3-diphenyliso-benzofuran monitored at 410 nm in a 1 cm cell, (b) 1.5×10^{-4} M 1,3-diphenylisobenzofuran monitored at 435 nm in a 2 mm cell and (c) as (a) but with 5×10^{-3} M tetramethyl-ethylene added. Arrows indicate % T before flash.

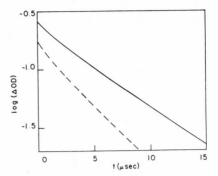

Fig. 2. First order decay plots of (a) ——— and (b) - - - - - of fig. 1.

that $\log \Delta OD = -2.3(1/\tau + k_A [A])\, t + \text{const.}$, where ΔOD is the optical density of the acceptor at time t minus the asymptotic value. Plots of ΔOD versus t at two acceptor concentrations provide values for both τ and k_A.

Photobleaching of 1,3-diphenylisobenzofuran under three sets of conditions is illustrated in fig. 1. In (a) a low acceptor concentration (2×10^{-5} M) results in a decay rate limited primarily by τ. The initial short lived increase in absorption arises from the methylene blue triplet—triplet absorption. In a deoxygenated solution no bleaching is observed and the methylene blue triplet decays at the characteristic rate. This eliminates the possibility of a direct reaction between the acceptor and triplet methylene blue. Increasing the acceptor concentration (1.5×10^{-4} M) predictably accelerates the decay rate by consumption of a signifi-cant portion of $^1\Delta$ as is demonstrated in (b). The ef-fect of adding a second oxidizable acceptor is depicted in (c). Addition of a sufficient amount of tetramethyl-ethylene (5×10^{-4} M) results in a decrease in both the duration and yield of 1,3-diphenylisobenzofuran photo-oxidation in accordance with expectations [compare amounts of bleaching in (a) and (c)].

First order plots of the decay curves in figs. 1(a) and (b) appear in fig. 2. The slight curvature reflects the consumption of acceptor as photo-oxidation pro-gresses. From the slopes we obtain values of $\tau = 7 \pm 1$ μsec and $k_A = 7 \pm 1 \times 10^8$ M^{-1} sec^{-1} *. This yields

* Foote, in ref. [8], quotes unpublished results of F. Wilkin-son which gave a lifetime of approximately 10 μsec.

a β value of 2×10^{-4} for 1,3-diphenylisobenzofuran reasonably close to that observed in 1,1,2-trichloro-trifluorethane [9]. The absolute reaction rate constant for tetramethylethylene (TME) is calculated to be $k_A = 4 \pm 1 \times 10^7$ M^{-1} sec^{-1}, β being in agreement with the previously determined value [1]. The absolute values of the rate constants for 1,3-diphenylisobenzo-furan and TME can now be used with tabulated β values to obtain absolute rate constants for other ac-ceptors.

In the present experiments where the transient singlet oxygen concentration is considerably greater than in steady state photo-oxidations, $^1\Delta - ^1\Delta$ anni-hilation might possibly affect the observed decay. This is ruled out by the observation that a reduction of $[^1\Delta]$ by a factor of two, produced by decreasing the methylene blue concentration, produces no change in the singlet oxygen lifetime. Since air saturated and oxygen saturated solutions give similar lifetimes, quenching of $^1\Delta$ by $^3\Sigma$ is also ruled out as a competi-tive decay mode.

In retrospect the agreement between the $^1\Delta$ life-time measured directly and that estimated by Foote substantiates the proposal that the quenching of $^1\Delta$ by β-carotene is diffusion controlled [8]. Lifetime measurements in a variety of other solvents are current-ly being conducted and this information, along with additional quenching rate constants, will appear in a subsequent more detailed publication.

The support of the U.S. Public Health Service (Grant No. CA 11459) is most gratefully acknowledged.

We are indebted to Dr. Robert Nilsson for many useful discussions.

References

[1] C.S.Foote, Accounts Chem. Res. 1 (1968) 104.

[2] K.Gollnick, Advan. Chem. Ser. 77 (1968) 78.

[3] C.S.Foote and S.Wexler, J. Am. Chem. Soc. 86 (1964) 3879.

[4] E.J.Corey and W.C.Taylor, J. Am. Chem. Soc. 86 (1964) 3881.

[5] K.Kawaoka, A.U.Khan and D.R.Kearns, J. Chem. Phys. 46 (1967) 1842.

[6] C.K.Duncan and D.R.Kearns, unpublished results.

[7] S.J.Arnold, M.Kubo and E.A.Ogryzlo, Advan. Chem. Ser. 77 (1968) 133.

[8] C.S.Foote, R.W.Denny, L.Weaver, Y.Chang and J.Peters, Ann. Acad. Sci. N.Y. 171 (1970) 139.

[9] I.B.C.Matheson and J.Lee, Chem. Phys. Letters 7 (1970) 475.

73

Reprinted from *J. Chem. Soc., Faraday Trans. II*, **68**, 586–593 (1972)

Lifetime of Singlet Oxygen in Liquid Solution

By D. R. Adams and F. Wilkinson

School of Chemical Sciences, University of East Anglia, Norwich

Received 4th November, 1971

A laser photolysis system has been used to obtain unimolecular rate constants for the decay of singlet oxygen in liquid solution. The decay constants were derived from a kinetic analysis of the disappearance of an oxidizable acceptor, diphenylisobenzofuran, which reacts with singlet oxygen produced by energy transfer from several electronically excited sensitizers. The singlet oxygen decay constant obtained is independent, within experimental error, of the sensitizer employed and direct excitation of aromatic hydrocarbons is shown to give the most reliable values. The lifetime of singlet oxygen, which shows no temperature dependence in the range 0-23°C, was measured using four different solvents and the values obtained varied from 5 μs (methanol) to 12 μs (benzene).

In recent years there has been a surge of interest in the properties of the excited singlet states of oxygen both in the gas phase [1] and in solution.[2] The rates of decay of the $^1\Sigma_g^+$ and $^1\Delta_g$ states of oxygen and the quenching constants for various additives have been determined in the gas phase from studies of the emission from these singlet states.[1, 3] In condensed phases, the singlet oxygen molecules relax by radiationless processes and no direct measurements of the decay constants have hitherto been possible. The present paper describes a method for determining the lifetime of the singlet oxygen species in solution.

As early as 1939, studies of dye-sensitized photo-oxygenation reactions with molecular oxygen led Kautsky to propose the following scheme to explain his findings [4]

$$S^* + O_2 \rightarrow S + O_2^*$$

$$O_2^* + M \rightarrow MO_2$$

where S is a sensitizer, M an oxygen acceptor, and an asterisk denotes an electronically excited state. Much evidence has been accumulated in support of the original suggestion by Kautsky that O_2^* is the $^1\Delta_g$ state of molecular oxygen.[5] Furthermore it has been shown that S* is the triplet state of the sensitizer and that singlet oxygen is produced solely by energy transfer from triplet states.[6]

In the present experiments, singlet oxygen was generated following laser photolysis of a suitable sensitizer in the presence of oxygen and an oxidizable acceptor. The concentration of the acceptor was monitored as a function of time while it reacted with singlet oxygen. From a kinetic analysis of the rate of disappearance of the acceptor, the decay rate of the singlet oxygen can be determined.

EXPERIMENTAL

APPARATUS

The short lifetime of singlet oxygen and the correspondingly brief duration of its reaction with any acceptor required the use of a laser flash photolysis system. Experiments attempting to apply this method using a conventional flash photolysis apparatus exciting with a flash of ~ 10 μs half-peak width proved unsuccessful. The laser photolysis system was designed so that concentrations could be monitored using a spectroscopic beam at right angles to, or colinear with, the exciting pulse as shown schematically in fig. 1a and b.

The excitation pulse was obtained by passing the output from a Laser Associates model 211A ruby laser, passively Q-switched, through a Wratten 39 gelatin filter, F_1, which removed any light scattered from the flash lamp that pumped the ruby rod. The laser pulse then entered an ammonium dihydrogen phosphate (ADP) crystal whereupon about 5 % conversion to the second harmonic was achieved, giving a pulse of about 80 mJ at 347 nm with a half-peak width of 20 ns. A 15 mm cell containing 0.05 M aqueous copper sulphate solution, F_2, was used to remove 75 % of the red light whilst transmitting more than 85 % of the second harmonic. This minimized possible inhomogeneities in the excited solution brought about by local heating.

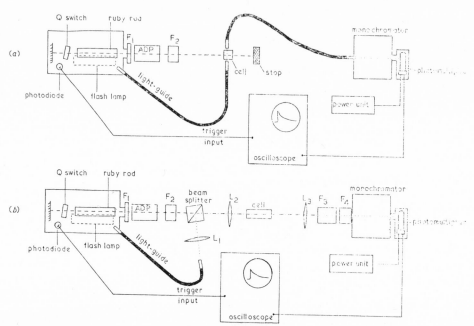

FIG. 1.—Schematic diagram of the laser photolysis system with (a) right angle and (b) colinear analysis of transients.

The monitoring beam utilized light from the comparatively long-lived ($\sim$ 2 ms) high intensity discharge lamp used to pump the ruby rod. The monitoring beam was conducted through a hole in the side of the laser cavity which held a light-guide of optical fibres. For right angle analysis the monitoring beam from this light-guide was passed through one side of the absorption cell and diametrically opposite this a second light guide picked up the monitoring light and conveyed it to the Bausch and Lomb high intensity grating monochromator (see fig. 1a). The alternative arrangement with colinear analysis to the exciting beam is represented schematically in fig. 1b. The monitoring beam, rendered parallel by lens, L_1, and the exciting beam were mixed by means of the beam splitting device and condensed slightly by the 25 cm focal length lens, L_2, placed 10 cm from the reaction cell. Any remaining 347 and 694 nm irradiation were removed by filter F_3 comprising a 3 cm cell containing 3.7×10^{-4} M biphenylene in benzene and by F_4 which has a path length of 25 mm and contained 0.4 M aqueous copper sulphate. The lens, L_3, focused the monitoring light onto the slits of the monochromator.

The photomultiplier was an RCA 931-A tube, fitted with a 110 Ω load resistor and with ten 22 kΩ resistors forming the dynode chain. The normal working range of 440-550 V, supplied to the tube by a Fluke 415B power supply unit, allowed a maximum anode current of about 0.2 mA to be drawn from the tube. Small capacitors were incorporated between the last four dynodes to prevent overloading by fluorescence, so that valid measurements could be taken immediately after the laser pulse. The voltage signal was carried to the Tektronix 585A oscilloscope via a Tektronix P6045 high-impedance probe. The oscilloscope was fitted with a type 82 plug-in amplifier unit, and the total rise-time of the detector system was less than 5 ns. The oscilloscope was triggered externally by a signal from a photodiode placed in the laser cavity and protected by suitable attenuating filters.

PROCEDURE

Kinetic information concerning singlet oxygen was obtained by monitoring the decrease in the acceptor concentration resulting from reaction with singlet oxygen as a function of time. For accurate kinetic analysis the transmittance change due to removal of the acceptor had to be at least 10 %. Using the laser photolysis system, this method is thereafter restricted to acceptor molecules which react with singlet oxygen with rate constants $\sim 10^8$ l. mol^{-1} s^{-1}. Furthermore, an ideal acceptor should not absorb any of the exciting light and it is convenient if the molecule has an absorption band which is not overlapped by sensitizer absorption.

In the present studies, the acceptor used was 1,3-diphenylisobenzofuran (DPBF) which has a bimolecular rate constant for reaction with $^1O_2^*$ of about 10^8 l. mol^{-1} s^{-1}. The molecule reacts with singlet oxygen as follows [7]:

DPBF has an absorption band maximum at 410 nm. Kinetic analysis was made at 390 nm where there is no absorption by any sensitizer except for methylene blue which absorbs slightly in this region.

DPBF slowly reacts with oxygen even in the absence of light [8] and therefore fresh solutions were made up prior to each experiment. Solutions usually contained $\sim 5 \times 10^{-5}$ M DPBF. The concentrations of the sensitizers were chosen such as to absorb ~ 80 % of the exciting light. Triplet naphthalene which does not absorb at 347 nm was produced by energy transfer from benzophenone. Methylene blue absorbs strongly at 694 nm as well as at 347 nm, and in this case experiments were made with and without frequency doubling of the ruby laser output.

The reaction vessel was a 1 cm^2 Pyrex fluorimeter cell. Air saturated solutions were used throughout except in one series of experiments where oxygen saturated solutions were investigated. Room temperature was $23 \pm 1°$C and in two series of experiments the temperature of the solution was varied.

Absorption spectra of mixtures of DPBF and the sensitizers were simply superimpositions of the spectra of the separate components. Absorption spectra taken before and after laser photolysis showed that the sensitizers were not consumed in the oxygenation process. Fluorescence spectra of the mixtures showed no new emission peaks which could have been assigned to exciplexes, etc.

MATERIALS

Anthracene and naphthalene were B.D.H. microanalytical grade. Pyrene was zone refined by Koch-Light, 1,2-benzanthracene was Koch-Light puriss grade and phenanthrene was freed from anthracene with maleic anhydride in purified benzene and re-crystallized from spectroscopic ethanol. Methylene blue and benzophenone were re-crystallized. All solvents were spectroscopic grade.

RESULTS AND DISCUSSION

Recent evidence strongly supports the following mechanism for many photo-sensitized oxidation reactions.[5, 6]

	reaction	rate
	$S + h\nu \rightarrow {}^1S^* \rightarrow {}^3S^*$	$I_a\phi_T$
(1)	${}^3S^* \rightarrow {}^1S$	$k_1[{}^3S]$
(2)	${}^3S^* + {}^3O_2 \rightarrow {}^1O_2^* + {}^1S$	$k_2[{}^3S][{}^3O_2]$
(3)	${}^1O_2^* \rightarrow {}^3O_2$	$k_3[{}^1O_2^*]$
(4)	${}^1O_2^* + M \rightarrow MO_2$	$k_4[{}^1O_2^*][M]$

where I_a is the rate of absorption of photons by the sensitizer and ϕ_T is the quantum yield of production of triplet sensitizer in the presence of oxygen. When 1,3-diphenyl-isobenzofuran (DPBF) is the oxidized species, M, reactions due to excited states of M need not be considered since no transients are observed when DPBF alone is subjected to laser photolysis. Although triplet energy transfer from many of the triplet sensitizers is possible, i.e., the process

(5) ${}^3S^* + {}^1M \rightarrow {}^3M^* + {}^1S$

does occur, reaction (5) is of negligible importance relative to reaction (2) under the experimental conditions employed since the concentration of oxygen was at least 30 times that of DPBF. k_5 was measured in the case of triplet 1,2-benzanthracene in de-aerated ethanol solution and a value of 4.4×10^9 l. mol^{-1} s^{-1} was obtained. This can be compared with the value of k_2 for triplet 1,2-benzanthracene in ethanol which is 3×10^9 l. mol^{-1} s^{-1}. In the de-aerated solution sensitized absorption was observed below 500 nm presumably due to the triplet state of DPBF but no consumption of DPBF occurred due to any reaction of ${}^3M^*$ in the absence of oxygen. In the presence of oxygen no absorption by ${}^3M^*$ was observed. Less than 4 % of ${}^3S^*$ molecules transfer energy to M in the presence of oxygen and the small amount of ${}^3M^*$ produced is probably quenched by oxygen to produce ${}^1O_2^*$, i.e., the overall effect is equivalent to reaction (2).

According to the mechanism given above the rate of change in the concentration of DPBF is given by

$$-d[M]/dt = k_4[{}^1O_2^*][M]. \qquad (1)$$

At times greater than 1 μs after the excitation pulse the decay of all excited sensitizers is complete and hence there is no further production of ${}^1O_2^*$. The rate of disappearance of singlet oxygen is then given by

$$-d[{}^1O_2^*]/dt = (k_3 + k_4[M])[{}^1O_2^*]. \qquad (2)$$

Under conditions such that $k_3 \gg k_4[M]$ equation (2) becomes

$$-d[{}^1O_2^*]/dt = k_3[{}^1O_2^*]$$

and therefore

$$[{}^1O_2^*] = [{}^1O_2^*]_0 \exp(-k_3 t) \qquad (3)$$

where $[{}^1O_2^*]_0$ is the concentration of singlet oxygen at the time from which the kinetic analysis is commenced (usually 1 μs after excitation). Substitution of equation (3) into (1) gives upon integration

$$\ln([M]/[M_0]) = (k_4[{}^1O_2^*]_0/k_3)(e^{-k_3 t} - 1). \qquad (4)$$

From equations (1), (3) and (4) it follows that

$$k_3 \ln M = B - d(\ln M)/dt \qquad (5)$$

where B is a constant.   The absorbance of M at the analysis wavelength, A, is equal
to $\varepsilon_M[M]l$ and therefore eqn (5) becomes

$$k_3 \ln A = B' - d(\ln A)/dt \qquad (6)$$

where B' is a consta   .us if $k_3 \gg k_4[M]$ a plot of $-d(\ln A)/dt$ against $\ln A$
should be linear with a slope equal to k_3.

Fig. 2 shows a plot of $\ln A$ against t for photosensitized oxidation of DPBF with
methylene blue in 95 % ethanol.   Tangents can be drawn to curves of this type and
plots of $-d(\ln A)/dt$ against $\ln A$ obtained which demonstrate that eqn (6) is well
obeyed (see fig. 3).   Analysis of different traces for this mixture gave slopes which

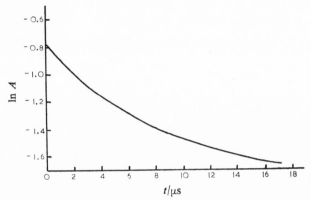

FIG. 2.—Change in absorbance, A, at 390 nm of 5×10^{-5} M diphenylisobenzofuran solution in
methanol due to reaction with singlet oxygen produced following laser photolysis with 10^{-4} M
methylene blue as sensitizer.

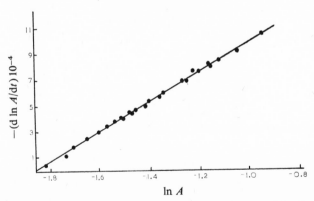

FIG. 3.—Data taken from analysis of three separate oscilloscope traces of the decrease in absorbance
of diphenylisobenzofuran due to reaction with $^1O_2^*$ plotted according to eqn (6) (sensitizer, methylene
blue).

agreed to within 10 %.   This graphical analysis is rather laborious and not particularly accurate and therefore a least squares computer fit of the data was attempted
using the VBO1A subroutine from the Atlas Computer Laboratories.   Values
obtained by this method are compared with those obtained by the graphical method

in table 1. The values obtained from both methods agree within experimental error and the precision from the computer fit is somewhat better (compare standard deviations in table 1). The least squares analysis was therefore used for all subsequent experiments.

TABLE 1.—COMPARISON OF GRAPHICAL AND COMPUTER LEAST SQUARES FIT METHODS FOR DETERMINATION OF THE DECAY CONSTANTS OF SINGLET OXYGEN, k_3 (STANDARD DEVIATIONS IN PARENTHESES)

| | | $10^{-5} k_3/s^{-1}$ | |
solvent	sensitizer	graphical	least squares fit
ethanol	methylene blue	1.1 (0.3)	1.3 (0.2)
methanol	methylene blue	1.4 (0.6)	1.6 (0.4)
benzene	anthracene	1.1 (0.2)	0.8 (0.1)

Values of k_3 obtained with different sensitizer-solvent systems are given in table 2. Each value was the average of several analyses and standard deviations are given. Repeated photolysis of a single solution in the presence of oxygen gave the same values of k_3 as the acceptor was consumed showing that the products of photo-oxygenation did not quench $^1O_2^*$. As shown in table 2 the values obtained show little or no temperature dependence. Also in a series of experiments with 1,2-benzanthracene as sensitizer the value of k_3 obtained was shown to be independent of the analyzing wavelength employed in the range 390-425 nm. In the case of methylene blue as sensitizer the identical values of k_3 were obtained with and without frequency doubling, i.e., excitation at 347 and 694 nm gave identical results. Changing the acceptor concentration from 5×10^{-6} to 5×10^{-5} M had no effect on the values of k_3 obtained, neither did a tenfold change in anthracene concentration (from 5×10^{-4} to 6×10^{-3} M) or methylene blue concentration (from 5×10^{-5} to 7×10^{-4} M). Right angle and colinear monitoring gave identical values of k_3.

TABLE 2.—UNIMOLECULAR RATE CONSTANTS, k_3, FOR SINGLET OXYGEN DECAY AT 23°C OBTAINED IN A VARIETY OF SOLVENTS USING DIFFERENT SENSITIZERS

| | $10^{-5} k_3/s^{-1}$ | | | |
sensitizer	benzene	95 % ethanol	methanol	dimethylformamide
anthracene	0.8 (0.1)	1.7 (0.3)	1.9 (0.3)	1.4 (0.1)
	0.9 (0.3) ^{a}			
phenanthrene	0.7 (0.1)	1.9 (0.3)	1.9 (0.5)	1.4 (0.2)
	0.8 (0.3) ^{a}			
pyrene	0.9 (0.1)	1.4 (0.4)	2.0 (0.4)	1.4 (0.2)
1,2-benzanthracene	0.7 (0.1)	1.6 (0.3)	2.1 (0.3)	1.4 (0.2)
	0.8 (0.2) ^{b}			
	0.8 (0.2) ^{c}			
benzophenone	1.3 (0.5)	1.6 (0.4)	1.7 (0.4)	2.3 (0.2)
benzophenone with 10^{-1} M naphthalene	1.3 (0.6)	1.5 (0.3)	1.8 (0.4)	2.0 (0.2)
methylene blue	1.0 (0.2) ^{e}	1.3 (0.2)	1.6 (0.4)	2.5 (0.5)
		1.4 (0.2) ^{b}		
		1.3 (0.1) ^{d}		

^{a} perdeuterated sensitizer; ^{b} oxygen saturated; ^{c} 5°C; ^{d} 0°C; ^{e} 4 : 1 benzene/methanol.

Table 2 demonstrates that, within experimental error, k_3 is independent of the sensitizer. This is especially true when sensitization arises from direct excitation of aromatic hydrocarbons. The fact that the rate constant for singlet oxygen decay is independent of the energy of the triplet sensitizer is consistent with rapid internal conversion of any $^1\Sigma_g^+$ states formed to give $^1\Delta_g$ oxygen which, in the absence of an acceptor, decays via a first order radiationless transition in agreement with the conclusions of previous workers.[6]

With benzophenone as sensitizer the scatter is much higher but the fact that the mean values of k_3 in benzene and dimethylformamide are higher seems significant. Gollnick et al.[9] have shown that sensitizers which have a lowest triplet state which is $^3(n,\pi^*)$ in character give some free radical oxidation of substrates. Small amounts of DPBF consumed in any free radical reaction could affect the kinetic analysis adversely. It is interesting to note that the k_3 values obtained using a mixture of benzophenone with 10^{-1} M naphthalene as sensitizer have identical values to those obtained from benzophenone alone. Since in this case triplet benzophenone is completely quenched to produce triplet naphthalene which then transfers energy to oxygen, this suggests that any side reactions which adversely affect the kinetic analysis arise from singlet state reactions of benzophenone. With methylene blue as sensitizer the value of k_3 obtained in dimethylformamide is somewhat high. Both Foote[10] and Gollnick[9] have found independently that hydrocarbon sensitizers lead to the same product distribution in stereoselective oxidation reactions regardless of the sensitizer whilst dye molecules show different distributions dependent on the dye concentration. Some complexing of $^1O_2^*$ with methylene blue in this polar solvent possibly occurs with a consequential increase in the decay of $^1O_2^*$.

All these considerations together with the elegant work of Stevens and Algar[6] convinces us that aromatic hydrocarbons constitute a class of compounds which photosensitize oxidation solely via $^1O_2^*$, and that the best values of k_3 are those obtained from the measurements with aromatic hydrocarbons as sensitizers. These values are tabulated in table 3 together with some physical properties of the solvents used. The values vary slightly with solvent but the reasons for this are unclear. There is no correlation with dielectric constant or viscosity but the decay constant is higher in alcoholic solvents where there is considerable hydrogen bonding within the solvent.

TABLE 3.—UNIMOLECULAR DECAY CONSTANTS FOR $O_2^*(^1\Delta_g)$ IN VARIOUS SOLVENTS

solvent	$10^{-5} k_3/\text{s}^{-1}$	viscosity/cP	dielectric constant
benzene	0.8 ± 0.05	0.60	2.3
methanol	2.0 ± 0.2	0.55	32.6
95 % ethanol	1.8 ± 0.2	1.1	24.3
dimethylformamide	1.4 ± 0.1	0.80	36.7

However, the values of k_3 may not depend on bulk properties of the solvent. They might equally well be considered as pseudo first order rate constants resulting from specific quenching of $O_2^*(^1\Delta_g)$ by solvent molecules. For example taking the concentration of benzene in benzene solutions as ~ 11 M, $k_3 \approx 11k_q$ yielding $k_q \approx 7 \times 10^3$ l. mol^{-1} s^{-1}. This compares with the gas phase value for quenching of $O_2^*(^1\Delta_g)$ by benzene[11] of 3.2×10^3 l. mol^{-1} s^{-1}. Unfortunately gas phase values for other solvents are not available. No mechanism for quenching of $^1O_2^*$ in the gas phase has been elucidated. However, quenching by amines which are quite efficient appears to be dependent on their ionization potentials.[12]

Quenching of $O_2^*(^1\Delta_g)$ by ground state oxygen has a value of 1.2×10^3 l. mol^{-1} s^{-1} in the gas phase[3] and the fact that k_3 was unaffected when solutions were saturated with oxygen suggests that in solution the quenching constant is $< 10^6$ l. mol^{-1} s^{-1}. This conflicts with the value of $10^3 k_3$, i.e., $\sim 10^8$ l. mol^{-1} s^{-1} reported by Koizumi, Usui and Iwanaga.[13]

Foote et al.[14] have shown that the sensitized photo-oxygenation of 2-methylpent-2-ene is reduced markedly in the presence of β-carotene indicating very rapid quenching of $^1O_2^*$ by β-carotene. By assuming that this quenching is diffusion controlled with a rate constant of 3×10^{10} l. mol^{-1} s^{-1} and from their measured value of k_3/k_4 for 2-methylpent-2-ene they obtain a decay rate for singlet oxygen in benzene of 10^5 s^{-1}. This is in striking agreement with the values obtained in this work.

There are many literature values of k_3/k_4 and k_4/k_q determined from measurement of sensitized photo-oxygenation yields as a function of the concentration of various acceptors and quenchers. The values of k_3 given in table 3 can therefore be used to determine absolute rate constants for k_4 and k_q. For example the value of k_3/k_4 for 2-methylpent-2-ene in methanol is reported to be 0.18 M and therefore $k_4 = 1.2 \times 10^6$ l. mol^{-1} s^{-1}. Previously only relative values were obtainable, e.g., see ref. (2).

[1] R. P. Wayne, *Adv. Photochem.*, 1969, **7**, 311.
[2] K. Gollnick, *Adv. Photochem.*, 1969, **6**, 1.
[3] K. H. Becker, W. Groth and U. Schurath, *Chem. Phys. Letters*, 1971, **8**, 259.
[4] H. Kautsky, *Trans. Faraday Soc.*, 1939, **35**, 216.
[5] D. R. Kearns, *Chem. Rev.*, 1971, **71**, 395.
[6] B. Stevens and B. E. Algar, *Ann. N.Y. Acad. Sci.*, 1970, **171**, 50.
[7] C. Dufraisse and S. Ecary, *Compt. rend.*, 1946, **223**, 735.
[8] A. de Berne and R. Ratsimbazafy, *Bull. Soc. chim. France*, 1963, 229.
[9] K. Gollnick, Th. Franken, G. Schadde and G. Dorhofer, *Ann. N.Y. Acad. Sci.*, 1970, **171**, 89.
[10] C. S. Foote, *Ann. N.Y. Acad. Sci.*, 1970, **171**, 105.
[11] F. D. Findlay, C. J. Fortin and D. R. Snelling, *Chem. Phys. Letters*, 1969, **3**, 204.
[12] E. A. Ogryzlo and C. W. Tang, *J. Amer. Chem. Soc.*, 1970, **92**, 5034.
[13] M. Koizumi, Y. Usui and C. Iwanaga, *Bull. Chem. Soc. Japan*, 1969, **42**, 1231.
[14] C. S. Foote and R. Denny, *J. Amer. Chem. Soc.*, 1968, **90**, 6233.
[15] K. Gollnick, *Adv. Chem. Series*, 1968, **77**, 78.

74

Reprinted from *J. Amer. Chem. Soc.*, **94**, 1032–1033 (1972)

Chemistry of Singlet Oxygen. XVI. Long Lifetime of Singlet Oxygen in Carbon Disulfide[1]

Sir:

Two recent studies concluded that neither the rate of reaction of singlet oxygen with substances (k_A) nor its decay rate (k_d) vary much in organic solvents.[2,3] Both of these studies were limited by the fact that only the ratio of k_d to another rate, either k_A[2,3] or the calculated diffusion-controlled rate of singlet oxygen quenching by β-carotene (k_Q),[3] could be determined. By the latter technique, k_d in benzene–methanol (4:1) was calculated to be 10^5 sec^{-1},[3,4] a value which has recently been confirmed by direct measurement.[5]

Highly anomalous effects on β (k_d/k_A) in CS$_2$ have been reported for self-sensitized oxygenations of anthracene derivatives; decreases in β of a factor of 12–60 on going from aromatic solvents to CS$_2$ were found.[6] We now report that these effects are real and not caused by possible effects of CS$_2$ on any processes in anthracene, and that the effect of CS$_2$ is to increase the lifetime of singlet oxygen by an order of magnitude compared to benzene or methanol.

Sensitized photooxygenation of anthracene was carried out in various solvents; consumption of anthracene was monitored by loss of uv absorption at 387 nm. Self-sensitization was prevented by filtering the exciting light through anthracene solutions, when sensitizer was dinaphthalenethiophene (DNT), or through K$_2$Cr$_2$O$_7$ solutions, when sensitizer was tetraphenylporphine (TPP). In addition, quenching of 2-methyl-2-pentene (2M2P) photooxygenation by β-carotene and by DABCO (diazabicyclo[2.2.2]octane) in various solvents was measured by previously reported techniques.[4,7] Results are shown in Table I.

The values of β for anthracene in CS$_2$ and benzene, 2.6 and 47 × 10^{-3} M, are the same within experimental error as those reported previously (2.0–3.6 and 45 × 10^{-3} M)[6] for self-sensitized oxygenation; thus, the complicated behavior, including fluorescence quenching

and λ$_{max}$ changes which anthracene (as a sensitizer) undergoes in CS$_2$,[6] do not cause the increase in reactivity in CS$_2$; in any case, β is independent of sensitizer.[7]

It is obvious that all the ratios of decay rate to reaction or quenching rate are much smaller in CS$_2$ than C$_6$H$_6$ or CH$_3$OH. The only reasonable explanation for these changes is that the major cause is a decrease in k_d by an order of magnitude in CS$_2$, since k_A and k_Q for the four substrates probably involve several different mechanisms and need not all behave in a precisely parallel fashion in going from C$_6$H$_6$ or CH$_3$OH to CS$_2$. In particular, k_Q for β-carotene, which is already diffusion controlled in benzene,[4,5] cannot increase by the factor of 20 in CS$_2$ which would be required if k_d were constant.

This conclusion is confirmed by the results of Merkel and Kearns, who find an increase of k_d in CS$_2$ by factors of 8 and 28 in C$_6$H$_6$ and CH$_3$OH, respectively.[8] These values agree well with the changes in β in Table I, if allowance is made for probable small changes in k_A and k_Q in the various solvents. These results are sufficient to account for the anomalous effects of CS$_2$ on self-sensitized oxidation of anthracenes[6] and the reported effects of CS$_2$ on other photooxidation rates.[9]

A further solvent anomaly on the self-sensitized oxidation rates of anthracene is that addition of a few per cent of other solvents to CS$_2$ causes large changes in β.[6] Table II shows that this effect is also found in the

Table II. Effect of Mixtures of CS$_2$ and C$_6$H$_6$ on β and λ$_{max}$ for Anthracene

Mol % C$_6$H$_6$	β × 10^3, M	Anthracene λ$_{max}$, nm
0	2.5^a	387
3.5	27^a	386
7	52^a	386
25	94^b	385
50	77^b	383
100	46^a	379

^{a} DNT sensitized. ^{b} TPP sensitized.

Table I. Values of β (k_d/k_A) or β^Q (k_d/k_Q)

Solvent	Anthracene^a β × 10^3 M	2M2P β × 10^2 M	β-Carotene β^Q × 10^7 M	DABCO^b β^Q × 10^4 M
C$_6$H$_6$	47	10^c	33^f	9.6
CH$_3$OH		16^c	61^a	65
CS$_2$	2.6	2.2^{d,e}	1.5^d	1.70
β(C$_6$H$_6$)/β(CS$_2$)	18	4.6	22	5.7
β(CH$_3$OH)/β(CS$_2$)		7.3	40	38

^{a} DNT sensitized. ^{b} Zinc TPP sensitized. ^{c} Reference 2. ^{d} TPP sensitized. ^{e} Average of four independent determinations, range ±0.004; value in ref 2 is in error. ^{f} C$_6$H$_6$–CH$_3$OH, 4:1 (ref 4). ^{a} Reference 3.

sensitized oxygenation, and is even more bizarre than originally thought; β for anthracene goes through an enormous increase on addition of 3.5% C$_6$H$_6$ to CS$_2$; a broad maximum occurs at about 25% C$_6$H$_6$.

Although pronounced changes in λ$_{max}$ of anthracene also occur, they do not seem to parallel the change in β, and are probably not related. In fact, the shift in λ is linear with mole per cent of C$_6$H$_6$. Like previous authors,[6] we are unable to account for the effect of small amounts of added solvents to CS$_2$; the effect of CS$_2$ on lifetime of singlet oxygen could be related to the fact that sulfides form adducts with ^{1}O$_2$,[10] but further work will be required to establish this.

(1) Paper XV: C. S. Foote, T. T. Fujimoto, and Yew C. Chang, *Tetrahedron Lett.*, 45 (1972); supported by National Science Foundation Grants No. GP-8293 and 25,790, and Public Health Service Grant No. AP00681.
(2) C. S. Foote and R. W. Denny, *J. Amer. Chem. Soc.*, **93**, 5168 (1971).
(3) R. H. Young, K. Wehrly, and R. L. Martin, *ibid.*, **93**, 5774 (1971).
(4) C. S. Foote, Y. C. Chang, and R. W. Denny, *ibid.*, **92**, 5216 (1970).
(5) P. B. Merkel and D. R. Kearns, *Chem. Phys. Lett.*, **12**, 120 (1971).
(6) (a) E. J. Bowen, *Discuss. Faraday Soc.*, **14**, 143 (1953); (b) E. J. Bowen, *Advan. Photochem.*, **1**, 23 (1963); (c) E. J. Bowen and D. W. Tanner, *Trans. Faraday Soc.*, **51**, 475 (1955); (d) R. Livingston in "Autoxidation and Antioxidants," W. O. Lundberg, Ed., Vol. I, Wiley-Interscience, New York, N. Y., 1961, p 249.
(7) C. S. Foote, R. W. Denny, L. Weaver, Y. Chang, and J. Peters, *Annu. N. Y. Acad. Sci.*, **171**, 139 (1970).
(8) P. B. Merkel and D. R. Kearns, *J. Amer. Chem. Soc.*, **94**, 1029 (1972). We thank Professor Kearns for a prepublication manuscript and discussion of his results.
(9) K. Gollnick, *Advan. Photochem.*, **6**, 1 (1968); E. J. Forbes and J. Griffiths, *Chem. Commun.*, 427 (1967). The self-sensitized oxidations of anthracenes are accelerated additionally by the increased intersystem crossing yields in CS$_2$; this is not a complication in the studies reported here, since a sensitizer was used and the kinetic treatment factors out sensitizer effects.[7]
(10) C. S. Foote and J. W. Peters, *J. Amer. Chem. Soc.*, **93**, 3795 (1971).

Christopher S. Foote,* Elaine R. Peterson, Kyu-Wang Lee
Contribution No. 2936
Department of Chemistry, University of California
Los Angeles, California 90024

Received December 6, 1971

Reprinted from *J. Amer. Chem. Soc.*, 94(3), 1029-1030 (1972)

Remarkable Solvent Effects on the Lifetime of $^1\Delta_g$ Oxygen[1]

Sir:

Knowledge of the lifetime of singlet oxygen in solution plays an important role in interpreting many of the photooxidation studies now being carried out. There now appears to be a widespread belief that the lifetime of singlet oxygen ($^1\Delta_g$) is approximately solvent independent.[2-7] This is based on experimental observations that the "β" value for a given oxygen acceptor, the ratio of the rate of the decay of singlet oxygen to the rate of reaction with the acceptor, does not depend strongly on the nature of the solvent. In most of the recent collections of data we find that β values for some of the common acceptors vary by at most a factor of 4 in different solvents.[2,3,5] In contrast to these more recent findings, Bowen[7] and Livingston and Owens[8] found relatively large solvent effects on the *rate* of photooxidation of anthracene (and hence presumably on its β value), but the significance of these observations was questioned because of the possible solvent effects on the fluorescence of anthracene.[2]

Recently we developed a spectroscopic method which has permitted us to directly measure the lifetime of singlet oxygen in solution.[9] We have used this method to measure the lifetime of singlet oxygen in some of the solvents which are commonly used in photooxidation studies. Contrary to previous conclusions, we find that the lifetime of singlet oxygen is remarkably sensitive to the nature of the solvent.

In our experiments singlet molecular oxygen ($\sim 5 \times 10^{-5}$ M) was generated in $\sim 10^{-7}$ sec by energy transfer from methylene blue triplets which were excited by a 20-nsec ruby laser pulse.[9] Singlet oxygen was then monitored spectroscopically by following its reaction with the colored acceptor, 1,3-diphenylisobenzofuran, to form a colorless product.

The results of these experiments, which are presented in Table I, demonstrate that the lifetime of $^1\Delta_g$ varies by a factor of 100 in going from water to CS_2. It is not surprising that there are large solvent effects on the $^1\Delta_g$ lifetime since it represents a good example of what Robinson[10] and others,[11-13] would classify as a small

Table I. Solvent Effects on the Lifetime of $^1\Delta_g$

Solvent	Viscosity, cP[a]	Dielectric constant[a]	$\tau^1\Delta_g$, μsec
H_2O	0.80[b]	80	2
CH_3OH	0.51	33	7
C_6H_6	0.56	2.2	24
$CH_3C(=O)CH_3$	0.29	~ 20	26
CS_2	0.35	2.6	200

[a] "Handbook of Chemistry and Physics," 35th ed, Chemical Rubber Publishing Co., Cleveland, Ohio. [b] Extrapolation based on data using 1:1 H_2O–CH_3OH.

molecule in their theoretical treatments of radiationless transitions in molecules. The solvent-dependent quenching of $^1\Delta_g$ oxygen to its ground state is to be contrasted with the somewhat analogous relaxation of triplet state organic molecules which (aside from heavy atom effects) is relatively independent of solvent because large molecules have sufficiently large numbers of internal vibrational modes to act as their own heat sinks.[10]

Even at this preliminary stage of investigation it is clear that certain solvent properties are unimportant in affecting the lifetime of singlet oxygen. The fact that the lifetime is shortest in water might suggest that the dielectric constant or dipole moment of the solvent molecules is an important parameter, but this possibility is eliminated by the fact that the lifetime is almost the same in both benzene and acetone. Comparison of the data in Table I further demonstrates that there is no correlation between viscosity and lifetime. Foote has also arrived at similar conclusions.[2]

Our observation that the lifetime of singlet oxygen is quite sensitive to the nature of the solvent appears to be at variance with conclusions reached by other workers, and it is worthwhile to see how this discrepancy might have arisen. Many of the previous studies of dye-sensitized photooxidation involved mixed solvents, and using the numbers in Table I we compute that the lifetime of $^1\Delta_g$ in a 4:1 mixture of benzene and methanol (a common mixture)[2,6] would be 16 μsec. This is considerably reduced from the lifetime in neat benzene, so it is easy to see that incorporation of small amounts of methanol in varous solvents would prevent observation of large solvent effects on the lifetime of $^1\Delta_g$. Furthermore, reaction rate constants of certain acceptors may be solvent dependent (the rate constant for 1,3-diphenylisobenzofuran in CS_2 is about one-half its value in methanol) and therefore obscure lifetime changes.

We are continuing our investigation of the various factors which influence the lifetime of $^1\Delta_g$ in solution, and a more detailed experimental and theoretical analysis of these observations follows.[14]

(1) This work was supported by a grant (CA 11459) from the U. S. Public Health Service.
(2) C. S. Foote and R. W. Denny, *J. Amer. Chem. Soc.*, 93, 5168 (1971).
(3) K. Gollnick, *Advan. Photochem.*, 6, 1 (1968).
(4) R. Higgins, C. S. Foote, and H. Cheng, *Advan. Chem. Ser.*, No. 77, 102 (1968).
(5) C. S. Foote, *Accounts Chem. Res.*, 1, 104 (1968).
(6) R. H. Young, C. K. Wehrly, and R. Martin, *Amer. Chem. Soc., Div. Petrol. Chem.*, 16, A89 (1971); *J. Amer. Chem. Soc.*, 93, 5774 (1971).
(7) E. J. Bowen, *Advan. Photochem.*, 1, 23 (1963).
(8) R. Livingston and K. E. Owens, *J. Amer. Chem. Soc.*, 78, 3391 (1956).
(9) P. B. Merkel and D. R. Kearns, *Chem. Phys. Lett.*, 12, 120 (1971).
(10) W. Robinson, *J. Chem. Phys.*, 47, 1967 (1967).
(11) M. Bixon and J. Jortner, *ibid.*, 48, 715 (1968).
(12) J. Jortner, S. A. Rice, and R. M. Hochrasser, *Advan. Photochem.*, 7, 149 (1969).
(13) G. W. Robinson and R. P. Frosch, *J. Chem. Phys.*, 37, 1962 (1962); 38, 1187 (1963).
(14) P. B. Merkel, R. Nilsson, and D. R. Kearns, *J. Amer. Chem. Soc.*, 94, 1030 (1972).

Paul B. Merkel, David R. Kearns*
Department of Chemistry, University of California
Riverside, California 92502
Received October 22, 1971

76

Reprinted from *J. Amer. Chem. Soc.*, **94**(21), 7244–7253 (1972)

Radiationless Decay of Singlet Molecular Oxygen in Solution. An Experimental and Theoretical Study of Electronic-to-Vibrational Energy Transfer

Paul B. Merkel and David R. Kearns*

Contribution from the Department of Chemistry, University of California, Riverside, California 92502. Received January 29, 1972

Abstract: A Q-switched ruby laser has been used to spectroscopically measure the lifetime of singlet ($^1\Delta$) oxygen in solution. The nature of the solvent is found to have a remarkable effect on the lifetime of singlet oxygen with values ranging from 2 μsec in water to 700 μsec in CCl_4. A simple theory has been developed to account for the quenching of $^1\Delta$ by the solvent in terms of intermolecular electronic-to-vibrational energy transfer. According to this theory the quenching efficiency can be quantitatively related to the intensities of infrared overtone and combinational absorption bands of the solvent in the energy regions of $^1\Delta \rightarrow {}^3\Sigma$ transitions and especially near 7880 and 6280 cm^{-1}, the respective $0 \rightarrow 0$ and $0 \rightarrow 1$ oxygen transition energies. Direct calculation of the quenching rates using this theory with no adjustable parameters yields a better than order of magnitude agreement with experimental results. An analysis of gas phase and solution quenching rate constants indicates rate constants obtained in one phase may be used to compute quenching constants in the other phase. This applies to the quenching of $^1\Sigma$ as well as $^1\Delta$. The apparent lack of a heavy atom effect on the $^1\Delta$ lifetime can be accounted for. Large deuterium effects are predicted and observed. The theory indicates that the quenching involves a second-order indirect mixing of $^3\Sigma$ and $^1\Delta$ states through $^1\Sigma$ by interaction with the solvent, in which $^3\Sigma$ and $^1\Sigma$ are mixed *via* an intramolecular spin–orbit coupling. Mixing of $^1\Sigma$ and $^1\Delta$ is also important in the quenching of $^1\Sigma$, and the matrix elements used to account for the observed quenching rate constants for $^1\Delta$ automatically lead to rate constants for the quenching of $^1\Sigma$ which are in good agreement with the experimental values. The theory thus appears to be internally consistent. The techniques used to measure the solvent-controlled decay of singlet oxygen have also been used to evaluate the absolute rate constants for the reaction of singlet oxygen with various acceptors. A quantum yield of 0.9 ± 0.1 was measured for the formation of $^1\Delta$ from quenching of triplet state methylene blue. The quenching of singlet oxygen by β-carotene and a polymethene pyrylium dye (I) has been investigated. We confirm the earlier suggestion by Foote that the β-carotene quenching of singlet oxygen is nearly diffusion controlled (2 × 10^{10} M^{-1} sec^{-1} in benzene). The rate constant for the quenching of $^1\Delta$ by I is found to be 3 × 10^{10} M^{-1} sec^{-1} in acetonitrile. The discrepancy between our observation that the lifetime of $^1\Delta$ is extremely solvent sensitive and earlier indications that the lifetime is relatively solvent independent is resolved. Our findings indicate the factors which may permit $^1\Delta$ lifetimes of greater than 1 msec in appropriate solvents to be obtained.

Recognition of the importance of singlet molecular oxygen as an intermediate in the photooxidation of unsaturated hydrocarbons has stimulated much of the current interest in the chemical and physical properties of this species.[1–4] One of the most common methods of generating singlet oxygen is *via* energy

transfer from an excited triplet sensitizer to ground-state oxygen. With sensitizers having a lowest triplet state above 13,100 cm^{-1}, both $^1\Sigma$ and $^1\Delta$ oxygen are produced.[5–9] However, for solution-phase reactions

(1) C. S. Foote and S. Wexler, *J. Amer. Chem. Soc.*, **86**, 3879 (1964).
(2) E. J. Corey and W. C. Taylor, *ibid.*, **86**, 3881 (1964).
(3) K. Gollnick, *Advan. Chem. Ser.*, No. 77, 78 (1968).
(4) C. S. Foote, *Accounts Chem. Res.*, **1**, 104 (1968).

(5) K. Kawaoka, A. U. Khan, and D. R. Kearns, *J. Chem. Phys.*, **46**, 1842 (1967).
(6) C. K. Duncan and D. R. Kearns, *ibid.*, **55**, 5822 (1971).
(7) D. R. Kearns, A. U. Khan, C. K. Duncan, and A. H. Maki, *J. Amer. Chem. Soc.*, **91**, 1039 (1969).
(8) E. Wasserman, V. J. Kuck, W. M. Delevan, and W. A. Yager, *ibid.*, **91**, 1040 (1969).
(9) E. W. Abrahamson, *Chem. Phys. Lett.*, **10**, 113 (1971).

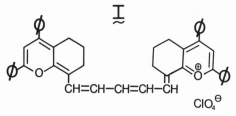

Figure 1. Polymethene pyrylium dye (I).

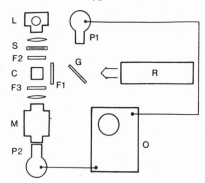

Figure 2. Schematic of laser photolysis apparatus: R (ruby laser), P1 (triggering photomultiplier), L (analyzing lamp), M (monochromator), P2 (monitoring photomultiplier), C (sample cell), S (shutter), G (glass plate), F1 (Corning 2–64 filter), F2 (5–60 filter), F3 (4–96 filter), O (oscilloscope).

only $^1\Delta$ is expected to be important due to the short lifetime ($\lesssim 10^{-10}$ sec) estimated[10] for $^1\Sigma$.

Under usual photooxygenation conditions the two important competitive decay modes for $^1\Delta$ are physical quenching to the ground $^3\Sigma$ state

$$^1\Delta \xrightarrow{1/\tau} {}^3\Sigma \qquad (1)$$

and chemical reaction of $^1\Delta$ with an acceptor A

$$^1\Delta + A \xrightarrow{k_a} AO_2 \qquad (2)$$

where AO_2 denotes the peroxide products. Steady-state measurements of photooxidation quantum yields as a function of concentration have allowed the determination of $\beta = (1/\tau k_a)$.[4] Physically, β represents the concentration of acceptor required to intercept half of the singlet oxygen produced, and thus it serves as an index of relative acceptor reactivities, but it does not permit either the absolute value of the reaction rate constant or the $^1\Delta$ lifetime to be determined unless one is known from other experiments.

We have recently reported a direct measurement of the lifetime of singlet oxygen and of absolute acceptor rate constants in methanol[11] as well as in other solvents.[12]

In this paper we have carried out a more comprehensive experimental and theoretical study of solvent effects on the lifetime of $^1\Delta$. We find that there are pronounced solvent effects on the decay of singlet oxygen, with lifetimes varying over almost three orders of magnitude. A strong correlation is noted between the singlet oxygen lifetime and the infrared absorption intensity of the solvent overtone and combinational bands in near resonance with the $^1\Delta \rightarrow {}^3\Sigma$ transition frequencies. Our experimental observations and theoretical considerations lead us to propose that the major pathway for radiationless decay of the $^1\Delta$ state involves conversion of electronic energy of oxygen directly into vibrational excitation of the solvent. The behavior of oxygen, a prototype small molecule in which the rate of intramolecular electronic relaxation is very slow, is to be contrasted with the behavior of large molecules where intramolecular relaxation of electronically excited states is rapid.

Experimental Section

Materials. With the exception of distilled water, deuterium oxide (Mallinckrodt 99.8%), perdeuterioacetone (Stohler Isotope Chemicals 99.5%), and absolute ethanol, solvents were Matheson

Coleman and Bell Spectroquality. Methylene blue (MB) was obtained also from Matheson Coleman and Bell. Aldrich 1,3-diphenylisobenzofuran (DPBF) was purified by recrystallization from methanol–water. Rubrene, also from Aldrich, was used without further purification. Fresh samples of 2,5-dimethylfuran (99%), 2,4-dimethylbutene (99%), 1,3-dimethylcyclohexene (98%), and 2,5-dimethyl-2,4-hexadiene (99%) from Chemical Samples Co. were used as received. Sigma Chemical β-carotene was purified by recrystallization from a mixed benzene-methanol solvent. The polymethene pyrylium dye (I) shown in Figure 1 was a gift from J. Williams of Eastman Kodak.

Methods. A ruby laser pulse was used to excite the sensitizer methylene blue (10^{-5}–10^{-4} M) in oxygenated or air-saturated solutions. With the 1-J laser pulse, conversion to triplets is essentially complete. The monomer and dimer triplets are quenched by ground-state oxygen in <0.5 μsec forming singlet oxygen with an efficiency near unity.[13] As a probe of the singlet oxygen concentration the acceptor 1,3-diphenylisobenzofuran ($\beta \sim 10^{-4}$)[14,15] was added to the solution, and the rate of decrease in absorption ($\lambda_{max} \sim 410$ nm) as a result of peroxide formation following a laser pulse was monitored.

The combination of MB and DPBF is well suited for these laser kinetic experiments since MB exhibits strong absorption at the fundamental ruby wavelength, while the absorption of DPBF occurs in a region in which MB is nearly transparent. An acceptor with a very low β value is required since the maximum fractional bleaching can be shown to equal $[^1\Delta]_0/\beta$, $[^1\Delta]_0$ being the concentration of singlet oxygen produced by the light pulse. Furthermore, in the investigation of the effect of addition of another acceptor on the decay of singlet oxygen, it is desirable that the concentration of the acceptor remain nearly constant during the bleaching of DPBF. This condition requires that β for DPBF be substantially smaller than that of the second acceptor.

A schematic diagram of the experimental apparatus is given in Figure 2. A Q-switched (cryptocyanine passive dye cell) ruby laser, constructed in this laboratory, was used to deliver 694-nm pulses of $\sim$20-nsec duration. Transient absorption produced in a 1-cm square cell was monitored at 90° to the exciting laser beam. The analyzing unit consisted of a 650-W tungsten–iodine lamp, a Bausch and Lomb high intensity monochromator, an RCA IP28 photomultiplier (P2), and a Tektronix 545A monitoring oscilloscope, equipped with a Type 1A1 plug-in unit. The oscilloscope was triggered synchronously with the laser pulse by a second photomultiplier (P1) which viewed a portion of the laser light.

(10) S. J. Arnold, M. Kubo, and E. A. Ogryzlo, *Advan. Chem. Ser.*, No. 77, 133 (1968).

(11) P. M. Merkel and D. R. Kearns, *Chem. Phys. Lett.*, 12, 120 (1971).

(12) P. B. Merkel and D. R. Kearns, *J. Amer. Chem. Soc.*, 94, 1029 (1972).

(13) Second-order rate constants for oxygen quenching of dimer and monomer triplets are of the order of 3×10^9 M^{-1} sec^{-1} (e.g., 2×10^9 M^{-1} sec^{-1} for monomer quenching in methanol), and lifetimes of monomer and dimer triplets in deoxygenated solution while solvent and concentration dependent in the neighborhood of 10 μsec (e.g., 7 μsec for monomer in a 10^{-4} M solution of MB in methanol).

(14) I. B. C. Matheson and J. Lee, *Chem. Phys. Lett.*, 7, 475 (1970).

(15) R. H. Young, K. Wherly, and R. Martin, *J. Amer. Chem. Soc.*, 93, 5774 (1971).

7246   *Journal of the American Chemical Society* / *94:21* / *October 18, 1972*

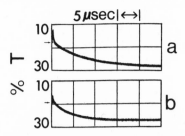

Figure 3. Bleaching in an oxygenated solution of methylene blue (10^{-4} M) in methanol of (a) 1.5×10^{-5} M 1,3-diphenylisobenzofuran monitored at 410 nm in a 1-cm cell and (b) 1.0×10^{-4} M 1,3-diphenylisobenzofuran monitored at 435 nm in a 2-mm cell. Arrows indicate $\%$ T before the pulse.

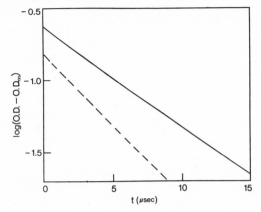

Figure 4. First-order decay plots of (a) ——— and (b) - - - - - of Figure 3.

Under normal circumstances the signal-to-noise ratio in the photoelectric measurement of transient absorption is limited by the photocathode current. The maximum value of the photocathode current is in turn fixed by the gain of the photomultiplier and the space charge limited anode current. Reduction of the photomultiplier gain allows an increase in photocathode current without overloading the anode. This was accomplished by using only the first seven dynodes of the 1P28, the last of which constituted the collector. Dynodes 8 and 9 and the anode were connected to ground through a resistor. A tenfold increase in the signal-to-noise ratio was obtained with this arrangement.

Excitation of MB by the analyzing light can itself produce significant bleaching of DPBF. This problem was virtually eliminated by positioning a shutter, which was manually opened immediately prior to triggering the laser, and a Corning 5-60 band pass filter between the lamp and sample cell.

Steady-state measurements of methylene blue sensitized photobleaching of rubrene were carried out in acetone and perdeuterioacetone in a manner reported previously.[16]

Results

1. Measurement of $^1\Delta$ Lifetime and Solvent Effects. Under conditions where [A] does not change substantially during the course of an experiment, solution of the kinetic equations describing processes 1 and 2 yields

$$[AO_2] - [AO_2]_\infty = \frac{k_a[A][^1\Delta]_0}{(1/\tau) + k_a[A]} e^{-(1/\tau + k_a[A])t} \quad (3)$$

where $[AO_2]_\infty$ is the concentration of peroxide after all singlet oxygen produced in a pulse has decayed. From this it follows that

$$-2.3 \log ([A] - [A]_\infty) = ((1/\tau) + k_a[A])t + \text{constant}$$

which can more conveniently be expressed as

$$-2.3 \log (OD - OD_\infty) = ((1/\tau) + k_a[A])t + \text{constant}$$

where OD is the optical density of the acceptor at time t and OD_∞ is the asymptotic value. Plots of (OD − OD_∞) $vs.$ t for two acceptor concentrations yield values for both τ and k_a.

Laser-induced photobleaching of DPBF is illustrated in Figure 3. In Figure 3a an acceptor concentration of 1.5×10^{-5} M results in a decay rate fixed primarily by τ. The initial short-lived transient absorption arises from MB triplet–triplet absorption. In a deoxygenated solution no bleaching is observed and the MB triplet decays at the same rate as observed in the absence of DPBF, thus eliminating the possibility of direct reaction between acceptor and sensitizer. In-

(16) P. B. Merkel, R. Nilsson, and D. R. Kearns, *J. Amer. Chem. Soc.*, **94**, 1029 (1972).

creasing the acceptor concentration to 10^{-4} M noticeably decreases the duration of photobleaching by consumption of a significant portion of $^1\Delta$ as is demonstrated in Figure 3b.

First-order plots of the decay curves in Figures 3a and 3b appear in Figure 4. The slight curvature reflects the consumption of acceptor as photooxidation progresses. From the slopes of these two plots we calculate $\tau(CH_3OH) = 7$ μsec and $k_a(DPBF) = 8 \times 10^8$ M^{-1} sec^{-1}.

The same procedures were used to measure the $^1\Delta$ lifetimes in ten additional solvents. Concentrations of methylene blue were chosen in each case such that fractional bleaching was small enough to prevent substantial departures from first-order linearity.

To solubilize the methylene blue, 1% methanol was added to the carbon disulfide and carbon tetrachloride solutions, while cyclohexane and benzene contained 2% methanol. The decay component due to the methanol was extracted from the observed decay by subtracting the appropriate fraction of the decay constant $(1/\tau)$ for pure methanol. Although this procedure may only be approximately correct, the correction factor is relatively large only in the case of carbon tetrachloride.

The lifetime for water was extrapolated from the data for pure methanol and 50% methanol–50% water. This procedure was necessitated by the insolubility of DPBF in water and by the marked decrease in the reaction rate constant for DPBF in solutions containing more than $\sim$70% water. The latter phenomenon probably arises from dimerization of the DPBF (the absorption spectrum red shifts and fluorescence disappears).

Bleaching of DPBF in 50% water–50% methanol and in carbon disulfide is shown in Figures 5a and 5b, respectively. These two decay curves illustrate the strong dependence of the lifetime of singlet oxygen upon solvent. The decay measurements are summarized in Table I. Lifetimes range from 2 μsec in water to approximately 700 μsec in carbon tetrachloride. They are estimated to be accurate to within 20% except for CS_2 ($\sim$30%) and CCl_4 ($\sim$50%). Solvent optical densities (1-cm cell) at 7880 and 6280 cm^{-1} are

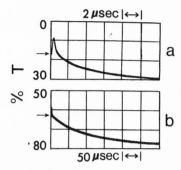

Figure 5. The decay of $^1\Delta$ as monitored by DPBF in (a) oxygenated 50% H_2O–50% CH_3OH with 1.5×10^{-5} M DPBF and 5×10^{-5} M MB (at 410 nm), and in (b) air-saturated CS_2 (containing 1% CH_3OH) with 0.5×10^{-5} M DPBF and 2×10^{-5} M MB (at 415 nm). Note the different time scales for (a) and (b).

Table I. Lifetimes of Singlet Oxygen in Various Solvents

Solvent	$\tau^1 \Delta$, μsec	$OD_{7880 cm^{-1}}$ (1 cm)	$OD_{6280 cm^{-1}}$ (1 cm)
H_2O	2^a	0.47	3.4
CH_3OH	7	0.18	3.9
50% D_2O–50% CH_3OH	11	0.09	1.0
C_2H_5OH	12	0.14	2.0
C_6H_{12}	17	0.09	0.08
C_6H_6	24	0.009	0.11
CH_3COCH_3	26	0.015	0.08
CH_3CN	30	0.016	0.14
$CHCl_3$	60	0.002	0.01
CS_2	200	<0.0005	0.00
CCl_4	700	<0.0005	0.00

^{a} Extrapolated from data obtained using 1:1 mixture of H_2O and CH_3OH.

also listed in Table I. The significance of these quantities will be explained in the Discussion and in the Theoretical Section.

Lifetimes measured in oxygen saturated ($\sim10^{-2}$ M) solutions were not noticeably different from those obtained in air-saturated solutions. This sets an upper limit of 10^6 M^{-1} sec^{-1} for the bimolecular rate constant for the quenching of $^1\Delta$ by $^3\Sigma$ oxygen in solution. In some cases slight increases in photooxidation yields were observed with oxygen bubbling due to the more effective competition of oxygen quenching of dye triplets with other nonradiative decay processes (unimolecular decay, triplet–triplet annihilation, dimer quenching of monomer triplets, and quenching of singlet oxygen itself by triplet-state sensitizer).[17]

Mutual annihilation of two $^1\Delta$ molecules provides another possible decay channel.[10] The absence of a second-order component in the singlet oxygen decay curves and lack of effects arising from changes in $[^1\Delta]_0$ (achieved by varying the dye concentration) indicate that the annihilation rate constant in solution is <10^9 M^{-1} sec^{-1}.

2. Deuterium Effects on the Lifetime of Singlet Oxygen. The observation that there are large solvent effects on the lifetime of singlet oxygen[12] suggested that it would be worthwhile to investigate further the effect of deuter-

(17) C. K. Duncan and D. R. Kearns, *Chem. Phys. Lett.*, **10**, 306 (1971).

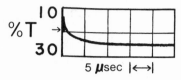

Figure 6. Same conditions as Figure 3a but with 5×10^{-3} M 2,4-dimethyl-2-butene added.

ation of the solvent on τ. Comparison of photooxidation efficiencies in nondeuterated and deuterated solvents under a fixed set of conditions has been used to obtain $^1\Delta$ lifetimes in the latter.[16] In the present studies the efficiency of rubrene photooxidation was observed to be the same (within an experimental accuracy of $\pm20\%$) in acetone and perdeuterioacetone. This indicates a singlet oxygen lifetime in perdeuterioacetone of approximately 26 μsec. These results along with those reported previously[16] are summarized in Table II. An interpretation of the deuterium effects will be presented later, and the optical densities which are listed in Table II are important in this respect.

Table II. Deuterium Effects on the Lifetime of Singlet Oxygen

Solvent	$\tau^1\Delta$, μsec	$OD_{7880 cm^{-1}}$ (1 cm)	$OD_{6280 cm^{-1}}$ (1 cm)
H_2O	2	0.47	3.4
D_2O	20	0.06	0.27
$H_2O:CH_3OH$, 1:1	3.5	0.33	3.7
$D_2O:CH_3OH$, 1:1	11	0.10	1.0
$D_2O:CD_3OD$, 1:1	35	0.03	0.30
CH_3COCH_3	26	0.015	0.08
CD_3COCD_3	26	0.002	0.13

3. Quantum Yield for Formation of $^1\Delta$ from Triplet Sensitizer. In dilute oxygenated solutions where only monomer dye is present, oxygen quenching is the only important decay mode for ^3MB. If the [DPBF] is kept low enough so as not to influence the decay of $^1\Delta$, then $[^1\Delta]_0 = \beta[AO_2]_\infty/[A]$. Comparison of $[^1\Delta]_0$ and the initial [^3MB] under such circumstances allowed us to determine that $^1\Delta$ is formed (either directly or from $^1\Sigma$) with an efficiency of 0.9 $\pm$ 0.1 when methylene blue triplets are quenched by oxygen. This result is in agreement with the theoretical prediction that for a molecule like MB oxygen quenching by energy transfer will predominate over quenching by enhancement of intersystem crossing to the ground state.[5]

4. Absolute Reaction Rate Constants. The disappearance of DPBF in a solution containing a second chemical or physical quencher of singlet oxygen, Q, having a quenching constant k_Q can be expressed in terms of

$$\log (OD - OD_\infty) = -2.3^{-1}((1/\tau) + k_A[A] + k_Q[Q])t + \text{constant}$$

Since τ and k_a(DPBF) have been measured as described previously, k_Q can be calculated from a first-order decay plot for a given [DPBF] and [Q].

If Q is an acceptor which quenches solely by reacting with singlet oxygen, then k_Q is the reaction rate constant k_a. The effect of 5×10^{-4} M 2,4-dimethyl-2-

butene on the decay of singlet oxygen is depicted in Figure 6. Both the duration and yield of DPBF photooxidation are decreased (compare with Figure 3a). An acceptor rate constant of k_a(2,4-dimethyl-2-butene) = 4×10^7 M^{-1} sec^{-1} is calculated from the first-order plot of this bleaching curve. Absolute rate constants in methanol obtained in this manner for a series of common acceptors are given in Table III. Probable error limits are $\pm 25\%$. Where com-

Table III. Absolute Acceptor Reaction Rate Constants in Methanol

Acceptor	k_a, M^{-1} sec^{-1}
1,3-Diphenylisobenzofuran	8×10^8
2,5-Dimethylfuran	4×10^8
2,4-Dimethyl-2-butene	4×10^7
1,2-Dimethylcyclohexene	1×10^7
2,5-Dimethyl-2,4-hexadiene	2×10^6

parison is possible, the resulting β values are in reasonable agreement with those obtained in methanol by steady-state methods.[15,18] It should be noted that as in the steady-state methods the presence of a small component in rate constant due to purely physical quenching cannot be ruled out.

Reaction rate constants for DPBF were not observed to be highly solvent dependent, varying only by about a factor of 2 over the range of solvents studied (excepting water).

5. Quenching of $^1\Delta$ by β-Carotene and I. Some time ago Foote and coworkers found β-carotene to be an extremely effective inhibitor of singlet oxygen reactions.[19,20] By assuming that β-carotene physically quenches singlet oxygen at a diffusion-controlled rate, they have been able to estimate a lifetime for $^1\Delta$ of 10^{-5} sec in a 4:1 benzene–methanol solution.[20] We have now been able to directly measure the β-carotene quenching of $^1\Delta$ and find $k_Q = 2 \times 10^{10}$ M^{-1} sec^{-1} in benzene. The polymethene pyrylium dye (I) is found to be a similarly efficient quencher of $^1\Delta$. In acetonitrile a value of $k_Q = 3 \times 10^{10}$ M^{-1} sec^{-1} is obtained for this dye.

Discussion

1. Solvents Effects on $^1\Delta$ Lifetime. Prior to the present work the lifetime of singlet oxygen was commonly believed to be nearly independent of solvent. This view was based on observations that β values for some common acceptors were not strongly dependent upon the nature of the solvent.[4,21,22] In contrast, large solvent effects (of the same order as our directly measured lifetimes would indicate) were observed on the rate of photooxidation of anthracene.[23] These results have recently been confirmed by Foote.[24] The results of our measurements, presented in Table I, demonstrate that there are very large solvent effects on the $^1\Delta$ life-

(18) E. Koch, *Tetrahedron*, **24**, 6295 (1968).
(19) C. S. Foote and R. W. Denny, *J. Amer. Chem. Soc.*, **90**, 6233 (1968).
(20) C. S. Foote, R. W. Denny, L. Weaver, Y. Chang, and J. Peters, *Ann. N. Y. Acad. Sci.*, **171**, 139 (1970).
(21) K. Gollnick, *Advan. Photochem.*, **6**, 1 (1968).
(22) C. S. Foote and R. W. Denny, *J. Amer. Chem. Soc.*, **93**, 5168 (1971).
(23) E. J. Bowen, *Advan. Photochem.*, **1**, 23 (1963).
(24) C. S. Foote, E. H. Peterson, and K. W. Lee, *J. Amer. Chem. Soc.*, **94**, 1032 (1972).

time. The apparent discrepancy between our work and the earlier studies may be traced in part to the use of mixed solvent systems and to solvent effects on reaction rate constants which tend to obscure lifetime variations in the steady-state experiments. Furthermore, the extremes in our measurements, water and carbon tetrachloride, have not previously been studied.

Examination of the data presented in Table I clearly indicates that the $^1\Delta$ lifetime is uncorrelated with most of the usual solvent properties. Contrary to the suggestion of Young and coworkers[25] no correlation between the lifetime of singlet oxygen and solvent polarity is indicated. For example, the lifetime of $^1\Delta$ in nonpolar cyclohexane is an order of magnitude greater than the lifetime in water but approximately half that in highly polar acetonitrile. Likewise, viscosity does not appear to play a prominent role in the radiationless decay of singlet oxygen. While carbon disulfide has *ca.* one-third the viscosity of water, the $^1\Delta$ lifetime is two orders of magnitude *greater* in the former. Similarly, polarizability, ionization potential, and oxygen solubility appear to be unimportant factors in the radiationless decay of singlet oxygen.

There is, however, one solvent property which does correlate well with the $^1\Delta$ lifetime, and this is infrared absorption intensity at 7880 and 6280 cm^{-1}. The theoretical basis of this relationship will be developed in detail in the next section.

2. Deuterium Effect on the $^1\Delta$ Lifetime. Although it hardly alters most of the properties of the solvent, we find that deuteration can have a very pronounced effect on the lifetime of singlet oxygen as the data in Table II demonstrate. Photooxidation rate measurement indicate that the lifetime of singlet oxygen is approximately ten times longer in D_2O than in H_2O. Similar large deuterium effects are observed in the water–methanol solvent mixtures. Interestingly deuteration of a solvent does not always lead to a large change in the singlet oxygen lifetime. There is very little difference between the lifetime of singlet oxygen in acetone and perdeuterioacetone. The effect of solvent deuteration on the lifetime of singlet oxygen clearly suggests that C–H and O–H vibrations are somehow important in relaxing singlet oxygen.

The striking deuterium effects on the lifetime of singlet oxygen is an additional salient experimental observation which is accounted for by the theoretical treatment in the following section.

3. Comparison of Solution and Gas-Phase Quenching of $^1\Delta$. It is interesting to compare quenching rates in solution with gas-phase values. Reciprocal lifetimes for water and benzene give quenching constants of 5×10^5 and 4×10^4 sec^{-1}, respectively. Gas-phase second-order rate constants for deactivation of $^1\Delta$ have been measured as $\sim 3 \times 10^3$ M^{-1} sec^{-1} for both water and benzene.[26]

To compare gas phase and solution quenching, consider the following scheme

$$^1\Delta + Q \underset{k_{-1}}{\overset{k_1}{\rightleftarrows}} (^1\Delta + Q) \overset{k_2}{\longrightarrow} {}^3\Sigma + Q$$

(25) R. H. Young, N. Chinh, and C. Mallon, *Ann. N. Y. Acad. Sci.*, **171**, 130 (1970).
(26) F. D. Findlay and D. R. Snelling, *J. Chem. Phys.*, **55**, 545 (1971).

where k_1 is the rate constant for formation, k_{-1} is the rate constant for dissociation, and k_2 is the first-order quenching constant of the $(^1\Delta + A)$ collision complex. A kinetic analysis assuming steady-state for $(^1\Delta + Q)$ and $k_2 \ll k_{-1}$ gives

$$-\frac{d[^1\Delta]}{dt} = \frac{k_1 k_2}{k_{-1}}[^1\Delta][Q] = k_Q[^1\Delta][Q]$$

where k_Q is the experimentally observed second-order quenching constant. In the gas phase where the collision frequency is $\sim 10^{11}$ sec^{-1} at 1 M and the estimated pair lifetime is 10^{-12} sec, we find $k_2 \sim 10 k_Q$. From the observed gas-phase values for k_Q, we calculate $k_2 \simeq 3 \times 10^4$ sec^{-1} for both water and benzene. To the extent that gas-phase and solution-state collision complexes are the same, k_2 should be the same for both phases.

In solution each singlet oxygen molecule is in constant contact with N nearest-neighbor solvent molecules and we can write $1/\tau \sim N k_2$. If we estimate N to be 5 for water and 2 for benzene, then from the observed values in solution, k_2 is calculated to be $\sim 10^5$ and $\sim 2 \times 10^4$ sec^{-1} in the respective solvents in reasonable agreement with the values derived from gas-phase data.

Considerations of this nature are not unique to $^1\Delta$ but apply to intermolecular deactivation of any molecule when $k_2 \ll k_{-1}$. For example, from the vapor phase rate constant for quenching of $^1\Sigma$ oxygen by water[27, 28] we can estimate the lifetime of $^1\Sigma$ in liquid water to be $\sim 10^{-11}$ sec $(k_2 \sim 2 \times 10^{10})$.

4. Absolute Reaction Rates. Absolute acceptor rate constants and absolute quenching rate constants may now be calculated from published β values for the solvents in which singlet oxygen lifetimes have been measured (refer to Table I). A knowledge of solvent influences on reaction reate constants should be helpful in elucidating reaction mechanisms.

5. Dye Quenching of $^1\Delta$. In retrospect, the assumption of Foote and coworkers that the quenching of singlet oxygen by β-carotene is diffusion controlled appears to have been correct. Using their quenching data in a 4:1 benzene–methanol solution[20] and our measured value of $k_Q = 2 \times 10^{10}$ M^{-1} sec^{-1} for β-carotene in benzene (as opposed to their estimate of 3×10^{10} M^{-1} sec^{-1} based on Ware's measurement of oxygen quenching of the fluorescence of naphthalene in benzene),[29] we obtain a $^1\Delta$ lifetime of 15 μsec. From the data in Table I a lifetime of 16 μsec can be calculated for this same solvent mixture.

The quenching efficiency of β-carotene can most plausibly be accounted for by a spin-allowed transfer of electronic excitation energy from singlet oxygen to β-carotene, provided the lowest triplet state of β-carotene is either nearly degenerate with or energetically below the $^1\Delta$ level. Reactive quenching cannot explain the observed efficiency since β-carotene is not consumed in the quenching process.[20] The observation[30] that oxygen quenches β-carotene triplets with a rate constant of $\sim 10^9$ M^{-1} sec^{-1} does not conflict with

the above interpretation, since for low-lying triplets, oxygen quenching by enhancement of intersystem crossing (as opposed to energy transfer from the triplet to produce singlet oxygen) becomes important.[5, 31]

Dyes of the type used in Q-switching neodymium lasers exhibit allowed optical absorption in the region of 9000 cm^{-1}.[32] If these dyes have normal singlet–triplet splittings, then they almost certainly have triplet states which lie below $^1\Delta$. This is borne out by the high quenching efficiency of the polymethene pyrylium dye, a member of this category.[33] While the quenching constant for I is slightly higher than that for β-carotene, this probably is due to the lower viscosity of the acetonitrile solvent.

Theoretical Section

We noted in the previous section that the lifetime of $^1\Delta$ is uncorrelated with most of the common properties of the solvent. There is, however, a striking parallel between the $^1\Delta$ lifetime and the intensity of the solvent absorptions near 7880 and 6280 cm^{-1}, resonant with $0 \to 1$ and $^1\Delta \to {}^3\Sigma$ transitions. This correlation is reminiscent of the electric dipole resonant electronic energy transfer first treated by Förster[34] and expanded by Dexter,[35] except that the present case would require electronic-to-vibrational energy transfer (the conversion of electronic excitation of oxygen into vibrational excitation of the solvent). The substantial deuterium effects observed also suggest that solvent vibrations play a prominent role in the decay of singlet oxygen. At this point we wish to develop the theoretical framework which will permit us to account for the various experimental features of the $^1\Delta$ decay and to make predictions regarding the behavior of singlet oxygen in solutions which have not yet been studied.

Radiationless decay in large molecules has usually been treated in terms of a primarily intramolecular mechanism.[36-39] In this approach only small amounts of energy are taken up by indirectly coupled low-frequency vibrations of the medium. In the absence of heavy atom effects relaxation is thus expected to be essentially independent of the nature of the solvent. In contrast to the behavior of large molecules the solvent can be expected to play a major role in radiationless transitions of small molecules, such as oxygen, where internal vibrational modes cannot provide the required set of vibrational states nearly degenerate with the initial state.[38, 39]

General Formulation. In our theoretical treatment of the radiationless decay of singlet oxygen we will make use of the general formalism developed by Robinson and Frosch.[36] Radiationless transitions can be considered to arise from interaction of zero-order nonstationary Born–Oppenheimer initial, ψ_i, and final,

(27) S. V. Filseth, A. Zia, and K. H. Welge, *J. Chem. Phys.*, **52**, 5502 (1970).

(28) F. Stuhl and H. Niki, *Chem. Phys. Lett.*, **7**, 473 (1970).

(29) W. R. Ware, *J. Phys. Chem.*, **66**, 455 (1962).

(30) E. J. Land, S. Sykes, and T. G. Truscott, *Photochem. Photobiol.*, **13**, 311 (1971).

(31) D. R. Kearns, *Chem. Rev.*, **71**, 395 (1971).

(32) F. P. Schäfer, *Angew. Chem.*, **9**, 9 (1971).

(33) J. L. R. Williams and G. A. Reynolds, *J. Appl. Phys.*, **39**, 5327 (1968).

(34) T. Förster, *Ann. Phys., Phys. Chem.*, **2**, 55 (1948); *Discuss. Faraday Soc.*, **27**, 7 (1959).

(35) D. L. Dexter, *J. Chem. Phys.*, **21**, 836 (1953).

(36) G. W. Robinson and R. P. Frosch, *ibid.*, **37**, 1962 (1962); **38**, 1187 (1963).

(37) B. R. Henry and M. Kasha, *Annu. Rev. Phys. Chem.*, **19**, 161 (1968).

(38) G. W. Robinson, *J. Chem. Phys.*, **47**, 1967 (1967).

(39) J. Jortner, S. A. Rice, and R. M. Hochstrasser, *Advan. Photochem.*, **7**, 149 (1969).

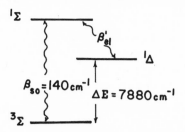

Figure 7. A schematic representation of the coupling scheme responsible for the relaxation of $^1\Delta$ oxygen. β_{so} is a spin–orbit coupling matrix element responsible for coupling the $^1\Sigma$ and $^3\Sigma$ states, and β_{el}' is a matrix element describing the solvent-induced mixing of the $^1\Sigma$ and $^1\Delta$ states.

ψ_f, states under the influence of an interaction Hamiltonian $\mathcal{H}'$. In this approach the rate constant for the transition from the initial state i to a particular final state f may be written as

$$k_{if} = \frac{2\pi}{\hbar \alpha}|\beta_{if}|^2 \simeq \frac{2\pi \tau_{vib}}{\hbar^2}|\beta_{if}|^2 \qquad (4)$$

where α is the energy of interaction between the final state and the molecules of the medium, τ_{vib} is the vibrational relaxation time of the solvent, and $\beta_{if} = \langle \psi_i/\mathcal{H}'/\psi_f \rangle$. If transitions can occur to a number of different final states, then the net quenching rate constant will involve a sum over all possible final states.

With neglect of electron exchange the interaction between O_2 and some solvent molecule with a dipole moment μ can be approximated by

$$\mathcal{H}' = \sum_i \frac{q_i \mu \cos \theta_i}{R_i^2} \qquad (5)$$

where q_i is the charge on an electron or oxygen nucleus located at a distance R_i from the center of the solvent dipole and θ_i is the angle formed between $\mathbf{R}_i$ and $\boldsymbol{\mu}$. If we express wave functions as products of oxygen (ψ) and solvent (Ω) wave functions and substitute the above expression for $\mathcal{H}'$ into eq 2, we obtain the following expression for β_{if}.

$$\beta_{if} = \sum_{mn}\left\langle \psi^0{}_{1\Delta}\Omega^0 \left| \sum_i \frac{q_i \mu \cos \theta_i}{R_i^2} \right| \psi^m{}_{3\Sigma}\Omega_n \right\rangle \equiv \sum_{mn}\beta_{mn} \qquad (6)$$

where subscripts on ψ refer to electronic states, and m and n to vibrational levels related by conservation of energy.

Making use of the Born–Oppenheimer separability we can further factor the matix elements and obtain the following expression

$$\beta_{mn} = \beta_{el} F_m{}^{1/2}\langle \Omega_0|\mu|\Omega_n \rangle \qquad (7)$$

in which

$$\beta_{el} = \left\langle \psi(^1\Delta)\left| \sum_i \frac{e \cos \theta_i}{R_i^2} \right| \psi(^3\Sigma) \right\rangle \qquad (8)$$

where ψ denotes purely electronic oxygen wave functions, Ω_0 and Ω_n are purely vibrational wave functions for the ground electronic state of the *solvent*, and $F_m = |\langle \chi_0(^1\Delta)|\chi_m(^3\Sigma)\rangle|^2$ is the Franck–Condon overlap between the vibrational wave functions, χ, for the specific initial and final electronic states of oxygen indicated.

Since $\psi(^1\Delta)$ and $\psi(^3\Sigma)$ are orthogonal and not explicit functions of nuclear coordinates, all terms in β_{el} involving the nuclear coordinates of oxygen will vanish.

From the nature of the electronic wave functions for oxygen,[31] we immediately conclude that in the absence of some spin-dependent perturbation $^1\Delta$ cannot couple with $^3\Sigma$. Consequently, in nonheavy atom solvents there is no solvent-induced *direct* mixing of the $^3\Sigma$ and $^1\Delta$ states. We know from studies of spin–orbit coupling in oxygen, however, that the $^1\Sigma$ and $^3\Sigma$ states are strongly coupled by a matrix element of ~ 140 cm^{-1}.[40,41] Nonheavy atom solvents can, thus, induce *indirect* mixing of $^1\Delta$ and $^3\Sigma$ by causing mixing of $^1\Sigma$ and $^1\Delta$ states as indicated schematically in Figure 7. Application of second-order perturbation theory[42] to this mixing scheme yields the following modified expression for β_{el}

$$\beta_{el} = \beta_{el}' \frac{\beta_{so}}{\Delta E} \qquad (9)$$

in which

$$\beta_{el}' = \left\langle \psi(^1\Delta)\left| \sum_i \frac{e \cos \theta_i}{R_i^2} \right| \psi(^1\Sigma) \right\rangle \qquad (10)$$

and

$$\beta_{so} = \langle \psi(^1\Sigma)|\mathcal{H}_{so}|\psi(^3\Sigma)\rangle = 140 \text{ cm}^{-1} \qquad (11)$$

For reasons which will become evident, it is desirable to expand μ as is commonly done in treatments of infrared vibrational selection rules[43] and write

$$\mu = \mu_0 + \sum_j \left(\frac{\partial \mu}{\partial Q_j}\right)_{Q_{j=0}} \cdot Q_j + \cdots \qquad (12)$$

where μ_0 is the permanent ground-state dipole moment of the solvent molecule and Q_j is a normal coordinate. Substituting we obtain

$$\beta_{mn} = \beta_{el}' \times \frac{\beta_{so}}{\Delta E} F_m{}^{1/2} M_n \qquad (13)$$

where

$$M_n = \left\langle \Omega_0 \left| \mu_0 + \sum_j \left(\frac{\partial \mu}{\partial Q_j}\right)_{Q_{j=0}} Q_j \right| \Omega_n \right\rangle \qquad (14)$$

When energy is transferred into internal vibrations of the solvent molecules, $n \neq 0$ and thus the term containing μ_0 vanishes due to the orthogonality of the solvent vibronic wave functions.

Using the above expressions, we can express the quenching rate constant as

$$k_{if} \simeq \frac{2\pi \tau_{vib}}{\hbar^2}\left| \beta_{el}' \times \frac{\beta_{so}}{\Delta E} \right|^2 \sum_{mn} F_m |M_n|^2 \qquad (15)$$

Variation of Lifetime with Solvent Infrared Band Intensities. At this point, we note that the quantity M_n which appears in the above quenching rate constant expression is the same factor which determines the intensity of $0 \rightarrow n$ transitions of the solvent in the infrared region,[43] and consequently we can obtain M_n directly from spectral data for the solvent.

(40) K. Kayama and J. C. Baird, *J. Chem. Phys.*, **46**, 2604 (1967).
(41) O. Zamani-Khamira and H. F. Hameka, *ibid.*, **55**, 2191 (1971).
(42) W. Kauzmann, "Quantum Chemistry," Academic Press, New York, N. Y., 1957, p 175.
(43) K. S. Pitzer, "Quantum Chemistry," Prentice-Hall, New York, N. Y., 1953, p 372.

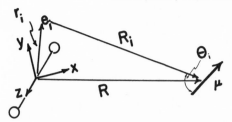

Figure 8. A schematic representation showing the orientation of a solvent molecule with dipole moment μ relative to oxygen at the origin.

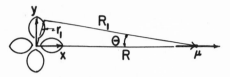

Figure 9. A schematic representation depicting a special orientation of the solvent molecule relative to oxygen. The molecular axis of oxygen is perpendicular to the plane of the paper.

The quantity F_m is the Franck–Condon factor for a transition from the 0th vibrational level of the $^1\Delta$ state to the mth vibrational level of the ground $^1\Sigma$ electronic state. Because the potential energy curves for oxygen in its $^1\Delta$ and $^1\Sigma$ state are so similar, both in regard to average internuclear separation (1.207 *vs.* 1.215 Å) and force constants ($\omega = 1580$ *vs.* $\omega = 1509$ cm^{-1}),[44] the Franck–Condon factors decrease extremely rapidly as m increases. For the isolated molecule, the values decrease as $\sim 1:10^{-2}:10^{-4}:2 \times 10^{-7}$ as m increases from 0 to 3.[45] Complete intramolecular conversion of energy from $^1\Delta$ into vibration of $^3\Sigma$ requires excitation to $m = 5$. Because of the rapid decrease in F_m, the relaxation of $^1\Delta$ oxygen is expected to be most rapid when *large* amounts of electronic energy can be transferred to vibrational excitation of the solvent. If, for example, all of the $^1\Delta$ excitation energy is transferred to the solvent, theory predicts that the $^1\Delta$ decay rate will be directly proportional to the solvent ir absorption intensity at ~ 7880 cm^{-1}. If the solvent absorption bands at this energy are not particularly strong, then quenching in which $^3\Sigma$ is formed in its $m = 1$ vibrational state and the solvent picks up (7880–~ 1600) cm^{-1} of vibrational excitation will also occur, and solvent absorption in the region of 6280 cm^{-1} becomes important.

Quantitative Test of Theory. It is evident that theory can at least qualitatively account for the dramatic variation of the life-time of singlet oxygen in various solvents and for the fact that there is a good correlation between infrared absorption intensity and quenching rate constants. As a quantitative test of the theory, let us try to compute the lifetime of $^1\Delta$ in a particular solvent. Since experimental values of M_n are available, the only quantity that we will actually have to compute at this point is β_{el}'. To do this, we choose a coordinate system with an oxygen molecule at its center and the z axis as the molecular axis as illustrated in Figure 8.

Noting that $R = R_i + r_i$ or $R_i^2 = (R^2 - r_i^2)$, where r_i is the distance of the ith oxygen electron from the origin, we can write

$$\mathcal{H}' = -\sum_i \frac{e\mu \cos \theta_i}{R^2[1 - (r_i/R^2)]} \quad (16)$$

A Maclaurin series expansion of $[1 - (r_i^2/R^2)]^{-1}$ yields

$$\mathcal{H}' = -\sum_i \frac{e\mu \cos \theta_i}{R^2}\left(1 + \frac{r_i^2}{R^2} + \ldots\right) \quad (17)$$

(44) G. Herzberg, "Spectra of Diatomic Molecules," Van Nostrand, Princeton, N. J., 1950.
(45) M. Halmann and I. Laulicht, *J. Chem. Phys.*, **43**, 438 (1965).

Referring to eq 10 we can write

$$\beta_{el}' = -\frac{e}{R^2}\sum_i \left\langle \psi(^1\Delta) \left| \cos \theta_i \left(1 + \frac{r_i^2}{R^2} + \ldots\right)\right| \psi(^1\Sigma)\right\rangle \quad (18)$$

If we consider only the outermost pair of oxygen electrons explicitly, then the zero-order electronic wave functions for $^1\Delta$ and $^1\Sigma$ can be expressed in the form[31]

$$\psi(^1\Delta) = \frac{1}{\sqrt{2}}\{|\pi_x(1)\bar{\pi}_x(2)| - |\pi_y(1)\bar{\pi}_y(2)|\}$$

$$\psi(^1\Delta) = \frac{1}{\sqrt{2}}\{|\pi_x(1)\bar{\pi}_y(2)| - |\bar{\pi}_x(1)\pi_y(2)|\} \quad (19)$$

$$\psi(^1\Sigma) = \frac{1}{\sqrt{2}}\{|\pi_x(1)\bar{\pi}_x(2)| + |\pi_y(1)\bar{\pi}_y(2)|\}$$

where π_x and π_y are antibonding π orbitals occupied by unpaired electrons (1) and (2) and a bar indicates spin β. From the character of these functions it is evident that only interactions involving the unpaired electrons will be effective in mixing, and that a one-electron perturbation of the form $V = V(1) + V(2)$ which destroys axial (z) symmetry will suffice, since

$$\langle\psi(^1\Delta)|V|\psi(^1\Sigma)\rangle = \frac{1}{2}\{\langle\pi_x(1)|V(1)|\pi_x(1)\rangle +$$

$$\langle\pi_x(2)|V(2)|\pi_x(2)\rangle - \langle\pi_y(1)|V(1)|\pi_y(1)\rangle -$$

$$\langle\pi_y(2)|V(2)|\pi_y(2)\rangle\} \quad (20)$$

In the present situation $V(1)$ and $V(2)$ have the same form and we can write

$$\beta_{el}' = \langle\pi_x|V(1)|\pi_x\rangle - \langle\pi_y|V(1)|\pi_y\rangle \quad (21)$$

$$V(1) = -\frac{e}{R^2}\cos \theta_1\left(1 + \frac{r_1^2}{R^2} + \ldots\right) \quad (22)$$

To make a crude estimate of the magnitude of β_{el}', consider the special situation in which the solvent dipole points along the x axis as in Figure 9. Using only the first term in (22) we have

$$\beta_{el}' = -\frac{e}{R^2}(\langle\pi_x|\cos \theta_1|\pi_x\rangle - \langle\pi_y|\cos \theta_1|\pi_y\rangle) \quad (23)$$

For an electron in the π_x orbital $\cos \theta_1 \simeq 1$, while in the π_y orbital $\cos \theta_1 = (1 - \sin^2 \theta_1)^{1/2} \simeq 1 - (Y_1^2/2R^2)$. Thus

$$\beta_{el}' \simeq -\frac{e}{2R^4}\langle\pi_y|Y_1^2|\pi_y\rangle \quad (24)$$

Using $R = 5$ Å and $\langle\pi_y|Y_1^2|\pi_y\rangle = 1$ Å^2, we obtain $\beta_{el}' \simeq 4 \times 10^3$ esu/cm^2 and $\beta_{el} \simeq -80$ esu/cm^2. The contribution to β_{el}' from the r_i^2/R^2 term in $V(1)$

will be reduced by approximately an order of magnitude. Consequently, our neglect of this term and other higher order terms is justifiable.

The interaction matrix element can now be expressed as

$$\beta_{mn} \simeq 80 \ (\text{esu/cm}^2) F_m{}^{1/2} M_n \qquad (25)$$

If we assume the O–O oxygen transition (for which $F_0 \sim 1$) to be most important, then $|M_n|^2$ can be obtained from the solvent infrared absorption intensity in the region of 7880 cm^{-1} through use of the relationship[46]

$$|M_n|^2 = \frac{10^{-38}}{\bar{\nu}} \int \epsilon d\bar{\nu} \qquad (26)$$

The molar extinction coefficient for water at 7880 cm^{-1} is $\simeq 10^{-2}$. Integrating over 200 cm^{-1} (kT) gives $|M_n|^2 \simeq 2 \times 10^{-42}$ esu^2 cm^2, from which we calculate $\beta_{0n} \simeq -10^{-19}$ erg $\simeq -5 \times 10^{-4}$ cm^{-1}. Finally, assuming $\tau_{\text{vib}} \simeq 10^{-11}$ sec,[36] we obtain a value for the rate constant for the radiationless decay of singlet oxygen in water of

$$k_{\text{if}} \simeq \frac{2\pi\tau_{\text{vib}}|\beta_{0n}|^2}{\hbar^2} \simeq 6 \times 10^5 \ \text{sec}^{-1} \qquad (27)$$

In view of the approximations, the agreement with the experimental estimate of $k_2 \simeq 10^5$ sec^{-1} is remarkably good. We have considered a particularly favorable orientation of the solvent dipole relative to the oxygen molecule and the average orientation would not produce ·as strong an interaction. On the other hand, at the small distances considered the dipole approximation is not strictly valid and leads to an underestimation of β_{el}'. A more accurate, but involved treatment would require a consideration of the electrostatic interaction of a system of point charges. .

In order to properly include the effects of the $0 \rightarrow 1$, $0 \rightarrow 2$, and other nonadiabatic transitions, we need the Franck–Condon factors appropriate to each. Isolated molecule Franck–Condon factors are probably not appropriate for oxygen molecules in solution, since there are experimental data which indicate that these may be rather sensitive to solvent perturbations. For example, Evans has observed $0 \leftarrow 0 : 0 \leftarrow 1$ intensity ratio of $\sim 1 : 0.3$ for the $^1\Delta \leftarrow {}^3\Sigma$ absorption of oxygen at high pressure in 1,1,2-trichlorotrifluoroethane.[47] If we assume that in the present situation $F_1 \simeq 0.1$ (intermediate between the isolated molecule and high pressure cases), then with reference to the data in Tables I and II we obtain the following empirical expression relating the $^1\Delta$ lifetime to solvent ir absorption.

$$\frac{1}{\tau} \ (\mu\text{sec})^{-1} \simeq 0.5(\text{OD}_{7880}) + 0.05(\text{OD}_{6280}) +$$

$$\text{higher terms} \qquad (28)$$

where OD_{7880} and OD_{6280} are optical densities of 1 cm of solvent at 7880 and 6280 cm^{-1}, respectively.[48]

Examination of the data in Tables I and II indicates that for most solvents higher terms in expression 28

(46) J. N. Murrell, "The Theory of the Electronic Spectra of Organic Molecules," Wiley, New York, N. Y., 1963, p 9.
(47) D. F. Evans, *Chem. Commun.*, 367 (1969).
(48) The previous theoretical treatment indicates that k_{if} will be proportional to oscillator strength multiplied by the number of coupled solvent oscillators or roughly to optical density.

can be neglected. Lifetimes in benzene, chloroform, carbon disulfide, and carbon tetrachloride, however, are substantially shorter than those calculated using only the 0–0 and 0–1 terms, but the correlation can be improved by inclusion of terms involving transitions to higher vibrational levels of $^3\Sigma$. The $0 \rightarrow 2\ {}^1\Delta \rightarrow {}^3\Sigma$ transition energy is $\simeq 4700$ cm^{-1}. Benzene has a very strong absorption in this region (OD of 1 cm $\simeq 7.0$). A Franck–Condon factor of $F_2 \simeq 5 \times 10^{-3}$ (*vs.* 10^{-4} for an isolated oxygen molecule) would bring the predicted and observed $^1\Delta$ lifetimes in benzene into good agreement. Chloroform and carbon disulfide exhibit moderate absorption within $\simeq kT$ of 4700 cm^{-1}, and thus radiationless decay of $^1\Delta$ to the $m = 2$ level of $^3\Sigma$ is expected to be important in these solvents also. Of the remaining solvents studied only acetone shows an absorption at 4700 cm^{-1} which is strong (OD $\simeq 3.0$ in a 1-cm cell) relative to that at 7880 and 6280 cm^{-1}. With an F_2 of 5×10^{-3} the value of τ in acetone calculated from expression 28 is closer to the measured lifetime. Transitions further into the infrared evidently must be included to account for the $^1\Delta$ lifetime in carbon tetrachloride.

When oxygen Franck–Condon factors in various solvents become available, more accurate calculations of radiationless decay rates will be permitted. Singlet oxygen lifetimes are expected to be longest in solvents such as nitrogen and argon which lack infrared intensity. It is worth noting that for a particularly strong solvent–oxygen interaction, transitions to the $m = 5$ vibrational level of $^3\Sigma$ may become more allowed. Hence transfer of excitation energy to the solvent would be less important, and the static dipole moment of the solvent could become significant in radiationless decay.

Deuterium Effects. Since the infrared overtone bands of common solvents in the regions of interest are usually due to C–H or O–H vibrations, the theory predicts that there can be large solvent deuterium effects on the lifetime of singlet oxygen. The lifetime of $^1\Delta$ in water is 2 μsec, and a comparison of the infrared absorption intensities of H_2O and D_2O (Table II) suggests that the lifetime in the latter solvent should increase by approximately a factor of 9. Our experimental studies of the photooxidation efficiencies in H_2O and D_2O indicate that there is actually a tenfold change.[16] A similar tenfold increase in the singlet oxygen lifetime is predicted, and observed, in going from a 1:1 mixture of $H_2O:CH_3OH$ to $D_2O:CD_3OD$.

Solvent deuteration will not necessarily lead to an increase in the lifetime of singlet oxygen as the case of acetone illustrates. The reduction in absorption intensity at 7880 cm^{-1} upon deuteration of acetone is almost completely compensated for by the intensity increase at 6280 cm^{-1}. Although deuteration will in general lead to a decrease in near-ir absorption intensities, it is possible that a situation might arise in which deuteration shifts would create resonances that would actually enhance $^1\Delta$ relaxation.

Because deuteration involves a very minor perturbation of the solvent, the deuterium effect on the lifetime of singlet oxygen can be used as a powerful diagnostic tool for investigating the role of singlet oxygen in various chemical and physical processes. Recently, for example, we have used the deuterium effect to demonstrate that the methylene blue sensitized photooxidation of tryptophan occurs *via* reaction with singlet oxygen.[16,49] This technique should have widespread application in areas such as photodynamic action.

Order of magnitude increases in the $^1\Delta$ lifetime upon deuteration are predicted for methanol and chloroform and are consistent with some preliminary experimental observations. Thus, these will also be useful solvents for characterizing singlet oxygen reactions.

Heavy Atom Effects. Since the $^1\Delta \rightarrow {}^3\Sigma$ transition is formally spin forbidden, one might have expected to observe large external heavy atom solvent effects on the relaxation rate. Our theoretical analysis indicates that this may not be the case, however, in the relaxation of $^1\Delta$ for the following reason. There is a large spin–orbit mixing of the $^1\Sigma$ state with the $^3\Sigma$ ground state, and mixing of $^1\Delta$ with $^3\Sigma$ is indirectly achieved by an electrostatically induced mixing of $^1\Delta$ with $^1\Sigma$. In order for there to be a specific heavy atom solvent effect on the relaxation, the heavy atoms would have to provide some new route for mixing $^1\Delta$ and $^3\Sigma$. An entirely similar process is responsible for the external heavy atom enhancement of the radiationless relaxation of an excited triplet-state molecule to its ground state, and we can use triplet data to place an upper limit on the importance of this mechanism in the oxygen case. For many of the aromatic hydrocarbons in a bromine-containing solvent (tetrabromozene), the radiative lifetimes are on the order of 50 msec and nonradiative transition rates are estimated to be much smaller than $10^2/\text{sec}$.[50] If we assume that similar electronic matrix elements will be obtained in the case of oxygen, then heavy atom solvents are not expected to enhance the relaxation of $^1\Delta$, since nonheavy atom solvents already give relaxation rates of $\sim 10^5/\text{sec}$.

Amine Quenchers. The amines and other molecules with low ionization potentials may introduce new mixing routes involving charge-transfer states. As developed, our theory does not include this effect, but it could be extended to treat such cases.

(49) R. Nilsson, P. B. Merkel, and D. R. Kearns, *Photochem. Photobiol.*, **16**, 117 (1972).
(50) G. G. Giachino and D. R. Kearns, *J. Chem. Phys.*, **52**, 2964 (1970).

Relation between Quenching of $^1\Sigma$ and $^1\Delta$. In spite of the large spin–orbit coupling of $^1\Sigma$ and $^3\Sigma$, direct relaxation of $^1\Sigma$ to $^3\Sigma$ is expected to be small because of the extremely unfavorable oxygen Franck–Condon factors for those $^1\Sigma \rightarrow {}^3\Sigma$ transitions which coincide with solvent infrared bands of measurable intensity. The β_{el}' used in our treatment of $^1\Delta$ relaxation is precisely (without reduction by β_{so}/Δ_E) the quantity which determines relaxation of $^1\Sigma$ to $^1\Delta$, and thus it can be used to compute the rate constants for quenching of $^1\Sigma$ by various solvents. As in the relaxation of $^1\Delta$ to $^3\Sigma$, we expect relaxation of $^1\Sigma$ to $^1\Delta$ without energy transfer to the solvent to be hindered by the small Franck–Condon factors. For the adiabatic $^1\Sigma$ to $^3\Sigma$ transition the solvent will have to take up approximately 5200 cm^{-1} of vibrational excitation. For water the molar extinction coefficient in this region is ~ 1.0 from which we calculate a value of $\beta_{0n} \simeq 0.4$ cm^{-1}. Substitution of this value into eq 27 gives a value of $k_{if} \simeq 3 \times 10^{10}$ sec^{-1} for the quenching rate constant of water which should be compared with our earlier estimate from experimental data of $k_2 \simeq 2 \times 10^{10}$ sec^{-1}.

Application to Other Systems. Radiationless decay of other simple molecules may also fall within the scope of the present theoretical treatment. Molecules with low-lying excited states for which $|M_n|^2$ for the solvent is large in regions of favorable F_m would be the most likely candidates. Cases intermediate between the small molecule limit and the primarily intramolecular relaxation characteristic of large molecules may also exist.

The solvent dependence of radiationless decay of $^1\Delta$ is similar to that for europium(III) which was recently characterized in detail by Hass and Stein.[51] Preliminary studies[52] suggest that radiationless transitions in rare earth ions are another system to which the above theory might be applied.

Acknowledgments. The support of the U. S. Public Health Service (Grant GM 10449) is most gratefully acknowledged. We thank Dr. J. Williams for the gift of compound I.

(51) Y. Hass and G. Stein, *J. Phys. Chem.*, **75**, 3668, 3677 (1971).
(52) C. Long and D. R. Kearns, unpublished results.

77

Copyright © 1973 by the American Chemical Society

Reprinted from *J. Amer. Chem. Soc.*, **95**(2), 375–379 (1973)

The Determination of Rate Constants of Reaction and Lifetimes of Singlet Oxygen in Solution by a Flash Photolysis Technique

Robert H. Young,* David Brewer, and Richard A. Keller

Contribution from the Department of Chemistry, Georgetown University, Washington, D. C. 20007, and National Bureau of Standards, Washington, D. C. 20234. Received June 12, 1972

Abstract: A flash photolysis technique, employing a dye-laser as the flash excitation source, was developed to determine, directly, the rate constant of decay (k_d) of singlet oxygen ($O_2{}^1\Delta_g$) and rate constant of reaction (k_{rx}) between singlet oxygen and 1,3-diphenylisobenzofuran, DPBF, in a variety of solvents. Some results are the following: methanol, $k_d = (9.0 \pm 0.6) \times 10^4$ sec^{-1}, $k_{rx} = (1.3 \pm 0.1) \times 10^9$ sec^{-1} M^{-1}, $k_d/k_{rx} = 7.2 \times 10^{-5}$ M; *n*-butyl alcohol, $k_d = (5.2 \pm 0.8) \times 10^4$ sec^{-1}, $k_{rx} = (0.80 \pm 0.2) \times 10^9$ sec^{-1} M^{-1}, $k_d/k_{rx} = 6.5 \times 10^{-5}$ M; *tert*-butyl alcohol, $k_d = (3.0 \pm 0.4) \times 10^4$ sec^{-1}, $k_{rx} = (0.57 \pm 0.08) \times 10^9$ sec^{-1} M^{-1}, $k_d/k_{rx} = 5.3 \times 10^{-5}$ M. Values in benzene:methanol (4:1), bromobenzene:methanol (4:1), pyridine, dioxane, methanol:water (1:1), methanol:glycol (1:1), and glycol were also obtained. The resulting β values (k_d/k_{rx}) are in good agreement with β values determined in other ways. The technique was also shown useful for determination of absolute rate constants of reaction with or quenching of singlet oxygen by compounds such as 2,5-diphenylfuran and *N,N*-dimethylaniline.

Due to the wide interest in and application of singlet oxygen chemistry, it is of importance to know the lifetime of singlet oxygen in various solvents.

Early attempts to determine the lifetime of the reactive intermediate, singlet oxygen, in photooxidation reactions resulted in a lower limit of a few hundred nanoseconds.[1] More recently, several research workers have used β-carotene to quench singlet oxygen, and from the results and with the assumption that the quenching action occurs at the diffusion limit, they have calculated the lifetime of singlet oxygen in various solvents.[2–4] The results are dependent upon the knowledge of the rate constants of diffusion which vary considerably for a given solvent.[5,6]

Although determination of β values (k_d/k_{rx}) for reactions between singlet oxygen and an acceptor suffer from some limitations, the values are becoming more reliable and are now easily obtained.[3] In many cases it is desirable to know the variation of the lifetime of singlet oxygen and its rate constant of reaction as a function of some parameter (*i.e.*, solvent polarity). Variations in these individual rate constants can be hidden if only their ratio is measured. As an example of this limitation, Foote found no relation between β values and the polarity or viscosity of the solvent.[7] On the other hand, Young and coworkers found that there was a relationship between solvent polarity and relative rate constants of reaction for substituted furans and other compounds.[3,4,8] The assumptions inherent

in the technique and interference with the excited states of the dye-sensitizers could introduce errors into some of these results.

Merkel and Kearns developed a method for determining the rate constant for the decay of singlet oxygen and the rate constant for the rate of reaction of singlet oxygen with 1,3-diphenylisobenzofuran (DPBF).[9] A modification of this technique was used to determine k_d and k_{rx} in a variety of solvents. In addition, the rate constant for quenching of singlet oxygen by *N,N*-dimethylaniline and the rate constant of reaction of 2,5-diphenylfuran with singlet oxygen were determined.

The modified technique involved the use of an organic dye laser to produce a high concentration of singlet oxygen in less than 1 μsec. The rate of decay of singlet oxygen was measured by monitoring the rate of the reaction of singlet oxygen with 1,3-diphenylisobenzofuran, DPBF. Analysis of the data yielded the rate constant of the reaction of DPBF with singlet oxygen (k_{rx}) and the rate constant of the decay of singlet oxygen (k_d). The following kinetic equations illustrate the technique

$$^1\text{Dye-Sens}_0 \xrightarrow[\text{laser}]{h\nu} {}^1\text{Dye-Sens}_1{}^* \xrightarrow{\text{ISC}} {}^3\text{Dye-Sens}_1{}^* \quad (1)$$

$$^3\text{Dye-Sens}_1{}^* + {}^3O_2 \longrightarrow {}^1\text{Dye-Sens}_0 + {}^1O_2 \quad (2)$$

$$^1O_2 \xrightarrow{k_d} {}^3O_2 \quad (3)$$

$$^1O_2 + \text{DPBF} \xrightarrow{k_{rx}} \text{DPBF-O}_2 \longrightarrow \text{products} \quad (4)$$

where DPBF is assumed to undergo reaction with $O_2{}^1\Delta$ without quenching of $O_2{}^1\Delta$ by comparison with dimethylfuran[3] and ISC refers to intersystem crossing from the singlet manifold to the triplet manifold.

Steps 1 and 2 are fast with respect to steps 3 and 4. Additional rate constants can be studied by including eq 5 in the kinetic scheme. Intense irradiation is neces-

$$^1O_2 + X \xrightarrow{k_q{}'} \text{products} \quad (5)$$

(1) G. O. Schenck and E. Koch, *Z. Elektrochem.*, **64**, 170 (1960).

(2) C. S. Foote and R. W. Denny, *J. Amer. Chem. Soc.*, **90**, 6233 (1968); C. S. Foote, Y. C. Chang, and R. W. Denny, *ibid.*, **92**, 5216, 5218 (1970).

(3) R. H. Young, K. Wehrly, and R. L. Martin, *ibid.*, **93**, 5774 (1971).

(4) R. H. Young, R. Martin, K. Wehrly, and D. Feriozi, Abstracts, 162nd National Meeting of the American Chemical Society, Washington, D. C., Sept 1971, No. PETR 31.

(5) W. R. Ware, *J. Phys. Chem.*, **66**, 455 (1962).

(6) B. Stevens and B. E. Algar, *ibid.*, **72**, 2582 (1968); L. K. Patterson, G. Porter, and M. R. Topp, *Chem. Phys., Lett.* **7**, 612 (1970).

(7) C. S. Foote and R. W. Denny, *J. Amer. Chem. Soc.*, **93**, 5168 (1971).

(8) R. H. Young, N. Chinh, and C. Mallon, *Ann. N. Y. Acad. Sci.*, **171**, 130 (1970); R. H. Young, R. L. Martin, N. Chinh, C. Mallon, and R. H. Kayser, *Can. J. Chem.*, **50**, 932 (1972).

(9) P. B. Merkel and D. R. Kearns, *Chem. Phys. Lett.*, **12**, 120 (1971); P. B. Merkel and D. R. Kearns, *J. Amer. Chem. Soc.*, **94**, 1029, 1030 (1972).

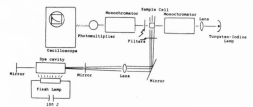

Figure 1. Schematic of the flash photolysis apparatus.

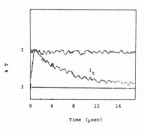

Time (μsec)

Figure 2. Trace of an oscilloscopic display of the photooxidation reaction of DPBF ($I = 34\%$ transmittance (expl 35%)) and ($I_\infty = 49\%$ transmittance (exptl 50%)).

sary to induce a detectable depletion in the DPBF concentration. The direct method has the advantage that it permits measurements of interactions between singlet oxygen and organic compounds without any interference from interactions of the dye sensitizer with the organic compound.

Experimental Section

A schematic of the flash photolysis apparatus is given in Figure 1. The reaction mixture was irradiated with a $1/4$ J, coaxial dye laser with a flash lifetime of $1/2$ μsec and a band pass of 10 nm. The dye used in the laser was either Rhodamine B (610 nm) or Rhodamine 6G (583 nm). The homogeniety of the laser radiation at the reaction cell was checked by observing burn spots on carbon paper and polaroid film. The monochromators were adjusted to pass 410-nm radiation with a band pass of 3.2 nm. Scattered light from the flashlamp was removed by the use of filters and the second monochromator. In many cases a differential preamplifier was used in the oscilloscope to increase the observability of small changes in the intensity of the 410-nm radiation.

The dye sensitizers, Methylene Blue and Rose Bengal, were obtained from Fisher Chemical Co. and used as received. 1,3-Diphenylisobenzofuran was obtained from Aldrich Chemical Co. and was used as received. N,N-Dimethylaniline was obtained from Fisher Chemical Co. while 2,5-diphenylfuran was obtained from Dr. A. Trozzolo of Bell Laboratories. Methanol and 1-butanol were spectrograde quality; other solvents were commercial grade and used as received. The experimental procedure involved dissolving Methylene Blue (optical density = 0.55 at 610 nm) or Rose Bengal) optical density = 0.14 at 583 nm) and DPBF (from 10^{-5} to 1.5×10^{-4} M) in 1 ml of solvent and saturating the solution with oxygen. The concentration of DPBF was determined from the optical density of the solution and its extinction coefficient at 410 nm (ϵ 23,500 l. mol^{-1} cm^{-1} in methanol with slight variations for the other solvents).

Analysis of Data

The coupled differential equations describing the reaction scheme outlined in eq 1–4 are

$$-\frac{d[DPBF]}{dt} = k_{rx}[DPBF][{}^1O_2] \qquad (6)$$

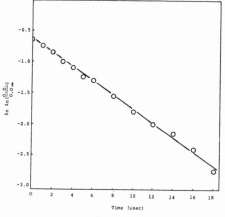

Figure 3. An example of the flash photooxidative decay results of DPBF (data from Figure 2) for the determination of α.

$$-\frac{d[{}^1O_2]}{dt} = k_d[{}^1O_2] + k_{rx}[DPBF][{}^1O_2] \qquad (7)$$

An approximate solution (see Appendix A) to these coupled equations is

$$\ln \ln \left[\frac{[DPBF]_t}{[DPBF]_{t=\infty}} \right] = \ln \frac{k_{rx}[{}^1O_2]_{t=0}}{\alpha} - \alpha t \qquad (8)$$

where

$$\alpha = k_{rx}[DPBF] + k_d \qquad (9)$$

Inspection of the above equation shows that the slope of a linear plot of $\ln \ln [[DPBF]_t/[DPBF]_{t=\infty}]$ *vs.* time is equal to α.

The concentration of DPBF was determined from the per cent transmission of 410-nm light. In all cases the initial and final optical densities at 410 nm were checked with an absorption spectrophotometer. Good agreement between optical densities calculated from the oscilloscope trace and spectrophotometer optical densities was obtained. A slight difference in the final optical density was obtained on occasion. This may have been due to the flash area being slightly less than 1 cm² or having a slight unevenness in intensity over the cell. The size of the flash area (or unevenness in intensity) did not deviate by more than 20% based on agreement of final calculated and spectrophotometric optical densities. This would result in no more than 5% error in the value of α. No other species present in the solution had significant absorption at this wavelength.

A typical oscilloscope trace is shown in Figure 2. The data obtained from Figure 2 were analyzed and plotted as shown in Figure 3. A value for α was obtained from a straight line drawn through these points. In a similar manner other values of α at various concentrations of 1,3-diphenylisobenzofuran, DPBF, were obtained. A plot of α *vs.* [DPBF] resulted in a line with a slope equal to the rate constant of the reaction between singlet oxygen and DPBF (k_{rx}) and an intercept equal to the rate of decay of singlet oxygen (k_d) as shown in Figures 4a and 4b. The concentration of

Table I. Rate Constants of Decay and Reaction of Singlet Oxygen in Various Solvents[a]

Solvent	Dye sensitizer	$10^4 k_d$, sec^{-1}	$10^{-9} k_{rx}$, M^{-1} sec^{-1}	$10^5 \beta$, M ($\beta_{lit.}$)[b]	$10^6 \tau_0$, sec
Methanol	MB[c]	8.8 ± 0.4	1.3 ± 0.1	$6.7 \pm 0.3 \, (7.3 \pm 0.7)$	11.4 ± 0.6
Methanol	RB[d]	9.7 ± 1.1	1.2 ± 0.4	$8.1 \pm 1.5 \, (7.3 \pm 0.7)$	10 ± 1.0
n-Butyl alcohol	MB	5.2 ± 0.8	0.8 ± 0.2	$6.5 \pm 1.0 \, (7.1 \pm 0.7)$	19 ± 3
tert-Butyl alcohol	MB	3.0 ± 0.4	0.57 ± 0.08	$5.3 \pm 0.7 \, (4.9 \pm 0.5)$	34 ± 4
Benzene:methanol (4:1)	MB	3.8 ± 0.8	0.91 ± 0.2	4.1 ± 0.8	26 ± 5
Bromobenzene:methanol (4:1)	MB	4.3 ± 0.7	0.70 ± 0.13	6.2 ± 1.0	23 ± 4
Pyridine	MB	3.1 ± 1.4	2.1 ± 0.3	1.5 ± 0.7	33 ± 15
Dioxane	MB	2.9 ± 1.0	1.21 ± 0.16	2.4 ± 0.8	32 ± 10

[a] Reported errors are to 95% confidence level. [b] See ref 3. [c] MB refers to Methylene Blue. [d] RB refers to Rose Bengal.

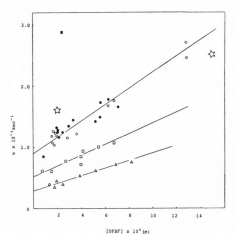

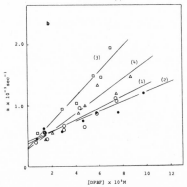

Figure 4 (a). A plot of α vs. [DPBF] for the determination of the lifetime of singlet oxygen (and rate constant of reaction) in methanol (O, with Methylene Blue and ● with Rose Bengal), n-butyl alcohol (□) and tert-butyl alcohol (△) (☆ from ref 9). (b) A plot of α vs. [DPBF] for the determination of the lifetime of singlet oxygen (and rate constant of reaction) in (1) benzene:methanol (4:1) (O), (2) bromobenzene:methanol (4:1) (●), (3) dioxane (□), and (4) pyridine (△).

DPBF used in these plots was the average value (see Appendix A).

Another route for singlet oxygen decay results when an additional compound is added to the reaction solution. From a variation of the kinetic equations it can be shown that in this case

$$\alpha = k_{rx}[\text{DPBF}] + k_d + k_q[\text{Q}] \qquad (10)$$

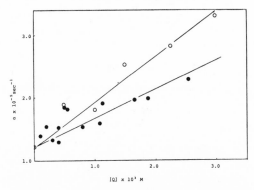

Figure 5. A plot of α vs. [Q] for Q = N,N-dimethylaniline (O) and 2,5-diphenylfuran (●) for the determination of the rate constant of quenching (reaction) of (with) singlet oxygen.

where Q is any compound that intercepts singlet oxygen either by quenching or reaction. At constant [DPBF] a plot of α vs. [Q] will give a straight line with a slope equal to the rate constant of quenching or reaction. In this manner the rate constant of quenching, or reaction, of different compounds with singlet oxygen can be determined. This was done for N,N-dimethyl-aniline and 2,5-diphenylfuran with Rose Bengal as the dye sensitizer (Figure 5).

Results

The results for the determination of the lifetime of singlet oxygen ($1/k_d$), the rate constant of reaction of singlet oxygen with 1,3-diphenylisobenzofuran, DPBF (k_{rx}), and the resultant β values (k_d/k_{rx}) are given in Tables I and II. The larger percentage error in pyridine and dioxane is the result of the small magnitude of the intercept in these compounds. For compounds with long lifetimes, small errors in the measured points introduce large percentage errors in the intercept.

An estimate of the lifetime can also be obtained by combining β values determined by other methods with a value of α at one concentration of DPBF. Lifetimes calculated by this method are listed in Table II.

Lifetimes of singlet oxygen were measured as a function of the concentration of N,N-dimethylaniline and 1,3-diphenylfuran. The results of these measurements are summarized in Table III.

Discussion

Lifetimes of Singlet Oxygen in Solution. Lifetimes of singlet oxygen were obtained in several alcohol and

379

Table II. Rate Constants of Decay of Singlet Oxygen in Various Solvents Determined by an Abbreviated Method[a]

Solvent	$10^5 \beta$, M	10^5, sec^{-1}	10^5[DPBF], M	$10^6 \tau_0$, sec[b]	$10^{-9} k_{rx}$, sec^{-1} M^{-1}
Methanol:water (1:1)	5.5^c	3.7 ± 1.2	1.7	3.6 ± 1.2	5.1
Methanol:glycol (1:1)	6.3^d	1.6 ± 0.1	1.9	8.3 ± 1	1.9
Glycol	31.6^d	5.1	1.7	2.1	1.5

[a] Reported errors are from average of two measurements. [b] Calculated from ([DPBF] + β)/$\alpha\beta$. [c] Determined from the method described in ref 8. [d] Determined from the method described in ref 3.

Table III. Rate Constants for Quenching/Reaction of Singlet Oxygen by *N,N*-Dimethylaniline and 2,5-Diphenylfuran[a]

Compound	k_{rx}' or k_q, sec^{-1} M^{-1}
N,N-Dimethylaniline	$(7.3 \pm 0.8) \times 10^7$
2,5-Diphenylfuran	$(4.6 \pm 0.78) \times 10^7$

[a] Reported errors are to a 95% confidence level.

mixed alcohol solvents. In all cases the experimental decay times gave very satisfactory results. Our lifetime measurement in methanol (11.1 ± 0.7 μsec) agrees well with the value obtained by Kearns (7 ± 1 μsec)[9] and Wilkinson (approximately 10 μsec).[10]

In addition, very good agreement was obtained between the β values from the direct method and our previous results as shown in Table I. This is perhaps the best evidence for the reliability of the data. For example, the direct method yielded β values of 7.2 $\times$ 10^{-5}, 6.5 $\times$ 10^{-5}, and 5.3 $\times$ 10^{-5} M for methanol,[11] *n*-butyl alcohol and *tert*-butyl alcohol. These can be compared with the previous results for β of 7.3 $\times$ 10^{-5}, 7.1 $\times$ 10^{-5}, and 4.9 $\times$ 10^{-5} M for the same solvents, respectively.[3]

Solvent Effect on the Lifetime of Singlet Oxygen. Although the lifetimes determined in this paper differ slightly from our results obtained using β-carotene,[3] it is of interest to note that the relative lifetimes are in quite good agreement as can be seen from Table IV.

Table IV. Comparison of the Relative Lifetimes Determined by a Quenching Technique[3] with Those from Table I

Solvent	τ_{0rel} (from ref 3)	τ_{0rel} (this work)
Methanol	1.00	1.00
n-Butyl alcohol	1.65	1.79
tert-Butyl alcohol	2.46	3.11
Benzene:methanol (4:1)	1.51 (1.8)^a	2.45

[a] Calculated from data in paper by E. R. Peterson, K. W. Lee, and C. S. Foote, *J. Amer. Chem. Soc.*, **94**, 1032 (1972).

This means that the solvent effects for the relative rate constants given in the previous work[3,4] are probably quite reasonable and confirms that there are only small solvent effects on the rate constant of reaction of singlet oxygen with most organic compounds.

We had previously suggested that the rate of decay of singlet oxygen in solution is related to the polarity of the solvent.[4] From an analysis of the results of Tables I and II it can be easily seen that there is no

relationship with the viscosity of the solvent, and we concur with Kearns and coworkers[9] on this point. However, it does appear that there may be a correlation with the polarity of the solvent. This trend is best seen when comparing alcohol and mixed alcohol solvents. The decrease in lifetime with increasing solvent polarity is not unusual. Fluorescence quantum yields are known to decrease in more polar solvents, perhaps through an "encounter complex" between the solvent and the excited solute.[12] Other factors such as hydrogen bonding should play an important part in such "encounter complexes." For this reason complete comparisons between the solvents considering *only* polarity is somewhat misleading.

It is worthwhile noting that there is a negligible heavy-atom effect on the lifetime of singlet oxygen as can be seen by comparing the lifetime in benzene:methanol (4:1), 26 μsec, with bromobenzene:methanol (4:1), 23 μsec. This confirms Foote's previous observation based on relative β values in these solvents.[7]

Rose Bengal *vs*. Methylene Blue. One of the real advantages of the dye laser is the wide range of monochromatic excitation light available from such a flash source. This means that different dye-sensitizers for the photooxidation reactions may be employed which can be a very practical option. For example, the two sensitizers used in this work were excited with different wavelengths: Methylene Blue, 610 nm; Rose Bengal, 583 nm. Good agreement between the lifetimes (11.4 ± 0.6 and 10 ± 1 μsec), rate constants of reaction [$(1.3 \pm 0.1) \times 10^9$ and $(1.2 \pm 0.4) \times 10^9$ M^{-1} sec^{-1})], and β values [$(6.7 \pm 0.3) \times 10^{-5}$ and $(8.1 \pm 1.5) \times 10^{-5}$ M)] were obtained for the two systems with different sensitizers.

The possibility that the species responsible for photooxidation reactions is a complex between oxygen and the dye sensitizer has been discussed by several workers.[13,14] In addition to their evidence, this intermediate can be eliminated because neither the lifetime of the reactive species nor the rate constant of the reaction with DPBF are a function of the dye sensitizer (see Table I). It is unlikely that any form of a sensitizer-O$_2$ complex would have the same lifetime and rate of reaction with different sensitizers. This confirms that the reactive species is singlet oxygen in dye-sensitized photooxidation reactions.

Conclusions

A new technique for the determination of lifetimes of singlet oxygen in solution and the determination of

(10) See comment in paper by C. S. Foote, R. W. Denny, L. Weaver, Y. Chang, and J. Peters, *Ann. N. Y. Acad. Sci.*, **171**, 139 (1970).

(11) Results in methanol were determined from a plot of all the α values with both Methylene Blue and Rose Bengal used as sensitizers.

(12) D. M. Hercules, "Fluorescence and Phosphorescence Analysis," Interscience, New York, N. Y., 1966, p 107.

(13) E. J. Corey and W. C. Taylor, *J. Amer. Chem. Soc.*, **86**, 3881 (1964); C. S. Foote and S. Wexler, *ibid.*, **86**, 3879 (1964).

(14) K. Gollnick, T. Franken, G. Schade, and G. Dorhofer, *Ann. N. Y. Acad. Sci.*, **171**, 89 (1970).

absolute rate constants of reaction between singlet oxygen and various organic acceptors was developed. Good agreement between the β values found in this manner and those reported in the literature was obtained emphasizing the reliability of the method. In addition, a variation of the technique enabled the determination of the rate constant of quenching (N,N-dimethylaniline)[10] and the rate constant of reaction (DPF[15]) of an organic compound with singlet oxygen. This may well be the method of choice for the determination of such rate constants when the interaction between the organic species and the dye sensitizer is unknown. Extensions of the technique for the determination of rate constants of interaction (reaction and/or quenching) will be given elsewhere.

Appendix A

The coupled differential equations describing the decay of $^1O_2$ are listed in the text (eq 6 and 7) and repeated here for clarity.

$$\frac{d[DPBF]}{dt} = -k_{rx}[^1O_2][DPBF] \qquad (1A)$$

$$\frac{d[^1O_2]}{dt} = -[k_{rx}[DPBF] + k_d][^1O_2] \qquad (2A)$$

Let

$$\alpha = k_{rx}[DPBF] + k_d \qquad (3A)$$

Assume that α is constant. This assumption is essentially true if [DPBF] is small or if the amount of the reaction is small. The error introduced by this assumption will be discussed below. The solution of eq 2A is

$$[^1O_2]_t = [^1O_2]_{t=0}e^{-\alpha t} \qquad (4A)$$

When this solution is substituted into eq 1A the solution of eq 1A is

$$\ln\left[\frac{[DPBF]_t}{[DPBF]_{t=\infty}}\right] = \frac{k_{rx}[^1O_2]_{t=0}e^{-\alpha t}}{\alpha} \qquad (5A)$$

(15) DPF is assumed to undergo reaction with $O_2{}^1\Delta$ without quenching of $O_2{}^1\Delta$ by comparison with dimethylfuran.[3]

This equation is equivalent to eq 8 in the main text and was used for the analysis of the data.[16]

In order to evaluate the errors introduced by assuming α is a constant, the coupled equations eq 1A and eq 2A were solved by a Runge–Kutta[17] technique on a computer. When reasonable values of the parameters were used, even when large amounts of DPBF were consumed (as much as 50%), the difference between a "true" α and the α derived from the linear plot described in the text was less than 15%, if the *average* concentration of DPBF was used in the calculation of the "true" α. For example, in methanol the change in DPBF concentration was about 30% at low concentrations (10^{-5} M) which results in a 5% change in α. At higher concentrations of DPBF (up to 1.5×10^{-4} M) the reaction resulted in about a 10–15% change in [DPBF] or a 10–12% change in the value of α. Use of average concentrations of [DPBF] in these cases reduced the error to less than 10%.

(16) The equation used by Kearns, *et al.*,[9] is essentially equivalent to eq 5A. The additional assumptions used by them introduce no more than 10% error under the conditions used in these experiments.

(17) H. Morgenau and G. M. Murphy, "Mathematics of Physics and Chemistry," Van Nostrand, Princeton, N. J., 1956, p 486.

(18) NOTE ADDED IN PROOF. Two recent publications by Kearns[19] and Wilkinson[20] illustrate variations of the technique presented in this paper. Their results for the lifetime of singlet oxygen ($^1O_2$) in comparable solvents agree reasonably well with our results. Their values for other solvents in addition to ours give a large number of solvent-lifetime results. However, we do advise some caution in the use of these results. The reactivity of the acceptor, DPBF, varies in different solvents. In most cases the reaction of $^1O_2$ with DPBF contributes significantly to the decay rate of $^1O_2$ at the concentrations of DPBF used in the experiments. This leads to low values for the lifetime of $^1O_2$ unless this factor is taken into consideration. Indeed if Wilkinson's values for the lifetime of $^1O_2$ in methanol ($\tau = 6$ μsec) and benzene:methanol (4:1, $\tau = 10$ μsec) are corrected for reaction by DPBF using the β (k_d/k_{rx}) ratios given in this paper, the agreement of the lifetimes with our results is extremely good (corrected results 10.5 μsec (ours 11 μsec) and 23 μsec (ours 26 μsec) for methanol and benzene:methanol, respectively). This makes it imperative that the effect of variation of the concentration of the acceptor (DPBF) be considered carefully. For the solvents which have very long lifetimes this becomes more important. If a solvent has both a small k_d and a large k_{rx} (low β value), then the experimental method used here is not capable of determination of the lifetime with a reasonable degree of accuracy as the intercept of the plot of α vs. [DPBF] passes too close to zero. This means that accurate β values should be obtained where possible in order to confirm the potential accuracy of the experimental results for all solvents.

(19) P. B. Merkel and D. R. Kearns, *J. Amer. Chem. Soc.*, **94**, 7244 (1972).

(20) D. R. Adams and F. Wilkinson, *J. Chem. Soc., Faraday Trans.*, 586 (1972).

Editor's Comments
on Paper 78

78   **BLOSSEY** et al.
      Polymer-Based Sensitizers for Photooxidations

HETEROGENEOUS PHOTOSENSITIZERS

Although many sources of singlet oxygen are now available, the photosensitized formation of singlet oxygen using various dyestuffs remains the method of choice for most synthetic and mechanistic applications. However, there have been several limitations to the photooxidation method: (1) the sensitizer (dye) must be soluble in the reaction solvent, which limits the dye–solvent combinations that can be used; (2) the dye is often bleached over long reaction times; (3) the dye can interact with the substrates and/or the products, and (4) the separation of the dye from the products can be difficult.

The preparation of an insoluble polymer-bound dye and the use of this heterogeneous sensitizer for the photochemical generation of singlet molecular oxygen is described in Paper 78. Polymer-bound sensitizers have several advantages over free sensitizers in solution. They can be used in solvents in which the unbound dye is both insoluble and therefore unable to sensitize singlet oxygen formation efficiently. They are significantly more stable toward bleaching than are the free sensitizers. The polymer-bound sensitizers can be easily removed at the end of the reaction by filtration, and can be reused with little or no loss in efficiency.

78

Reprinted from *J. Amer. Chem. Soc.*, **95**(17), 5820–5822 (1973)

Polymer-Based Sensitizers for Photooxidations

Sir:

Insoluble polymer supports were introduced several years ago by Merrifield[1] and by Letsinger[2] to facilitate polypeptide synthesis. The technique involves the use of an insoluble styrene–divinylbenzene copolymer bead to provide a foundation upon which successive chemical transformations can be carried out.

For some time we have been interested in the use of insoluble polymer supports in photochemical reactions. In this report, we describe the preparation and use of the first example of a synthetically applicable, polymer-based photosensitizer. The reagent, polymer-based Rose Bengal (Ⓟ-Rose Bengal), is utilized to sensitize the generation of singlet molecular oxygen. Rose Bengal[3] is attached to a chloromethylated polystyrene support *via* the following procedure: Rose Bengal, 2.0 g (2.1 mmol), was stirred at reflux in 60 ml of reagent grade dimethylformamide with 2.0 g of chloromethylated styrene–divinylbenzene copolymer beads (1.38 mequiv of CH_2Cl, 50–100 mesh). After 20 hr, the polymer (now dark red) was filtered and washed successively with 150-ml portions of benzene, ethanol, ethanol–water (1:1), water, methanol–water (1:1), and methanol. After these washings, the final filtrate was colorless. The polymer beads[4] were dried in a vacuum oven to a final weight of 2.17 g.

Singlet molecular oxygen exhibits three modes of reaction with alkenes:[5] 1,4-cycloaddition with conjugated dienes to yield cyclic peroxides, an "ene" type reaction to form allylic hydroperoxides, and 1,2 cycloaddition[6] to give 1,2-dioxetanes which cleave thermally to carbonyl-containing products. Examples of all of

(1) R. B. Merrifield, *Science*, **150**, 178 (1965).
(2) R. L. Letsinger, M. J. Kornet, V. Mahedevon, and D. M. Jerina, *J. Amer. Chem. Soc.*, **86**, 5163 (1964).
(3) Rose Bengal

(4) The Rose Bengal is probably attached to the polymer as the carboxylate ester.
(5) (a) C. S. Foote, *Accounts Chem. Res.*, **1**, 104 (1968); (b) D. R. Kearns, *Chem. Rev.*, **71**, 395 (1971); (c) K. Gollnick and G. O. Schenck in "1,4-Cycloaddition Reactions," J. Hammer, Ed., Academic Press, New York, N. Y., 1967, p 255; (d) K. Gollnick, *Advan. Photochem.*, **6**, 1 (1968); (e) W. R. Adams in "Oxidation," Vol. 2, R. L. Augustine and D. J. Trecker, Ed., Marcel Dekker, New York, N. Y., 1971, p 65; (f) J. T. Hastings and T. Wilson, *Photophysiology*, **5**, 49 (1970).
(6) (a) A. P. Schaap and G. R. Faler, *J. Amer. Chem. Soc.*, **95**, 3381 (1973); (b) N. M. Hasty and D. R. Kearns, *ibid.*, **95**, 3380 (1973).

Table I. Photoxidations with Ⓟ-Rose Bengal

Singlet oxygen acceptor	Product	% yield (isolated)
1	2	95
3	4	69
5	6	82

these three reaction types have been carried out utilizing Ⓟ-Rose Bengal as a sensitizer (see Table I).

To a solution of 140 mg (0.6 mmol) of 1,2-diphenyl-*p*-dioxene (**1**) in 6 ml of CH_2Cl_2 was added 200 mg of sensitizer beads. The resultant mixture contained in a Pyrex vessel was vigorously stirred at 10° under O_2 and irradiated with a 500-W tungsten–halogen lamp through a uv-cutoff filter. Gas chromatography indicated complete oxidation of **1** after 6 hr. Removal of the sensitizer beads by filtration of the reaction mixture through a sintered glass disk[7] and removal of the solvent under vacuum gave colorless crystals of **2** in 95% yield. The photooxidation product was compared with an authentic sample of **2**.[8] Absorption spectra of the reaction solution before and after photolysis indicated that no Rose Bengal or other sensitizer is leached into the reaction solution.

The following control experiments indicate that the conversion of **1** to **2** is a singlet oxygen-mediated reaction. The reaction is inhibited by the addition of 10 mol % (based on **1**) of 1,4-diazabicyclo[2.2.2]octane (DABCO), a singlet oxygen quencher.[9] The photooxidation of **1** can be carried out in the presence of 10 mol % of 2,6-di-*tert*-butylcresol, a free radical inhibitor. The conversion of **1** to **2** can also be effected by photooxidation with 562-nm radiation using a Bausch and Lomb grating monochromator and SP-200 mercury light source.[10] It should also be noted that a suspension of solid Rose Bengal in CH_2Cl_2 is relatively ineffective in photosensitizing the generation of singlet oxygen.

1,3-Cyclohexadiene (**3**) and tetramethylethylene (**5**) undergo the 1,4-cycloaddition and ene reactions, respectively, with singlet oxygen produced by Ⓟ-Rose Bengal sensitization. The reactions were carried out as described for the photooxidation of **1**. Products

(7) The dried beads can be reused with no detectable decrease in efficiency.
(8) E. J. Bourne, M. Stacey, J. C. Tatlow, and J. M. Tedder, *J. Chem. Soc.*, 2976 (1949).
(9) C. Ovannès and T. Wilson, *J. Amer. Chem. Soc.*, **90**, 6527 (1968).
(10) Rose Bengal: $\lambda_{max}^{(CH_2)_2CO}$ 562 nm.

4[11] and 6[12] were isolated by distillation under vacuum and compared with authentic samples.

One criterion for the generation of free singlet oxygen from various sources has been the product distribution obtained from 1,2-dimethylcyclohexene (7). Photooxidation of 7 with polymer-based Rose Bengal yields a similar distribution of the two possible ene products 8 and 9 (see Table II).

Table II. Oxidation of 1,2-Dimethylcyclohexene (7) Using Various Singlet Oxygen Sources

	8	9	10
Sources			
(P)-Rose Bengal[a]	87	13	0
Photooxidation (soluble sens.)[b]	89	11	0
OCl⁻–H₂O₂[b]	91	9	0
(C₆H₅O)₃PO₃[c]	96	4	0
K₃CrO₈[d]	82	18	0
Radical autoxidation[b]	6	39	54

[a] Products from this reaction were analyzed by gas chromatography as the alcohols obtained by triphenylphosphine reduction of 8 and 9. [b] See ref 5a. [c] R. W. Murray and J. W.-P. Lin, *Ann. N. Y. Acad. Sci.*, **171**, 121 (1970). [d] J. W. Peters, J. N. Pitts, Jr., I. Rosenthal, and H. Fuhr, *J. Amer. Chem. Soc.*, **94**, 4348 (1972).

Dihydropyran (11) is a singlet oxygen acceptor that also yields two products: 13 is obtained by thermal cleavage of the 1,2-dioxetane 12 and 15 is formed upon dehydration under the reaction conditions of the ene product 14. Photooxidation of 11 in CH₂Cl₂ with tetraphenylporphine gives 73% 13 and 27% 15.[13]

With the (P)-Rose Bengal sensitizer in CH₂Cl₂, the

(11) G. O. Schenck and D. E. Dunlap, *Angew. Chem.*, **68**, 248 (1956).
(12) C. S. Foote and S. Wexler, *J. Amer. Chem. Soc.*, **86**, 2879 (1964).
(13) The product distribution from 11 is independent of the sensitizer used but a function of the solvent employed for the reaction: P. D. Bartlett, G. D. Mendenhall, and A. P. Schaap, *Ann. N. Y. Acad. Sci.*, **171**, 79 (1970).

photooxidation of 11 gives an identical product distribution.

We conclude, on the basis of the experiments described in this report, that free singlet oxygen is efficiently formed by energy transfer from (P)-Rose Bengal to oxygen. The possible uses for an insoluble, easily recovered sensitizer in preparative photochemical reactions are obvious. Insoluble polymer-based sensitizers may also be useful in mechanistic investigations in which the particular sensitizer is itself insoluble in the solvent of choice. Experiments with other types of (P) sensitizers are in progress.

Acknowledgment. Financial support to D. C. N. from the Research Corporation (Cottrell Research Grant) and the National Science Foundation (GP-33566), to E. C. B. from the National Institutes of Health for a special post-doctoral fellowship, and to A. P. S. from the Research Corporation (Cottrell Research Grant) and the U. S. Army Research Office—Durham is gratefully acknowledged. The authors wish to thank Dow Chemical Co. for a gift of styrene–divinylbenzene copolymer beads.

(14) On leave from Rollins College, 1972–1973.
(15) Fellow of the Alfred P. Sloan Foundation, 1971–1973.

Erich C. Blossey,[14] **Douglas C. Neckers***[15]
Department of Chemistry, The University of New Mexico
Albuquerque, New Mexico 87106

Arthur L. Thayer, A. Paul Schaap*
Department of Chemistry, Wayne State University
Detroit, Michigan 48202
Received June 15, 1973

SELECTED READINGS

REVIEWS ON SINGLET MOLECULAR OXYGEN

Abe, T., M. Sukigara, and K. Honda, "Singlet Oxygen and its Photochemical Reactions," *Seisan-Kenkyu*, **26**, 71–77 (1974) (Japanese).

Adam, W., "Singlet Molecular Oxygen and its Role in Organic Peroxide Chemistry," *Chem. Z.*, **99**, 142–155 (1975).

Adams, W. R., "Photosensitized Oxygenations" in *Oxidation*, Vol. 2, R. L. Augustine and D. J. Trecker (eds.), Marcel Dekker, Inc., New York, 1971, pp. 65–112.

Ando, W., and J. Suzuki, "Singlet Oxygen, Its Formation and Reactions," *Yuki Gosei Kagaku Kyokai Shi*, **30**, 391–400 (1972) (Japanese).

Arbuzov, Y. A., "The Diels-Alder Reaction with Molecular Oxygen as Dienophile, "*Russ. Chem. Rev.*, **34**, 558–574 (1965) (English translation).

Denny, R. W., and A. Nickon, "Sensitized Photooxygenation of Olefins" in *Org. Reactions*, **20**, 133–336 (1973).

Duynstee, E. F. J., "De rol van singlet zuurstof in de organische chemie," *Chem. Weekblad*, **67**(37), 21–24 (1971) (Dutch).

Foote, C. S., "Mechanisms of Photosensitized Oxidation," *Science*, **162**, 963–970 (1968).

Foote, C. S., "Photosensitized Oxygenations and the Role of Singlet Oxygen," *Accounts Chem. Res.*, **1**, 104–110 (1968).

Foote, C. S. "Mechanism of Addition of Singlet Oxygen to Olefins and Other Substrates," *Pure Appl. Chem.*, **27**, 635–645 (1971).

Foote, C. S. "Photosensitized Oxidation and Singlet Oxygen: Consequences in Biological Systems," in *Free Radicals in Biology*, W. A. Pryor (ed.), Academic Press, Inc., New York, 1975.

Gollnick, K., "Type II Photooxygenation Reactions in Solution," *Advan. Photochem.*, **6**, 1–122 (1968).

Gollnick, K., and G. O. Schenck, "Oxygen as a Dienophile," in *1, 4-Cycloaddition Reactions*, J. Hamer (ed.), Academic Press, Inc., New York, 1967, pp. 255–344.

Kaplan, M. L., "Singlet Oxygen," *Chem. Technol.*, **1**, 621–626 (1971).

Kearns, D. R., "Physical and Chemical Properties of Singlet Molecular Oxygen," *Chem. Rev.*, **71**, 395–427 (1971).

Kearns, D. R., and A. U. Khan, "Sensitized Photooxygenation Reactions and the Role of Singlet Oxygen," *Photochem. Photobiol.*, **10**, 193–210 (1969).

Muset, P. P., "Peroxidaciones, Radical Superoxido y Oxigeno Singlete," Proceedings of Real Academia de Farmacia de Barcelona, A-Sesion Inaugural (1975) (Spanish).

Oda, R., "Oxidation by Singlet Oxygen," *Yuki Gosei Kagaku Kyokai Shi*, **32**, 926–932 (1974) (Japanese).

Ogryzlo, E. A., "Physical Properties of Singlet Oxygen," *Photophysiology*, **5**, 35–47 (1970).

Pitts, J. N., Jr., "Photochemical Air Pollution: Singlet Molecular Oxygen as an Environmental Oxidant," *Advan. Environ, Sci.*, **1**, 289–337 (1969).

Politzer, I. R., G. W. Griffin, and J. L. Laseter, "Singlet Oxygen and Biological Systems," *Chem. Biol. Interactions*, **3**, 73 (1971).

Rabek, J. F., "Singlet Oxygen Mechanism of Photooxidation Reactions of Organic Compounds I.," *Wiadomości Chemiczne*, **25**, 293–381 (1971) (Polish).

Rigaudy, J., "Photooxydation des Derives Aromatiques," *Pure Appl. Chem.*, **16**, 169–186 (1968) (French).

Stevens, B., "Kinetics of Photoperoxidation in Solution," *Accounts Chem. Res.*, **6**, 90–96 (1973).

Wayne, R. P., "Singlet Molecular Oxygen," *Advan. Photochem.*, **7**, 311–371 (1969).

Wilson, T., and J. W. Hastings, "Chemical and Biological Aspects of Singlet Excited Molecular Oxygen," *Photophysiology*, **5**, 49–95 (1970).

PROCEEDINGS OF MEETINGS AND SYMPOSIA ON SINGLET MOLECULAR OXYGEN

"International Oxidation Symposium on Ozone Chemistry, Photo and Singlet Oxygen Oxidations, and Biochemical Oxidations," arranged by the Stanford Research Institute; general chairman, F. R. Mayo; session chairmen, C. S. Foote and K. Gollnick; San Francisco, Calif., August 28–September 1, 1967; proceedings published in *Advan. Chem. Ser.* **77**, 78–168 (1968).

"International Conference on Singlet Molecular Oxygen and Its Role in Environmental Sciences," held by the New York Academy of Sciences; conference cochairmen, A. M. Trozzolo and R. W. Murray; New York, N.Y., October 23–25, 1969; proceedings published in *Ann. N.Y. Acad. Sci.*, **171**, 1–302 (1970).

"Oxidations by Singlet Oxygen," sponsored by Division of Petroleum Chemistry, American Chemical Society; chairman, C. S. Foote; Washington, D.C., September 12–17, 1971; proceedings published in *Petrol. Preprints*, **16**,(4), A9–91 (1971).

AUTHOR CITATION INDEX

Hirt, A., 92
Hochstetler, A. R., 110, 297
Hochstrasser, R. M., 366, 372
Hock, H., 72
Hoffman, R. W., 299, 300
Hoitjink, G. J., 140
Holleman-Wiberg, 73
Hollinder, G. A., 349
Hollins, R. A., 113, 175, 218, 222, 223, 231, 258, 264, 272
Honda, K., 385
Hopfield, H. S., 87
Hornbeck, G. A., 87
Horning, E. C., 114
Horwitt, M. K., 329
Houben-Weyl, 111, 270
Houpillart, J., 280
Howard, J. A., 269, 276, 279
Howell, L. G., 282
Hoytink, G. J., 96
Hsu, J., 334
Huber, J., 295, 298, 300
Hudson, R. L., 88, 117, 118
Huie, R. E., 349
Huisgen, R., 295, 299, 346
Humphries, F. S., 113
Hunig, S., 118
Hunten, D. M., 264

Imamura, M., 140
Imotu, M., 218
Ingersoll, H. G., 322
Inglett, G. E., 329, 334, 337
Ingold, C. K., 154, 175, 269, 276, 279
Iori, G., 295
Iredale, T., 264
Ireland, R. E., 154
Ishikawa, H., 222
Iverson, L., 334
Iwanaga, C., 364
Izod, R. P. J., 264

Jackson, G., 184
Jarmin, W. R., 96
Jerina, D. M., 383
John, F., 159
Johnson, H. L., 85
Johnston, R. G., 148
Jones, J., 178
Jorgensen, C. K., 322
Jortner, J., 366, 372
Jost, W., 217
Julien, P. L., 107

Kahn, A. H., 180
Kaiser, E. M., 115

Kallman, H. P., 218
Kane, V. V., 298
Kaplan, J., 87, 94
Kaplan, M. L., 217, 225, 269, 276, 321, 385
Kasche, V., 193
Kasha, M., 3, 58, 70, 89, 90, 91, 99, 103, 119, 140, 154, 174, 178, 180, 217, 279, 282, 285, 372
Katakis, D., 289
Kato, S., 337
Katz, T. J., 297
Kautsky, H., 32, 43, 44, 56, 58, 70, 73, 87, 101, 110, 118, 139, 140, 147, 154, 175, 178, 180, 193, 216, 217, 223, 264, 364
Kauzmann, W., 373
Kawaoka, K., 97, 175, 185, 222, 224, 264, 266, 356, 367
Kayama, K., 373
Kayser, R. H., 334, 377
Kearns, D. R., 3, 97, 113, 175, 185, 194, 218, 222, 223, 224, 225, 231, 257, 258, 264, 266, 268, 272, 291, 298, 299, 300, 308, 321, 326, 327, 340, 343, 347, 356, 364, 365, 366, 367, 368, 369, 370, 372, 376, 377, 381, 383, 385, 386
Kees, K., 268
Kelleher, P. G., 321
Kenney, R. L., 148
Kettner, K. A., 285
Keulemans, A. I. M., 154
Khan, A. U., 3, 89, 90, 97, 99, 103, 113, 119, 140, 154, 175, 178, 185, 217, 218, 222, 223, 224, 225, 231, 257, 258, 264, 266, 272, 279, 282, 285, 356, 367, 386
Khan, N. A., 329
Kharasch, M. S., 115, 175
Kim, B., 217
King, T. A., 178
Kinkel, K. G., 60, 63, 139, 216, 217
Kirrman, K. A., 90
Kirschke, K., 295
Kitagawa, T., 295
Klager, K., 61
Klein, E. A., 139, 140
Kline, O. L., 334
Klyne, W., 140
Kneisley, J. W., 269
Kneser, H. O., 3, 5, 88, 90, 217
Knobler, C. M., 87
Knopka, W. P., 295
Knotzel, H., 89, 113
Knotzel, L., 89, 113
Koblitz, W., 70, 139, 180
Kobuke, Y., 352
Koch, E., 99, 101, 113, 137, 147, 154, 175, 217, 218, 318, 371, 377

Seliger, H. H., 85, 87, 88, 89, 90, 91, 99
Sharp, D. B., 147, 154, 175, 347
Shaw, F. R., 154
Sheremetiev, G. D., 140
Shombert, D. J., 101, 193
Silverman, S., 87
Simamura, O., 105
Sinha, B., 118
Slater, J., 95
Small, R. D., Jr., 337
Smith, E. B., 225, 285
Smith, J. R., 352
Smith, W. F., 321, 334
Snelling, D. R., 223, 225, 258, 263, 266, 364, 371
Solo, R. B., 289
Sondermann, J., 193
Southern, P. F., 103, 279
Sperling, W., 193
Spikes, J. D., 3, 295, 317
Spruit, C. J. P., 85
Spruit-Van Der Burg, A., 85
Sprunch, G. M., 218
Sprung, J. L., 263
Stacey, M., 383
Stanier, R. Y., 317
Staudinger, H. S., 304
Stauff, J., 87, 91
Steadman, F., 70, 139
Steer, R. P., 263
Stefani, A. P., 346
Stefanic, A. P., 154
Stein, G., 376
Steiner, W., 92, 217
Steinmetz, R., 139, 217
Steinmetzer, H. C., 291
Stephens, E. R., 285
Stephenson, L. M., 268
Stevenick, J. van, 334
Stevens, B., 178, 181, 321, 334, 337, 338, 340, 350, 364, 377, 386
Stevens, M. P., 100, 111, 270, 277
Stewart, E. T., 156
Stewart, T. D., 8
Stief, L. J., 289
Stiles, M., 269
Stoll, A., 139
Straight, R., 295, 317
Strating, J., 308
Strickland, S., 282
Stuhl, F., 264, 372
Subba Rao, V., 101, 112, 175
Suell, P. S., 154
Sukigara, M., 385
Sullivan, P. A., 282
Summerbell, R. K., 100, 273
Suprunchuk, T., 310, 326, 334, 337

Sussenbach, J. S., 295
Sutton, D. A., 76
Suurmond, D., 334
Suzuki, J., 385
Svreshnikov, B. Y., 140
Swern, D., 154
Sykes, A., 319, 326
Sykes, S., 372
Symonds, M., 279
Szwarc, M., 154

Tamelen, E. E. van, 118
Tang, C. W., 314, 334, 364
Tanner, D. W., 139, 175, 178, 180, 344, 365
Tappeiner, H. von, 65, 73
Tappel, A. L., 334, 337
Tate, J. T., 117
Tatlow, J. C., 383
Tatsuno, Y., 322
Taube, H., 87, 95, 289
Taylor, J. A., 264
Taylor, L. S., 85
Taylor, O. C., 285
Taylor, W. C., 109, 110, 112, 119, 156, 163, 165, 175, 178, 180, 217, 223, 257, 269, 276, 277, 279, 311, 356, 367, 380
Teale, F. W. J., 140
Tedder, J. M., 383
Terenin, A. N., 140, 216
Thakar, M. S., 217
Thayer, A. L., 268
Theiling, E., 61, 63
Theimer, E. T., 274
Thiele, J., 218
Thier, W., 118
Thomas, M., 337
Thomaz, M. F., 178
Thompson, Q. E., 119, 269, 278, 308
Thrush, B. A., 88, 119
Tillman, P., 193
Timmons, R. B., 349
Timmons, R. J., 118
Ting, I. P., 285
Toepel, T., 61
Topp, M. R., 350, 377
Totter, J. R., 282
Trozzolo, A. M., 3, 334, 337, 340
Truscott, T. G., 317, 319, 326, 372
Tschurdy, D. P., 334
Tsubomura, H., 140, 180, 208
Tuleen, D. L., 299
Turner, A. M., 280
Turro, N. J., 159, 291

Umstätter, H., 217
Unger, I., 140

SUBJECT INDEX

About the Editor

A. PAUL SCHAAP is Associate Professor of Chemistry at Wayne State University. His areas of research include the reactions of singlet molecular oxygen and organic chemiluminescent reactions. Professor Schaap received his A.B. degree (summa cum laude) from Hope College in 1967. The last semester of his undergraduate education was spent as a research associate with Professor Hans Wynberg at the University of Groningen, The Netherlands. He obtained his Ph.D. degree in 1970 under the direction of Professor Paul Bartlett at Harvard University. He is currently an Alfred P. Sloan Research Fellow.